Coasts, Estuaries and Lakes

N. Jayaraju • G. Sreenivasulu
M. Madakka • M. Manjulatha
Editors

Coasts, Estuaries and Lakes

Implications for Sustainable Development

Springer

Editors
N. Jayaraju
Department of Geology
Yogi Vemana University
Kadapa, Andhra Pradesh, India

G. Sreenivasulu
Department of Geology
Sri Venkateswara University
Tirupati, Andhra Pradesh, India

M. Madakka
Department of Biotechnology
and Bioinformatics
Yogi Vemana University
Kadapa, Andhra Pradesh, India

M. Manjulatha
Department of Floriculture
National Institute of Horticultural
and Herbal Science
Seonghwan, Republic of Korea

ISBN 978-3-031-21643-5 ISBN 978-3-031-21644-2 (eBook)
https://doi.org/10.1007/978-3-031-21644-2

This Springer imprint is published by the registered company Springer Nature Switzerland AG
The registered company address is: Gewerbestrasse 11, 6330 Cham, Switzerland

Dedicated to the family members of the Editors

Foreword

Congratulations on putting together such an impressive set of papers that examine our coasts, estuaries, and lakes! Please accept this brief foreword that aims to synthesize and shine light upon this important book:

Coasts, estuaries, and lakes are critical zones globally where humans interact and often conflict with nature. These Earth features are complex in morphology and inherently dynamic as a result of a multitude of processes that shift their form and change their functioning. Communities depend on the water and shoreline for sustenance, revenue, and countless aspects of life. The challenges of living in coastal zones, of both lakes and marine areas, are ever increasing as shoreline populations continue to climb and demand for food and financial sustainability increases. Moreover, climate change is creating new threats and perturbations that require humanity and governments to react, often rapidly, to water surpluses or shortages or shifting land and seascapes. Research is required to advance our understanding of these complicated systems and human impacts, and around the world, people must strive to share insights and to learn from each other. Beyond research, communities need wise planning and smart public and private action that is rooted in and driven by science. In addition to reading and studying about Earth processes and systems, scientists and students are encouraged to engage with governments and communities to help them understand our evolving world and guide careful decision-making.

The book *Coasts, Estuaries and Lakes: Implications to Sustainable Development* is an excellent collection of 29 chapters in 7 parts that delve into critical science and related challenges. From processes to pollutants, the chapters examine fundamental aspects, but are too numerous and detailed to review or attempt to synthesize succinctly here. Led by Dr. N. Jayaraju from the Department of Geology at Yogi Vemana University, the experienced editorial team includes Drs. Sreenivasulu, Madakka, and Manjulatha. Part I addresses two immediate and important topics, flooding and pollution. Water quality and sedimentary aspects are explored in Parts II and III, respectively. Parts IV and V address biodiversity and climate change which are key concerns for coasts, estuaries, and lakes, and Parts VI and VII examine socioeconomic considerations and geospatial tools that are essential to and will inform sustainable development. The compilation of papers is truly impressive, covering a wealth of information and including examples at varying scales – from local and national studies in India, Kuwait, and elsewhere to global insights from past and recent work.

In summary, the book collection will provide an excellent resource for years to come, and the various chapters will serve as invaluable references for experts and students around the world. The unfortunate reality is that our coasts, estuaries, and lakes are facing unprecedented challenges. Science must be used in partnership with continuous stakeholder engagement and wise governance to allow our shores and seas to sustain society for future generations.

J. P. Walsh
Director, Coastal Resources Center & Professor
Graduate School of Oceanography, URI-GSO Bay Campus,
Narragansett, RI, USA

Foreword

Coasts, estuaries and brackish lakes support an important biodiversity and large human populations, and they deliver many ecosystem services from which society gains goods and benefits. However, at the same time, they are the sites of many hazards which give rise to risks and threats to natural and human populations. These threats have been summarised as a "triple whammy" – that the areas are subject to three major threats: an increasing urbanisation and industrialisation; an increasing use of resources such as fish, water and space; and a decreasing resistance and resilience to external, wide-ranging factors such as climate change. These threats and risks may be even greater in developing countries which may have a lesser capability to deal with them.

Given the above features, there is the need to summarise the characteristics of all areas and to gain better information and data to help us manage the areas sustainably. In this way, both nature and society benefit from a healthy environment. This requires natural and social scientists to work together and to work with policymakers and policy implementers. It requires gathering knowledge and data at local, regional, national and global levels and summarising these for a greater benefit. It requires us to be clear about why we need information and how it can be used, often based on studying and monitoring the areas.

Against this background, in its five parts, this book integrates the science and management and gives attention to various features for many Indian areas, including hot spots within those, although other geographical locations are also mentioned. The editors and many authors have done an excellent job of bringing together information and data on local and regional processes and dynamics. These cover sediment contamination and its effects, and coastal flooding, thereby covering physico-chemical interactions, water quality and hydrological processes for the coastal and catchment areas. Using evidence from various regions, and as well as on different spatial scales, it covers temporal scales including long-term natural and anthropogenic forcing on marine, coastal and estuarine ecosystems and multi-year biogeochemical observations.

It is axiomatic that we cannot understand the ecological structure and functioning of an area unless we understand the physical and chemical structure and interactions. Hence the book covers saline water intrusion, and sediment physical, chemical and geochemical features. The seasonal and spatial features of local beaches, the substratum and suspended sediment dynamics and mineral chemistry give a background to a greater understanding. This in turn will enable environmental management which has to be based on fit-for-purpose science and good analytical techniques, which are also described.

The physical and chemical background then allows an interrogation of the biological and ecological features. As such, the contributors give the background with greater information on biodiversity and ecology. The chemical conditions then result in contamination of the biota, reflected here in studies on a range of taxa from foraminifera to molluscs and fishes. This contamination often leads to pollution effects, defined as a decline in the health of organisms, their populations and communities, and indeed a reduction in their fitness-for-survival. Understanding these aspects requires us to define and apply bioindicators and follow this with detailed monitoring. As shown here, chemical contamination can have biochemical effects in organisms and lead to the sequestration of the contaminants in shells and other hard structures of organisms.

As shown in this book, the features of an area may be affected by pressures, defined as the mechanisms of change on the natural and human systems. These may relate to local factors and activities, such as industrial discharges, leading to what may be termed endogenic managed pressures, where the causes and consequences of environmental change both occur in a given area being managed. However, all areas are also exposed to what may be called exogenic unmanaged pressures, in which the causes of change are outside the area being managed but the consequences have to be managed locally. The most important example of this is global climate change which may be manifest as alterations in weather patterns, storminess, sea-level rise, flooding and erosion, all of which increase the vulnerability of our coasts, estuaries and brackish lakes. Hence, although the causes of climate change have to be managed on a global scale, the consequences have to be managed locally.

The editors and contributors summarise these local, regional and global effects and look at the implications of climate change for the geographical regions covered by the book. These features are important and so have an interest beyond the local areas. In particular, it is important for all natural and social scientists and policy-makers to understand such changes in wider global areas. Because of this, the book will be of interest to coastal and estuarine scientists worldwide.

Chair in Estuarine & Coastal Sciences, Research Professor, Mike Elliott
Department of Biological and Marine Sciences
The University of Hull
Hull, UK

Formerly Director, the former Institute
of Estuarine & Coastal Studies (IECS)
Hull, UK

Director, International Estuarine
& Coastal Specialist Ltd
Beverley, UK

Co-Editor-in-Chief, Estuarine Coastal
& Shelf Science
Hull, UK

UNIVERSITY
OF HULL

University of Hull
Hull, HU6 7RX, UK
Mob. +44 (0)7946 533938
University Tel. +44 (0)1482 466773
Email Mike.Elliott@hull.ac.uk
Alternative Email Mike.Elliott@iecs.ltd

Foreword

In 2015, the United Nations General Assembly established its position on Sustainable Development Goals (SDG). These 17 inter-linked global goals are designed to be a "blueprint to achieve a better and more sustainable future for all" and should be in place by 2030. While some of the goals (*e.g.*, gender equality, peace, justice and strong institutions) are exceptionally broad in outlook, several relate to environmental issues, especially No. 13 (Climate Action). This goal is designed to "take urgent action to combat climate change and its impacts". Geoscientists have, for many years, been presenting evidence to the Inter-Governmental Panel on Climate Change (IPCC) that there was a clear separation of "natural climate change" from modern, "anthropogenic" climate change. Meetings, such as COP26 (Glasgow, 2021), have identified a number of changes consequent on a global temperature rise of 1.5–1.8 °C (*e.g.*, sea level rise, coastal erosion, reduction of dissolved O_2 in the oceans, ocean acidification, coral bleaching, loss of biodiversity and extreme weather events). Many of these outcomes of climate change require geoscience awareness to resolve the potential problems, sustain ecosystem services and provide leadership in areas such as flooding and coastal zone management.

Many of these topics feature in this volume, where sustainable development is considered from a geoscience perspective. Our lakes, estuaries, and coastal seas are archives of the Pleistocene, Holocene and "Anthropocene" and provide an understanding of the rates of change involved. This volume is also a contribution to SDG No. 4 (Quality Education) in communicating the sustainability issues.

To some, geoscience *is the problem* (*e.g.*, mining and mining pollution, oil exploration, and coalfield development) but it is also clear that, in many aspects of environmental understanding, *geoscience knowledge will also be part of the solution.*

Emeritus Professor of Micropalaeontology
University of Plymouth
Plymouth, UK

Malcolm Hart

University of Plymouth
Plymouth, U.K.

UNIVERSITY OF PLYMOUTH
School of Geography, Earth and Environmental Sciences

Foreword

This volume titled *Coasts, Estuaries and Lakes: Implications for Sustainable Development* covers a wide range of topics in coastal science, primarily in reference to the Indian subcontinent. The studies cover topics including coastal and estuarine water quality, suspended sediments and trace elements, beach morphology and sediments, biodiversity and marine pollution in coastal waters, the impacts of climate change on coasts in this region, consideration of sustainable development of coastal resources, and the usefulness of geospatial tools for monitoring and investigation in the coastal zone.

The wide range of topics, study locations, methods and authors involved in this volume is indicative of an extremely vibrant community of coastal researchers working in this field and who care deeply about preservation and future management of coastal and estuarine environments around the Indian subcontinent. The editors have done an excellent job in pulling this volume together, and I expect that it will become essential reading for researchers interested in the breadth of work being undertaken within coastal science in this region.

Sarah Woodroffe

President of the INQUA Commission,
(INQUA= International
Union for Quaternary Research)
Coastal and Marine Processes, Department
of Geography, Durham University,
Durham, UK

Durham
University
Department of Geography

Inspiring the extraordinary

Foreword

Coastal communities are prone to a range of hazards that can result in the loss of life, damage to infrastructure, economic hardship, and the degradation of ecosystems. While coastal issues of this nature are influenced by a range of biological, chemical, and physical processes, and increasing anthropogenic pressure, many scientific works focus on a subset of these contributors to coastal change. The book *Coasts, Estuaries and Lakes: Implications to Sustainable Development* is a timely and multidisciplinary synthesis of physical and biological processes impacting coastal environments and examines the role they play in sustainable development. The chapters cover a broad, yet well-integrated, spectrum of Earth science disciplines ranging from coastal environments, Earth system processes, and the link between socioeconomics and sustainable development. That sets this volume apart from related books is the multidisciplinary aspect, not only regarding the book as a whole but also the individual chapters. The case studies focus mainly on the vulnerable coastlines of India, but they can be applied globally as they explore issues and solutions that are both relevant and informative to many of the world's coastlines. As co-leader of IGCP Project 725 "Forecasting Coastal Change," I am particularly pleased to see that this book provides a comprehensive understanding of many of the drivers, processes, and scale of coastal change and how this improved understanding enhances a community's ability to manage the coastal zone and better support sustainable development. Researchers, students, practitioners, and enthusiasts alike will find value in this book that is carefully organized and edited by Drs. N. Jayaraju, G. Sreenivasulu, M. Madakka, and M. Manjulatha.

Jessica Pilarczyk

Co-Leader, UNESCO/IGCP 725&, Tier II Canada
Research Chair
Department of Earth Sciences, Simon Fraser University,
Burnaby, BC, Canada

SFU FACULTY OF SCIENCE | DEPARTMENT OF EARTH SCIENCES

Preface

Global warming and climate change continue to pose a severe threat to the coasts and the fragile marine marginal fringes. Coastlines need to be managed effectively and responsibly. Considering the fact that the Earth's surface consists of about 70% water, oceans are crucial for the existence of life on earth. It is indispensable for sustaining all life forms, hence conservation of coastlines is key towards protection of the vulnerable ecosystem as well as human life.

The Earth has around 620,000 km of coastline. With the growing human population and an increasing number of people living near the coasts, it is all the more reason to conserve and look for new ways to protect the natural ecosystems, which is often home to a wide range of spectacular and species-specific biodiversity. On land, they harbor key ecosystems such as freshwater or marine marginal wetlands, which are considered to be vital for migratory bird sanctuaries and other terrestrial biota. In wave-protected zones, they house mangroves, salt marshes, deltas, and sea grasses, among others, all of which can support replenishment and reproduction of aquatic biosphere. Rocky shores along coasts support for a wide range of both benthic and sessile species and several types of seaweed. Colonies of coral reefs are usually found at depths of around 50 m along the shores of tropical regions in clear and warm water.

Other coastal attractions like beaches, tourist islands, and seaside resorts spin a huge economy through coastal tourism. The coastal fragile ecosystems generally provide protection against erosion, rise in sea level, and natural tsunamis, if planted and protected with mangroves and marine vegetation. In several countries, mangroves are the main source of firewood for fuel and domestic usage. Ecosystems such as mangroves and sea grasses have a much bigger ability for carbon sequestration over several well-known terrestrial ecosystems. Thus, they can play a vital role to mitigate, monitor, and manage the impact of climate change by uptake of anthropogenic carbon dioxide pumping. However, because of how economically significant coasts are, many of these communities are susceptible to climate change, which increases the frequency of extreme weather events, sea level rise, and related problems like beach erosion, saltwater intrusion, and marine transgression, which includes flooding. The anthropogenic uses of the coast are further complicated and its vulnerable marginal coastal zones are threatened by a number of additional chronic coastal challenges, including marine trash, microplastic waste, coastal development activities, and marine ecosystem dispersion.

Globally, coastal ecosystems are in decline as a result of the combined effects of climate change, bio habitat imbalance, uncontrolled overfishing, and water pollution (microplastic pollution). This decade (2021–2030) has been designated as the UN Decade on Ecosystem Restoration by the UN. Rare attention has been paid to the rehabilitation of delicate marine habitats. This is due to the fact that coastlines are always changing, making it impossible to calculate their precise perimeter. Today, however, coastal regions are home to the majority of the world's population. Forty-four percent of the world's population, according to the United Nations' atlas, lives about 160 km from the ocean. Many huge cities are situated near harbors and have lavish access to port facilities. Some countries defend and safeguard their coasts against military attacks, marine smugglers, and illegal migrants. This has added a crucial importance to the coasts. Therefore, research on this fragile and economically vital part is mandatory at the present juncture.

The Intergovernmental Oceanographic Commission (IOC) of UNESCO, which focuses on ocean observations, data, services, and associated capacity building. The International Decade of Ocean Science for Sustainable Development, or 2021–2030, has been designated by the United Nations. The 2030 Agenda, which the United Nations endorsed in 2015, was closely related to the idea of an ocean decade. It is intended that the decade would inspire the ocean community to take action on crucial issues including the preservation and usage of the ocean, including research and technology advancements in oceanography. In order to support this sincere and scientific endeavor, we have taken the responsibility of bringing out this edited book to mark the IOC–UNESCO Ocean decade celebrations.

With this background, a humble beginning has been made by us to compile and edit research papers on several burning issues related to the marine marginal bodies including marine pollution. This endeavor has attracted several investigations dealing with the most basic to critical issues like coastal resources, pollution, erosion, restoration, and so on.

We hope that the research, recommended study, and suggestions will provide the needed baseline data for the researchers, stakeholders, policy makers, students, and others who are in constant touch with the coastal zone for science, survival, sightseeing, tourism, economy, and preservation of fragile marine ecosystems. Ocean protection and responsibility at regional and global level is paramount for healthy sustenance of life in the waters. It is imperative to prevent further ecological degradation that in turn affects human well-being and the environment at large.

Kadapa, Andhra Pradesh, India N. Jayaraju
Tirupati, Andhra Pradesh, India G. Sreenivasulu
Kadapa, Andhra Pradesh, India M. Madakka
Jeonju, South Korea M. Manjulatha

Contents

About the Editors

N. Jayaraju is a professor of Geology, Yogi Vemana University, India. His research interests include of environmental geosciences, marine ecology, marine pollution, coastal zone resources, sediments of marine marginal water bodies, viz., lakes and estuaries. He was awarded gold medal for being a university topper in his master's degree course and later received University Merit Fellowship for pursuing PhD. He worked as PDF for UGC, CSIR, DST, and all central government prestigious funding agencies. He has published more than 60 research papers in various national and international journals. Jayaraju has attended more than 50 international conferences abroad and 40 in India. He is an active member in many scientific associations/organizations. Moreover, he is serves on editorial boards and reviewer boards for many prestigious journals. Jayaraju has travelled widely on his research endeavor, visiting over 60 countries like Australia, China, Japan, Germany, France, Italy, Middle East, the UK, and the USA.

G. Sreenivasulu received his MSc (with university top rank and gold medal) and PhD from the Department of Geology, Yogi Vemana University, Kadapa, India. He worked as a Research Associate at the National Centre for Earth Science Studies, Thiruvananthapuram, India. Currently, he is working as Dr. D. S. Kothari Postdoctoral Fellow in the Department of Geology at Sri Venkayeswara University, Tirupati, India. He was awarded DST-INSPIRE, a prestigious fellowship, for 5 years for PhD program by the Department of Science and Technology (DST), Government of India. Thrust areas of his research include micropaleontology, marine pollution, coastal morphodynamics, and application of geospatial tools. He has attended many scientific events related to his relevant area of specialization both within India and abroad. He has published more than 25 papers in international and national journals. He is an active member in many scientific associations/organizations. Moreover, he serves on editorial boards and reviewer boards of many prestigious journals.

M. Madakka is currently an associate professor in the Department of Biotechnology and Bioinformatics, Yogi Vemana University, Kadapa, AP, India. She received her MSc and PhD degrees in microbiology from Sri Krishnadevaraya University, Anatapur, India. Her research interests are in the fields of environmental biotechnology and environmental microbiology. She was awarded Rajiv Gandhi fellowship for PhD from UGC, New Delhi, India. She secured DST-SERB fast track young scientist fellowship during 2012–2015, New Delhi. She has over 17 years of research and 14 years of teaching experience in the university. She has published more than 40 research papers in peer-reviewed national and international scientific journals and presented several papers in national/international conferences.

M. Manjulatha, Senior Research Associate, is currently affiliated to the National Institute of Horticultural and Herbal Sciences, RDA, South Korea. She obtained her PhD from the Department of Biotechnology, Sri Krishna Devaraya University, in collaboration with the Department of Crop Physiology, University of Agricultural Sciences, GKVK, Bangalore. She worked as a research associate in DBT-supported Center of Excellence program and ICAR FIST-DST niche area of excellence for drought research in UAS, GKVK, Bangalore. She has 14 years of research experience in the area of plant molecular biology, biotechnology, transgenics, gene editing, and molecular markers. She has published 25 research papers in peer-reviewed journals and 4 book chapters. She has been a reviewer for various international journals.

Part I
Coastal, Estuarine and Lake (Brackish) Environments: Introduction, Definition, Processes and Dynamics

Chapter 1
Bibliometric Analysis of the Literature on Coastal Sediment Pollution

Nezha Mejjad, Abdelmourhit Laissaoui, Bouabid El Mansouri, Ahmed Fekri, Aniss Moumen, Khalid El Khalidi, and Ouafa El Hammoumi

Abstract This study addresses a bibliometric analysis of the global research literature and the temporal research record on coastal sediment pollution. Two thousand five hundred publications were exported from the Dimensions database and analyzed by the VOSviewer software tool. The results showed a notable increasing trend in scientific productivity related to coastal sediment pollution recorded since 2007, while the highest number of publications are in 2014 and 2013. Biological, environmental, and earth sciences were the top investigated subject categories. The Marine Pollution Bulletin journal was ranked as the lead in coastal sediment pollution research publications. The analysis based on term occurrences indicates the shift of research interest from studying heavy metals pollution in marine sediments to microplastic pollution. The bibliographic coupling analysis of countries indicated that China is the most active and productive country based on the publication weight (813), citations number (6365), and total link strength (583,176). These analytical

N. Mejjad (✉)
Faculty of Sciences, Ibn Tofail University, Kenitra, Morocco

Faculty of Sciences – Chouaïb Doukkali University, El Jadida, Morocco

National Center for energy, Sciences and Nuclear Techniques, Rabat, Morocco

A. Laissaoui
National Center for energy, Sciences and Nuclear Techniques, Rabat, Morocco

B. El Mansouri
Faculty of Sciences, Ibn Tofail University, Kenitra, Morocco

A. Fekri · O. El Hammoumi
Faculty of Sciences Ben M'sik, Hassan II University, Casablanca, Morocco

A. Moumen
National School of Applied Sciences, Ibn Tofail University, Kenitra, Morocco

K. El Khalidi
Faculty of Sciences – Chouaïb Doukkali University, El Jadida, Morocco

N. Jayaraju et al. (eds.), *Coasts, Estuaries and Lakes*,
https://doi.org/10.1007/978-3-031-21644-2_1

investigations provide a global overview of academic publications related to marine sediments pollution, which will be helpful to the researchers in discerning the latest trends and progress in this research domain. However, future investigations are recommended to fill the knowledge gaps on coastal pollution management and its social and economic implications.

Keywords Coastal sediment pollution · Global research analysis · Heavy metals · Microplastic · Dimensions · VOSviewer · Bibliometric analysis

1.1 Introduction

Coastal sediment is an essential component of ecosystem cycles where the biogeochemical processes related to nutrients occur. It also plays a critical role as a habitat for marine organisms (Snelgrove 1997; Snelgrove 1999; Reise et al. 2009; Levin et al. 2001, Joensuu et al. 2018; Thrush et al. 2021) and acts as a sink for various naturally occurring and anthropogenic pollutants (Seeley et al. 2020; Li et al. 2021). Besides, sediment is the final receptor of pollutants and is used as a pollutant indicator and accumulator (Viguri et al. 2002; Nasr et al. 2006; Yousef et al. 2015). In this context, many studies served sediment to assess the health of the environment (e.g., Gang et al. 2018; Schintu et al. 2016; Chuan and Yunus 2019; Rezaei et al. 2021) and define the possible impact on marine organisms (Mejjad et al. 2020a; Bat et al. 2021; Tamilmani and Venkatesan 2021; Maghsodian et al. 2022, Yuksel et al. 2022). Besides, they were used to follow the pollutants' deposition history in an ecosystem, reconstruct the chronology of pollution, and identify the pollutant's sources (Mejjad et al. 2016; Laissaoui et al. 2018; Hassen et al. 2019; Lin et al. 2020; Martin et al. 2020; Benmhammed et al. 2021a; Dahl et al. 2021, Uddin et al. 2021; Shi et al. 2022; Mejjad et al. 2022).

As it provides a habitat for the marine organism, sediment is an indirect indicator of fish habitat quality (Joensuu et al. 2018; Thrush et al. 2021). Many aquatic species that play an essential role in climate regulation and nutrient generation live in sedimentary habitats; thus, the deterioration of their quality would impact the species living in and the environment.

In response to human activities growth, coastal land development has significantly impacted coastal water quality, causing an increase in sediment accumulation rate and pollutants amount flowing into the ocean (Mejjad et al. 2020b). Accordingly, different types of pollution could be introduced to coastal environments, such as heavy metals (Morillo et al. 2004; Buccolieri et al. 2006; Delshab et al. 2017; Mejjad et al. 2018; Benmhammed et al. 2021a) and radioactive elements (Urry 1949; Hetherington and Harvey 1978; Yii et al. 2009; Zare et al. 2012; Mejjad et al. 2016; Uddin and Behbehani 2018; Yümün et al. 2021). These chemicals are well known for their toxicity to living organisms, persistence in the environment, and their bioaccumulative nature, leading to the contamination of the food chain (Hazrat et al. 2019). Thus, the geochemical, sedimentological, and environmental analyses of coastal sediments are crucial for assessing and monitoring the potential environmental and health risks and helping in the management projects and strategies' implementation for sustainable development of coastal ecosystems.

In this study, a historical review and bibliometric analysis of research on coastal sediment pollution were carried out to understand the global research growth trend of the investigated topic. Thus, this paper will overview scientific literature development from 1892 to 2022 on coastal sediment pollution based on the Dimensions database.

1.2 Methods

1.2.1 Data Collection

Data were collected from Dimensions.ai (www.dimensions.ai, accessed on May 02 2022), one of the major bibliometric databases (Basson et al. 2022) that has higher broader coverage than Web of Sciences and Scopus (Herzog et al. 2020; Visser et al. 2022; Guerrero-Bote et al. 2021). This is due to the selection criteria of Dimensions, which is based mainly on a single variable, the presence of DOI (Digital Object Identifier) and because it uses Crossref, PubMed, and Open Citation, among other sources, to populate its database (Basson et al. 2022). This bibliometric database was launched in January 2018 by Digital Sciences (Adams et al. 2018). This database platform includes articles, chapters, books, policy documents, grants, patents, and clinical trials. In the present study, we limited our research to articles, chapters, edited books, preprints, and proceedings. The research parameters query is as follows:

- Date range: from 1892 to 2022
- Publication type: articles, chapters, edited books, preprints, and proceeding
- Query date: performed on May 02 2022
- Used terms: "Sediment AND coastal AND pollution"
- The number of extracted publications: 2500 publications relevant to the search query

1.2.2 Bibliometric Analysis Using VosViewer Software

VOSviewer is a software tool allowing constructing and visualizing bibliometric networks of data that may include researchers, journals, or individual publications (Van Eck and Waltman 2010). Using bibliometric databases such as Scopus, Web of Sciences, or Dimensions allows constructing analysis based on citation, co-citation, bibliographic coupling, or coauthorship relations. This software allows visualizing three types of maps: (i) the network visualization, (ii) the density visualization, and (iii) the overlay visualization. We used in the present study the latest version of VOSviewer (Version 1.6.18), released on Jan. 24 2022. Then, we performed the bibliometric mapping using the extracted data from the Dimensions database as an input to VOSviewer. Two thousand five hundred publications dated 2019, 2020, 2021, and 2022 were exported to carry out bibliometric mapping.

The first step of creating a map using VOSviewer consists of choosing the fields from where the terms will be extracted, so we extracted the terms from the title and abstract fields. In the second step, we excluded the terms with a number of occurrences inferior to 30 terms where from 45,697 terms, only 589 met the defined threshold. Then for each of the 589 terms, a relevance score was calculated, and based on this score, 353 most relevant terms were selected. Before visualizing the maps, the last step consisted of verifying the selected terms in the function of occurrence and relevance (Fig. 1.1).

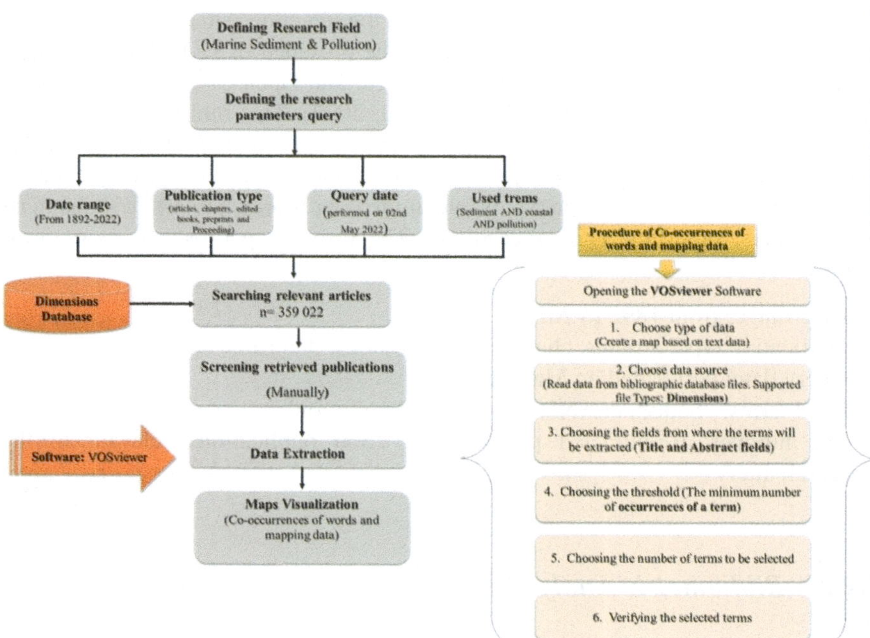

Fig. 1.1 Flowchart displaying the procedures of map visualization through VOSviewer software

1.3 Results and Discussions

1.3.1 Chronology of Publications from 1892 to 2022

1.3.1.1 Type of Publications

Our search query in Dimensions exhibited that 52.47% (*N* = 188,398) of documents are articles, 42.37% (*N* = 152,120) are chapters, 3.52% (*N* = 12,626) edited books, (*N* = 0.93%) 3335 preprints, and 0.71% (*N* = 2546) proceedings (Fig. 1.2). These publications related to coastal sediments pollution were carried out between 1892 and 2022; these publications were mostly published in "The Marine Pollution Bulletin journal" (*N* = 11,013) and in the "Sciences of the Total Environment journal" (*N* = 7955). This result is consistent with the Sun et al. 2012 study where the Marine pollution bulletin (MPB) journal was found to be the most active journal in estuary pollution-related research.

1.3.1.2 Research Category

Three lakh fifty-nine thousand and seventeen publications related to our research query were shown in Dimensions by May 02, 2022, most of them published in peer-reviewed journals with 11,013 publications in "Marine Pollution Bulletin" followed by 7955 publications in "The Sciences of the Total Environment"; the number of publication in the other source titles ranged between 4199 and 1. Figure 1.3 shows the main field of research related to our search query. The fields ranked the

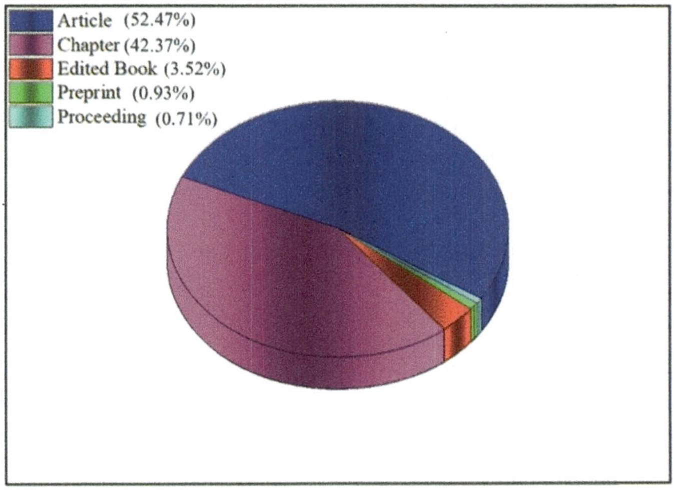

Fig. 1.2 Percentage of publications according to their type

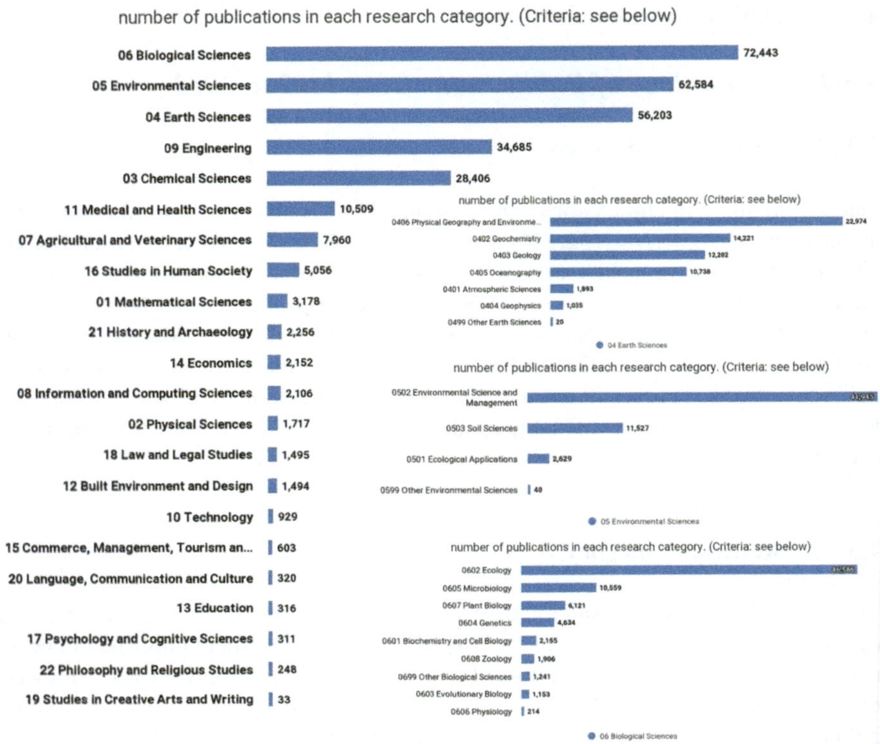

Fig. 1.3 Free screenshot showing the number of Publications versus research category. (*Exported on May 03, 2022; sourced from Dimensions at* www.dimensions.ai)

"Biological Sciences" as the first investigated field in relation to our search query with 2,304,654 citations, followed by "Environmental Sciences" and "Earth Sciences" in second and third places with 1,882,058 and 1,596,385 citations, respectively. Other research categories are also shown in Fig. 1.2. This suggests the increasing interest in studying the biological aspects, such as the ecological and microbiological characterization of sediments, as well as genetic aspects of sediments.

Indeed, Fig. 1.4 indicates the growing trend of the number of publications related to coastal sediment pollution between 1973 (51 publications) and 2021 (906 publications) (3 years after the creation of the MPB journal), reflecting the ongoing interest in the investigated subject. Biological sciences, environmental sciences, and earth sciences were found as the most popular fields of study related to our search query "coastal sediment pollution". This is due to the importance of sediment as a habitat for the benthic community, such as annelids, mollusks, worms, and crustaceans (Thrush et al. 2004). This close liaison between benthic organisms and sediments that act as a sink of various environmental pollutants increases the concern about the impacts of coastal sediments pollution on this community, leading to

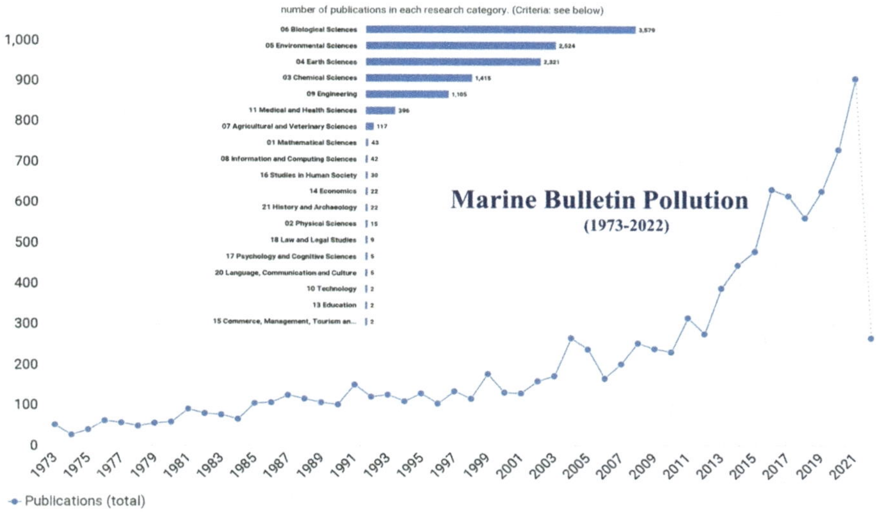

Fig. 1.4 Free screenshot showing the number of publications published in Marine Bulletin Pollution Journal between 1973 and 2022 and the number of publications according to the research category. (*Exported on May 04, 2022; sourced from Dimensions at* www.dimensions.ai)

growing interest in this area of research. Noting that the benthic community plays an essential role in controlling the water and sediment quality. These living organisms are relatively immobile, which indicates that the continued exposure to pollutants in their habitat in contact with polluted sediment would threaten them and affect the ecosystems (Santos et al. 2009). In this sense, sediments are an essential medium between environmental, biological, and chemical processes, which explain the highest number of publications recorded for these research categories.

1.3.1.3 Evolution of Publication Rate by Year

Coastal sediment pollution-related research has known notable progress in the past two decades; however, an evident variation in the publication number per year is shown in Fig. 1.5. A significant increasing trend in publication outputs has been detected since 2007, while the maximum number of publications was noted in 2014 and 2013 (95,392 publications), representing 26% of the total published publication between 1892 and 2022, according to the Dimensions analyzed database. The maximum number of publications recorded during 2013 and 2014 related to our search query could probably be linked to the increasing project trends calling for protecting and managing the ocean, where many initiatives and management plans were defined and launched between 2010 and 2020. For instance, at the margin of the 10th meeting of the Conference of the Parties to the Convention on Biological Diversity (COP10) held in Nagoya, Japan, in October 2010, the Sustainable Ocean Initiative (SOI) was born (SOI 2014).

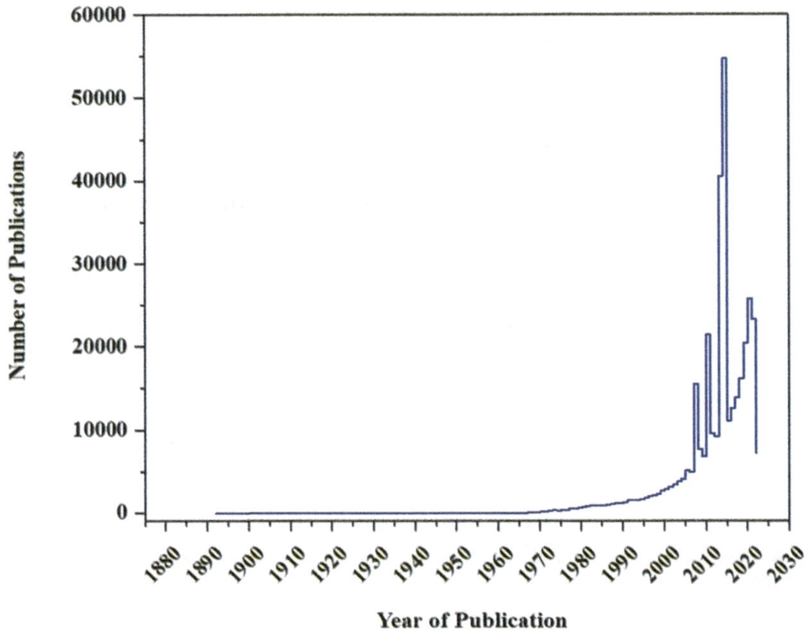

Fig. 1.5 Evolution of publications rate between 1982 and 2022

Indeed, according to White et al. (2017), the publication output trends in 2014 in peer-reviewed Sciences and Engineering journals, conference proceedings, and books have known an exceptional growth, reaching around 2.3 million globally. This recorded growth varied across the world according to data availability and technological capabilities, suggesting further investigations of the unbalanced growth of Scientific research input across the globe. Besides, multidisciplinary biology, environmental sciences, and management were the major research categories.

The observed low rate of publications in the fields of "coastal pollution" (<200 publications per year between 1892 and 1970) is probably because the issue of ocean pollution was not of major concern where communities around the world used to dump their waste into the ocean. Thus, little attention was given to the effects of dumping waste in the marine environment, while according to the United States Environmental Protection Agency (EPA), before 1970, various establishments used to dump their chemical by-products (e.g., industrial waste, sewage sludge, organic chemical wastes, heavy metals, and radioactive waste) into waterways to eliminate their waste. Then since 1970, various congresses were organized (Fig. 1.6) in response to the marine pollution increase, while in 1972, a convention called the "London Convention" was signed to protect the marine environment from dumping waste in the ocean. Noting also that the increasing interest in ocean pollution and its impact on the marine environment led scientists to investigate marine sediments because of their high adsorption capabilities of contaminants and because

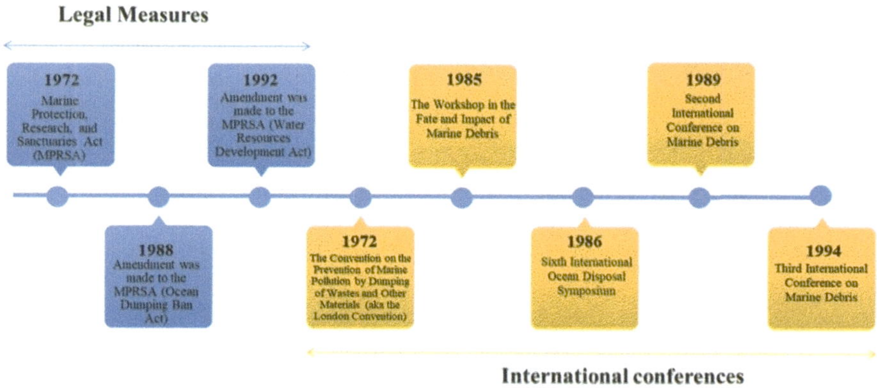

Fig. 1.6 The main legal measures and international conferences carried out between 1972 and 1994. (*MPRSA*, 1972; *EPA*, 1988; *PUBLIC LAW 102-580—OCT. 31*, 1992; *Workshop on the Fate and Impact of Marine Debris* (1984: *Honolulu, Hawaii*); *International Ocean Disposal Symposium*, 1986; *Proceedings of the Second International Conference on Marine Debris, 2–7 April* 1989; *Faris* 1994)

they represent a suitable tool to establish the history of contamination in a system (Valette-Silver 1993). Consequently, in the late 1970s, Goldberg and other Scientific researchers started using sediment cores to reconstruct the coastal pollution history by using radiometric-dating tools (Goldberg 1963; Goldberg and Bruland 1974, Goldberg et al. 1977, 1978). Accordingly, the low publication rate recorded before 1970 is because little was known about the marine environment and the effects of dumping waste in the ocean.

1.3.2 *Bibliometric Analysis Using VOSviewer*

1.3.2.1 Visualization Map/Network Visualization Based on Text Data

Two thousand five hundred from around 78,000 papers published in 2019, 2020, 2021, and 2022 sorted by relevance to the search query (Coastal AND Sediment AND Pollution) extracted from Dimensions were used for generating the visualization map. Thus, text data were used to establish a network of co-occurrence links between terms, while these terms were defined in the text data by utilizing natural language processing algorithms (Van Eck and Waltman 2022). In total, 333 terms or keywords were grouped into 4 clusters and presented in Fig. 1.7 in the form of colored circles linked by 33,030 links and a total link strength of 338,738. The map shows four colored clusters (Red, Blue, Green, and Yellow).

The main keywords co-occurrence for the red cluster (include 94 items) are heavy metals (occurrence: 762), metals (occurrence: 680), index (occurrence: 539), trace element (occurrence: 208), and surface sediment (occurrence: 389). Noting

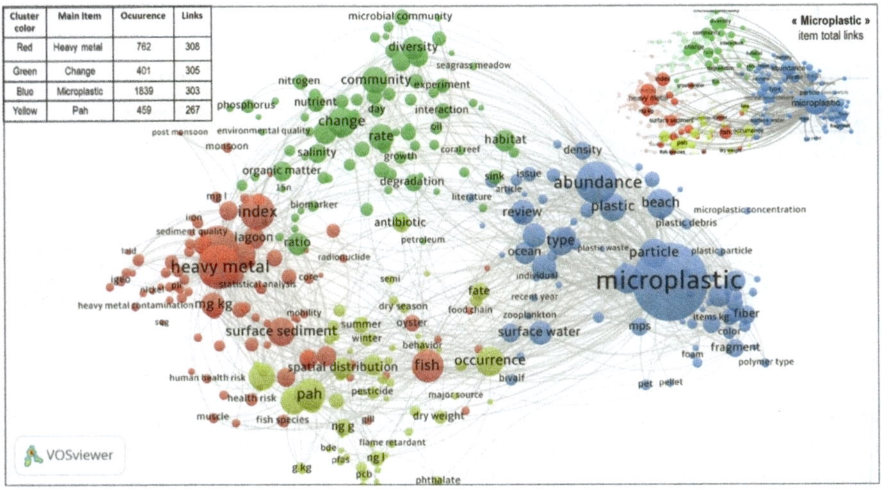

Fig. 1.7 Visualization map according to publications weights. *Each circle's size depends on the number of published studies related to the search query or items (keywords or terms); the line represents the linkage between every item. The link between every term/keyword is solid when the line is short*

that the term "heavy metals" was reported in many studies as a misused term (Chapman 2007; Batley 2012; Pourret et al. 2021) where the authors recommended replacing "heavy metals" with "metal", "trace metal", or metalloids, which explain the co-occurrence of two terms indicating the same group of elements. The occurrence number of the term "index" is related to the monitoring of pollutants' levels in sediments where environmental indexes are used, such as enrichment factor and geo-accumulation index (Benmhammed et al. 2021b) and ecological indexes (Maanan et al. 2018; Mejjad et al. 2022). The cluster group terms related to sediments analysis and pollution level assessment and monitoring using environmental indexes allow defining the potential risk (occurrence: 95), presenting these chemical elements on the studied system environment and sometimes on living organisms using sediment guideline values (occurrence: 65). The red cluster then covers topics related to chemical pollution evaluation using sediments (surface sediment and sediment core) and fish (occurrence: 484) and environmental assessment of pollution levels using environmental indexes.

The second cluster with 85 items covers the biological sciences research category. The main keywords include community (occurrence: 352), change (occurrence: 401), habitat (occurrence: 241), and diversity (occurrence: 278). The terms "change" in the visualization map represent 305 links; the links with the terms "Change" are strong with the following keywords "community" and "nutrient", "response", and "degradation" as they are close to each other. According to Van Eck and Waltman (2022), the distance between two items in the visualization map indicates the relatedness of topics in terms of research category, so the closer items present strong links. Also, the term "change" shows a weak link with "heavy metal",

"metal", and "index" terms, but it shows an existing connection of the terms and suggests that heavy metal accumulation could impact the generation of nutrients, impact the marine community, and cause degradation of the marine ecosystem. The increase in heavy metal concentrations raises the toxicity risk in the marine environment (Perumal et al. 2021). Indeed, the physicochemical composition of sediment can considerably influence its role as a secondary source of metals bioavailability to the water column (Miranda et al. 2021).

The blue cluster comprises 84 items and covers the microplastic occurrence and fate in the marine environment topic. The term "microplastic" presents the high number of occurrences (occurrence: 1839), then "abundance" (occurrence: 702) and "size" (occurrence: 394), reflecting the growing interest in microplastic pollution effects analysis and investigation. The significant occurrence of the term "size" is an essential parameter to define, measure, and study because smaller microplastic particles are considerably more consumed by marine living organisms (Azizi et al. 2022; Hankins et al. 2022; Liu et al. 2022). da Silva et al. (2022) have demonstrated that the smaller microplastic particles can severely affect the trophic chain suggesting that the most significant effect of the microplastic consumption derives from the lower trophic levels of the trophic web. The third cluster items present short lines reflecting strong links between each other, where the term "microplastic" indicates 303 links and 32,717 strength links with the blue cluster items and other cluster items.

The yellow cluster presents fewer items (70 items) and a low number of occurrences compared to the other clusters. It covers the topics related to the ecological effects of persistent organic pollutants such as Polycyclic aromatic hydrocarbons (Pah). The occurrence of terms "Pha" (occurrence: 469) and "ecological risk" (coccurrence: 371) in a small number of occurrences is because probably there is little attention given to the analysis of Pha ecological risk on coastal sediment in the last decades compared to heavy metal pollution and microplastic pollution.

Almost the four clusters cover topics related mainly to environment quality assessment and monitoring of the marine environment, while little attention is given to coastal systems' management and restorations topics.

1.3.2.2 Overlay Map/Overlay Visualization Based on Text Data

The overlay map is the same as the visualization map, where only the colors of items are different. Here, in Fig. 1.8, each item is defined by its score and represented by a specific color where by default, colors range from the lowest score (blue) to the highest score (green to yellow). The overlay map shows the average citations; the blue terms indicate average citations below 4, the green terms have average citations around 10, and the yellow keywords present average citations of 10 or higher. We can notice that "microplastic" presents the high average citations (16.66) together with all the closer keywords presenting the strongest link with the microplastics. Biological research category (Green cluster) and heavy metals (Red cluster) related terms presented the lowest score, indicating average citations ranged

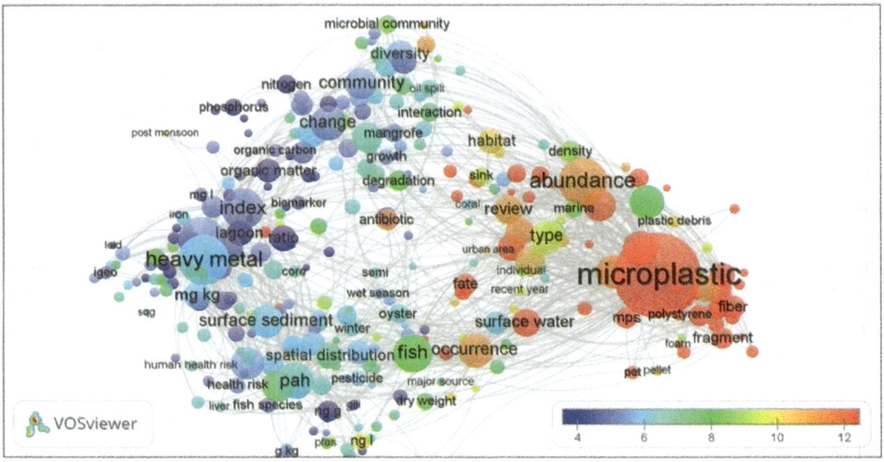

Fig. 1.8 Overlay visualization showing the average citations corresponding to each item. (*The blue color represents the lowest score while the red color indicates the highest score*). (Produced by VOSviewer 1.6.18)

from 4 to 6 compared to microplastic (blue cluster) related terms, which presented average citations higher than 10. This results in average citations network of 2500 publications and suggests a trend toward investigating and assessing microplastic pollution in the last 4 years. It was reported by Hu et al. 2019, that microplastic pollution in the terrestrial and aquatic environment has become a significant concern, especially this kind of pollutants presents long and poor degradability and could affect fish and living organisms when they ingest the microplastic particles (Jiang 2017; Wang et al. 2018). The obtained results are consistent with the Zhang et al. 2020 study focused on a bibliometric analysis of the global trend in microplastic research between 1986 and 2019. The authors reported that in 2018 and 2019, the publications related to microplastic pollution accounted for more than half of the total (at least 2500 papers published so far on microplastic). This is explained by the growing concern about microplastic pollution's impact on the environment, ecosystem services, and human health research. Also, the development of methods used to extract and identify microplastics in different matrices among the reason for the increasing number of related publications in the latest years. Similar results were reported in studies based on bibliometric analysis of microplastic pollution, confirming that the microplastic-related research input has grown in the latest years and expanded into many other related fields, as shown in Fig. 1.9 (Qin et al. 2020; Wong et al. 2020; Frias et al. 2021).

Figure 1.9 is a fragment of overlay visualization for the term "Microplastics", showing the links between "microplastic" and other terms. Noting that link strength between two circles presents the number of articles in which the two terms occurred together. The "microplastic" term occurred 1839 times in the extracted papers, while it presents 32,717 of total link strength, representing then the most

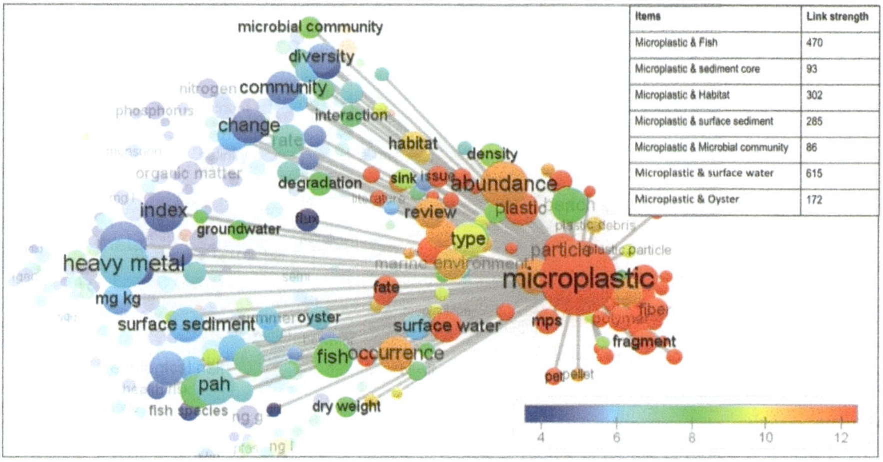

Items	Link strength
Microplastic & Fish	470
Microplastic & sediment core	93
Microplastic & Habitat	302
Microplastic & surface sediment	285
Microplastic & Microbial community	86
Microplastic & surface water	615
Microplastic & Oyster	172

Fig. 1.9 The fragment of overlay visualization for the term "Microplastics"; the table shows the strength links between "Microplastic" and other occurred terms. (*Produced by VOSviewer 1.6.18*)

high-frequency terms. The "Microplastic" and "Fish" link strength analysis shows that the two terms have occurred together in 470 articles from 2500. "Microplastic" and "Surface Sediments", "Microplastic" and "Surface Water", "Microplastic", and "Habitat" terms were respectively found together in 285, 615, and 302 articles (Fig. 1.9). Also, "microplastic" and "sediment cores" have occurred together in 93 articles. Indeed, sediment cores are used to follow the disposition history of microplastic in marine ecosystems, allowing a better understanding of microplastic pollution trends over the years (Uddin et al. 2021). Consequently, this analysis shows that the "Microplastic" research field has grown mainly by expanding into other fields.

1.3.2.3 Density Heat Map/Density Visualization Based on Text Data

The terms extracted from the titles and abstracts and keywords fields related to the search query "coastal sediment pollution" were summarized and counted. They are 45,697 terms in title and abstract and 353 terms meeting the threshold of 30 occurrences. Figure 1.10 displays the density heat maps of terms and clusters where the hot areas are represented by a warm red while the cool area is represented by blue color.

Density Visualization Map

Figure 1.10b shows the density visualization of the research hotspot. Each term density relies upon the number and weight of neighboring terms. Hence, a larger number of neighboring terms and shorter distances designate greater density,

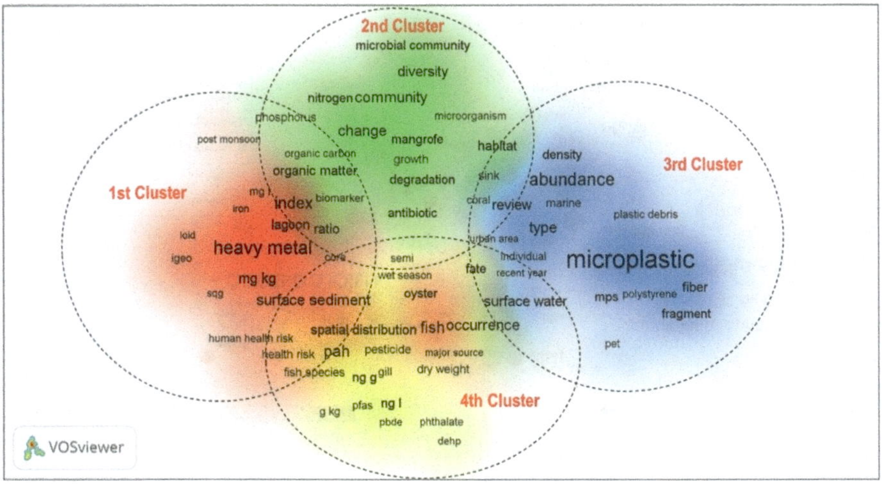

Fig. 1.10 Cluster density visualization map of 353 terms, each occurring 30 times or more in the title, abstract, and the list of the keywords

resulting in an intense color (Van Eck and Waltman 2022). The terms presenting high density are microplastic, Pha, heavy metal, community, index, lagoon, and change (Fig. 1.10). We can conclude that these terms are the core keywords in sediment coastal pollution studies.

Three hundred and fifty three keywords are grouped into four clusters showing four research fields. The red cluster mainly focused on heavy metals pollution assessment using environmental indexes as indicated by the keywords "heavy metal", "sediment", "index", and "Igeo". The second cluster is mainly related to marine biological studies as shown by the terms "community", "microorganism", "change", "mangrove", "microbial community", "nitrogen", and "diversity". The blue-colored cluster presents the most occurred term which is "microplastic"; other terms are shown in the map such as "abundance", "plastic debris", "fragment", "density", and "surface water", while the yellow cluster is showing terms mainly related to pesticides, Pha pollution, and the impact on fish, oyster, and human health. Among the terms, "Microplastic", "heavy metal", "abundance", "community" "lagoon", and "Pha" have intense colors suggesting that they are the main research hotspots.

1.3.2.4 Visualization Map and Density Map Based on Bibliographic Data

In order to follow the publications rate trend by country, we created maps based on bibliographical data, which allowed making bibliographic coupling and analyzed the bibliographic coupling links' strengths. Accordingly, bibliographic coupling analysis was chosen to identify the number of publications related to our search

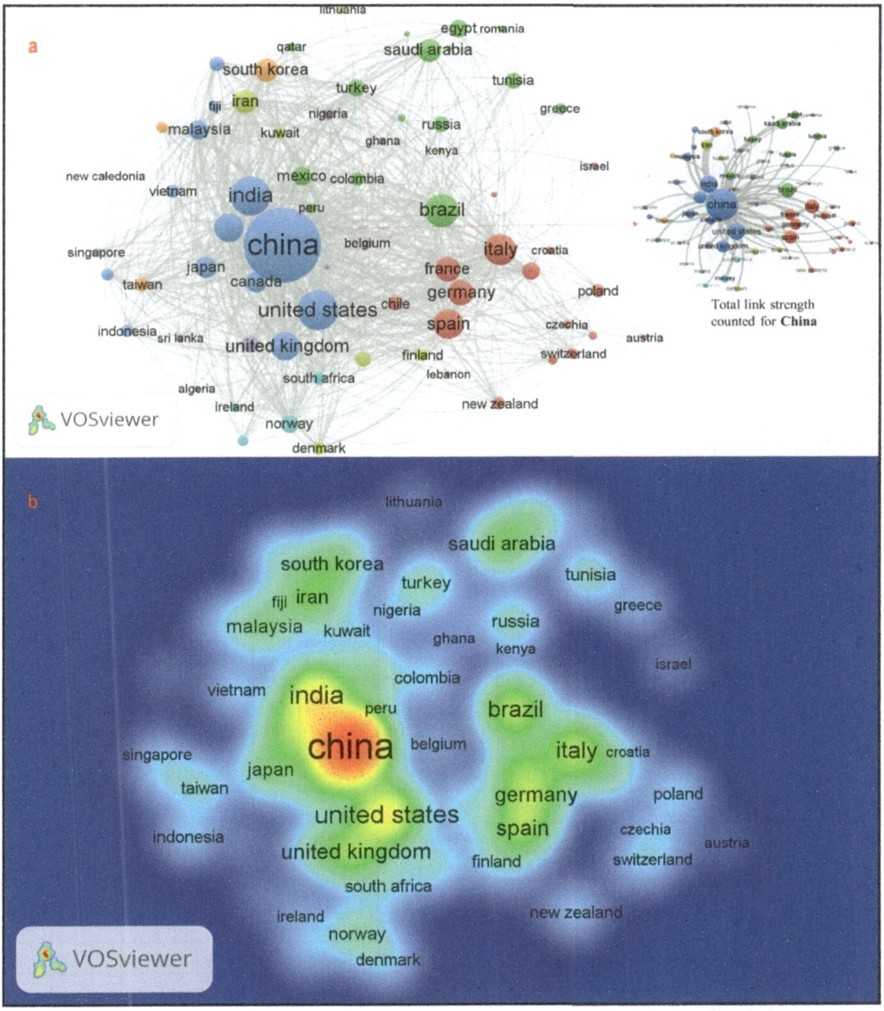

Fig. 1.11 Bibliographic coupling analysis of the countries based on documents weight (**a**) *network visualization*; (**b**) *density visualization*

query by country. We excluded the country with a number of publications inferior to 5. From the 120 countries, 68 meet the threshold. As shown in the visualization network, we obtained seven clusters from the analysis (Fig. 1.11a). The size of the circles in the figure rises in proportion to the number of publications. Different colors indicate different clusters showing links with each other. Countries showing in the same clusters are citing each other more regularly. Figure 1.11b shows the hotspots countries on this topic, where China come first (warm red), then India and the United States (yellow), followed by Brazil, Italy, the United Kingdom, Spain,

Germany, France (yellow to green), Saudi Arabia, South Korea, and Iran (green to blue).

The calculated total link strength of bibliographic coupling analysis shows that the country leader in the field of coastal pollution is China, with 813 publications, 6365 citations, and 583,176 total link strength (Table 1.1). Indeed, the number of Chinese academic articles on microplastic recorded in 2015 accounted for 17% of the world's total (Towards Osaka Blue Ocean Vision 2021). Countries with a number of publications higher than 100 are the United States (227), India (221), Brazil (155), Italy (143), Spain (123), Australia 120, and the United Kingdom (119). We can notice the unbalanced development of scientific research input across the world, which was explained by the variation in data availability and technological capabilities across the world (White et al. 2017). Table 1.1 compares the scientific input between developed countries and developing countries. Except for China, India, and Brazil, almost all developed countries meet the threshold where the number of publications exceeds five papers. Nothing that also, some other developing countries, such as Iran and Mexico, indicate significant scientific productivity in good accordance Gonzalez-Brambil et al. 2016, which indicated that the nation's size and their

Table 1.1 Number of documents per country per cluster based on the bibliographic coupling analysis

Cluster 1 (17 items)		Cluster 2 (16 items)		Cluster 3 (15 items)		Cluster 4 (6 items)		Cluster 5 (6 items)		Cluster 6 (5 items)		Cluster 7	
Country	Doc. Wgt	Country	Doc. Wgt	Country	Doc. Wgt	Country	Doc. Wgt	Country	Doc. Wgt	Country	Doc. Wgt	Country	Doc. Wgt
Austria	5	Brazil	**155**	Australia	**120**	Denmark	26	Argentina	41	Algeria	7	Pakistan	19
Chile	29	Colombia	25	Bangladesh	63	Finland	23	Belgium	15	Ireland	13	South Korea	76
Croatia	11	Egypt	48	Canada	65	Iran	**82**	Ecuador	6	Netherland	18	Taiwan	29
Czechia	10	Greece	19	China	**813**	Kuwait	21	Ghana	8	Norway	49	---	-
Estonia	12	Jordan	5	Fiji	16	Lithuania	9	Nigeria	19	South Africa	27	---	-
France	**86**	Kenya	6	India	**221**	Sweden	34	Sri Lanka	8	---	-	---	-
Germany	**95**	Mexico	53	Indonesia	23	---	-	---	-	---	-	---	-
Israel	9	Morocco	8	Japan	67	---	-	---	-	---	-	---	-
Italy	**143**	Peru	15	Malaysia	57	---	-	---	-	---	-	---	-
Lebanon	6	Qatar	13	New Caledonia	6	---	-	---	-	---	-	---	-
Montenegro	5	Romania	9	Singapore	12	---	-	---	-	---	-	---	-
New Zealand	23	Russia	37	Thailand	22	---	-	---	-	---	-	---	-
Poland	25	Saudi Arabia	77	United Kingdom	**119**	---	-	---	-	---	-	---	-
Portugal	45	Tunisia	40	United States	**227**	---	-	---	-	---	-	---	-
Slovenia	10	Turkey	44	Vietnam	27	---	-	---	-	---	-	---	-
Spain	**123**	United Arab Emirates	6	---	-	---	-	---	-	---	-	---	-
Switzerland	16	—	-	—	-	---	-	---	-	---	-	---	-

Developed Countries

Developing Countries

R&D investments are not always the key driver and indicator of efficiency. Besides, Table 1.1 shows us that the scientific input of African countries on "coastal sediment pollution" is too low compared to Asian, European, and American countries.

1.4 Conclusions

The scientific literature on coastal sediment pollution was investigated with sciento-metric software (VOSviewer), based on data extracted from the Dimensions. Accordingly, this study systematically presented the hotspots in the field of coastal sediment pollution. At the same time, the elaborated bibliometric analysis set the importance of the coastal sediment by summarizing and counting the number of publications and the main investigated research categories related to coastal sediment pollution.

The obtained analysis results suggest an increasing trend toward investigating and assessing microplastic pollution in the last 4 years compared to other environmental pollutants. The study revealed that the publication's subjects are mainly related to environmental quality assessment and marine environment monitoring, while little attention was given to the management and restorations side of coastal systems. Thus, there is a need to consider investigating the management tools and technologies to avoid the possible effects of sediments pollution on living organisms and coastal ecosystem services.

An increasing trend in the number of academic publications was recorded between 1892 and 2022. The growing concern on marine pollution since the 1970s was reported as the leading cause of the investigation of coastal sediments in the last five decades.

The bibliographic coupling analysis of countries indicated that China is the most active and productive country based on its publications weight (813), citations number (6365), and total link strength (583,176). In addition, the bibliographic coupling analysis indicates that the scientific input of African countries on "coastal sediment pollution" is too low compared to Asian, European, and American countries.

Also, the highlighted variation in academic publications productivity per year is probably related to data availability, technological capabilities, and development. Particularly in the fields of oceanography, technological issues remain the possible leading cause of the unbalanced growth of scientific research input across the globe. Accordingly, further investigations in terms of scientific productivity trends in developing and developed countries are needed.

These findings provide a global overview of academic publications related to marine sediments pollution, which will be helpful to the researchers in discerning the latest trends and progress in this research domain.

Acknowledgments The authors would like to highlight that this book chapter has been made possible, thanks to the Dimensions.ai database.

References

Adams J, Draux H, Jones P, Osipov I, Porter S, Szomszor M (2018) Dimensions -a collaborative approach to enhancing research discovery. Digital Science. https://www.digital-science.com/resources/portfolio-reports/dimensions-collaborative-approach-enhancing-research-discovery/

Azizi N, Nasseri S, Nodehi RN, Jaafarzadeh N, Pirsaheb M (2022) Evaluation of conventional wastewater treatment plants efficiency to remove microplastics in terms of abundance, size, shape, and type: a systematic review and meta-analysis. Mar Pollut Bull 177:113462

Basson I, Simard M-A, Ouangré ZA, Sugimoto CR, Larivière V (2022) The effect of data sources on the measurement of open access: a comparison of Dimensions and the Web of Science. PLoS One 17(3):e0265545. https://doi.org/10.1371/journal.pone.0265545

Bat L, Arici E, Öztekin A (2021) Threats to quality in the coasts of the Black Sea: heavy metal pollution of seawater, sediment, macro-algae and seagrass. In: Spatial modeling and assessment of environmental contaminants. Springer, Cham, pp 289–325

Batley GE (2012) "Heavy metal"—a useful term. Integr Environ Assess Manag 8(2):215–215. https://doi.org/10.1002/ieam.1290

Benmhammed A, Laissaoui A, Mejjad N, Ziad N, Chakir E, Benkdad A, AitBouh H, El Yahyaoui A (2021a) Recent pollution records in Sidi Moussa coastal lagoon (western Morocco) inferred from sediment radiometric dating. J Environ Radioact 227:106464

Benmhammed A, Laissaoui A, Mejjad N, Chakir EM, Ziad N, Benkdad A, El Yahyaoui A, Bouh HA (2021b) Assessment of chronological records of rare earth elements in Sidi Moussa lagoon sediment (North-Western Morocco). Environ Ecol Res 9(4):186–195

Buccolieri A, Buccolieri G, Cardellicchio N, Dell'Atti A, Di Leo A, Maci A (2006) Heavy metals in marine sediments of Taranto Gulf (Ionian Sea, southern Italy). Mar Chem 99(1–4):227–235

Chapman PM (2007) Heavy metal -music, not science. Environ Sci Technol 41(12):6C

Chuan OM, Yunus K (2019) Sediment and organisms as marker for metal pollution. In: Monitoring of marine pollution. IntechOpen. https://doi.org/10.5772/intechopen.85569

da Silva JVF, Lansac-Tôha FM, Segovia BT, Amadeo FE, Braghin LDSM, Velho LFM, Sarmento H, Bonecker CC (2022) Experimental evaluation of microplastic consumption by using a size-fractionation approach in the planktonic communities. Sci Total Environ 821:153045

Dahl M, Bergman S, Björk M, Diaz-Almela E, Granberg M, Gullström M, Leiva-Dueñas C, Magnusson K, Marco-Méndez C, Piñeiro-Juncal N, Mateo MÁ (2021) A temporal record of microplastic pollution in Mediterranean seagrass soils. Environ Pollut 273:116451. https://doi.org/10.1016/j.envpol.2021.116451

Delshab H, Farshchi P, Keshavarzi B (2017) Geochemical distribution, fractionation and contamination assessment of heavy metals in marine sediments of the Asaluyeh port, Persian Gulf. Mar Pollut Bull 115(1–2):401–411

Faris J (1994) Seas of debris: a summary of the third international conference on marine debris sea. Alaska Fisheries Science Center, Seattle, p 54

Frias JP, Ivar do Sul JA, Panti C, Lima ARA (2021) Editorial: microplastics in the marine environment: sources, distribution, biological effects and socio-economic impacts. Front Environ Sci 9:676011. https://doi.org/10.3389/fenvs.2021.676011

Gang Y, Won EJ, Ra K, Choi JY, Lee KW, Kim K (2018) Environmental assessment of contaminated marine sediments treated with solidification agents: directions for improving environmental assessment guidelines. Mar Environ Res 139:193–200. https://doi.org/10.1016/j.marenvres.2018.05.011

Goldberg ED (1963) Geochronology with ^{210}Pb.In radioactive dating. Report to the International Atomic Energy Agency, Vienna

Goldberg ED, Bruland KW (1974) Radioactive chronologies. In: Goldberg ED (ed) The sea. Wiley-Interscience, New York, pp 457–489

Goldberg ED, Gamble E, Griffin JJ, Koide M (1977) Pollution history of Narraganset Bay as recorded in its sediments. Estuar Coast Mar Sci 5:549–556

Goldberg ED, Hodge V, Koide M, Griffin J, Gamble E, Bricker OP, Matisoff G, Holdren GR, Braun R (1978) A pollution history of Chesapeake Bay. Geochim Cosmochim Acta 42:1413–1425

Gonzalez-Brambila CN, Reyes-Gonzalez L, Veloso F, Perez-Angón MA (2016) The scientific impact of developing nations. PLoS One 11(3):e0151328. https://doi.org/10.1371/journal.pone.0151328

Guerrero-Bote VP, Chinchilla-Rodríguez Z, Mendoza A, de Moya-Anegón F (2021) Comparative analysis of the bibliographic data sources dimensions and scopus: an approach at the country and institutional levels. Front Res Metr Anal 5. pmid: 33870055

Hankins C, Raimondo S, Lasseigne D (2022) Microplastic ingestion by coral as a function of the interaction between calyx and microplastic size. Sci Total Environ 810:152333

Hassen NEH, Reguigui N, Helali MA, Mejjad N, Laissaoui A, Benkdad A, Benmasour M (2019) Evaluating the historical sedimentation patterns in two different Mediterranean deep environments (Sardinia and Sicily channels). Mediterr Mar Sci 20(3):542–548

Hazrat A, Ezzat K, Ikram I (2019) Environmental chemistry and ecotoxicology of hazardous heavy metals: environmental persistence, toxicity, and bioaccumulation. J Chem 6730305:14. https://doi.org/10.1155/2019/6730305

Herzog C, Hook D, Konkiel S (2020) Dimensions: bringing down barriers between scientometricians and data. Quant Sci Stud 1:387–395

Hetherington JA, Harvey BR (1978) Uptake of radioactivity by marine sediments and implications for monitoring metal pollutants. Mar Pollut Bull 9(4):102–106

Hu Y, Gong M, Wang J, Bassi A (2019) Current research trends on microplastic pollution from wastewater systems: a critical review. Rev Environ Sci Biotechnol. https://doi.org/10.1007/s11157-019-09498-w

International Ocean Disposal Symposium, United States, Wood DH, Robertson A (1986) Sixth International Ocean Disposal Symposium: 21–25 April 1986, Asilomar Conference Center, Pacific Grove, California. The Symposium, Place of publication not identified

Jiang J (2017) Occurrence of microplastics and its pollution in the environment: a review. Sustain Prod Consum 13:16–23. https://doi.org/10.1016/j.spc.2017.11.003

Joensuu M, Pilditch CA, Harris R, Hietanen S, Pettersson H, Norkko A (2018) Sediment properties, biota, and local habitat structure explain variation in the erodibility of coastal sediments. Limnol Oceanogr 63(1):173–186

Laissaoui A, Mejjad N, Ziad N, Ait Bouh H, El Hammoumi O, Benkdad A, Fekri A (2018) Evidence for a recent increase in delivery of atmospheric 210Pb to Oualidia lagoon, coastal Morocco. Environ Monit Assess 190(11):1–9

Levin LA, Boesch DF, Covich A, Dahm C, Erséus C, Ewel KC, Kneib RT, Moldenke A, Palmer MA, Snelgrove P, Strayer D, Weslawski JM (2001) The function of marine critical transition zones and the importance of sediment biodiversity. Ecosystems 4(5):430–451

Li Y, Zhang Y, Chen G, Xu K, Gong H, Huang K, Yan M, Wang J (2021) Microplastics in surface waters and sediments from Guangdong Coastal Areas, South China. Sustainability 13(5):2691

Lin J, Xu XM, Yue BY, Xu XP, Liu JZ, Zhu Q, Wang JH (2020) Multidecadal records of microplastic accumulation in the coastal sediments of the East China Sea. Chemosphere. https://doi.org/10.1016/j.chemosphere.2020.128658

Liu Q, Liu L, Huang J, Gu L, Sun Y, Zhang L, Lyu K, Yang Z (2022) The response of life history defense of cladocerans under predation risk varies with the size and concentration of microplastics. J Hazard Mater 427:127913

London Convention (1972) Convention on the prevention of marine pollution by dumping of wastes and other matter. 6 May 2022. Available from: https://www.epa.gov/sites/default/files/2015-10/documents/lc1972.pdf

Maanan M, El Barjy M, Hassou N, Zidane H, Zourarah B, Maanan M (2018) Origin and potential ecological risk assessment of trace elements in the watershed topsoil and coastal sediment of the Oualidia lagoon, Morocco. Hum Ecol Risk Assess Int J 24(3):602–614

Maghsodian Z, Sanati AM, Tahmasebi S, Shahriari MH, Ramavandi B (2022) Study of microplastics pollution in sediments and organisms in mangrove forests: a review. Environ Res 208:112725

Marine Protection, Research, and Sanctuaries Act (MPRSA) of 1972, Pub. L. No. 92-532, tit. III, 86 Stat. 1052 (1972)

Martin C, Baalkhuyur F, Valluzzi L, Saderne V, Cusack M, Almahasheer H, Krishnakumar PK, Rabaoui L, Qurban MA, Arias-Ortiz A, Masqué P, Duarte CM (2020) Exponential increase of plastic burial in mangrove sediments as a major plastic sink. Sci Adv 6. https://doi.org/10.1126/sciadv.aaz5593

Mejjad N, Laissaoui A, El-Hammoumi O et al (2016) Sediment geochronology and geochemical behavior of major and rare earth elements in the Oualidia Lagoon in the western Morocco. J Radioanal Nucl Chem 309:1133–1143. https://doi.org/10.1007/s10967-016-4714-8

Mejjad N, Laissaoui A, El-Hammoumi O, Fekri A, Amsil H, El-Yahyaoui A, Benkdad A (2018) Geochemical, radiometric, and environmental approaches for the assessment of the intensity and chronology of metal contamination in the sediment cores from Oualidia lagoon (Morocco). Environ Sci Pollut Res 25(23):22872–22888

Mejjad N, Laissaoui A, Fekri A, Benmhammed A, El Hammoumi O, Cherif EK (2020a) Does human activities growth lead to biodiversity loss in the Moroccan coastal lagoons? A diagnostic comparison study. In: Proceedings of the 4th Edition of International Conference on Geo-IT and Water Resources 2020, Geo-IT and Water Resources 2020, pp 1–5

Mejjad N, Laissaoui A, Fekri A, Hassen NEH, Benmhammed A, El Hammoumi O, Benkdad A, Amsil H (2020b) Tracking natural and human impact on sediment dynamics using radiometric approach in Oualidia lagoon (Morocco). Int J Environ Anal Chem 102(16):4300–4315. https://doi.org/10.1080/03067319.2020.1782394

Mejjad N, Laissaoui A, Benmhammed A, Fekri A, El Hammoumi O, Benkdad A, Amsil H, Chakir EM (2022) Potential ecological risk assessment of rare earth elements in sediments cores from the Oualidia lagoon, Morocco. Soil Sediment Contam 31(8):941–958. https://doi.org/10.1080/15320383.2022.2027342

Miranda LS, Ayoko GA, Egodawatta P, Hu WP, Ghidan O, Goonetilleke A (2021) Physicochemical properties of sediments governing the bioavailability of heavy metals in urban waterways. Sci Total Environ 763:142984

Morillo J, Usero J, Gracia I (2004) Heavy metal distribution in marine sediments from the southwest coast of Spain. Chemosphere 55(3):431–442

Nasr SM, Okbah MA, Kasem SM (2006) Environmental assessment of heavy metal pollution in bottom sediments of Aden Port, Yemen. Int J Oceans Oceanogr 1(1):99–109

Ocean Dumping Ban Act of 1988. EPA press release – Nov. 21, 1988. https://archive.epa.gov/epa/aboutepa/epa-history-ocean-dumping-ban-act-1988.html

Perumal K, Antony J, Muthuramalingam S (2021) Heavy metal pollutants and their spatial distribution in surface sediments from Thondi coast, Palk Bay, South India. Environ Sci Eur 33(1):1–20

Pourret O, Bollinger JC, Hursthouse A (2021) Heavy metal: a misused term? Acta Geochim 40:466–471. https://doi.org/10.1007/s11631-021-00468-0. hal-03174937

Proceedings of the Workshop on the Fate and Impact of Marine Debris, 27–29 November 1984, Honolulu, Hawaii (1985)

Proceedings of the Second International Conference on Marine Debris, 2–7 April 1989, Honolulu, Hawaii. NOAA technical memorandum NMFS; NOAA-TM-NMFS-SWFSC, p 154

Qin F, Du J, Gao J, Liu G, Song Y, Yang A, Wang H, Ding Y, Wang Q (2020) Bibliometric profile of global microplastics research from 2004 to 2019. Int J Environ Res Public Health 17(16):5639. https://doi.org/10.3390/ijerph17165639

Reise K, Bouma TJ, Olenin S, Ysebaert T (2009) Coastal habitat engineers and the biodiversity in marine sediments. Helgol Mar Res 63:1–2. https://doi.org/10.1007/s10152-009-0147-x

Rezaei M, Mehdinia A, Saleh A, Modabberi S, Mansouri Daneshvar MR (2021) Environmental assessment of heavy metal concentration and pollution in the Persian Gulf. Model Earth Syst Environ 7(2):983–1003

Santos IR, Friedrich AC, Ivar do Sul JA (2009) Marine debris contamination along undeveloped tropical beaches from Northeast Brazil. Environ Monit Assess 148(1):455–462

Schintu M, Marrucci A, Marras B, Galgani F, Buosi C, Ibba A, Cherchi A (2016) Heavy metal accumulation in surface sediments at the port of Cagliari (Sardinia, western Mediterranean): environmental assessment using sequential extractions and benthic foraminifera. Mar Pollut Bull 111(1–2):45–56

Seeley ME, Song B, Passie R, Hale RC (2020) Microplastics affect sedimentary microbial communities and nitrogen cycling. Nat Commun 11(1):1–10

Shi C, He H, Xia Z, Gan H, Xue Q, Cui Z, Chen J (2022) Heavy metals and Pb isotopes in a marine sediment core record environmental changes and anthropogenic activities in the Pearl River Delta over a century. Sci Total Environ 814:151934

Snelgrove PV (1997) The importance of marine sediment biodiversity in ecosystem processes. Ambio 26:578–583

Snelgrove PVR (1999) Getting to the bottom of marine biodiversity: sedimentary habitats: ocean bottoms are the most widespread habitat on earth and support high biodiversity and key ecosystem services. Bioscience 49(2):129–138. https://doi.org/10.2307/1313538

SOI (2014) Action plan for the sustainable ocean initiative (2015–2020). Sustainable ocean initiative global partnership meeting (3–4 October, 2014, Seoul, Republic of Korea). Available from https://www.cbd.int/doc/meetings/mar/soiom-2014-02/official/soiom-2014-02-actionplan-en.pdf. 4 May 2022

Sun J, Wang MH, Ho YS (2012) A historical review and bibliometric analysis of research on estuary pollution. Mar Pollut Bull 64(1):13–21

Tamilmani A, Venkatesan G (2021) Assessment of trace metals and its pollution load indicators in water and sediments between Upper and Grand Anicuts in the Cauvery. Int J Environ Sci Technol 18(12):3807–3818

Thrush SF, Hewitt JE, Cummings VJ, Ellis JI, Hatton C, Lohrer A, Norkko A (2004) Muddy waters: elevating sediment input to coastal and estuarine habitats. Front Ecol Environ 2(6):299–306. https://doi.org/10.2307/3868405

Thrush S, Hewitt J, Pilditch C, Norkko A (2021) Ecology of coastal marine sediments: form, function, and change in the anthropocene. Oxford University Press, Oxford

Uddin S, Behbehani M (2018) Concentrations of selected radionuclides and their spatial distribution in marine sediments from the northwestern Gulf, Kuwait. Mar Pollut Bull 127:73–81

Uddin S, Fowler SW, Uddin MF, Behbehani M, Naji A (2021) A review of microplastic distribution in sediment profiles. Mar Pollut Bull 163:111973

Urry WD (1949) Radioactivity of ocean sediments. VI. Concentrations of the radio-elements in marine sediments of the southern hemisphere. Am J Sci 247(4):949

Valette-Silver NJ (1993) The use of sediment cores to reconstruct historical trends in contamination of estuarine and coastal sediments. Estuaries 16(3):577–588. https://doi.org/10.2307/1352796

Van Eck NJ, Waltman L (2010) VOSViewer: Visualizing Scientific Landscapes [Software]. Available from https://www.vosviewer.com

Van Eck J, Waltman L (2022) VOSviewer manual. (Manual for VOSviewer version 1.6.18). University Leiden, Lieden

Viguri J, Verde J, Irabien A (2002) Environmental assessment of polycyclic aromatic hydrocarbons (PAHs) in surface sediments of the Santander Bay, Northern Spain. Chemosphere 48(2):157–165

Visser M, van Eck NJ, Waltman L (2022) Large-scale comparison of bibliographic data sources: Scopus, Web of Science, Dimensions, Crossref, and Microsoft Academic. Quant Sci Stud 2:20–41

Wang F, Wong CS, Chen D et al (2018) Interaction of toxicchemicals with microplastics: a critical review. Water Res 139:208–219. https://doi.org/10.1016/j.watres.2018.04.003

Water Resources Development Act of 1992. 102d Congress. PUBLIC LAW 102-580—OCT. 31, (1992), 106 STAT. 4797

White KE, Robbins C, BKhan B, Freyman C (2017) Science and engineering publication output trends: 2014 shows rise of developing country output while developed countries dominate highly cited publications. Info brief. National Science Foundation. NSF 18-300

Wong SL, Nyakuma BB, Wong KY, Lee CT, Lee TH, Lee CH (2020) Microplastics and nanoplastics in global food webs: a bibliometric analysis (2009–2019). Mar Pollut Bull 158:111432

Yii MW, Zaharudin A, Abdul-Kadir I (2009) Distribution of naturally occurring radionuclides activity concentration in East Malaysian marine sediment. Appl Radiat Isot 67(4):630–635

Youssef M, El-Sorogy A, Al Kahtany K, Al Otiaby N (2015) Environmental assessment of coastal surface sediments at Tarut Island, Arabian Gulf (Saudi Arabia). Mar Pollut Bull 96(1–2):424–433

Yüksel B, Ustaoğlu F, Tokatli C, Islam MS (2022) Ecotoxicological risk assessment for sediments of Çavuşlu stream in Giresun, Turkey: association between garbage disposal facility and metallic accumulation. Environ Sci Pollut Res 29(12):17223–17240

Yümün Z, Kam E, Dinçer A, Önce M, Yümün S (2021) The investigation of toxic element pollution and radioactivity analyses of marine sediments in the Gulf of Gemlik (Bursa, Turkey). Appl Ecol Environ Res 19(2):881

Zare MR, Mostajaboddavati M, Kamali M, Abdi MR, Mortazavi MS (2012) 235U, 238U, 232Th, 40K and 137Cs activity concentrations in marine sediments along the northern coast of Oman Sea using high-resolution gamma-ray spectrometry. Mar Pollut Bull 64(9):1956–1961

Zhang Y, Pu S, Lv X, Gao Y, Ge L (2020) Global trends and prospects in microplastics research: a bibliometric analysis. J Hazard Mater 400:123110. https://doi.org/10.1016/j.jhazmat.2020.123110

Chapter 2
Coastal Flooding in India: An Overview

P. S. Swathy Krishna, L. Sheela Nair, and M. Ramesh

Abstract The continuously changing coastal zones have been under the threat of natural disasters such as tsunamis, cyclones, flooding, erosion, along with human interventions. Flooding is one of the major coastal disasters while changing climate and sea-level rise intensify its impact. Increasing trend in flooding events adversely affects inland in the following ways as salt water intrusion damages vegetation and soil fertility and disturbs the coastal livelihood and its socioeconomic impacts can be very relevant in India with densely populated coastal zones. However, India has a long history of coastal flooding; only few are studied while many went unnoticed, with most of the studies restricted to storm surges and wind waves. Hence, a proper analysis of wind, wave, tide, and coastal morphology along Indian coast is needful as India comes under the influence of seasonally reversing monsoons and increased frequency of tropical cyclones along with risk of rising sea level. This chapter provides details on the coastal flooding scenario along Indian coast so far, considering the contributing factors as a critical input in the further development of an early flood warning system and thereby appropriate adaptation measures can be taken.

Keywords Coastal flooding · Storm surge · Swells · Kallakkadal · Sea-level rise

2.1 Introduction

Shorelines and coastal process are continuously changing under the influence of rising sea levels, freak waves, tides, along with many more anthropogenic factors and are becoming risky (Narayana et al. 2012). Coastal flooding has become a serious issue in many parts of the world, which adversely affects the foreshore defense

P. S. Swathy Krishna (✉) · M. Ramesh
National Centre for Earth Science Studies, Thiruvananthapuram, Kerala, India

Cochin University of Science and Technology, Cochin, Kerala, India

L. Sheela Nair
National Centre for Earth Science Studies, Thiruvananthapuram, Kerala, India

N. Jayaraju et al. (eds.), *Coasts, Estuaries and Lakes*,
https://doi.org/10.1007/978-3-031-21644-2_2

mechanism and posing threat to hazard management in coastal and estuarine zones (Nicholls et al. 2007). Several factors contribute in coastal flooding, which comes from oceanographic, geophysical, atmospheric, or even astronomic sources (Pugh 1987; Gornitz 1995; Cazenave and Llovel 2010) and the influence is quantified based on their amplitude (Middleton and Thomson 1986). Higher annual variation in sea levels could contribute to nuisance flooding in low-lying coasts, which occurs with the combined effect of spring tides and high mean sea level, under the clear sky (Moftakhari et al. 2015). The long period swells along the English Channel coast and posed serious threat of coastal flooding in Atlantic basin, even though the severity was rarely found (Sibley et al. 2015). Such swell-dominated flooding event was explained by tracking the long distant extra tropical storms and was later concluded for the cause of "nuisance flooding."

The extremity in rainfall and its uneven spread, intensity of storm surges, and hazardous coastal floods are growing in numbers with alarming trends in global warming and the fluctuating sea level (Burton and Kates 1964). Apart from that, increased strength of atmospheric process tends to amplify the frequency of high magnitude floods and storms (Berz 1993). The interaction of high winds with spring tides elevates the mean and fluctuating sea levels, floods the coastal plain, thus abolishing the coastal structures and affecting the coastal community (Hunt 2005). Climate change-induced subsidence has significant influence on increasing the risk due to coastal floods and the impact of inundations on coastal community can be of much relevance in ocean front countries, especially for a subcontinent like India. This chapter provides a review of various types of coastal flooding events, its impacts, and the status of flood risk in India along with some recently identified flooding events along the southwest (SW) coast of India.

2.2 Coastal Flood Risk Along Indian Coast: Monsoon, Climate, and Cyclones

India has a coastline of 7517 km comprising the Andaman and Nicobar Islands and the Lakshadweep Islands (IYB 2007). The wave climate of the IO depicts the multimodel spectrum as waves evolve from varying geographic areas with distinct atmospheric driving forces, especially, of the monsoon driven waves in the Arabian Sea (AS). However, over Bay of Bengal (BoB), the increased number of cyclones determines most of the wave climate in this basin. The coastal cyclones during 2006–2020 were analyzed and disclosed the influence of SW monsoon in creating 29 cyclonic disturbances near to the cost and 27 cyclonic turbulences associated to North East monsoon and 5 during the non-monsoon period (Mahadevan and Latha 2001). The Indian Ocean Tsunami during December 2004 was the most devastating Tsunami in modern times, flooding the entire coast, caused deaths of nearly 8835 people along with widespread destruction along Indian mainland (Georges 2011). It had significant impact on the southern peninsular region of India that severely affected the coastal states of union territories of Pondicherry and Kerala. Coastal flood hazard in India is classified as high (Fig. 2.1) with the extremity in mean sea level expected to increase in the coming years.

Fig. 2.1 Coastal flood risk map. (Source: https://data.amerigeoss.org/dataset/ss-global-muis-2)

2.3 Classification of Coastal Flooding in India

Coastal erosion and storm surges were the serious coastal hazards studied so far, whereas only a few cases of coastal flooding were studied in India, when compared to other parts of the world. Kurian et al. (2009) were the first to classify the coastal flooding phenomena in Indian context, based on the causative factors as:

(a) Synoptic scale flooding—due to storm surge where the forcing is from large-scale weather systems.
(b) Mesoscale flooding—known as "Rissaga," which is a high amplitude sea-level oscillation and occurs quite frequently.
(c) Remote forcing—"kallakkadal," distinguished with its frequency and local impact.

2.3.1 Flooding Due to Storm Surges

The occurrence of high amplitude waves in the Indian Ocean is often due to stronger storm events, with 7% of the global TC occurring in the NIO (Gray 1985). Coastal flooding owing to TC-generated storm surge in NIO often occurs with severity during post and pre-monsoon seasons. In fact, the disastrous flooding happens when the peak storm surge combines with high spring tide, which caused considerable damage and property loss (Bertin et al. 2014; Shaji et al. 2014; Marasinghe and Wijetunge 2015). During the landfall of cyclone, the sea water inundates the low-lying coastal regions, thereby floods the entire region leaving behind an eroded beach and most often damaging the vegetation and reducing soil fertility. The vulnerability of Indian coastline to coastal flooding was categorized into four zones in view of surge potential, of which coastal zones of BoB along with its off-shore islands are found more prone to storm-surge (~10–13 m). However, a major part of Orissa coast is under the Very High-Risk Zone (VHRZ) (~5–7 m) (ESSO-INCOIS-Indian National Centre for Ocean Information Services). The flooding due to surge is more noticeable in the East coast as BoB is considered one of the cyclogenesis areas of the world oceans. In areas with a wide continental shelf, a traveling external surge can have significant interaction with locally generated waves and tide (Mahadevan and Latha 2001; Wolf 2009). There a decline in frequency of the cyclonic disturbances from about six to seven per hundred years was reported, whereas the frequency of cyclonic storms from about one to two per hundred years over BoB and AS in the monsoon season, respectively (Singh 2001). Seasonal wise, the frequency of TC was found to be high in the post-monsoon and greater intensities during the pre-monsoon period (Li et al. 2013). Occurrences of TC are comparatively less in AS basin to Bob and exhibit a frequency ratio of 4:1 (Dube et al. 1997). On an average, 4.7 cyclonic storm days per year evolved over the AS from 1979 to 2008, with more than 15 cyclonic storm days during 1998 and 2004 (Evan and Camargo 2011).

Fig. 2.2 (**a**) Cyclone Hudhud at Uppada beach, Andhra Pradesh. (**b**) Cyclone Tauktae damages the beach roads in Alappuzha, Kerala

Vulnerability of Andhra Pradesh coast to TC and storm surges is reported by newspaper (https://timesofindia.indiatimes.com/city/visakhapatnam/three-million-at-risk-from-storms-along-aps-cyclone-prone-coast/articleshow/86818958.cms) and a file photo of cyclone impact is shown in Fig. 2.2a. Similarly, on the west coast, Cyclone Tauktae, the fifth strongest storm since 1998, in the AS, persisted for over 4 days along the West coast from May 14 to 17, 2021 (https://india.mongabay.com/2021/05/cyclone-tauktae-exposes-vulnerabilities-along-indias-west-coast/), is shown in Fig. 2.2b.

2.3.2 Sea-Level Rise (SLR) and Associated Chances of Coastal Flooding

Coastal areas are characterized as “at-risk” regions due to climate change through SLR (Deborah et al. 2019) and the water level variability at the coast was critical for coastal flooding assessment. The general mean sea level (GMSL) rose to 100 m below the present GMSL in 15,000 years (Clark et al. 2016) and would continue to increase and expected to exceed 1 m by 2100 (Church et al. 2013). The flood magnitude and frequency were affected by SLR and enduring changes in the wave climate (Fig. 2.3), thus the areas lying under MSL will experience higher frequency of floods due to under SLR (Vitousek et al. 2017). Sea-level changes in the Indian Ocean were not considerably noted for the flooding studies. However, a positive trend of mean sea level in the NIO was noticed, with an average growth of about 4 mm year^{-1} for 1993 to 2001 (Chowdary and Behra 2015) and the contribution of sea-level rise on the flooding phenomenon along Indian coast must also be taken into account.

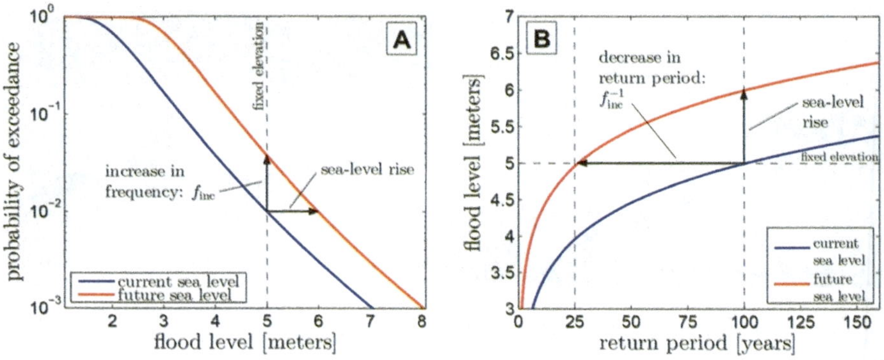

Fig. 2.3 A 1 m increase in SL increases the frequency (**a**) lowers the return period (**b**) of the 5 m flood level (Vitousek et al. 2017)

2.3.3 Remote Forcing: Swell-Induced Coastal Flooding

Swells hold the particular characteristics of the nature and intensity of wind at the generation zone which are pioneer to distant storms, tropical cyclones, or monsoon (Bhowmick et al. 2011). Apart from the storms and seasonal reversing monsoon winds, the SIO swells also determine the wave regime over NIO and along Indian coast (Chowdary et al. 2019). The eastward propagating swells from the South Atlantic spread energy all over the entire IO, with three swell generation zones, namely, tropical NIO, tropical SIO, and extra tropical SIO (Alves 2005). Such swells generated from the Southern Ocean belt took only 4 days to reach the South Indian coast (Fig. 2.4) and with least dissipation (Nayak et al. 2013).

The presence of Southern Ocean swells in the NIO could be found during the fair season and thus NIO is swell dominated (Glejin et al. 2013; Sandhya et al. 2015), where the amplitude of wind seas is quite lower in comparison with swells (Anoop et al. 2015). During monsoon, the swell average time period is found comparatively less in the eastern AS (Sanil Kumar et al. 2010), as they are generated within NIO domain due to a higher southwest wind condition (11–12 m/s). Such long period waves would reach to the Indian coast without much dissipation and most of the times resulted in flooding along the coast especially along the SW coast of India.

2.3.3.1 Kallakkadal

Kallakkadal is an intense wave activity, occurring along the SW coast of India that continues for several days without any change in local weather as in a flash flooding nature and causes panic and severe destruction to coastal communities. One of most noticed and studied Kallakkadal was 2005-May event (Murty and Kurian 2006; Kurian et al. 2009; Remya et al. 2016). According to fisherman, it occurs almost every year, where the coastal inundation due to Kallakkadal invites the attention of

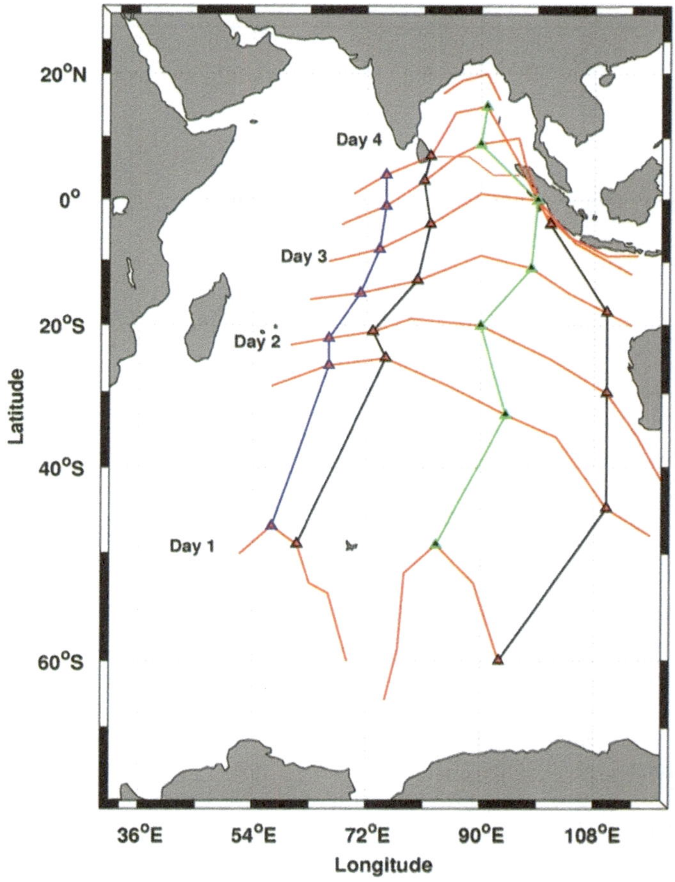

Fig. 2.4 Swell generation and advancement through Southern to Indian mainland (Nayak et al. 2013)

the government and media only when it affects the local population and causes damage to their property. So far, two theories were proposed for the 2005 kallakkadal event. First, the flooding is related to the seismic activities in Sumathra—Andaman Nicobar regions and observed a substantial flooding of TS canal during the 2005 kallakkadal event that was way similar to 2004 tsunami (Narayana and Tatavarti 2005) and the second theory was, these high wave activities are probably due to the storm occurred on the south west coast of Australia. The storm formed around 30° S and traveled toward south of Madagascar, then dissipated on 22 May (Baba 2005). Later, analysis of 2005 event confirmed that these remotely generated swells from Southern Ocean may interact with coastal current directed southward in the west coast of India and waves get amplified, leading to Kallakkadal with an inundation limit of 435 m along the Adimalathura coast (Kurian et al. 2009) (Fig. 2.5). The flooding impacts along the coast vary based on land topography and become more

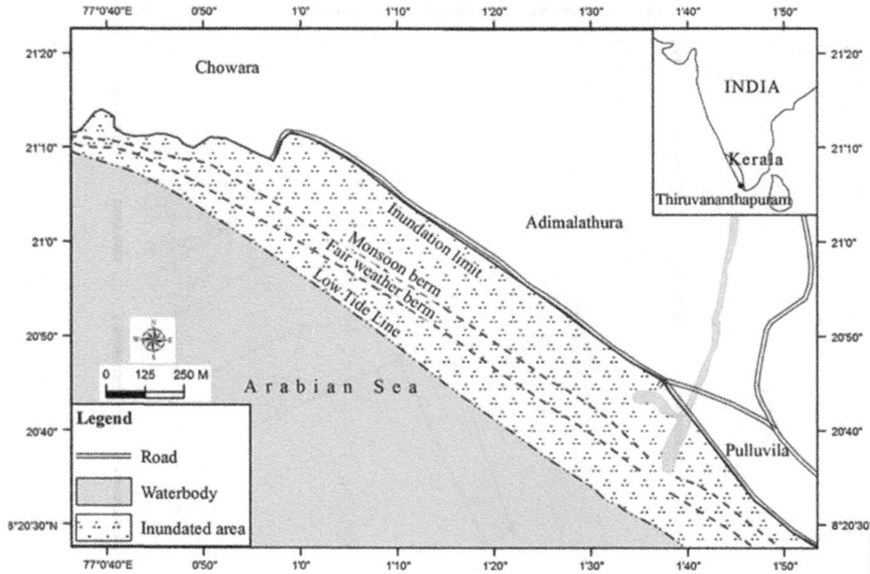

Fig. 2.5 Inundation limit of May 2005 Kallakkadal at Adimalathura, Thiruvananthapuram (Kurian et al. 2009)

severe when associated with high tide. Similar events appear severe and higher in frequency over southern Indian coast than in north mainly due to the orientation of the coast line.

Intense wind system that provides larger fetch for longer duration is required for the generation and subsequent propagation of long period swells, like SIO swells (Munk et al. 1963). Such strong surface winds and associated ocean turbulence exhibit Cut of lows (COL), that initiates high waves (McInnes and Hubbert 2001) and these long period swells sometimes cause flash floods along the coast (Lupo and Smith 1995; Wiedenmann et al. 2002). Such planetary scale wind patterns associated with the strong westerly winds prevailing over the Southern Ocean, as blocking patterns, acted as a strong wind system for the generation of these SIO swells (Murty and Kurian 2006; Remya et al. 2016). Another noticeable Kallakkadal event occurred during 2012 September (Fig. 2.6) (The Hindu, Alappuzha, Kerala, 2012, Sep. 4).

Both the cases (2005 and 2012) had similar characteristics of tsunami where, the sea receded several meters landward and cause large amount of sediment transport with severe erosion. However, the Kallakkadal event is observed only along the SW coast of India and is mostly found along South Kerala coast.

Fig. 2.6 Flooding event during September, 2012 along Kerala coast

2.4 Recent Coastal Flooding Along SW Coast of India

Some of the identified coastal flooding events during the years 2017 to 2019 along the SW coast of India are given in Fig. 2.7. The flooding associated to cyclone OCKHI (Fig. 2.7a) during 1 December 2017 along the Trivandrum coast was the first reported flooding event of 2017 due to cyclone. During 24 April 2018 (Fig. 2.7b), waves of period more than 20s, that persisted for more than a day, caused flooding and could be identified as a swell-induced coastal flooding of pre-monsoon. Wind, wave, and tide have a combined effect on the coastal flood and Fig. 2.7c indicates the erosion after the event. During the calm conditions of March 2019, the coast witnessed the flash flood that lasted for a day (Fig. 2.7d), as an effect of three cyclonic storms occurred over the SIO (30° S). The increased frequency of coastal floods along with monsoonal rough seas and cyclones is taking a toll on the livelihoods of coastal community especially on small-scale fishermen.

Fig. 2.7 Different flooding events along the SW coast. (**a**) High wave activity on the Trivandrum coast during the cyclone Ockhi of 1 December 2012. (**b**) High period swells lashing over the Valiyathura-Shangumugham coast on 24 April 2018. (**c**) The flooding signatures at the Valiyathura beach during 28 September 2018. (**d**) High intensity waves during 20 March 2019 along Valiyathura that caused flooding and severe erosion

2.5 Conclusions

This chapter explains various types of coastal flooding events in Indian context and concludes that BoB is more influenced by synoptic scale flooding while the AS is experiencing more flooding due to remote forcing. The flooding due to storm surges is indeed due to the impact of high frequency and high amplitude waves which lasts on an average of 2–3 days leaving behind an erosive beach along with devastating destructions. The high period long waves (peak wave period more than 18 s) with an average H_s nearly 1.5–3 m were found responsible for flooding due to remote forcing and last from few hours to days extending from few kilometers to thousand kilometers. So far, the meso-scale flooding named rissaga has not been identified in Indian coasts. This chapter also emphasizes on the major drawback of all the previous studies regarding coastal flooding along the Indian coast, in which most of the studies discussed about cyclones and Southern Indian Ocean swells and no one could explain the proper reasons behind it. All the studies are mainly related to originating and propagating of Southern Indian Ocean swells, but none of these studies could answer clearly what is the forcing factor behind Kallakkadal and why this is reporting only along Kerala coast. Apart from the SIO swells, increased number of cyclones over AS and predicted sea-level rise along the Indian Ocean in the near future are also creating an alarming situation of coastal flooding. Hence, it is concluded that coastal flooding may also have other features rather than kallakkadal, especially on the increasing tendency of storm events in the tropical Indian Ocean and rising sea-level warnings and it is in high demand to establish a continuous monitoring network and therefore developing an early warning system for flash floods along Indian coast.

References

Alves JHGM (2005) Numerical modelling of ocean swell contributions to the global wind-wave climate, 1463-5003/$ – see front matter. 2005 Elsevier Ltd. All rights reserved

Anoop TR, Sanil Kumar V, Shanas PR, Johnson G (2015) Surface wave climatology and its variability in the North Indian Ocean based on ERA-Interim reanalysis. J Atmos Oceanic Tech 32:1372–1385

Baba M (2005) Occurrence of "swell waves" along the southwest coast of India from southern Indian Ocean storm. J Geol Soc India 66:248–249

Bertin X, Li K, Roland A, Zhang YJ, Breilh JF, Chaumillon E (2014) A modeling-based analysis of the flooding associated with Xynthia, central Bay of Biscay. Coast Eng 94:80–89

Berz GA (1993) Global warming and the insurance industry. Interdiscip Sci Rev 18(2):120–125

Bhowmick SA, Kumar R, Chanudhuri S, Sarkar A (2011) Swell propagation over Indian Ocean region. Int J Ocean Clim Syst 2:87–99

Burton I, Kates RW (1964) The precipitation of natural hazards in response management. Nat Resour J 3:412–433

Cazenave A, Llovel W (2010) Contemporary Sea level rise. Ann Rev Mar Sci 2:145–173

Chowdary P, Behra MR (2015) A study on regional sea level variation along the Indian coast. Procedia Eng 116:1078–1084. 8th International Conference on Asian and Pacific Coasts, 2015

Chowdary P, Behra MR, Reeve DE (2019) Wave climate projections along the Indian coast. Int J Climatol 39:4531–4542

Church JA, Clark PU, Cazenave A, Gregory JM, Jevrejeva S, Levermann A, Merrifield MA, Milne GA, Nerem RS, Nunn PD, Payne AJ, Pfeffer WT, Stammer D, Unnikrishnan AS (2013) Sea level change. In: Stocker TF, Qin D, Plattner G-K, Tignor M, Allen SK, Boschung J, Nauels A, Xia Y, Bex V, Midgley PM (eds) Climate change 2013: the physical science basis. Contribution of Working Group I to the fifth assessment report of the Intergovernmental Panel on Climate Change. Cambridge University Press, Cambridge

Clark PU, Shakun JD, Marcott SA, Mix AC, Eby M, Kulp S, Levermann A, Milne GA, Pfister PL, Santer BD, Schrag DP, Solomon S, Stocker TF, Strauss BH, Weaver AJ, Winkelmann R, Archer D, Bard E, Goldner A, Lambeck K, Pierrehumbert RT, Plattner G-K (2016) Consequences of twenty first century policy for multi-millennial climate and sea-level change. Nat Clim Change 6:360–369

Deborah I, Xavier B, Philip T, Mark DP (2019) Interactions between mean sea level, tide, surge, waves and flooding: mechanisms and contributions to mean sea level. Surv Geophys 40:1603–1630

Dube SK, Rao AD, Sinha PC, Murty TS, Bahulayan N (1997) Numerical modelling of storm surges in the Arabian Sea. Appl Math Model 9:289–294

Evan AT, Camargo SJ (2011) A climatology of Arabian Sea cyclonic storms. J Clim 24:140–158

Georges R (2011) Impact of 2004 tsunami in the islands of Indian Ocean: lessons learned. Emerg Med Int:920813. https://doi.org/10.1155/2011/920813
Glejin J, Sanil Kumar V, Balakrishnan Nair TM, Singh J (2013) Influence of winds on temporally varying short and long period gravity waves in the near shore regions of Eastern Arabian Sea. Ocean Sci 9:343–353
Gornitz V (1995) Monitoring sea level changes. Clim Chang 31:515–544
Gray WM (1985) Technical document WMO/TD no. 72. WMO, Geneva, pp 3–19
Hunt JCR (2005) Inland and coastal flooding: developments in prediction and prevention. Philos Trans R Soc 363:1475–1491
IYB (2007) India year book. Government of India, New Delhi
Kurian NP, Nirupama N, Baba M, Thomas KV (2009) Coastal flooding due to synoptic scale meso-scale and remote forcing. Nat Hazards 48:259–273
Li Z, Yu W, Li T, Murty VSN, Tangang F (2013) Bimodal character of cyclone climatology in the Bay of Bengal modulated by monsoon seasonal cycle. J Clim 26:1033–1046
Lupo AR, Smith PJ (1995) Climatological features of blocking anticyclones in the Northern Hemisphere. Tellus Ser A 47:439–456
Mahadevan R, Latha G (2001) Influence of coastal flooding on surge estimates along the east coast of India. Indian J Mar Sci 30:115–122
Marasinghe CK, Wijetunge JJ (2015) A deterministic hazard assessment for storm surge induced flooding: a case study. In: Moratuwa Engineering Research Conference (MERCon), 2015, pp 34–39. https://doi.org/10.1109/MERCon.2015.7112316
McInnes KL, Hubbert GD (2001) The impact of eastern Australian cut-off lows on coastal sea levels. Meteorol Appl 8:229–244
Middleton JF, Thomson KR (1986) Return periods of extreme sea levels from short periods. J Geophys Res 91(C10):11707–11716
Moftakhari HR, AghaKouchak A, Sanders BF, Feldman DL, Sweet W, Matthew RA, Luke A (2015) Increased nuisance flooding along the coasts. Geophys Res Lett 42:9846–9852
Munk WH, Miller GR, Snodgrass FE, Barber NF (1963) Directional recording of swell from distant storms. Philos Trans R Soc Lond Ser A 255:505–584
Murty TS, Kurian NP (2006) A possible explanation for the flooding several times in 2005 on the coast of Kerala and Tamil Nadu. J Geol Soc India 67:535–536
Narayana AC, Tatavarti R (2005) High wave activity on the Kerala coast. J Geol Soc India 66:249–250
Narayana S, Hanson S, Nicholas RJ, Clarke D, Willems P, Ntegeka V, Monbaliu J (2012) A holistic model for coastal flooding using system diagrams and source pathway receptor (SPR) concept. Nat Hazards Earth Syst Sci 12:1431–1439
Nayak S, Bhaskaran PK, Venkatesan R, Dasgupta S (2013) Modulation of local wind-waves at Kalpakkam from remote forcing effects of Southern Ocean swells. Ocean Eng 64:23–35
Nicholls RJ, Hanson S, Herweijer C, Patmore N, Hallegatte S, Corfee-Morlot J, Cha^teau J, Muir-Wood R (2007) Ranking port cities with high exposure and vulnerability to climate extremes—exposure estimates. OECD environmental working paper no. 1. Organisation for Economic Co-operation and Development (OECD), Paris
Pugh DT (1987) Tides, surges and mean sea-level, vol 68(2). J. Wiley, Chichester, p 472
Remya PG, Vishnu S, Praveen Kumar B, Balakrishnan Nair TM, Rohith B (2016) Teleconnection between the North Indian Ocean high swell events and meteorological conditions over the Southern Indian Ocean. AGU J Geophys Res Oceans 121:11723
Sandhya KG, Remya PG, Balakrishnan Nair TM, Arun N (2015) On the co-existence of high-energy low-frequency waves and locally-generated cyclone waves off the Indian east coast. Ocean Eng 111:148–154
Sanil Kumar V, Sajiv Philip C, Balakrishnan Nair TN (2010) Waves in shallow water off west coast of India during the onset of summer monsoon. Ann Geophys 28:817–824
Shaji D, Sajal K, Vishal T (2014) Storm surge studies in the North Indian Ocean: a review. Indian J Geo-Mar Sci 43:125–147

Sibley A, Cox D, Titley H (2015) Coastal flooding in England and Wales from Atlantic and North Sea storms during the 2013/2014 winter. Weather 70(2):62–70

Singh OP (2001) Long term trends in the frequency of monsoonal cyclonic disturbances over the North Indian Ocean. Mausam 52:655–658

Vitousek S, Barnard PL, Fletcher CH, Frazer N, Erikson L, Storlazzi CD (2017) Doubling of coastal flooding frequency within decades due to sea-level rise. Sci Rep 7:1399

Wiedenmann JM, Lupo AR, Mokhov II, Tikhonova EA (2002) The climatology of blocking anti-cyclones for the Northern and Southern Hemisphere block intensity as a diagnostic. J Climatol 15:3459–3473

Wolf J (2009) Coastal flooding: impacts of coupled wave–surge–tide models. Nat Hazards 49(2):241–260

Part II
Water Quality/Hydrological Processes

Chapter 3
Appraisal of Coastal Water Quality of Two Hot Spots on Southwest Coast of India: A Case Study of Multi-Year Biogeochemical Observations

B. Upendra, M. Ciba, V. Arun, R. Sreelesh, and K. Anoop Krishnan

Abstract This study apprises the spatial and seasonal trends of coastal water quality factors using the multi-year physical, chemical, and biological observational data from the two hot spots on Southwest coast of India, the Kochi (creek, 0.5 km, 2 km, and 5 km regions), and the Mangalore (0.5 km, 2 km, and 5 km regions) stations. It also highlights the eutrophication status of coastal region using trophic index (TRIX index) that reveals the primary productivity status, while the statistical analysis of the water quality data reveals the source of pollution. The parameters such as temperature, pH, Total Suspended Solids (TSS), Dissolved Oxygen (DO), Salinity, Alkalinity, Ammonia (NH_4), Nitrite (NO_2), Nitrate (NO_3), Total Nitrogen (TN), Inorganic Phosphate (IP), Total Phosphorus (TP), Silicate (SiO_4), Chlorophyll a, Total Viable Count (TVC), Streptococcus faecalis (SFLO), and Escherichia coli (ECLO) are used for the analysis of coastal water characteristics. The spatial variation of dissolved parameters in the coastal waters shows a high concentration of parameters near the shore than offshore and the seasonal analysis shows that most of the higher concentrations are reported during the pre-monsoon season, while the lower concentrations are noted during the monsoon season except in few cases indicating the terrestrial influx, riverine influx, and evaporation activity in study regions. The biological, microbiological, and trophic index suggest a higher primary productivity and eutrophication stage of coastal water during study period indicating deteriorating quality of water in these coastal waters.

B. Upendra (✉)
National Centre for Earth Science Studies (NCESS), Thiruvananthapuram, Kerala, India

Department of Applied Chemistry, Cochin University of Science and Technology (CUSAT), Kochi, Kerala, India

M. Ciba · V. Arun · R. Sreelesh · K. Anoop Krishnan
National Centre for Earth Science Studies (NCESS), Thiruvananthapuram, Kerala, India

N. Jayaraju et al. (eds.), *Coasts, Estuaries and Lakes*,
https://doi.org/10.1007/978-3-031-21644-2_3

Keywords Coastal water quality · Southwest coast of India · Biogeochemical observations · TRIX index · Water quality indices · Eutrophication · Seasonal influence

3.1 Introduction

Coastal-near shore environments are one of the most dynamic and vulnerable ecosystems in the world that are more sensitive to human interferences due to economic developments in many aspects. Coastal regions are essential apart from their intense ecologic, cultural, and aesthetic values, a considerable section of the coastal community relies on this system for their survival. The nature of coastal environment depends upon natural and manmade activities around the coastline as well as the inputs from the backwater and riverine systems (Field et al. 2014). Numerous pollutants enter the marine environment through oil spills, industrial, and domestic waste, ballast water, agricultural, and ship waste, and overfishing destroys many marine habitats. The most powerful drivers of change in marine and coastal ecosystems are fisheries, invasive species, pollution, nutrient loading (eutrophication), land-use change and habitat loss, and global climate change. Eutrophication is one of the major ecosystem issues confronting the World's coastal regions as a result of increasing anthropogenic activities, particularly in India (Peng 2015). Hence, the assessment of contaminants from different sources is necessary for the proper management of coastal environments (Peng 2015). Satisfactory environmental conditions are of great importance to maintain the life and productivity of the ocean in addition to ensuring human well-being.

Coastal area of India covers around 7500 km and the coastal zone of country is diversified with coral reefs, mangroves, sea grass beds, wetlands, creeks, estuaries, and rich in natural resources. Since 1991, under the Ministry of Earth Sciences (MoES), New Delhi, a nationwide coastal monitoring programme, namely, Sea Water Quality Monitoring (SWQM), has been launched in order to identify ecologically sensitive areas in terms of coastal pollution through chemical, biological, and marine biological aspects. Among the various coastal ocean monitoring research sites, this study covered two important and environmentally sensitive coastal hotspots in the Arabian Sea, west coast of India, namely, Kochi and Mangalore during the period, 1 April 2012–31 March 2015. The physio-chemical and biological parameters in the overlying waters of these environmental hotspots are monitored seasonally to assess the current status of environmental health of this part of the coastal regions. These two hot spots are involved in rapid socio-economic developments and a study is utmost essential not only to understand the local biogeochemical processes occurring in the coastal waters and its global relevance, but also to lay down strategies for mitigating the ill-effects of human interferences including pollution of the coastal waters. This study includes 36-hour continuous monitoring in Kochi (creek and 0.5 km) and Mangalore (0.5 km) stations. Near shore and offshore studies include 2 km and 5 km transects at both Kochi and Mangalore hot spots.

This study tries to determine the spatial and seasonal trends of sea water quality factors and in terms of chemical, biological, and microbiological characteristics of coastal waters. It also highlighted the eutrophication status of coastal region using trophic index (TRIX index) that reveals the primary productivity status of the coastal environment and identifies the sources of pollutants, economically sensitive areas with respect to pollution, and suggests steps for abatement of pollution.

3.2 Methodology

3.2.1 Study Area

The hotspot monitoring studies are carried out at coastal waters of Kochi (9°55′ to 9°59′N and 76°11′ to 76°14′E) and Mangalore (12°49′ to 12°51′N and 74°46′ to 76°49′E). The details of the study area and sampling locations are given in Table 3.1 and presented in Fig. 3.1.

3.2.1.1 Kochi: Weather, Rainfall, Demography, and Land Use Pattern

Kochi is one among the largest harbour cities in India. Moreover, it is considered as the commercial capital of Kerala. Kochi estuary is one of the largest estuaries on the West coast of India which is permanently connected with the Arabian Sea by a gut, about 450 m wide which forms main entrance to Kochi harbour which is situated in and around the Willington Island. The city is situated at the sea mouth of seven major rivers that start from the high hills of Western Ghats and pass through the Kerala state midlands, lowlands, and coastal areas, to drain into the Arabian Sea. To the North and South, the harbour is continuous with shallow brackish water areas receiving the waters from several rivers. Five rivers flow into these backwaters, of which two, the Periyar on the north and the Pamba on the south, are the largest. The bordering areas and a couple of Islands in the estuary are the centres of intense

Table 3.1 Details of the sampling stations of the hot spots Kochi and Mangalore

SWQM – details of sampling stations			
Sampling station	Station name	Latitude	Longitude
Kochi	Kochi creek	09°58′16″N	76°14′43″E
	Kochi 0.5 km	09°56′17″N	76°14′02″E
	Kochi 2 km	09°56′17″N	76°13′13″E
	Kochi 5 km	09°56′17″N	76°11′36″E
Mangalore	Mangalore 0.5 km	12°50′49″N	74°49′17″E
	Mangalore 2 km	12°50′49″N	74°47′42″E
	Mangalore 5 km	12°50′49″N	74°46′05″E

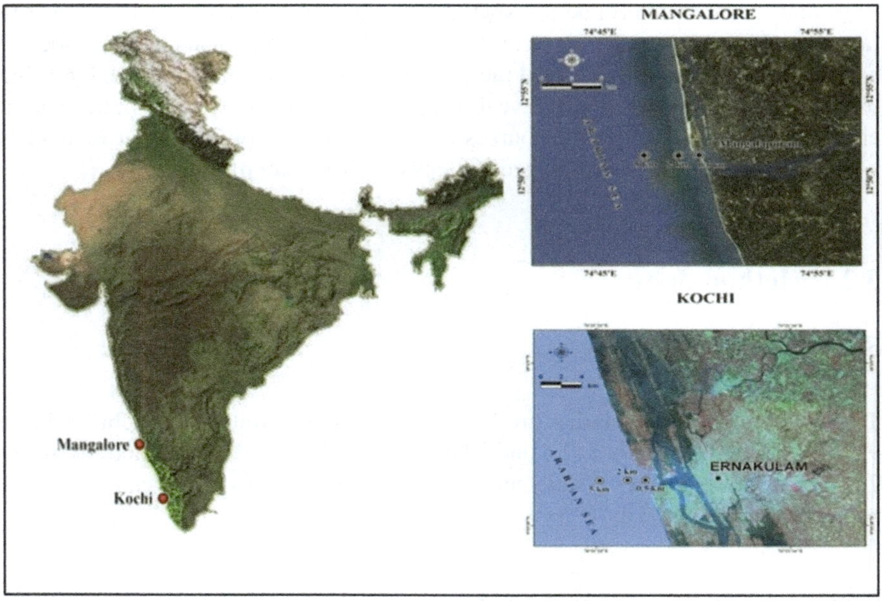

Fig. 3.1 Study area and sampling locations

human activity. Tropical monsoon climate is the peculiarity of Kochi station. The current recorded highest temperature is 38 °C and lowest is 17 °C. South west monsoon brings major portion of rainfall followed by North east monsoon to the Kochi and there are 132 average rainy days per year with a mean annual rainfall about 350 cm. Predominance of wetland and water bodies is the characteristic features of land use pattern in Kochi. But now majority of canals have deteriorated as mere drainage channels and getting possibility of flood in suburb areas. Coconut and paddy are the major plantations of Kochi. During last 20 decades, Kochi witnessed rapid population growth associated with socio-economic developments with an average local population growth of about 7.83%.

3.2.1.2 Mangalore: Weather, Rainfall, Demography, and Land Use Pattern

Mangalore is the largest coastal urban centre in Karnataka with a population of 684,785 as per the 2011. It is now a city bustling with trade and commerce. Netravathi and Gurupur rivers drain into Mangalore estuary. These rivers effectively encircle the city, with the Gurupur flowing around the north and the Netravathi flowing around the south of the city. Though the coastal areas of Mangalore are extensively exploited for marine resources, some area is also being used for dumping industrial and domestic wastes. Mangalore is influenced by the southwest monsoon and has the tropical monsoon climate. It receives most of (about 95%) its total annual rainfall during the 6 months period from May to October, while the remaining period from December to March is hot and dry. The average annual rainfall in Mangalore is 380 cm with 138 rainy days per year. The coastal area is covered with coconut gardens. The topography of the city varies from plain hilly regions to undulating with natural valleys within the city. The ambient temperature varies from a minimum of 17 °C to a maximum of 37 °C. Agriculture is the main activity of the people in the city with sown area comprising 28% of the total geographical area. Major crops are cashew nut, rubber, paddy, arecanut, coconut, and vegetables. Groundwater irrigates about 75% of the irrigated area and the remaining is by surface water sources. The predominant geological formations comprise Granites, Charnockite, laterites Gneisses, Schists, and alluvium with major soil types including alluvium and lateritic soils.

3.2.2 Sampling Locations and Periodicity

The samples are collected during both low and high tides at the mouths of creeks, shore (0.5 km), near shore (2 km), and offshore (5 km) for water quality analysis. Collection of surface waters from all the three zones (shore, near shore, and offshore) and bottom waters from near and offshore transects is mandatory. Samplings are carried out for 36 hrs continuously at an interval of 3 hours including night hours at river/backwater/ creek mouths at 2 km on either side of the shore station (along the shore) during both low and high tides. The samples are collected for 3 seasons including monsoon, pre-, and post-monsoon seasons with an interval of 3 months for each sampling with separate collection of samples for low and high tides. Seasonal sampling indicating month, date, and tidal condition is needed to be maintained for comparison of data during past year. All the samples are analysed for different physical, chemical, and biological parameters. The methodology used to analyse these parameters is described in the Table 3.2.

3.2.3 Data and Materials Used

In order to analyse the quality status of coastal waters, the chemical, biological as well as microbiological components have been determined using the methods described in Table 3.2. The chemical, biological, and microbiological data obtained from the above analysis are used to deduce meaningful conclusions regarding the coastal water quality status of the Arabian Sea. The obtained data are used to perform the statistical analysis such as factor analysis and correlation in order to relate the suitable variables within the data.

Table 3.2 Methods used to determine chemical, biological, and microbiological parameters

Sl. No	Parameters	Methodology
1	Temperature	Certified Thermometer
2	pH	Water Quality Analyser (PCD 650)
3	Dissolved Oxygen (DO)	Water Quality Analyser (PCD 650)/Winkler method
4	Total Suspended Solids (TSS)	Gravimetric Method
5	Salinity	Argentometric Titration /Water Quality Analyser (PCD 650)
6	Alkalinity	Acid-Base Titration
7	Nitrate (NO_3)	Using column reductor and measuring Spectrophotometrically at 542 nm. (In Variant Spectrophotometer – Carry Bio 50)
8	Nitrite (NO_2)	Spectrophotometrically at 542 nm (In Variant Spectrophotometer – Carry Bio 50)
9	Ammonia (NH_4)	Phenate method, Spectrophotometric analysis at 630 nm.
10	Total Nitrogen (TN)	Oxidation followed by reduction using column reductor and measuring Spectrophotometrically at 542 nm. (In Variant Spectrophotometer – Carry Bio 50)
11	Inorganic Phosphate (IP)	Spectrophotometrically at 882 nm (In Variant Spectrophotometer – Carry Bio 50)
12	Total Phosphorus (TP)	Oxidation method and measuring Spectrophotometrically at 882 nm. (In Variant Spectrophotometer – Carry Bio 50)
13	Silicate (SiO4)	Spectrophotometrically at 810 nm. (In Variant Spectrophotometer – Carry Bio 50)
14	Chlorophyll a	Spectrophotometric Method
16	Total Viable Count (TVC)	Standard Plate Count Method
17	Streptococcus faecalis (SFLO)	Standard Plate Count Method
18	Escherichia coli (ECLO)	Standard Plate Count Method

3.2.4 Assessment of Eutrophication Status

Different contaminants cause different ecological risk, for example, the higher concentration of nitrogen and phosphorus indicates the risk of eutrophication in coastal waters. Eutrophication has become a remarkable global problem and catastrophically affects aquaculture and local economies (Peng 2015). Hence, the ecological risks of nitrogen and phosphorus are assessed by using the trophic index (TRIX), which is estimated as follows (Vollenweider et al. 1998):

$$\mathrm{TRIX} = \frac{\log 10\left(Chl\text{-a} \times \%\mathrm{DO} \times \mathrm{DIP} \times \mathrm{DIN}\right) + k}{m}$$

Table 3.3 Trophic (TRIX) level by Peng (2015)

TRIX value	Trophic level
<4	High water quality, scarcely primary productivity, less eutrophication
4 < TRIX < 5	Good water quality, moderate productivity, moderate eutrophication
5 < TRIX <6	Fair water quality, high productivity, high eutrophication
>6	Poor quality water, strongly primary productivity, high eutrophication

Where, the *Chl*-a, DIP, and DIN are the concentrations of chlorophyll-a, dissolved inorganic phosphorus, and dissolved inorganic nitrogen, respectively, in μg/L whereas % DO is oxygen percent saturation, k and m are constants where, $k = 1.5$ and $m = 1.2$. The values of TRIX vary from 0–10 and are classified into trophic level as described in Table 3.3.

3.3 Results and Discussion

The seasonal and spatial hydrochemical variations of pH, salinity, and alkalinity and nutrients such as NO_2, NO_3, NH_4, TN, TP, and SiO_4 are described below for the two observational sites.

3.3.1 Kochi Hotspot (CH)

pH It is the most important parameter of aquatic environment for determining the quality of water and universally used to express the intensity of acid or alkaline condition of a solution. Spatially pH varied from 7.22 (creek) to 7.81 (5 km), while the seasonal variation of pH ranges from 7.23 (CH1, creek, monsoon) to 8.03 (CH4, post-monsoon, 5 km). Comparatively, the offshore region recorded mild rise in pH range (7.78–8.03) during pre-monsoon season. The recorded high pre-monsoon pH values could be due to the influence of seawater penetration and high biological activity (Balasubramanian and Kannan 2005) and also due to the occurrence of high photosynthetic activity (Sridhar et al. 2006).

Salinity The salinity varied spatially from 23.96 PSU (creek) to a maximum of 29.85 PSU (offshore, 2 km away from land) during this study. In this study, seasonal variation of salinity shows a range of 16.46 PSU to 32.43 PSU and it was higher during the pre-monsoon seasons. This could be due to the continuous evaporation of water from the study area especially during these seasons as also observed earlier by Sampathkumar and Kannan (1998). Since the salinity is the limiting factor for

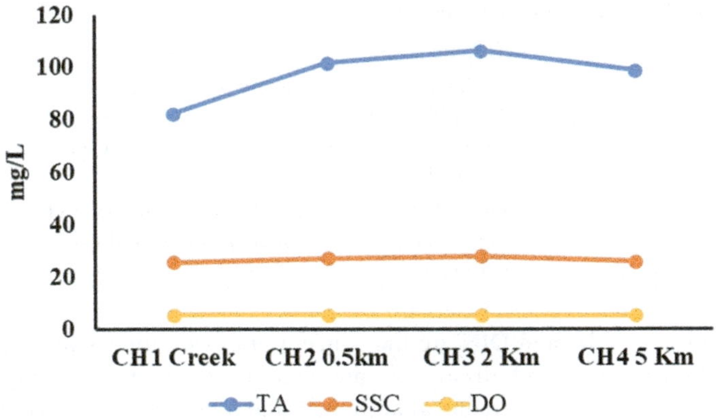

Fig. 3.2 Spatial variation of TA, SSC, and DO

distribution of living organisms, variation in the salinity caused by evaporation and dilution process might affect the faunal distribution strongly.

Total Alkalinity (TA) The total alkalinity determines the neutralization ability of the marine water to the acidic pollutants from rainfall or waste water. During the study period, the value of alkalinity spatially varied from 82.25 (creek) to 106.03 mg/L and are also high in offshores (Fig. 3.2). Seasonal variation (Fig. 3.3) shows that the values are observed high during pre-monsoon season and range from 54.01 to 118.07 mg/L. The 2 km site in monsoon found with maximum alkalinity and minimum value was recorded during post-monsoon at creek region (Fig. 3.3). High value at station 2 km may be due to cumulative effect of sea water invasion (Anitha and Kumar 2013) and pollution from various sources.

Dissolved Oxygen (DO) The DO plays a vital role in maintaining aerobic conditions of overlying water column. DO level gives general assessment of productivity of aquatic system and is a good indicator of pollution status. The DO spatially varies from an average of 5.28 mg/L to a maximum of 5.4 mg/L during the study period (Fig. 3.2), while seasonal variation value ranges from 4.33 to 6.18 (Fig. 3.3).

Suspended Sediment Concentration (SSC) The SSC ranges from 25.80 to 28.007 mg/L spatially and is maximum reported at the CH2 location. The seasonal average ranges between 18.49 and 36.26 mg/L and lowest and highest value reported during post-monsoon and monsoon season, respectively, at CH1 creek and CH2 0.5 km offshore. Moreover, comparatively higher concentration of SSC is observed during monsoon season in all location which can be attributed to the resuspension of sediments due to river influx during monsoon season (Naik et al. 2020).

Nutrients The nutrients such as NO_2, NO_3, NH_4, TN, IP, TP, and SiO_4 are essential for the primary productivity of ocean waters ranging spatially from a 56.10 to

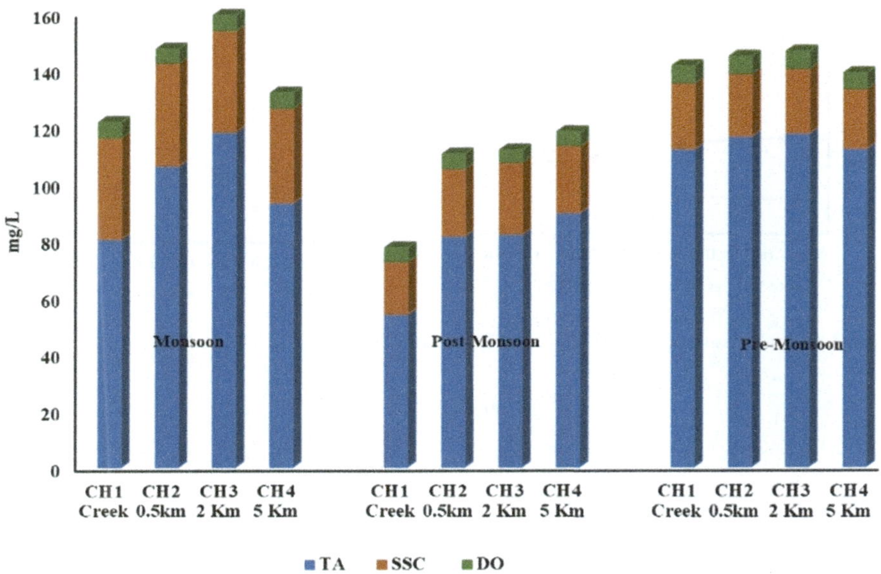

Fig. 3.3 Seasonal variation of TA, SSC, and DO at Cochin hot spot

94.82 µg/L, 618.09–965.85 µg/L, 13.01–17.19 µg/L, 274.70–444.32 µg/L, 47.51–81.74 µg/L, and 1289.63–2045.54 µg/L, respectively (Fig. 3.4a, b). Seasonal analysis of NO_2 shows that they are highly (111.40 µg/L) reported during pre-monsoon season due to evaporation. The lowest value reported is 45.38 µg/L during post-monsoon and compared to three sampling sites (creek, 0.5 km, and offshore); 0.5 km shows maximum variation (Fig. 3.5a). During the study period, minimum value for nitrite (417.97 µg/L) was recorded during pre-monsoon season at 5 km sampling site, while the maximum value observed as the 1410.81 µg/L during monsoon at the creek region. (Fig. 3.5b). Low values are noticed during summer and high value during monsoon seasons might be due to high freshwater inflow, high terrestrial flow, and high salinity (Murugan and Ayyakannu 1991). The ammonia (NH4) concentration varied from 12.50 µg/L (offshore, 2 km) to 19.139 µg/L (creek) during pre-monsoon (Fig. 3.5a). Total nitrogen (TN) fluctuated between 221.67 µg/L in post-monsoon at off shore region and 533.51 µg/L in post-monsoon at creek region during monsoon (Fig. 3.5b). Inorganic phosphate (IP) fluctuated between 18.39 µg/L in pre-monsoon at offshore region and 56.53 µg/L in post-monsoon at creek (Fig. 3.5a). The low TP value (28.30 µg/L) during pre-monsoon is reported from offshore area and the high concentration of 98.24 µg/L recorded at creek during monsoon (Fig. 3.5a). The high concentration of inorganic phosphate during post-monsoon season may be due to the input of domestic sewage water during the monsoon period. During the study time, seasonal variation in silicate values ranges from minimum 1083.31 µg/L in pre-monsoon at offshore to maximum 2839.04 µg/L in monsoon at creek area (Fig. 3.5b). The low values recorded during pre-monsoon

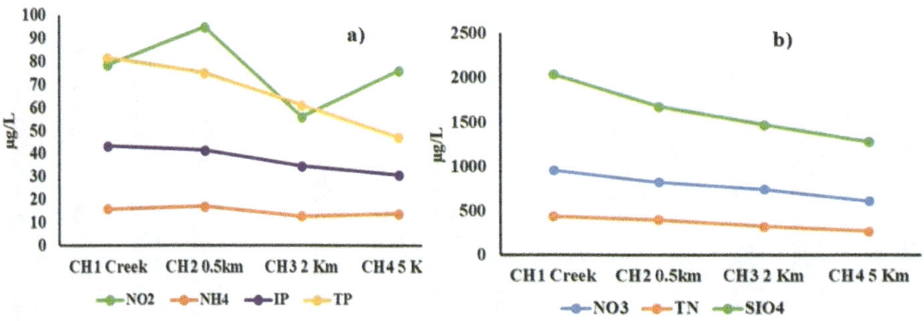

Fig. 3.4 Spatial variation of (**a**) NO_2^{3-}, NH_4^-, IP, and TP (**b**) NO_3^{2-}, TN, and SiO_4

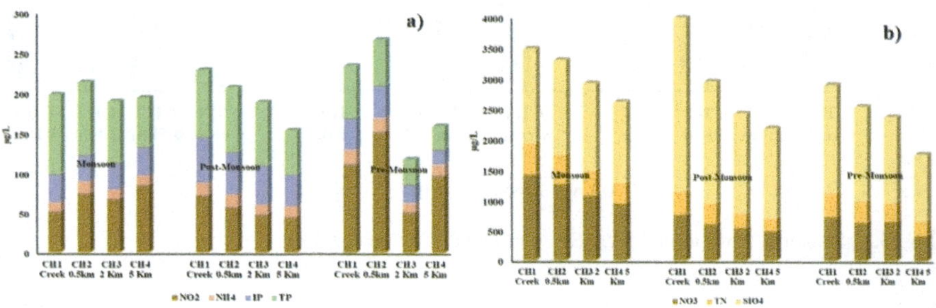

Fig. 3.5 Seasonal variation of (**a**) NO_2^{3-}, NH_4^-, IP, and TP (**b**) NO_3^{2-}, TN, and SiO_4

season must be due to silicate consumption by the phytoplankton for their biological activity (Mishra and Sujatha 1993). However, the lesser quantity of silicate was reported during monsoon season at different location due to the terrestrial runoff and river influx.

3.3.2 Mangalore Hotspot (MH)

pH The pH is ranged spatially between an average of 7.77 at 0.5 km and 8.07 at offshore, while the seasonal analysis shows that value varies between an average value of 7.49 and 8.15 during the study. The low values of pH noticed during monsoon season may be due to dilution and mixing of coastal waters by rain floods that leads to reduction in salinity and temperature and decomposition of organic matter (Soundarapandian et al. 2009; Rajendran and Kathiresan 1999).

Salinity Spatial analysis of salinity shows that the minimum average value of salinity is reported at 0.5 km (i.e. MH1) 29.65PSU and maximum value of 32.08PSU reported at offshore. The seasonal analysis shows an average salinity variation of 27.35–32.49PSU and is reported during monsoon season at 0.5 km and pre-monsoon at offshore, respectively. Salinity of surface water showed an increasing trend from estuarine to offshore region during the study period.

Dissolved Oxygen (DO) Spatial average of DO shows that minimum (4.75 mg/L) value reported at MH1 0.5 km location and maximum (4.87 mg/L) values were recorded at offshore region situated at 5 km (Fig. 3.6). The seasonal analysis of data suggests that the value range between 4.12 and 5.39 mg/L and high and low values are reported at monsoon and pre-monsoon season, respectively (Fig. 3.7). Generally, the oxygenation process in aquatic system is the result of degradation of organic matter, photosynthesis, and physicochemical properties of water (Aston 1980).

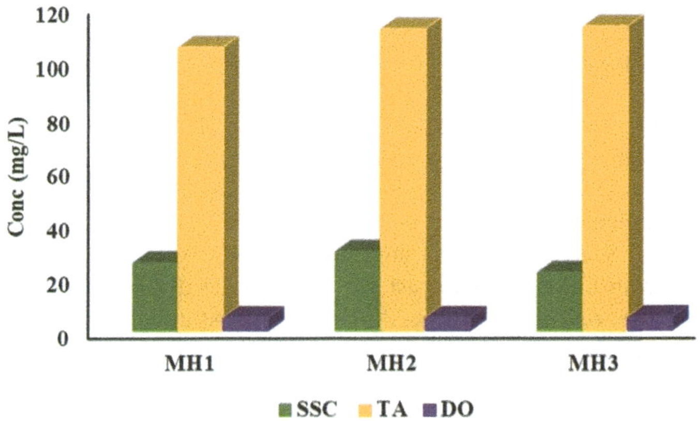

Fig. 3.6 Spatial variation of SSC, TA, and DO at Mangalore

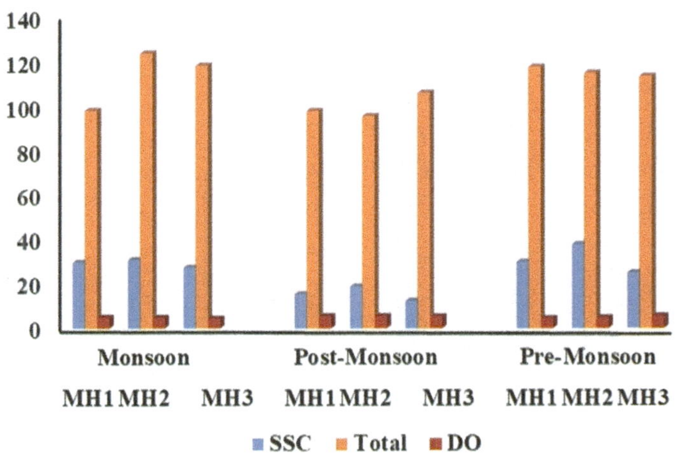

Fig. 3.7 Seasonal variation of SSC, TA, and DO at Mangalore hot spot

Alkalinity (TA) During the study period, the spatial value of alkalinity varied from 104.95 to 112.91 mg/L (Fig. 3.6); the seasonal value variation is between 98.514 and 124.19 mg/L. The minimum and maximum alkalinity is found during the monsoon season at MH1 (0.5 km) and MH2 (2 km) region, though comparatively highest value ranges are found in the pre-monsoon (Fig. 3.7).

Suspended Sediment Concentration (SSC) Spatially, the SSC ranges from 21.52 to 29.23 mg/L (Fig. 3.6), and contrast to CH, maximum reported at the shore creek region due to the shallowness as well as resuspension due to tidal mixing and riverine influx (Naik et al. 2020). The seasonal average ranges between 12.28 and 37.93 mg/L and lowest and highest value reported during post-monsoon and pre-monsoon season, respectively, at MH3 (5 km) and MH2 (2 km) offshore region (Fig. 3.7). Higher concentration of SSC during pre-monsoon season in all location can be attributed to the beach erosion during monsoon season (Naik et al. 2020).

Nutrients The concentration of NO_2, NO_3, NH_4, TN, IP, TP, and SiO_4 spatially varied in a range of 63.49–76.54 μg/L, 583.97–809.83 μg/L, 11.87–15.09 μg/L, 410.88–318.95 μg/L, 29.12–38.62 μg/L 51.24–65.06 μg/L, and 418.64–705.43 μg/L, respectively; higher concentrations are observed in MH1 (0.5 km) due to beach erosion (Fig. 3.8). Seasonal analysis suggests that NO_2 varies between 40.89 and 112.85 μg/L with highest and lowest values observed during monsoon and post-monsoon season, respectively (Fig. 3.9). The highest concentration of NO_3 was recorded at 1036.38 μg/L in post-monsoon at MH1 and minimum (247.33 μg/L) in pre-monsoon at offshore region. The large scale spatio-temporal variation of nitrate along the coastal milieu is attributed to the processes of quick assimilation by phytoplankton and enhancement by surface runoff (De Sousa 1983; Zepp 1997). The ammonia contents were negligible during the study period as its ranges between 8.261 and 17.98 μg/L. The variation of ammonia in the marine environ-

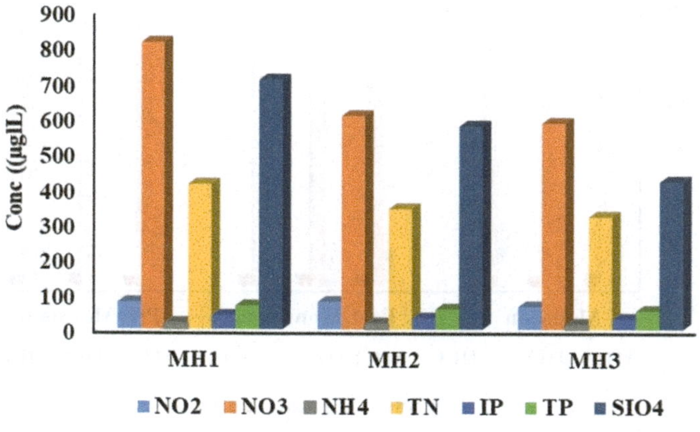

Fig. 3.8 Spatial variation of Nutrients at Mangalore hotspot

ment is mainly caused by two processes, excretory release by marine invertebrates and utilization by phytoplankton (Gilbert et al. 1982). Maximum TN was recorded (467.89 μg/L) in post-monsoon at MH1 (0.5 km) site and minimum TN value (235.15 μg/L) during pre-monsoon at offshore area. The sampling site MH1 (0.5 km) determined with maximum IP with 44.26 μg/L in monsoon and minimum value 25.21 μg/L in post-monsoon. During monsoon period, the highest TP values (73.67 μg/L) are recorded at MH (0.5 km) and lowest (48.95 μg/L) recorded at MH2 (2 km) during post-monsoon period. The higher concentration phosphate during monsoon is an indication of river flux contribution of phosphate. Nutrients like phosphate and nitrate are important in the primary production of marine environment and are higher during monsoon season which is an indication of river runoff, rapid oxidation of plankton detritus, and decomposing matter (Kalaiarasi et al. 2012). The silicate values varied from 248.83 to 1216.38 μg/L; highest values were recorded at 0.5 km during post-monsoon season and lowest at offshore during pre-monsoon season. The higher concentration of silica during post-monsoon can be attributed to river and stream addition to ocean during the monsoon season (Robin et al. 2011); lower concentration during pre-monsoon is attributed to utilization by phytoplankton and bottom algae along with the mixing bottom loss during its mixing with seawater. Comparatively, the lower concentration of SiO_4 at offshore regions than onshore regions in all through season is an indication of high biological productivity and biological removal by adsorption onto suspended sediments. The Fig. 3.9 represents the seasonal fluctuation values of nutrients in Mangalore hotspot.

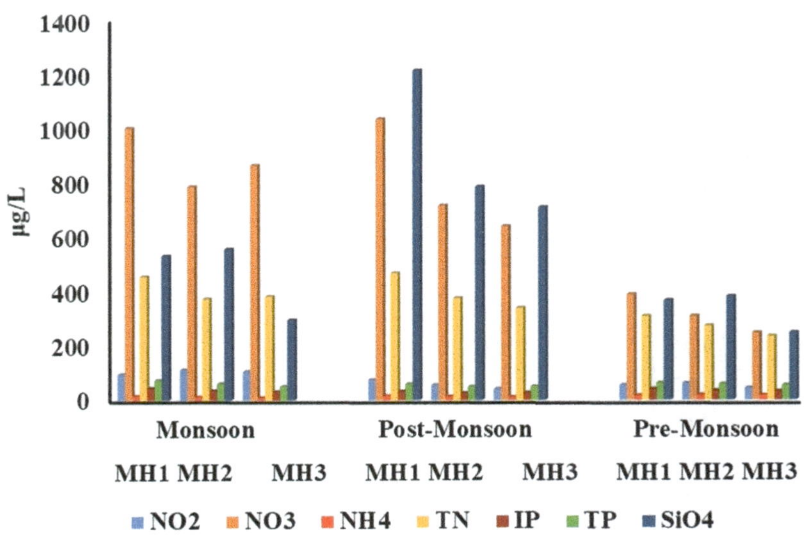

Fig. 3.9 Seasonal variation of Nutrients at Mangalore hot spot

3.3.3 *Marine Biology*

3.3.3.1 Kochi Hotspot

Chlorophyll Chlorophyll (*chl* a) concentration represents productivity of a particular region including phytoplankton biomass and plays a role in reducing light penetration in shallow water habitats; The *chl* a is direct indicator for the nutrient enrichment because its concentrations are a good measure of the phytoplankton biomass that utilizes nutrients for growth (Morsy et al. 2022). The recorded *chl* a during the study period spatially shows 1.68–1.89 mg m^{-3} and are showing high values in the creek region and low values in the CH2 region (0.5 km) (Fig. 3.10). As far as Kochi is concerned, mean peak chlorophyll concentration occurred around pre-monsoon season followed by monsoon with slight variation. During post-monsoon season, mean chlorophyll value was found to be 1.187 mg m^{-3} at Barmouth region, 1.31 mg m^{-3} at CH1 (0.5 km), 0.90 mg m^{-3} at CH2 (2 km), and 1.25 mg m^{-3} at offshore region. During the pre-monsoon season, mean chlorophyll values ranged from 1.5 mg m^{-3} (Offshore) to 2.84 mg m^{-3} (2 km). During monsoon season, CH4 (5 km) region reported highest mean value of 2.32 mg m^{-3} and offshore region recorded lowest value (1.71 mg m^{-3}). Among the four sampling seasons, post-monsoon of 2012 showed lower values of chlorophyll such as 3.25 mg m^{-3} at bar-mouth and 1.36 mg m^{-3} at offshore waters (Fig. 3.11). The general trend of decreasing chlorophyll values could be observed from bar mouth to offshore corresponding to the phytoplankton biomass and nutrient enrichment. Chlorophyll concentration in Kochi estuary is induced by the abundance of floating matter found in the estuary, which predominantly consists of Eichchornia crassipes (Water hyacinth). It has been reported that the overabundance of Eichchornia reduces phytoplankton species composition and which in turn affects species diversity (Morsy et al. 2022). Thus, in general a chlorophyll value of Kochi estuary strictly pronounces the mesotrophic condition of the area.

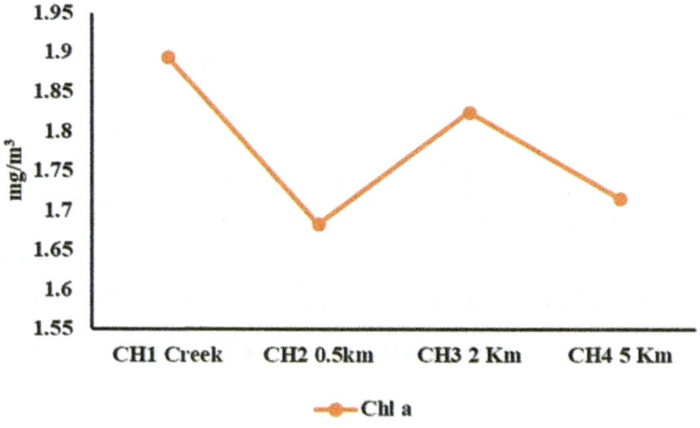

Fig. 3.10 Spatial variation of *Chl* a in Kochi

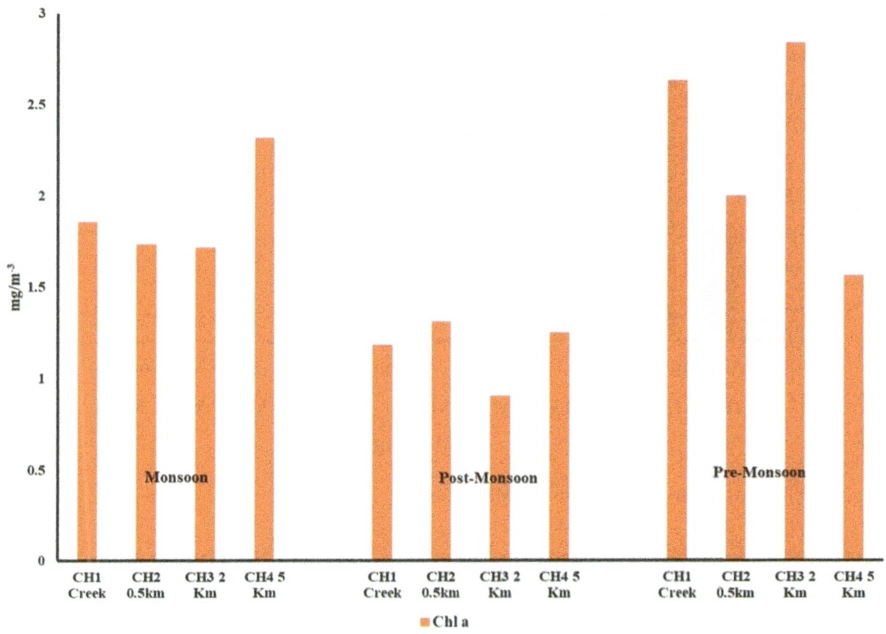

Fig. 3.11 Seasonal variation of *Chl* a in Kochi

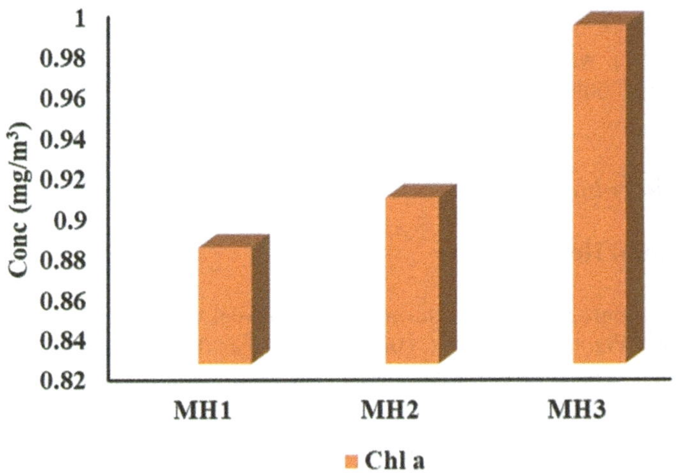

Fig. 3.12 Spatial variation of *Chl* a in Mangalore

3.3.3.2 Mangalore Hotspot

Chlorophyll Spatial average of the *chl* a reported for MH is range between 0.87 and 0.98 mg m^{-3} and increasing towards the offshore direction (Fig. 3.12). The seasonal *chl* a values range between 0.41 and 1.65 mg m^{-3} and are recoded low and

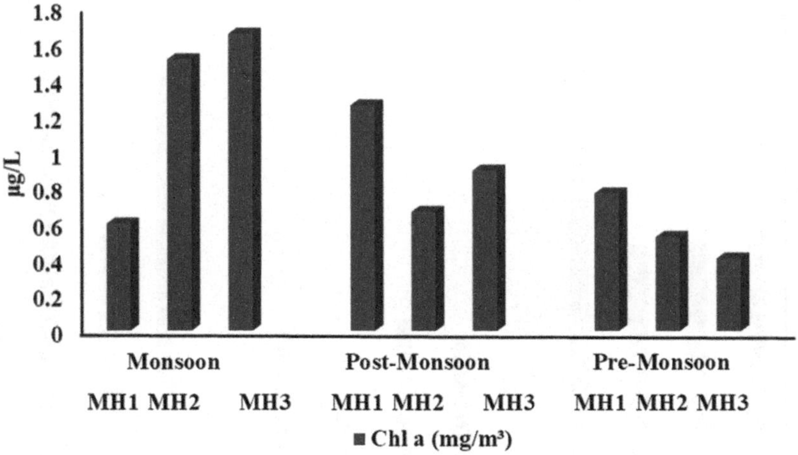

Fig. 3.13 Seasonal variation of *Chl* a in Mangalore

high during pre-monsoon season and monsoon season at offshore regions, respectively. The seasonal *chl* a variation is attributed in Fig. 3.13 and shows that higher concentration is observed during monsoon season followed by post-monsoon season due to coastal upwelling during monsoon. Temporal variation in chlorophyll values was attributed to phytoplankton productivity. During pre-monsoon season, the cooler nutrient water is tapped below due to the warm surface water and the vertical layers of the ocean are not mixed properly. Therefore, nutrients that have built up in deep water cannot reach surface water. This could be attributed to the lower values of chlorophyll during pre-monsoon season.

3.3.4 Microbiology

3.3.4.1 Kochi Hotspot

Continuous monitoring studies during seasons reveal that in Kochi hotspot reveals that the Total Viable Count (TVC) that distinguishes the concentration of microorganisms such as bacteria, yeast, or mould spores in a sample is high in creek region with a mean value of 24736.56 CFU/mL, while lower value (11184.44) observed in offshore region. The faecal Streptococci (SFLO) bacteria concentration varies between 12445.56 and 4928.33 CFU/mL and the highest value reported in offshore region CH4 and lowest at CH3. However, the E. coli (ECLO) comparatively shows lowest concentration. The highest mean value reported for ECLO is 752.78 CFU/mL at offshore region CH3 (2 km) and lowest reported as130CFU/mL at offshore region CH4 (5 km) (Fig. 3.14). The seasonal analysis of microbiological concentration suggests that the high concentration of TVC is observed in post-monsoon

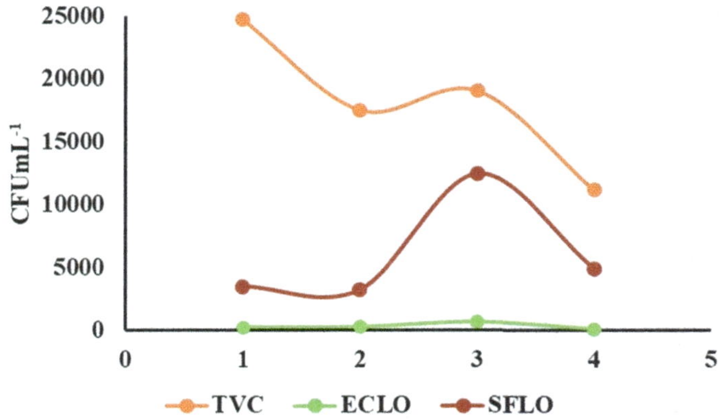

Fig. 3.14 Spatial variation of microbiology concentration in Kochi

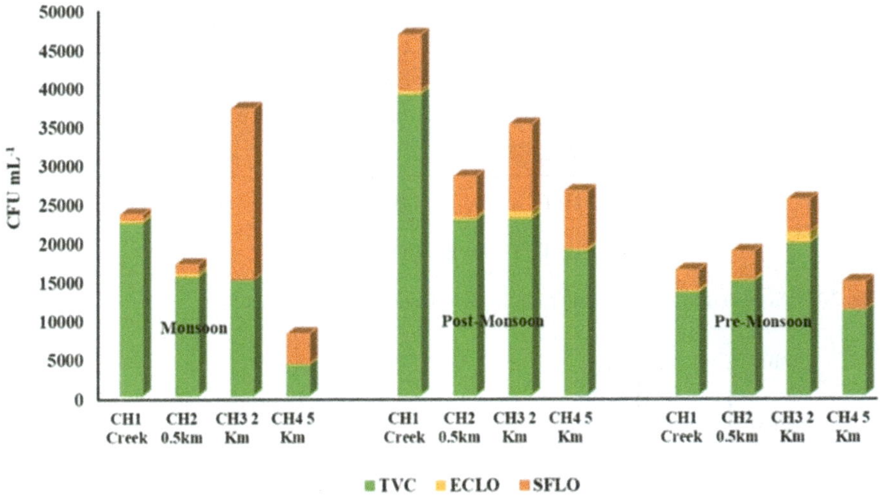

Fig. 3.15 Seasonal variation of microbiology concentrations in Kochi

(38741.94 CFU/mL) at CH1 (creek) and lowest (3920 CFU/mL) in monsoon at CH4. When comes to Kochi shore waters, the SFLO recorded a highest mean value of 22,000 CFU/mL during monsoon at CH3 (2 km) and lowest at CH1 (692.21 CFU/mL) during monsoon. However, the ECLO value is almost zero reported at CH3 during monsoon season and highest concentration (873.33 CFU/mL) observed at CH3 (2 km) during post-monsoon (Fig. 3.15). The higher faecal content during monsoon season and low concentration in offshore region suggest an increased land runoff to the Kochi coastal hotspot.

3.3.4.2 Mangalore Hotspot

The spatial average concentration of TVC, ECLO, and SFLO shows that the higher concentration of them is observed at MH2 (20253.61 CFU/mL), MH3 (164.25 CFU/mL), and MH2 (1229.36 CFU/mL), respectively (Fig. 3.16), while the lower concentration reported at the MH3 (13842.51 CFU/mL), MH1 (134.56 CFU/mL), MH1 (1062.91 CFU/mL), respectively. Seasonally, the TVC varies with value of 9126.66–29155.24 CFU/mL and the highest value observed at MH2 during monsoon season, lowest at offshore region during pre-monsoon (Fig. 3.17). The ECLO

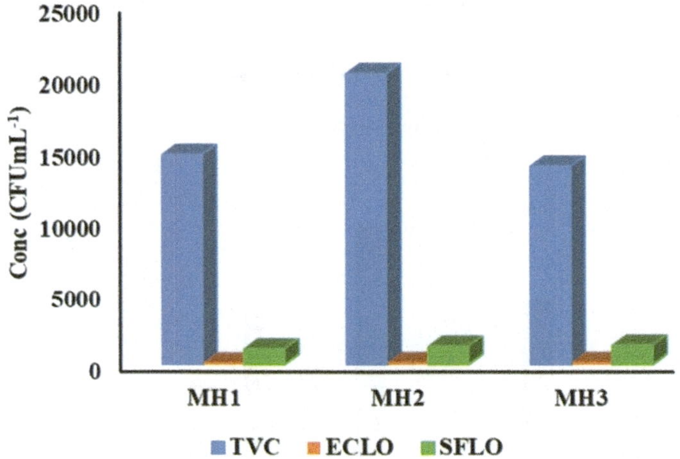

Fig. 3.16 Seasonal variation of microbiology concentrations in Mangalore

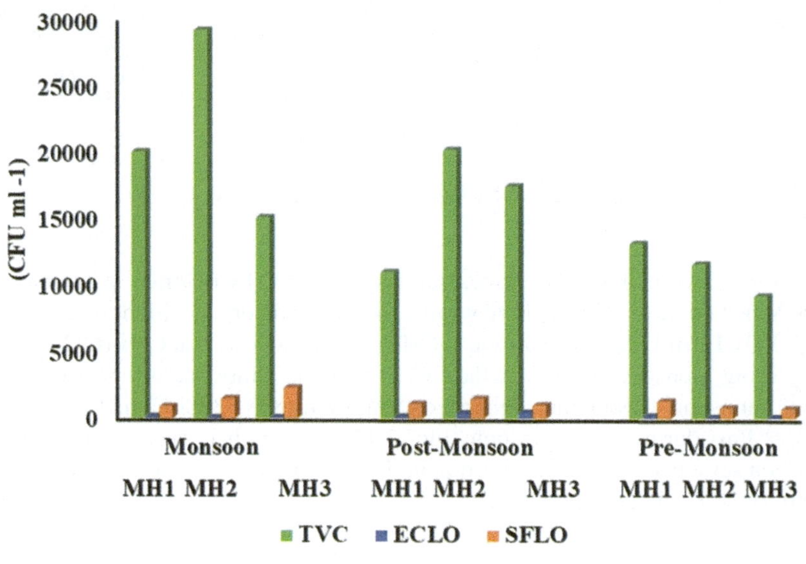

Fig. 3.17 Seasonal variation of microbiology concentrations in Mangalore

values of highest (414.44 CFU/mL) and Lowest (38.33 CFU/mL) value are observed at MH3 during post-monsoon and pre-monsoon, respectively. Similar to TVC, the highest SFLO value (2262.54 CFU/mL) is observed during monsoon and lowest SFLO value (681.67 CFU/mL) during pre-monsoon at MH3 (km) offshore location. The higher content TVC during monsoon is an indication of increased land runoff. The Central Pollution Control Board of India (CPCB) classifies the seawaters along the Indian coast as fit for commercial fishing, contact recreation, and bathing activities only when the FC is <100 CFU/mL. In case of both Kochi and Mangalore, generally, monsoon season shows the prominent increase in the faecal indicator bacteria (SFLO and ECLO) load throughout the study due to land runoff.

TRIX Index

TRIX index measures the primary productivity in the coastal waters as well as the eutrophication status; the annual values of oxygen saturation %, *Chl*-a, dissolved inorganic nitrogen (DIN), and dissolved inorganic phosphorus (DIP) are used to distinguish the index. The four trophic levels are defined as follows: high with TRIX < 4, good with 4 < TRIX < 5, fair with 5 < TRIX < 6, and poor with TRIX > 6 (Herrera-Silveira and Morales-Ojeda 2009).

3.3.4.3 Kochi Hotspot

TRIX index calculated for Kochi hotspot varies with 5.79–6.44 with an average value of 6.19. The spatial analysis suggests that the TRIX value of CH1 (creek), CH2 (0.5 km), CH3 (2 km), and CH4 (5 km) are 6.35, 6.27, 6.12, and 6.03, respectively (Fig. 3.18). Thus, the high TRIX value is observed in the creek region where the value ranges between 6.19 and 6.44. Similarly, the lowest variation of TRIX is observed at CH4 (5 km) offshore region where the value ranges between 5.79 and 6.37. The average TRIX seasonal value reported for Kochi is 6.34, 6.04, and 6.20

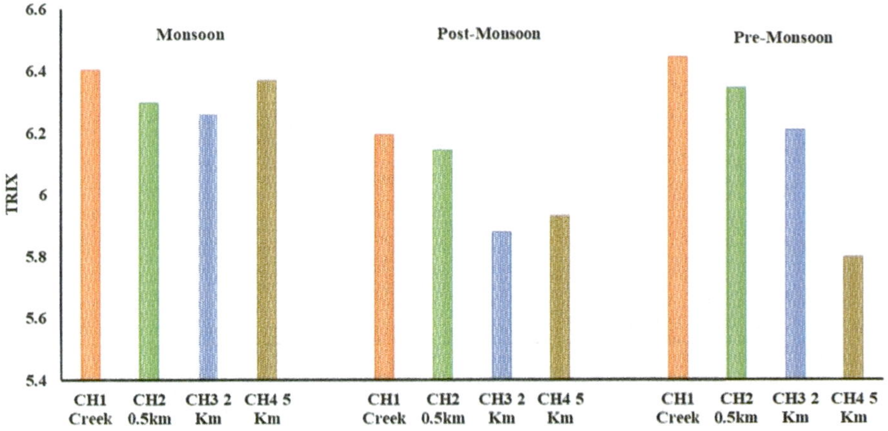

Fig. 3.18 Spatio-seasonal variation of TRIX variation in Kochi hotspot

during monsoon, post-monsoon, and pre-monsoon, respectively, indicating high productivity in the coastal environment. Although the higher value of TRIX is reported during pre-monsoon season at creek region close to ocean, the TRIX value >6 found mostly in monsoon season can be attributed to the land runoff as well as riverine flux. According to Herrera-Silveira and Morales-Ojeda (2009), the TRIX value of 6.19 calculated for Kochi hotspot is suggesting eutrophic condition of coastal water with less water quality. Here, the nutrient content as well as the primary productivity is higher. However, during post-monsoon season, the average value is almost close to 6 indicating fair condition (Peng 2015).

3.3.4.4 Mangalore Hotspot

The TRIX calculated for MH ranges between 4.9 and 5.71 with an average value of 5.37. The spatial calculation of TRIX suggests that highest value (5.51) is observed in the creek and lowest (5.29) found at offshore region (Fig. 3.19). Seasonal variation of TRIX value shows an average value of 5.62, 5.45, and 5.07 during monsoon, pre-monsoon, and post-monsoon season. The seasonal variation of TRIX shows that the higher value of TRIX (5.71) is observed during post-monsoon season at creek region, but the highest values (5.53–5.65) reported for different regions are observed in monsoon season. In the case of post- and pre-monsoon, seasonal variation of values between the regions is too high that the offshore regions show comparatively less TRIX. Since the average value of TRIX for MH is 5.37, the trophic status of coast falls in the fair condition proposed by Herrera-Silveira and Morales-Ojeda (2009). This suggest that Mangalore coastal waters are comparatively high in quality than Kochi, but the nutrient content as well as other primary productivity is higher in this region.

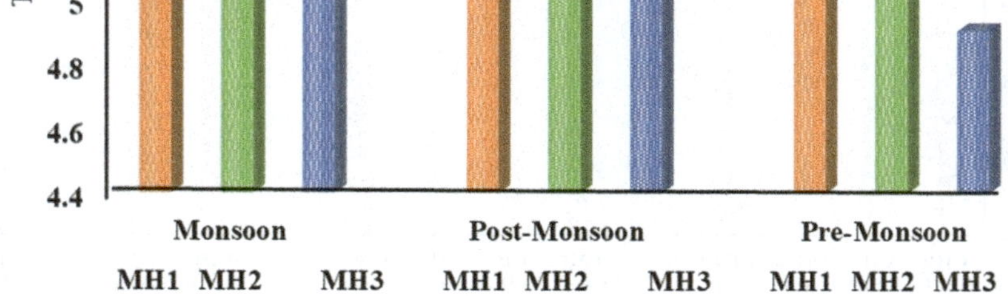

Fig. 3.19 Spatio-seasonal variation of TRIX variation in Mangalore hotspot

3.4 Conclusions

This study is carried out as a part of the nationwide coastal monitoring programme, namely, Sea Water Quality Monitoring (SWQM), that has been launched in order to identify ecologically sensitive areas in terms of coastal pollution through chemical, biological, and marine biological aspects. It attempts to appraise the spatial and seasonal trends of sea water quality factors and in terms of chemical, biological, and microbiological characteristics of coastal waters of the two hot spots from southwest coast of India, the Kochi and the Mangalore stations. It also highlighted the eutrophication status of coastal region using trophic index (TRIX index) that reveals the primary productivity status of coastal environment and identifies the sources of pollutants and economically sensitive areas with respect to pollution. The spatial trends of dissolved parameters in these coastal waters show a high concentration of parameters near the shore than offshore and the seasonal analysis shows that most of the higher concentrations are reported during the pre-monsoon season, while the lower values are noted for the monsoon season except in few cases, indicating the terrestrial influx, riverine influx, evaporation, and higher photosynthetic activity in both the study regions. Comparing the dissolved parameters, chemical and biological aspects, it is observed that the higher values of most parameters are observed in the Kochi hotspot than the Mangalore hotspot, suggesting strong influence of anthropogenic activity in the study region. Moreover, the biogeochemical parameters and the trophic index suggest a higher primary productivity and eutrophication stage of coastal water during study period, indicating deteriorating quality of water in these coastal regions.

References

Anitha G, Kumar SP (2013) Seasonal variations in physico-chemical parameters of Thengapattanam estuary, South west coastal zone, Tamilnadu, India. Int J Environ Sci 3(4):1253

Aston SR (1980) Nutrients, dissolved gases, and general biogeo-chemistry of estuaries. In: Olausson E, Cato I (eds) Chemistry and biogeochemistry of estuaries. John Wiley & Sons, Inc., Hoboken, New Jersey

Balasubramanian R, Kannan L (2005) Physicochemical characteristics of the coral reef environs of the Gulf of Mannar Biosphere Reserve, India. Int J Ecol Environ Sci 31(3):273–278

De Sousa SN (1983) Studies on the behaviour of nutrients in the Mandovi estuary during premonsoon. Estuar Coast Shelf Sci 16(3):299–308

Field CB, Barros VR, Dokken DJ, Mach KJ, Mastrandrea MD (2014) Climate Change 2014 Impacts, Adaptation, and Vulnerability (Working Group II Contribution to the Fifth Assessment Report of the Intergovernmental Panel on Climate Change) Coastal Systems and Low-Lying Areas. Cambridge University Press. https://doi.org/10.1017/CBO9781107415379 (Chapter 5), pp 361–410. https://doi.org/10.1017/CBO9781107415379.010

Gilbert PM, Lipschultz F, McCarthy JJ, Altabet MA (1982) Isotope dilution models of uptake and remineralization of ammonium by marine plankton 1. Limnol Oceanogr 27(4):639–650

Herrera-Silveira JA, Morales-Ojeda SM (2009) Evaluation of the health status of a coastal ecosystem in southeast Mexico: assessment of water quality, phytoplankton and submerged aquatic vegetation. Mar Pollut Bull 59:72–86

Kalaiarasi M, Paul P, Lathasumathi C, Stella C (2012) Seasonal variations in the physico-chemical characteristics of the two coastal waters of Palk-Strait in Tamil Nadu, India. Glob J Environ Res 6(2):66–74. https://doi.org/10.5829/idosi.gjer.2012.6.2.386

Mishra D, Sujatha P, Panigrahy RC (1993) Physico-chemical characteristic of the Bahuda estuary (Orissa), Eastern coast of India. Ind J Mar Sci 22:75–77

Morsy A, Ebeid M, Soliman A, Halim AA, Ali AE, Fahmy M (2022) Evaluation of the water quality and the eutrophication risk in Mediterranean Sea area: a case study of the Port Said harbour, Egypt. Environ Chall 7:100484

Murugan A, Ayyakannu K (1991) Ecology of Uppanar backwaters, Cuddalore-II nutrients. Mahasagar-Bull Natl Inst Oceanogr 24(2):103–108

Naik S, Mishra RK, Sahu KC, Lotliker AA, Panda US, Mishra P (2020) Monsoonal influence and variability of water quality, phytoplankton biomass in the tropical coastal waters – a multivariate statistical approach. Front Mar Sci 7:648. https://doi.org/10.3389/fmars.2020.00648

Peng S (2015) The nutrient, total petroleum hydrocarbon and heavy metal contents in the seawater of Bohai Bay, China: temporal–spatial variations, sources, pollution statuses, and ecological risks. Mar Pollut Bull 95(1):445–451

Rajendran N, Kathiresan K (1999) Seasonal occurrence of juvenile prawn and environmental factors in a Rhizophora mangal, southeast coast of India. Hydrobiologia 394:193–200

Robin RS, Muduli PR, Vardhan KV, Rao GN, Balasubramanian T (2011) Seasonal variation of nutrient concentration in a tropical estuary, southwest coast of India. PIET, Rourkela MIP

Sampathkumar P, Kannan L (1998) Seasonal variations in physico-chemical characteristics in the Tranquebar – Nagapattinam region, southeast coast of India. Pollt Res 17:397–402

Soundarapandian P, Premkumar T, Dinakaran GK (2009) Studies on the physico-chemical characteristic and nutrients in the Uppanar estuary of Cuddalore, South east coast of India. Curr Res J Biol Sci 1(3):102–105

Sridhar R, Thangaradjou T, Kumar SS, Kannan L (2006) Water quality and phytoplankton characteristics in the Palk Bay, southeast coast of India. J Environ Biol 27(3):561–566

Vollenweider RA, Giovanardi F, Montanari G, Rinaldi A (1998) Characterization of the trophic conditions of marine coastal waters with special reference to the NW Adriatic Sea: proposal for a trophic scale, turbidity and generalized water quality index. Environmetrics 9(3):329–357. https://doi.org/10.1002/(sici)1099-095x(199805/06)9:3<329::aid-env308>3.0.co;2-9

Zepp RG (1997) Interactions of marine biogeochemical cycles and the photodegradation of dissolved organic carbon and dissolved organic nitrogen. Water Sci. Technol Lib 25:329–352

Chapter 4
Assessment of Water Quality from the Gundlakamma Estuary, Andhra Coast, Southeast Coast of India

U. Suresh and B. C. Sundara Raja Reddy

Abstract Twenty-four marine water samples in three seasons (premonsoon-2016, monsoon-2016, and premonsoon-2017) from Gundlakamma estuary, Prakasam district, Southeast coast of India, were collected using shallow water sampler at 1 m below the surface. The physicochemical properties were determined. The results for the studied parameters revealed that the water composition of the study area is pH 7.0–8.9; TDS 25.16–64,972 ppm; Salinity 28.92–47.85 ppt; DO 4.04–40.04 ppm; Ca 412.50–1577.50 ppm; Cl 16,012–26487.50 ppm; Na 3023–17454.50 ppm; and K 85–186.5 ppm. Statistical treatment of Factor Analysis (FA) for the distribution of physicochemical parameters shows that they exhibit a strong affinity among several parameters of the study area. In the typical context, water composition may be fresher in some part of upstream and more concentrated near estuary. The impact of anthropogenic as well as natural input, as a source of physicochemical parameters, is lethal. However, a comprehensive investigation is warranted to assess the health of the fragile estuarine ecosystem.

Keywords Physicochemical parameters · Water quality · Gundlakamma estuary · Southeast coast of India

4.1 Introduction

Water is one of the most important components of the ecosystem (Priyanka et al. 2009). Water quality can be described by chemical and microbiological properties (Manjare et al. 2010). Significant correlation is possible among these studied variables, and thus be useful in indicating the health of the watery (Sreenivasulu et al.

U. Suresh (✉) · B. C. Sundara Raja Reddy
Department of Geology, Sri Venkateswara University, Tirupati, Andhra Pradesh, India

N. Jayaraju et al. (eds.), *Coasts, Estuaries and Lakes*,
https://doi.org/10.1007/978-3-031-21644-2_4

2015). Even when there is no pollution, the waters of rivers and lakes keep changing with the seasons and geographical areas. The specific ways that chemical constituents enter the marine environments are land runoff in the coastal zone via rivers and other anthropogenic effluents (Jayaraju et al. 2011). Pollution in a river first affects its chemical quality before destroying the environment and disturbing the delicate food chain. Pollution severely limits the diverse uses of rivers, and even polluters such as industry suffer as a result of rising river pollution (Dhirendra et al. 2009). As a result, a detailed understanding of water quality is required to determine the health status of the environment and its impact on biological fertility (Poonam and Rahul 2012).

The biochemistry of a water body is heavily influenced by physicochemical characteristics. Subtle changes in physical conditions can have a significant impact on the water quality of the studied system, affecting the geographical and temporal distribution of nutrients and/or biological populations. As a complex system, the marine environment is primarily influenced by physicochemical and biological processes. In recent years, industrial operations, economic upwelling, and urbanization in cosmopolitan city centers across the world have developed fast, and huge amounts of pollution have been pushed and flushed into rivers, estuaries, and marine marginal water bodies (Ravichandran and Manickam 2012). The scientific community is paying more attention to estuarine sediment contamination, which is known to contribute greater measure to the ecosystem health stress (Riba et al. 2002). Marine water, which serves as a habitat for many aquatic organisms, is contaminated with a variety of harmful and poisonous compounds, including heavy metals. As a result, they are regarded as key carriers and serve as sinks of pollutants (Ho et al. 2010). This reflects the system's current quality, thus providing information on the pollution status and sources. Thus, this serves as a natural archive of recent changes in environmental settings (Kruopiene 2007).

The present study involves the analysis of water described by its physicochemical variables of Gundlakamma estuary, Andhra coast, India. It is also aimed to know the quality of water sources with respect to physicochemical variables. In addition, the data sets were treated with statistical methods in order to understand the possible correlation among several environmental variables.

4.2 Study Area

The area under study is situated on the eastern part of Prakasam district. It is situated between latitude 15°31′ and 15°32′ N and longitude 80°12′ and 80°15′ E (Fig. 4.1). The river (Gundlakamma River) flows through the east-central region of Andhra Pradesh, India. It originates in the Nallamala Hills, a component of the Eastern Ghats. As it flows down the densely covered hills, it is joined by several mountain streams. Gundlakamma is the largest of all the rivers that originate from the Nallamalla Hills. The mean annual precipitation declines from 1270 mm (eastern extremity) to 762 mm (western extremity) of the basin. The NE monsoon arrives in October and leaves by November. The mean extreme temperature ranges from

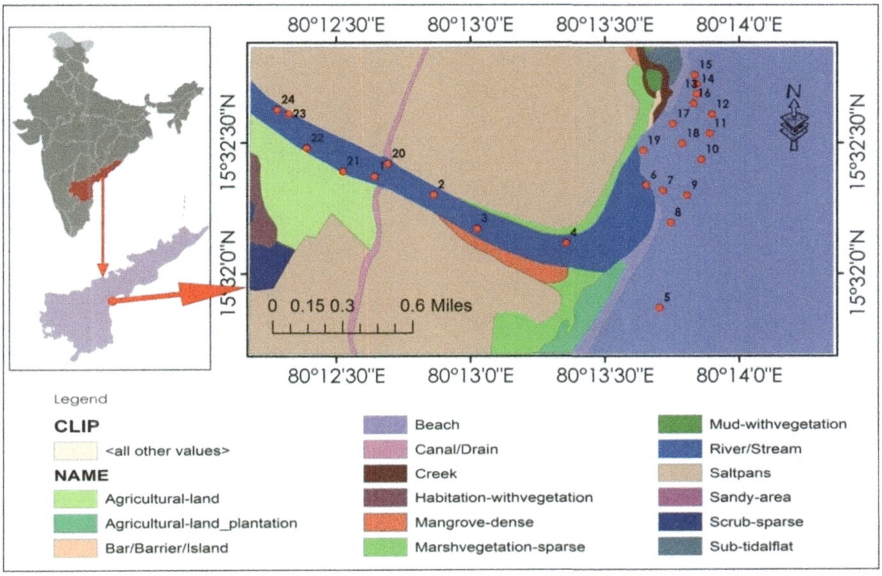

Fig. 4.1 Location map of the study area

30–32 °C, with the lowest temperature ranging from 22.5–25 °C in the catchment. The region around the study area is composed of deposits of recent origin such as sand bars, dunes, and mudflats. These are the outcome of regression of the sea and modified by in part by fluvial process.

4.3 Methodology

Twenty-four marine water samples in three seasons (premonsoon-2016, monsoon-2016, and premonsoon-2017) from Gundlakamma estuary, Prakasam district, Southeast coast of India, were collected using shallow water sampler at 1 m below the surface. The physicochemical properties were determined. The sampling stations were located precisely by using Global Positioning System (GPS) (Table 4.1). The bottom water samples were stored in polyethylene bottles that had already been cleaned by soaking in 5 M HCl for more than 3 days and thoroughly washing with distilled-deionized water before use. Salinity was measured using Atago refracto meter (Arumugam and Sugirtha 2014). Pye Unicam model 292 m (after standardization with buffer solution at pH 4.0, 7.0, and 9.0) was used for pH determination. Gravimetric (evaporation) method for TDS and EDTA Titrimetric method for Ca were used. Mohr titrimetric and chloride metric were used for Cl analysis and Flame photometer was used for Na and K. Dissolved Oxygen was measured using standard laboratory methods (APAH 1995). The data are obtained by statistically analyzing Factor loadings using XL STAT-2022.

Table 4.1 GPS locations of sampling stations

Sample	Longitude	Latitude
1	15°32′11.47″	80°12′ 38.55″
2	15°33′30.06″	80°12′51.76″
3	15°32′10.21″	80°13′01.45″
4	15°32′07.08″	80°13′21.32″
5	15°32′28.17″	80°13′38.56″
6	15°32′20.25″	80°13′39.21″
7	15°32′18.99″	80°13′42.96″
8	15°32′17.91″	80°13′48.32″
9	15°32′11.61″	80°13′44.65″
10	15°32′26.15″	80°13′51.52″
11	15°32′32.13″	80°13′53.36″
12	15°32′36.56″	80°13′53.86″
13	15°32′41.20″	80°13′50.44″
14	15°32′43.51″	80°13′46.38″
15	15°32′43.51″	80°13′46.38″
16	15°32′39.08″	80°13′45.08″
17	15°32′34.36″	80°13′45.12″
18	15°32′29.83″	80°13′00.48″
19	15°32′28.21″	80°13′38.56″
20	15°32′25.18″	80°12′41.54″
21	15°32′23.45″	80°12′31.46″
22	15°32′28.78″	80°12′23.36″
23	15°32′36.70″	80°12′19.36″
24	15°32′19.64″	80°12′24.01″

4.4 Results and Discussion

Physicochemical properties of marine water in three seasons (premonsoon-2016, monsoon-2016, and premonsoon-2017) were determined according to the standards of the American Public Health Association (APHA) and other standard methods. The common physicochemical parameters of water like pH, Total Dissolved Solids (TDS), Salinity, Dissolved Oxygen (DO), Calcium (Ca), Chloride (Cl), Sodium (Na), and Potassium (K) have been determined. The variation in significant physicochemical parameters along these transects is shown in Tables 4.2, 4.3, 4.4, and Fig. 4.2 has been explained as below;

Table 4.2 Physicochemical parameters during premonsoon-2016

Sample	pH	TDS (ppm)	Salinity (ppt)	DO (ppm)	Ca (ppm)	Cl (ppm)	Na (ppm)	K (ppm)
1	7.5	37,774	31.95	4.04	1222.5	17,687	3745	216
2	7.6	36,657	30.32	19.79	1382.5	16,787	3389	199
3	7.5	36,245	29.10	40.04	1247.5	16,112	3620	204
4	7.3	36,678	28.92	26.66	1257.5	16,012	3518	194
5	7.5	36,766	29.19	31.91	1387.5	16,162	3225	209
6	7.5	46,878	32.94	16.56	1237.5	18,237	3479	218
7	7.6	40,619	34.79	19.79	1112.5	19,262	3769	216
8	7.8	37,422	33.39	35.13	1152.5	18,487	3926	215
9	7.8	49,234	33.67	10.50	1192.5	18,637	3627	215
10	7.8	54,471	34.03	20.20	1097.5	18,837	3218	209
11	7.9	64,972	34.79	17.7	1107.5	19,262	3060	199
12	7.9	37,575	32.94	4.04	1107.5	18,237	3806	216
13	7.9	38,008	32.49	14.94	1122.5	17,987	3538	212
14	7.7	45,826	31.95	14.94	1137.5	17,687	3578	225
15	7.7	47,236	31.99	18.98	997.5	17,712	3676	217
16	7.7	35,669	33.39	22.22	987.5	18,487	3758	210
17	7.6	39,556	33.71	12.92	1182.5	18,662	3493	215
18	7.8	33,835	34.61	13.33	1172.5	19,162	3664	206
19	7.2	34,956	33.85	20.60	1052.5	18,737	3023	218
20	7.0	37,526	33.48	17.37	1112.5	18,537	5533	195
21	7.2	34,863	33.44	21.01	1187.5	18,512	5338	150
22	7.4	37,769	32.18	21.01	1232.5	17,812	5262	156
23	7.5	39,116	32.49	18.58	1167.5	17,987	4906	157
24	7.7	37,217	33.35	8.08	1227.5	18,462	4468	153

4.4.1 Potential Hydrogen (pH)

The pH represents the balance among various types of carbonic acid. These changes are accompanied by changes in other physicochemical parameters, which influence water quality. In the present study, pH ranged 7.0–7.9 during premonsoon-2016, whereas 8.6–8.9 and 7.9–8.4 during monsoon-2016 and premonsoon-2017, respectively. The average pH of sea water is 8.2; however, it can range from 7.5 to 8.5 based on the geographical conditions. Human activities, such as sewage overflows or runoff, can create severe short-term pH variations, with long-term consequences that can be extremely detrimental to plants and animals. Extreme pH shifts can stress local creatures, causing many species to abandon the area or die. Here, the pH values were increased during monsoon between 8.6 and 8.9. This may be due to rainwater runoff during the monsoon.

Table 4.3 Physicochemical parameters during monsoon-2016

Sample	pH	TDS (ppm)	Salinity (ppt)	DO (ppm)	Ca (ppm)	Cl (ppm)	Na (ppm)	K (ppm)
1	8.6	37.22	42.56	5.64	1557.5	23562.5	5910	175.5
2	8.7	37.56	47.85	14.94	1497.5	26487.5	5765	177.0
3	8.8	43.99	42.65	14.94	1327.5	23612.5	5873	177.0
4	8.9	44.29	41.70	39.60	1287.5	23087.5	5856	178.5
5	8.8	47.62	41.52	26.66	1392.5	22987.5	5834	180.0
6	8.6	45.94	41.80	21.01	1272.5	23137.5	6186	175.0
7	8.8	45.69	40.60	18.58	1222.5	22487.5	6265	179.5
8	8.8	41.13	43.10	8.08	1192.5	23862.5	5801	182.0
9	8.8	42.82	44.01	10.50	1362.5	24362.5	5831	179.5
10	8.7	34.65	42.65	20.20	1352.5	23612.5	5841	180.4
11	8.8	36.90	41.48	17.7	1307.5	22262.5	6063	182.3
12	8.8	48.30	39.31	4.45	1047.5	21762.5	5768	182.4
13	8.8	36.17	42.56	19.79	1052.5	23562.5	6070	181.2
14	8.6	37.33	47.17	18.98	1577.5	26112.5	6196	180.3
15	8.6	40.95	36.10	22.22	1167.5	19987.5	5655	182.5
16	8.7	40.97	35.79	21.01	1162.5	19812.5	6239	181.0
17	8.8	33.02	34.30	31.91	1227.5	18987.5	6075	180.8
18	8.7	44.35	31.68	16.56	1097.5	17537.5	6281	179.5
19	8.8	25.16	35.16	19.79	1282.5	19462.5	6471	183.6
20	8.8	55.20	32.72	35.13	1212.5	18112.5	6186	186.5
21	8.7	34.51	33.71	12.92	1177.5	18662.5	6046	183.6
22	8.7	33.77	34.16	13.33	1107.5	18912.5	6214	182.6
23	8.7	39.00	35.16	20.60	1327.5	19462.5	6217	183.0
24	8.8	46.07	35.88	17.37	1227.5	19862.5	6273	184.2

4.4.2 *Total Dissolved Solids (TDS)*

TDS is typically low for freshwater sources, at less than 500 ppm. Brackish water and seawater contain 30–40,000 and 500–30,000 ppm, respectively. TDS in water refers to dissolved organic matters and inorganic salts, including sodium, potassium, calcium, magnesium, chloride, bicarbonates, and sulfates (Zhang et al. 2010). Human activities such as agriculture, water use, industry processes, and mining can increase the TDS level in waterbody. Excess amount of TDS is hazardous to aquatic biosphere (Peng et al. 2020). In this study, TDS values ranged between 33,835 and 64,972 ppm during premonsoon-2016. 33.02–47.62 and 32.31–48.19 ppm was reported during monsoon-2016 and premonsoon-2017, respectively. The values are decreased drastically during monsoon and continued due to fresh water source from the seasonal rains.

Table 4.4 Physicochemical parameters during premonsoon-2017

Sample	pH	TDS (ppm)	Salinity (ppt)	DO (ppm)	Ca (ppm)	Cl (ppm)	Na (ppm)	K (ppm)
1	7.9	36.92	32.58	5.96	527.5	18037.5	13,535	103.9
2	8.1	45.11	33.21	7.04	447.5	18387.5	13,947	98.8
3	8.1	41.70	33.30	19.79	467.5	18437.5	5198	104.8
4	8.1	39.49	33.35	18.98	492.5	18462.5	16,352	100.7
5	8.2	36.38	33.48	19.79	502.5	18537.5	17,355	85.2
6	8.2	41.58	33.30	35.13	452.5	18437.5	13,865	106.7
7	8.3	33.19	33.35	12.92	442.5	18462.5	17,454	102.7
8	8.3	34.03	31.23	8.08	472.5	18287.5	16,303	96.7
9	8.2	34.46	32.94	10.50	552.5	17237.5	14,183	106.8
10	8.3	32.62	33.03	20.20	467.5	18287.5	17,231	95.9
11	8.3	32.31	33.58	17.7	532.5	18587.5	15,861	107.7
12	8.3	32.52	34.57	14.94	427.5	19137.5	13,801	99.6
13	8.4	43.13	32.72	14.94	482.5	18112.5	16,361	101.6
14	8.4	32.48	33.30	36.27	512.5	18437.5	12,096	98.1
15	8.3	33.78	32.99	26.66	482.5	18262.5	15,616	100.0
16	8.2	35.68	33.30	21.01	457.5	18437.5	13,600	95.0
17	8.3	34.42	33.12	18.58	4.7.5	18337.5	13,940	102.1
18	8.2	48.19	33.53	13.33	427.5	18562.5	14,433	100.4
19	8.1	46.77	33.03	20.60	422.5	18287.5	14,928	106.1
20	8.0	44.26	34.07	17.37	442.5	18862.5	13,426	104.1
21	8.2	34.52	33.48	22.22	427.5	18537.5	14,807	104.3
22	8.1	33.07	33.44	21.01	412.5	18512.5	13,218	108.0
23	8.0	36.85	33.58	31.91	427.5	18587.5	13,032	106.0
24	8.1	46.28	34.30	16.56	427.5	18987.5	15,992	107.3

4.4.3 Salinity

Salinity in surface waters is reasonably steady around 3.6 percent, regulated by precipitation and evaporation. It averages between 31.5 and 34.5 ppt (Tomascik et al. 1997). Due to heavy rains, the freshwater is added at the surface and dilute the denser seawater, thereby reducing the salinity obviously and the seawater becomes less denser. Seawater can also be less saline near shore coastal zones and estuaries. In the study area, the salinity ranged 28.92–34.79, 31.68–47.85, and 31.23–34.57 ppt during premonsoon-2-16, monsoon-2016, and premonsoon-2017, respectively. A variation in the ranges of salinity was observed from upstream to the estuary and premonsoon to monsoon.

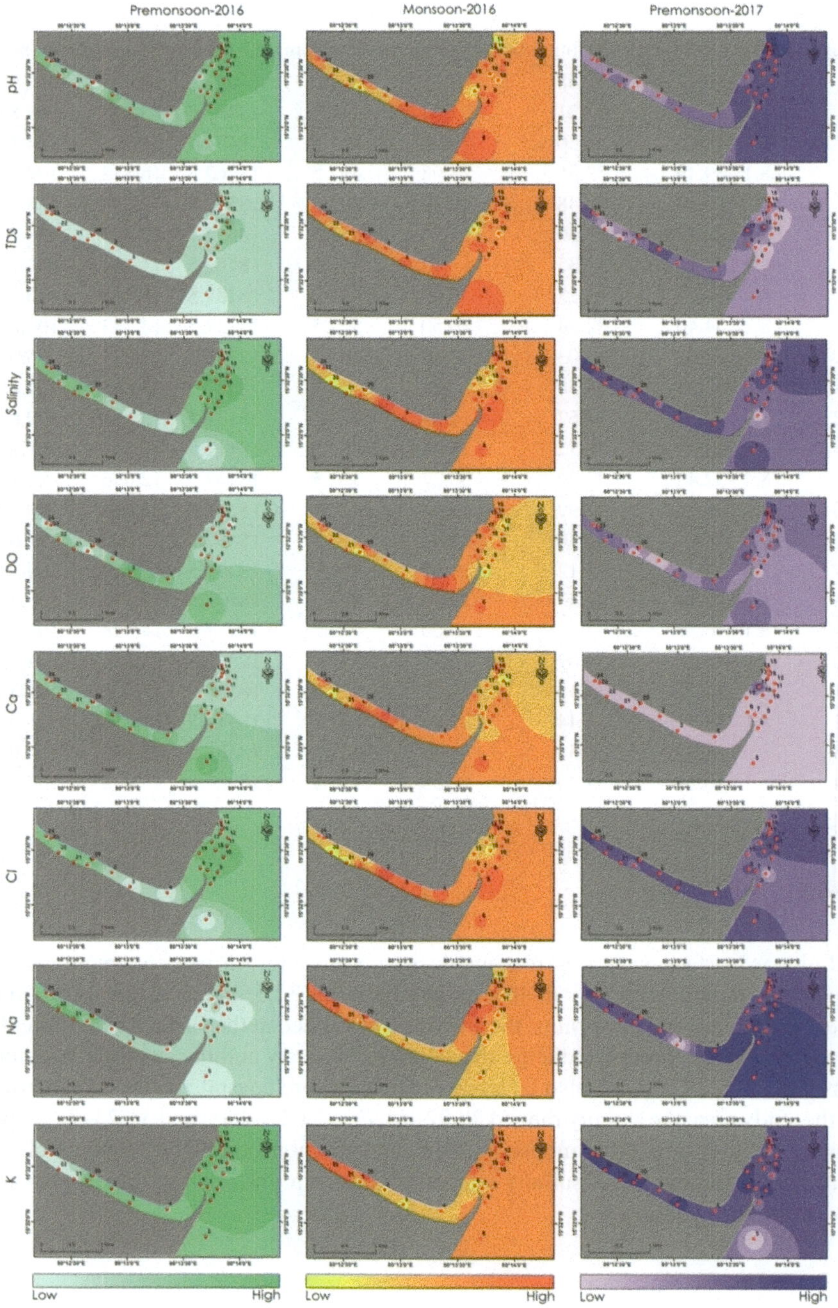

Fig. 4.2 Contour maps showing the vitiations in the distribution of physicochemical parameters in different seasons

4.4.4 Dissolved Oxygen (DO)

The seas contain several dissolved gases that are key to biosphere, particularly oxygen (O_2). Oxygen is required for respiration of biosphere including phytoplankton. They cannot survive for long in water with Dissolved Oxygen less than 5 ppm (Sreenivasulu et al. 2015). The low level of DO in water indicates contamination and is an important factor in determining water quality. Several investigations observed that 4–9 ppm of DO is the value. This will support a large extent of biosphere population (Abdus-Salam et al. 2010). However, at 2.5 ppm, the larvae of less sensitive species of crustaceans may start to die, and the growth of crab is hampered. But <2 ppm, some juvenile fish and crustaceans that cannot survive outside may die. Further, below 1 ppm, fish totally avoid the area or begin to die in huge stocks (USEPA 2000). The highest value of Dissolved Oxygen 40.04 ppm was reported from the study area during the premonsoon-2016. During premonsoon-2016, DO ranged between 4.04 and 40.04 ppm; 4.45–39.60 and 5.96–36.27 ppm were reported during monsoon-2016 and premonsoon-2017, respectively.

4.4.5 Calcium (Ca)

Seawater is reported to be almost saturated in calcium carbonate in the form of calcite or aragonite. Calcite is found in cooler seas, whereas aragonite is found in tropical conditions (Andersson et al. 2008). Ca is a conservative element in saltwater, which implies that its concentration in current seawater is rather stable. The calcium content of seawater is roughly 400 ppm. The natural occurrence of calcium in the earth's crust is one of the primary causes for its abundance in water. Calcium is another component of coral. Rivers typically contain 1–2 ppm Ca, but in lime areas, calcium values may reach to 100 ppm. In the present study, the calcium concentration was reported 987.50–1387.50, 1047.50–1577.50 and 412.50–552.50 ppm during premonsoon-2016, monsoon-2016, and premonsoon-2017 respectively. The concentration of Ca was increased from premonsoon-2016 to monsoon-2016 in 67% of the sampling stations. Decreasing trend with lowest values among the three seasons was observed during premonsoon-2017.

4.4.6 Chloride (Cl)

Chloride is primary inorganic anions (negative ions) in both salt and freshwater. It is formed by the dissolution of salts in water, such as NaCl or $CaCl_2$. A rise in chloride content can have a number of ecological consequences in both aquatic and terrestrial ecosystems. Salt has also been demonstrated to change the structure of microbial communities at low concentrations. However, the evolution of improved water desalination processes over the previous few decades has greatly lowered this barrier. Contamination of chlorides in bottom water may occur as a result of adjacent salt storage or salty rocks, mixing of freshwater with sea water, dissolving of salty industrial wastes, and other factors. The chloride ion concentration in seawater is around 19,400 ppm (a salinity of 35 ppt). Chloride levels in brackish water in tidal estuaries can range between 500 and 5000 ppm. The optimum limit of chloride in water is <250 ppm. During premonsoon-2016, chloride concentration was reported between 16,012 and 19,262 ppm. The highest concentration was reported from the estuary (stations 7 and 11), whereas the lowest values of 16,012 ppm from upstream. 17537.50–26487.50 and 17237.5–19137.50 ppm was reported during monsoon-2016 and premonsoon-2017, respectively. During monsoon-2016, the highest value of 26487.50 ppm was reported from the upstream (station 2), whereas lowest from the estuary (station 18).

4.4.7 Sodium (Na)

Sodium (Na) is a widely distributed cation in sea water. Usually, the salts of the seawater are greatly of sodium (Na^+) and chloride (CL^-) called NaCl. It constitutes about 85% of the salts in seawater. Sea water contains an average of 30.59% sodium as a percentage of its total salinity. The Na:Cl ratio is usually altered near estuaries. In this study, Na ranged between 3023 and 5332 ppm during premonsoon-2016. 5765–6471 ppm was reported during monsoon-2016. During premonsoon-2017, 5198–17,454 ppm of Na was reported during premonsoon-2017. A highest value of 17,454 ppm was reported during premonsoon-2017 from the estuary (station 7), whereas the lowest value of 3023 ppm was reported during premonsoon-2016.

4.4.8 Potassium (K)

Potassium is the fourth most abundant cation, but its concentration is only a few percent that of sodium. Although K is rarely measured directly, it appears to have a very consistent connection with chlorinity (Thompson and Robinson 1932). However, biological agents can influence potassium content because some organisms, notably large algae, concentrate potassium to a significant extent. Dilution with river water can also change the potassium to chlorinity ratio. The K reacts with clay and colloidal particles carried to the sea by freshwater outlets, influencing the ratio. From the study area, K ranged between 153 and 218 ppm during premonsoon-2016. Lowest value of 153 ppm was reported from upstream station 24. The highest value of 218 ppm was reported from two locations of the estuary (stations 6 & 19). During monsoon-2016, it showed the ranges between 175 and 186.5 ppm whereas 85–108 ppm during premonsoon-2017. During monsoon, the lowest and highest values of 175.5 and 186.5 ppm, respectively, were reported from upstream stations

(stations 1 and 20, respectively). The lowest value of 85.2 ppm was reported during premonsoon-2017 from the station 5.

4.5 Factor Analysis (FA)

The studied variables during three seasons, i.e., premonsoon-2016, monsoon-2016, and premonsoon-2017, were subjected to a statistical treatment such as Factor Analysis (FA). The analysis generated three factors for each season from the data of the study area.

During premonsoon-2016, the analysis generated three factors from the data of study area, namely, Factor 1 (39.23%), Factor 2 (26.65%), and Factor 3 (11.15%) (Fig.4.3). Factor 1 was represented by Salinity (0.769) and Cl (0.770) and was assigned as Salinity-Cl assemblage. Salinity and Cl showed lowest values of 28.92 and 16,012 ppm, respectively, during premonsoon-2016. Factor 2 was enveloped with Na (0.820) and K (0.566) and termed as Na- K assemblage. Na showed higher value of 5533 ppm at station 20 and the lower value of 4906 ppm at station 23. K showed higher value of 195 ppm at station 20, whereas lower (157 ppm) at station 23. In this factor, Na and K showed both lower and higher values at estuarine environment of the study area. Factor 3 was composed of DO (0.287) and Ca (0.287) and is called DO-Ca assemblage. DO showed highest value of 22.22 ppm, whereas the Ca showed lowest (987.5 ppm) at a station 16.

In monsoon-2016, factor loadings indicate three dominant assemblages: Factor 1 (41.96%), Factor 2 (18.38%), and Factor 3 (12.74%) (Fig. 4.4). Factor 1 envelops two distinct, Salinity (0.881) and Cl (0.886) named as Salinity-Cl assemblage. Salinity and Cl showed highest values of 47.85 and 26487.50 ppm, respectively, at station 2, whereas moderate at station 9. Factor 2 represented two districts, pH

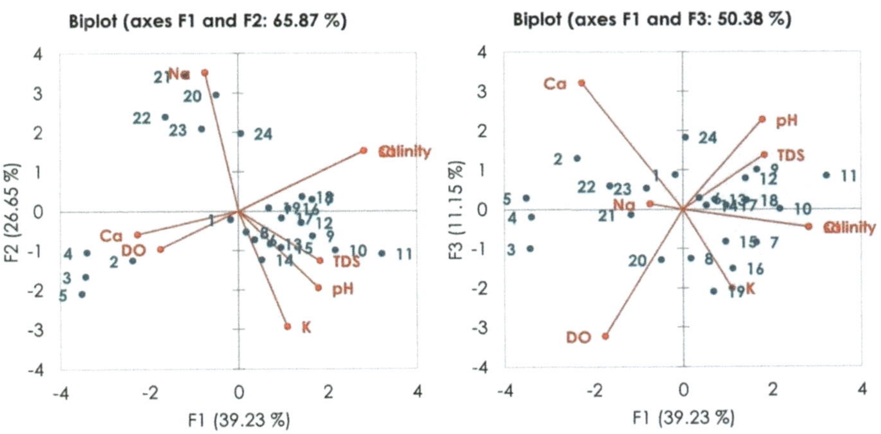

Fig. 4.3 Factor loading during premonsoon-2016 (F1, F2, and F3)

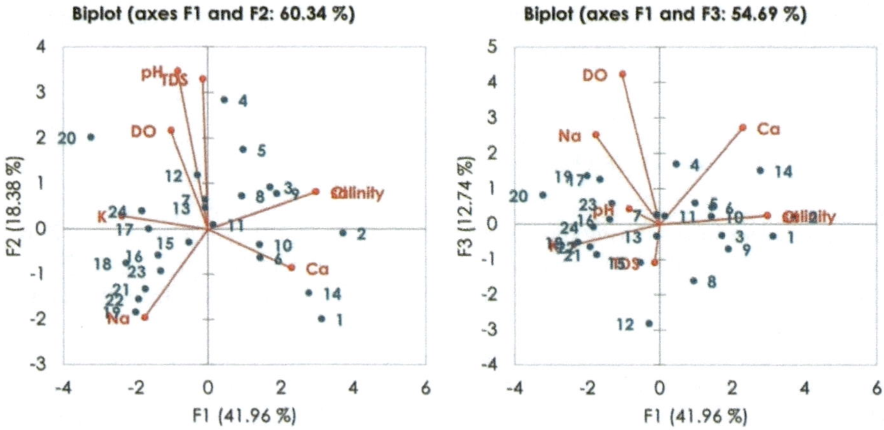

Fig. 4.4 Factor loading during monsoon-2016 (F1, F2, and F3)

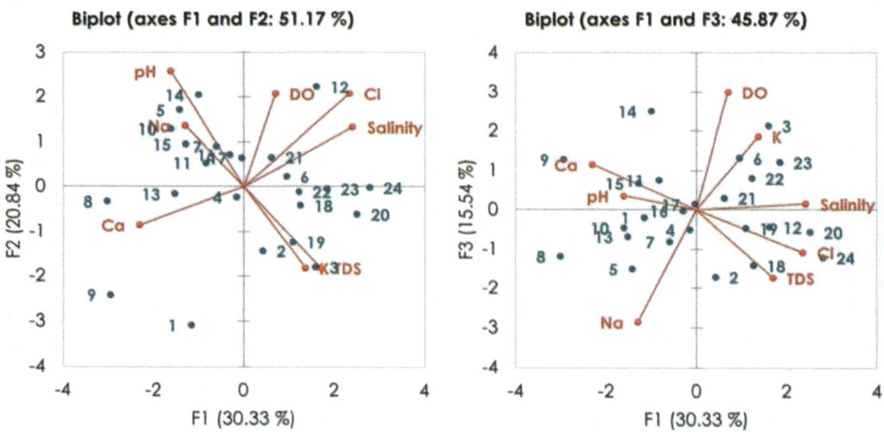

Fig. 4.5 Factor loading during premonsoon-2017 (F1, F2, and F3)

(0.527) and TDS (0.476). Therefore, it is termed as pH- TDS assemblage. pH showed highest value of 8.9 at station 4, whereas TDS showed highest value at station 5. Factor 3 includes DO (0.545) and Na (0.194) and it is called DO- Na assemblage. DO and Na showed lowest values of 4.04 and 5768 ppm at station 12 during monsoon-2016.

In premonsoon-2017, factor loadings indicate three dominant assemblages: Factor 1 (30.33%), Factor 2 (20.84%), and Factor 3 (15.54%) (Fig.4.5). Factor 1 is represented by two variables, Salinity (0.540) and Cl (0.513). Hence, it is named as Salinity-Cl assemblage. The assemblages were termed according to their most important physicochemical parameters. Salinity showed higher values (34.07 & 34.30) from the estuarine environment (stations 20 & 24). Factor 2 is represented by two variables, pH (0.426) and TDS (0.208). Hence, it is named as pH-TDS assemblage. pH showed both lower (7.9) and higher (8.3) values from the stations 1 and 12, respectively. TDS showed the lower values (32.52 ppm) at station 12. Factor 3 envelops two variables, DO (0.427) and Na (0.385). It is called DO-Na assemblage. DO show both lower and higher values from the stations 2 and 14, respectively. The distribution shows that physicochemical parameters have a strong affinity with marine water samples of the study area.

4.6 Conclusions

The health of the water in terms of its physicochemical parameters of Gundlakamma estuary, Andhra coast, Southeast coast of India, was assessed. The results for the studied parameters revealed that the water composition of the study area is pH 7.0–8.9; TDS 25.16–64,972 ppm; Salinity 28.92–47.85 ppt; DO 4.04–40.04 ppm; Ca 412.50–1577.50 ppm; Cl 16,012–26487.50 ppm; Na 3023–17454.50 ppm; and K 85–186.5 ppm. Statistical treatment of Factor Analysis (FA) for the distribution of physicochemical parameters shows that they exhibit a strong affinity with waters of the study area. In the typical context, water composition may be fresher in some part of upstream and more concentrated near estuary. The impact of anthropogenic as well as natural input, as a source of physicochemical parameters in the study area, is alarming. However, a detailed study is warranted to evaluate the status of health of the estuarine environment which will form a baseline data for future investigations.

References

Abdus-Salam N, Adekola FA, Apata AO (2010) A physico-chemical assessment of water quality of oil producing areas of Ilaje, Nigeria. Adv Nat Appl Sci 4(3):333–344

Andersson A, Mackenzie F, Bates N (2008) Life on the margin: implications of ocean acidification on Mg-calcite, high latitude and cold-water marine calcifiers. Mar Ecol Prog Ser 373:265–273

APHA/AWWA/WEF (1995) Standard methods for the examination of water and wastewater, 19th edn. American Public Health Association, Washington, DC

Arumugam A, Sugirtha PK (2014) Evaluation of physico-chemical parameters and nutrients in the Mangrove ecosystem of Manakudy Estuary, Southwest coast of India. Int J Latest Res Sci Technol 3(6):205–209

Dhirendra MJ, Kumar A, Agrawal N (2009) Studies on physico-chemical parameters to assess the water quality of River Ganga for drinking purpose in Haridwar District. Rasayan J Chem 2(2):195–203

Ho HH, Swennen R, Damme AV (2010) Distribution and contamination status of heavy metals in estuarine sediments near CuaOng Harbor, Ha Long Bay, Vietnam. Geol Belg 13(1–2):37–47

Jayaraju N, Sundara Raja Reddy BC, Reddy KR (2011) Heavy metal pollution of Tambaraparni River estuary, East Coast of India. J Chem Ecol 27(4):337–350

Kruopiene J (2007) Distribution of heavy metals in sediments of the Nemunas River (Lithuania). Pol J Environ Stud 16(5):715–722

Manjare SA, Vhanalakar SA, Muley DV (2010) Analysis of water quality using physico-chemical parameters of Tamdalge tank in Kolhapur district, Maharashtra. Int J Adv Biotechnol Res 1(2):115–119

Peng J, Kumar K, Gross M, Kunetz T, Wen Z (2020) Removal of total dissolved solids from wastewater using a revolving algal biofilm reactor. Water Environ Res 92:766–778. https://doi.org/10.1002/wer.1273

Poonam B, Rahul K (2012) Status of seawater quality at few industrially important coasts of Gujarat (India) off Arabian Sea. Indian J Geo-Mar Sci 41(1):90–97

Priyanka T, Amita B, Sukarma T (2009) Evaluation of water quality: physico-chemical characteristics of Ganga River at Kanpur by using correlation study. Nat Sci 1(6):91–94

Ravichandran R, Manickam S (2012) Heavy metal disrtibution in the coastal sediment of Chennai coast. IIOAB J 3(2):12–18

Riba I, DelValls TA, Forja JM, Go'mez-Parra A (2002) Influence of the Aznalco'llar mining spill on the vertical distribution of heavy metals in sediments from the Guadalquivir estuary (SW Spain). Mar Pollut Bull 44:39–47

Sreenivasulu G, Jayaraju N, Sundara Raja Reddy BC, Lakshmi Prasad T (2015) Physico-chemical parameters of coastal water from Tupilipalem coast, Southeast coast of India. J Coastal Sci 2(2):34–39

Thompson TG, Robinson RJ (1932) Chemistry of the sea. Physics of the earth. In: Oceanography, vol 5. National Research Council, Bull., no. 85, 1932, Washington, D.C., pp 95–203

Tomascik T, Mah AJ, Nontji A, Moosa MK (1997) The ecology of the Indonesian seas, the ecology of Indonesia series, vol 7. Periplus Editions (HK) Ltd., Singapore, 1388 p

USEPA (2000) Risk-based concentration table. United States Environmental Protection Agency, Philadelphia

Zhang S, Zhou Q, Xu D, Lin J, Cheng S, Wu Z (2010) Effects of sediment dredging on water quality and zooplankton community structure in a shallow of eutrophic lake. J Environ Sci 22:218–224

Chapter 5
Evaluation of Physicochemical Parameters of Coastal Water from Pennar River Estuary, East Coast of India: An Integrated Approach

B. Praveena, M. Pramod Kumar, T. Lakshmi Prasad, and N. Jayaraju

Abstract Pennar is one of the major rivers in south India and it has significant impact on water chemistry. This investigation intended to assess the water quality parameters of Pennar river estuary, Nellore, South East coast of India. A total of 20 bottom water samples were collected from chosen sample points during February 2020 and evaluated at fortnightly intervals to acquire baseline information on the estuarine water quality. Significant variations were observed in physicochemical parameters of water at different stations and these are as follows: pH (7.29–8.47), bottom water temperature (28.8–31.8 °C), air temperature (32.9–30.1 °C), EC (44.24–53.54 μmho/cm), Dissolved Oxygen (3.7–4.6 mg /l), Organic Matter (0.61–1.7 mg/l), Salinity (29.24–35.93 ppt), Nitrates (1.4–2.2 mg/l), and Sulphates (120–192 mg/l), respectively. Multiple statistical techniques such as Pearson Correlation Analysis, Factor Analysis (FA), and Hierarchical Cluster Analysis (HCA)were effectuated to the obtained datasets to determine the relationship among the selected parameters.

Keywords Pennar river estuary · Physicochemical parameters · Statistical techniques · East coast of India

B. Praveena · M. Pramod Kumar · T. Lakshmi Prasad (✉)
Department of Earth Sciences, Yogi Vemana University, Kadapa, Andhra Pradesh, India

N. Jayaraju
Department of Geology, Yogi Vemana University, Kadapa, Andhra Pradesh, India

N. Jayaraju et al. (eds.), *Coasts, Estuaries and Lakes*,
https://doi.org/10.1007/978-3-031-21644-2_5

5.1 Introduction

Water is a very important reliable natural resource and plays a significant role in the economy. It is primarily required for the Biosphere. The quality of water may depend upon the nature of the self-purification of water and the types of pollutants. Water has its own physical, chemical, and biological parameters that affect its quality. Water quality refers to show how well a body of water meets the demands of one or more biotic species, as well as any human requirement or purpose. The demand for water was continuously increasing in recent eras. Quality of water in rivers, lakes, and ponds may vary with geographic areas and seasons even if there is no pollution too. Due to the escalation of population, urbanization, and industrialization, large quantities of domestic and industrial effluents are released into water bodies. This has notably contributed to the contamination of the water bodies like lakes, rivers, estuaries, and ponds and degradation of the estuarine water quality (Wang et al. 2014; Jayaraju et al. 2021). It is estimated that around 90% of particulate matter transported by rivers will accumulate in coastal and estuary zones(Karen and Barid 2001; Abdus-Salam et al. 2010).

The Pennar river estuary is one of the most typical estuaries, with tidal oscillations and freshwater flows playing the most important roles and combined to have a high mixing mechanism with rapid urbanization. The catchment area of Pennar has aquacultural ponds where the effluents are discharged into the estuary which may lead to a constant increase in the quality of stormwater. The marine environment is a dynamic and complex aquatic system and is mainly influenced by physical, chemical, and biological processes when the river merges into the sea. It is a very difficult task to properly comprehend the biological phenomena of the water since the chemistry of water reveals a great deal about the metabolism of the ecosystem and explains the overall hydro-bio interaction. Pollution in a river primly affects its chemical purity, then gradually degrades the community, disrupting the delicate food web. Regular monitoring procedures are necessary to determine the geographical and temporal fluctuations in water quality.

Water features on the coast (marine environment) are rapidly changing with time and space, so there is a need of regular monitoring to retain their quality and quantity (Satheeshkumar and Khan 2012). Furthermore, the biological diversity and health status of Indian estuarine ecosystems are deteriorating gradually as a result of anthropogenic activities such as the disposal of domestic sewage and industrial effluents into estuaries. This has resulted in a drastic reduction of shallow water aquatic populations and the extinction of numerous flora and fauna. This heavy metal contamination in the marine environment penetrates the food chain via bioaccumulation processes, posing major environmental and health risks (Sreenivasulu et al. 2015). Many researchers not only in India but also in the world have studied the bottom and surface water chemistry and its seasonal fluctuations (Karen and Barid 2001; Damotharan et al. 2010; Sreenivasulu et al. 2015; Ajayan and AJIT 2016; Swain et al. 2021).

The main objective of the current investigation is to study the water quality of the Pennar river estuary with respect to physicochemical parameters. The data, thus generated, are subjected to multiple statistical analyses. The present study may serve as baseline information on water quality parameters and would form a useful tool for further ecological assessment and monitoring the quality of water.

5.2 Study Area

The present study area Pennar river (estuary) is one of the key rivers in south India, also known as "Uttara Pinakini" which originates in the Nandi hill ranges, Mysore, Karnataka, and empties into the Bay of Bengal at Utukur (V), Nellore district, Andhra Pradesh, India. It is situated in the survey of India topo sheets no: 66B/2 in the scale of 1: 50000 geographically located in between latitudes 14°40′0″ – 14°52′0″ N and longitudes 80°8′ 0″ – 80°14′ 0″ E. The maximum area of the river basin has six principal tributaries (Kunderu, Jayamangali and Sagileru from North and Chitravathi, Cheyyeruand Papagni from South) that are spread in the state of Karnataka and Andhra Pradesh and lie in the shadow region of the Eastern Ghats. The Pennar river's estuary stretches 7 km upstream from the Bay of Bengal. During the dry season, from November to June, due to tidal effects, saltwater reaches further upstream. Around the river mouth, sand dunes are grown as high as 7 mt. The 15-km-long Upputeru tidal creek and the 180-meter-long and 3-meter-high Isakapalli barrier island divide the lagoon from the Bay of Bengal. The Pennar river basin is formed by Archean rocks consisting of dolerites, granites, felsites, pegmatites, quartzite, peninsular gneisses, amphibolites, phyllites, and mica schist. Additionally, it consists of numerous intrusive metamorphic schists. The basin is ornamented with several soil types such as red, sandy, black, and mixed soils. Annual rainfall (average) is 1041 mm and the mean minimal and maximal temperatures are 20 and 39.6 °C, respectively. The location map of the study area with sampling stations was shown (Fig. 5.1).

5.3 Methodology

A total of 20 bottom water samples from different sites were collected at the depth of 1 meter by using a shallow water sampler from the current study area with reference to longitude and latitude (by using GPS) in 500 ml pre-cleaned and labeled plastic bottles during February 2020 (pre-monsoon season). A few of common physicochemical parameters of water, namely, pH, Temperature, Electrical Conductivity (EC), Salinity, Dissolved Oxygen (DO), Organic Matter (OM), Nitrates, and Sulphates, had been analyzed according to the standards of the American Public Health Association (APHA 1998). Some of the field parameters like pH, temperature, and Electrical Conductivity (EC) were measured (carried out) itself in the field (spot) by using a portable kit (Velsamy et al. 2013). DO and Salinity were determined by using an Atago refractometer (Sreenivasulu et al. 2015). DO, OM, and nitrates

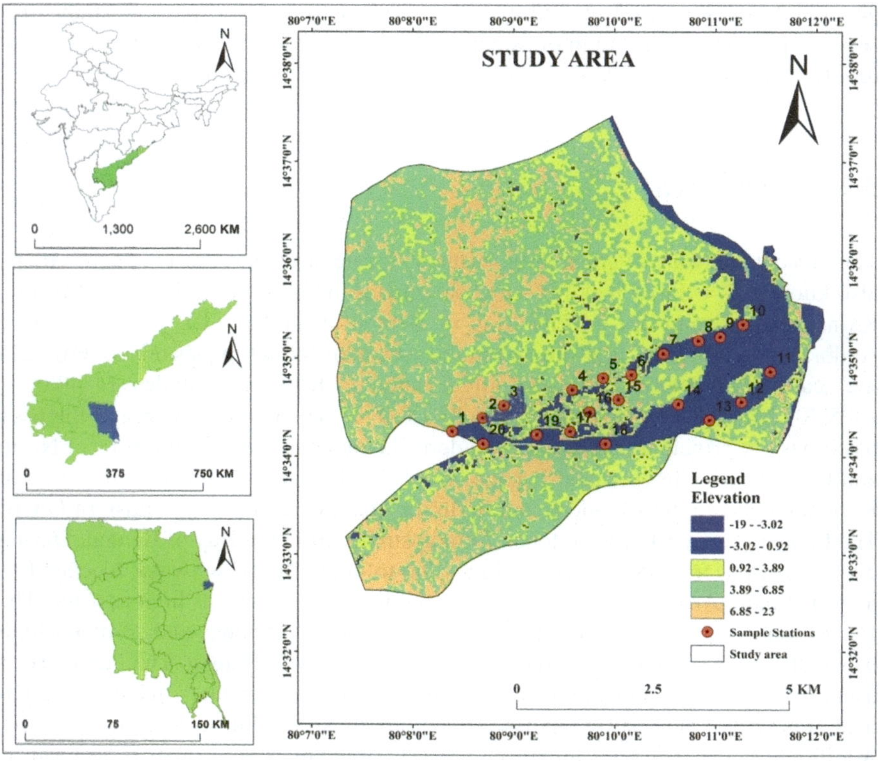

Fig. 5.1 Location map of the study area along with sampling stations

analyses were performed in the laboratory by using standard laboratory methods (APHA 1998). The contents of sulphates (SO_4^{2-}) were determined by UV spectrophotometric method. The obtained datasets were determined statistically employed with different analyses, namely, Pearson Correlation Analysis, Factor analysis (FA), and Hierarchical Cluster Analysis (HCA). Spatial distribution maps of different parameters were created by using ArcGIS 10.3 software with the Inverse Distance Weighted (IDW) interpolation technique (PramodKumar et al. 2020a, b) to point out the most affected stations. IDW interpolation is a technique for predicting a value from observed values to any unmeasured or un-sampled place. Weights are applied to the measured location points in such a way that measured values closer to the prediction location have a greater influence on the predicted value than those further away.

5.4 Results and Discussions

The current study deals with the physiochemical parameters of Pennar river estuary East coast of India. The prescribed parameters were analyzed according to the standards of the American Public Health Association (APHA), 2008. Analytical results

Table 5.1 Physicochemical Parameters of the study area

Sample No	pH	Temperature (°C) Bottom water	Air	EC (μmho/cm)	DO (mg/l)	OM (mg/l)	Salinity (ppt)	Nitrates (mg/l)	Sulphates (mg/l)
1	7.71	29.8	30.1	46.58	4.4	1.32	32.4	1.4	167
2	7.75	30.3	31.7	46.91	4	1.2	31.2	1.6	170
3	7.88	29.9	31.5	46.08	4.1	1.7	30.6	1.7	187
4	7.7	29.5	32.2	47.95	4.2	1.58	31.92	1.9	182
5	8.19	28.8	30.4	47.78	3.9	0.92	31.8	1.9	182
6	8.1	31	31.6	48.61	4	0.98	32.4	2	182
7	8.09	31.1	32.8	49.36	4.1	1.23	32.95	2.1	192
8	8.1	30.4	31.3	51.13	4.5	1.11	34.24	2	185
9	8.14	29.4	32.6	52.36	3.7	0.85	35.14	2.1	187
10	8.47	28.8	32.8	53.44	4.3	1	35.93	2	187
11	8.07	29.6	31.7	51.69	4.4	0.79	34.65	2	187
12	8.29	29.1	32.1	49.79	3.9	0.81	33.26	2	170
13	7.98	30.2	32.2	48.96	4.6	0.76	32.66	1.9	180
14	7.81	29.9	31.6	49.47	4.1	0.99	33.03	1.9	132
15	7.76	31.7	31.7	47.81	3.8	1.01	31.9	1.9	140
16	7.68	31.8	32.9	45.98	4.5	1.24	30.49	1.9	147
17	7.78	30.9	31.2	45.85	4	1.34	30.4	2	120
18	7.3	30.8	31.2	46.18	4.2	0.61	30.64	1.9	132
19	7.83	31.6	31.9	44.24	4.4	1	29.24	1.9	127
20	7.29	30.8	31.2	46.18	4	1	30.64	2.2	145
Max.	8.47	31.8	32.9	53.44	4.6	1.7	35.93	2.2	192
Min.	7.29	28.8	30.1	44.24	3.7	0.61	29.24	1.4	120
Avg.	7.89	30.27	31.71	48.36	4.15	1.079	32.30	1.904	164.22

of variations of different parameters were tabulated (Table 5.1). A graphical representation of the physicochemical parameters was shown (Fig.5.2). Spatial distribution maps of different parameters were shown (Figs. 5.3a–d and 5.4a–d). There is not much difference in spatial distribution maps of air and bottom water temperature, so the air temperature spatial distribution map has not been included.

5.4.1 pH

pH is one of the critical environmental factors for water quality monitoring as it governs most of the chemical processes that occur in soil and water and yields a significant piece of information in many sorts of solubility calculations or geochemical equilibrium (Gupta et al. 2008). Thus, it influences the aquatic production and the production of calcareous microfauna. According to Lloyd (1987), the high pH range is not immediately hazardous to freshwater fish. With a few exceptions, pH

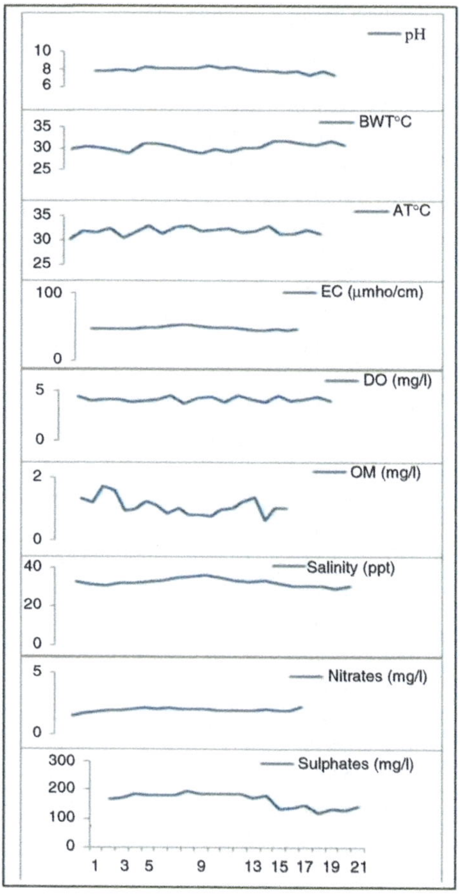

Abbreviations: BWT-Bottom Water Temperature, AT- Air Temperature, EC- Electrical Conductivity, DO – Dissolved Oxygen, OM – Organic Matter.

Fig. 5.2 Graphical representation of physicochemical parameters

values between 6.5 and 9.0 are long-term suitable for fish and other freshwater aquatic species. The constituents like temperature and air cause variations in the pH of water. Almost alkaline pH was observed at all the stations. All of the water samples tested were under the range of 7.29–8.47 and the mean pH was (7.89). Low pH was observed at station 20 (7.29) and high alkaline pH at Station 10 (8.47) might be because of the alkaline-rich environment prevailing in the estuary (Sundara Raja Reddy 2006).

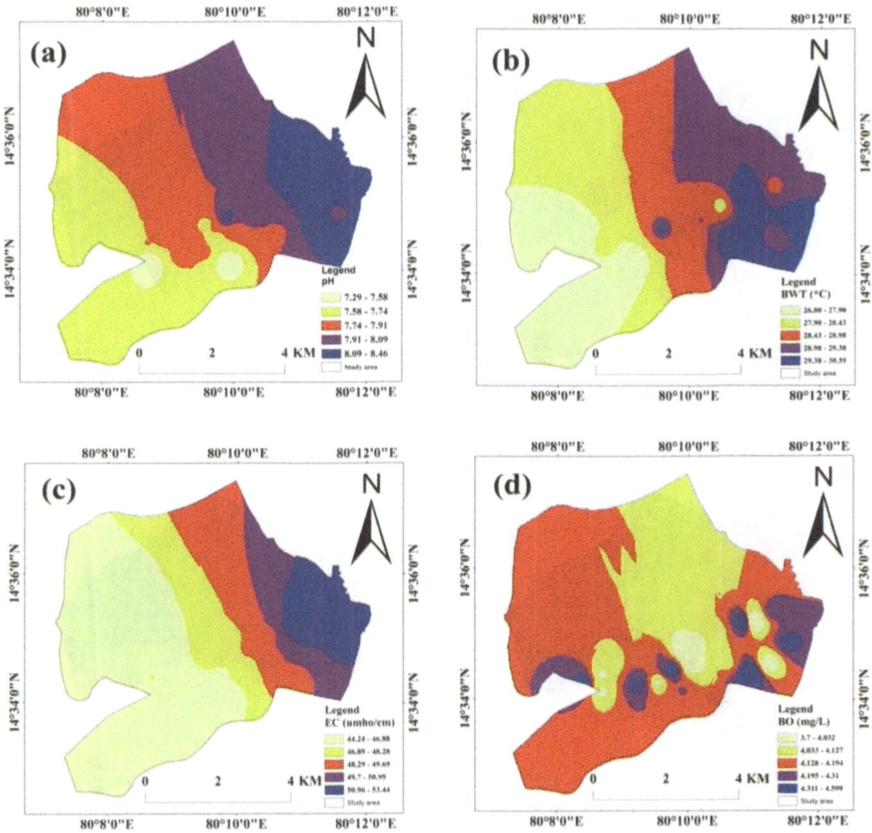

Fig. 5.3 (**a**) Spatial distribution map of pH. (**b**) Spatial distribution map of BWT. (**c**) Spatial distribution map of EC. (**d**) Spatial distribution map of DO

5.4.2 *Temperature (°C)*

Water temperature is a significant component that governs the chemical, biochemical, and biological capabilities of a water body(Manjare et al. 2010). As reported by Alabaster and Lloyd, (1980) in the tropics, fish are normally accustomed to temperatures ranging from 28 to 30. During the study, time mean temperature range in bottom water was observed from 28.8 to 31.8 and the highest temperature was observed from stations 8 and 13 and the lower bottomwater temperature was recorded at stations 5 and 10. The air temperature ranged between 30.1 and 32.9 and was recorded from stations 1 and 16. Change in temperature has an impact on aquatic life. Poikilothermic creatures, such as fish, have body temperatures that closely match the temperature of their surroundings. The highest value (32.9) may be due to the extreme shallowness of the bay and the "White Water" condition(Sundara Raja Reddy 2006). The fluctuations in the temperature may be due to a few factors,

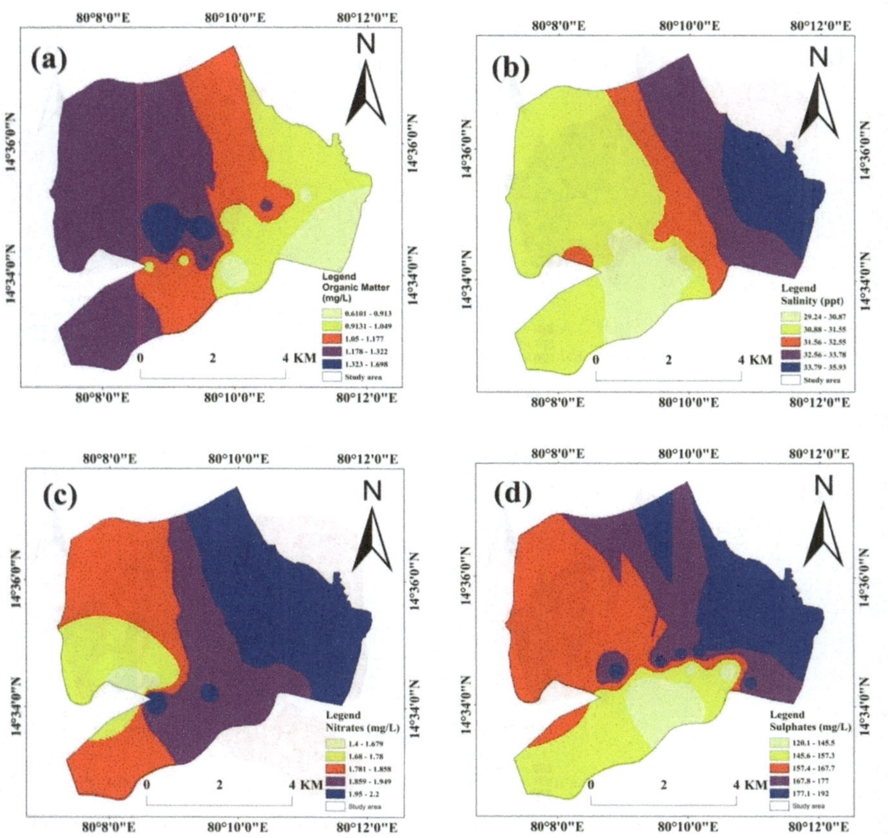

Fig. 5.4 (**a**) Spatial distribution map of OM. (**b**) Spatial distribution map of salinity. (**c**) Spatial distribution map of nitrates (**d**) Spatial distribution map of sulphates

e.g., the estuary/river temperature is heat exchange on the earth's surface under controlled radiation, groundwater movement, and thermonuclear phenomena functioning in the aquifers (Gupta et al. 2017).

5.4.3 *Electrical Conductivity (EC) (µmho/cm)*

EC is a measure of an aqueous solution's efficiency to transport an electron current, which is affected by the presence and total concentration of ions, their mobility, valence, and temperature (Young et al. 2005). In the study area, the range of values of EC in all sampling stations was 44.24–53.44 (mean 48.36). The maximal content of EC was observed at station 10, whereas minimal conductivity was observed at station 1. There were differences in EC across all of the stations, but these differences were not statistically significant. In this study, the maximal content of EC may be due to high ionic concentration, trophic levels, pollution status, some domestic effluents, and other OM in water at the particular station (Yadav et al. 2013).

5.4.4 *Salinity (ppt)*

Salinity is a significant factor and a slight change in the salinity can affect the conductivity-density relationship and influences the flora and fauna. It is a prime abiotic factor that controls the distribution of living organisms (Sobha Rani 2018). The fluctuations in salinity affect the biological characteristics of the environment. In brackish water, the range of salinity is 0.5–30 (Karleskint et al. 2009). This is mainly found in coastal, salt marshes, and estuaries. Some estuaries can have salinities as high as 30 depending on their source of fresh water and location. Salinity content in the collected water samples ranges from 29.24 to 35.93 and its average was 32.3. A high amount of salinity is recorded at station 10. This may be owing to decreased fresh water inflow. The low salinity was observed at station 19 which may be due to the large quantity inflow of fresh water at the particular station (Kaliyaperumal 1992).

5.4.5 *Dissolved Oxygen (DO) (mg/l)*

The quantity of DO in water is an indicator of current water quality because it supports a well-balanced aquatic life. It correlates with photosynthesis, bacterial activity, stratification, nutrient availability, and physical processes like temperature and water movement either directly or indirectly. In the current study, DO was found to be in the range of 3.7–4.6 with an average of 4.15. The obtained study reports revealed that the DO was higher at Station 2 and lower at station 9 as it reveals that the range of DO was suitable for the survival of many aquatic insects and aquatic organisms.

5.4.6 *Organic Matter (OM) (mg/l)*

OM is a soluble organic substance with a wide range of reactivity and ecological significance. It's important for nutrient transport and aquatic food web dynamics and naturally OM is present in all natural water. The composition of OM is determined by its sources like the plant, soil, algae, wastewater, and environmental processing such as biodegradation and photodegradation. In the present study, the level of DO range was noted between 0.61 and 1.7 from stations 3 and 8 with an average of 1.07. The high amount of OM may be due to few cumulative factors such as the supply of considerable volume of OM by water run-off and relatively rapid rate of accumulation of grained inorganic matter, a good amount of organic debris emptying from the overlying column, and low oxygen content of the waters immediately above the bottom sediments which would favor elevated OM in the bottom sediments (Sundara Raja Reddy 2006).

5.4.7 *Nitrates (mg/l)*

Generally, nitrates were considered as a limiting nutrient for primary production in the marine ecosystem. Nitrate in water is a reason for concern, not only because of its hazardous potential, but also because it might suggest that the water is polluted with additional contaminants. The concentration of nitrates in the study area ranged between 0.4 and 2.2 with an overall mean value of 1.9. Maximal values of Nitrates were observed at station 20, whereas minimal value was noticed at station 1. High levels of nitrates in water may be because of variations in the reduction of nitrates, excretion, and oxidation of ammonia.

5.4.8 *Sulphates (mg/l)*

Sulphates are widely distributed in nature and their concentration in bottom water. The sulphates content value was between 120 and 192 (average 164.2). The low value of sulphates(120) was recorded at station 17 and a sudden increase of sulphates(192) was recorded at station 7. The higher value of sulphates may be because of mineral dissolution and atmospheric deposition (Sharma and Kumar 2020).

5.5 Multiple Statistical Analysis

5.5.1 *Pearson's Correlation Analysis*

The relationship among the several parameters of bottom water samples was investigated by using Pearson's Correlation Co-efficient (Table 5.2). The analysis yielded a considerable positive correlation between pH – EC (0.721), pH – Salinity (0.703), pH – Sulphates (0.636), EC – Salinity (0.983), EC – Sulphates (0.619), and Salinity – Sulphates (0.631) (bolded in Table 5.2).

There were at least five affinity assemblages among all the physicochemical parameters from the study area. Affinity groups are EC – Salinity (0.983), pH – EC – Salinity (0.721, 0.703), Salinity – pH (0.636), EC – Sulphates (0.619), and AT-Nitrates (0.447). The majority of parameters were found to correlate with EC

Table 5.2 Pearson Correlation's Analysis for physicochemical parameters of Pennar river estuary

Parameters	pH	BWT	AT	EC	DO	OM	Salinity	Nitrates	Sulphates
pH	1								
BWT	−0.529	1							
AT	0.325	0.117	1						
EC	**0.721**	−0.575	0.389	1					
DO	−0.055	0.14	0.073	−0.045	1				
OM	−0.13	0.075	−0.03	−0.341	0.039	1			
Salinity	**0.703**	−0.603	0.299	**0.983**	−0.006	−0.305	1		
Nitrates	0.187	0.086	0.447	0.381	−0.23	−0.391	0.257	1	
Sulphates	**0.636**	−0.563	0.219	**0.619**	0.047	0.114	**0.631**	0.012	1

Table 5.3 Factor loadings for physicochemical parameters

Parameters	F1	F2	F3
pH	0.777	−0.087	−0.104
BWT	−0.621	0.475	−0.145
AT	0.407	0.641	−0.536
EC	0.985	0.084	0.122
DO	−0.055	−0.049	−0.189
OM	−0.255	−0.327	−0.51
Salinity	0.94	−0.043	0.111
Nitrates	0.323	0.639	0.19
Sulphates	0.708	−0.329	−0.307

Abbreviations: *BWT* bottom water temperature, *AT* air temperature, *EC* electrical conductivity, *DO* dissolved oxygen, *OM* organic matter

(Salinity, Sulphates, and Nitrates). Positive correlation among the parameters revealed that there was a strong linear relationship between the parameters, whereas low and negative correlation showcases that there was a poor relationship among the parameters. It also established that EC is the prime controlling factor of other parameters.

5.5.2 *Factor Analysis (FA)*

Obtained data of physicochemical parameters were subjected to statistical analysis by performing FA using XL STAT, 2021 software (Jayaraju et al. 2010; Sreenivasulu et al. 2015). This analysis exhibited that there are three different factors (Table 5.3, Factor 1, 2 & 3).

5.5.2.1 Factor – 1 (EC – Salinity – pH – Sulphates – BWT Assemblage)

The factor loading yielded 40.92% of the variance among the EC – Salinity – Ph – Sulphates – BWT.

In this factor, maximum number (5) of parameters are covered including EC (0.985), pH (0.940), Salinity (0.708), Sulphates (0.708), and BWT (−0.621). Thus, it can be called EC – Salinity – pH – Sulphates – BWT Assemblage. These assemblages are called by their most prominent characteristics (Reddy et al. 2009).

5.5.2.2 Factor – 2 (AT-Nitrates Assemblage)

This factor accounting 14.20% of the variance with AT (0.641) and nitrates (0.639). Hence, this factor is named as AT-Nitrates Assemblage.

5.5.2.3 Factor – 3 (DO-OM Assemblage)

In this factor, factor loadings accounted for 8.58% of the variance and include DO (−0189) and OM (−0.510). So the assemblage can be termed as DO-OM Assemblage. Therefore, a cross plot of Factors 1 and 2 and Factors 1 and 3 (Fig. 5.5a, b) gave perceptible enlightenment about the parameters. This also showed that Factor-1 is having strong variance.

5.5.3 Hierarchical Cluster Analysis (HCA)

HCA was performed to the selected stations to delineate the dendrogram of parameters with single linkage Euclidean and correlation coefficient distance. Majorly two clusters, i.e., cluster A and cluster B, are formed as shown (Fig. 5.5c). Long distance cluster depicts an insignificant relationship, whereas the low distance cluster indicates an insignificant relationship among the parameters (PramodKumar et al. 2020a, b). According to the dendrogram, Cluster A (blue series) is formed by stations S – 3, 6, 13, 5, 4, 7.8.11, 9, 10, 12, 1, and 2. The Cluster B is formed with stations S – 14, 17, 16, 20, 19, 15, and 18. The report states that cluster A has a more significant relationship when compared with cluster B.

5.6 Conclusion

This study focused on determining the water quality properties of bottom water at Pennar river estuary, South East coast of India. During this experimentation, a few physicochemical characteristics were investigated to assure good quality attributes

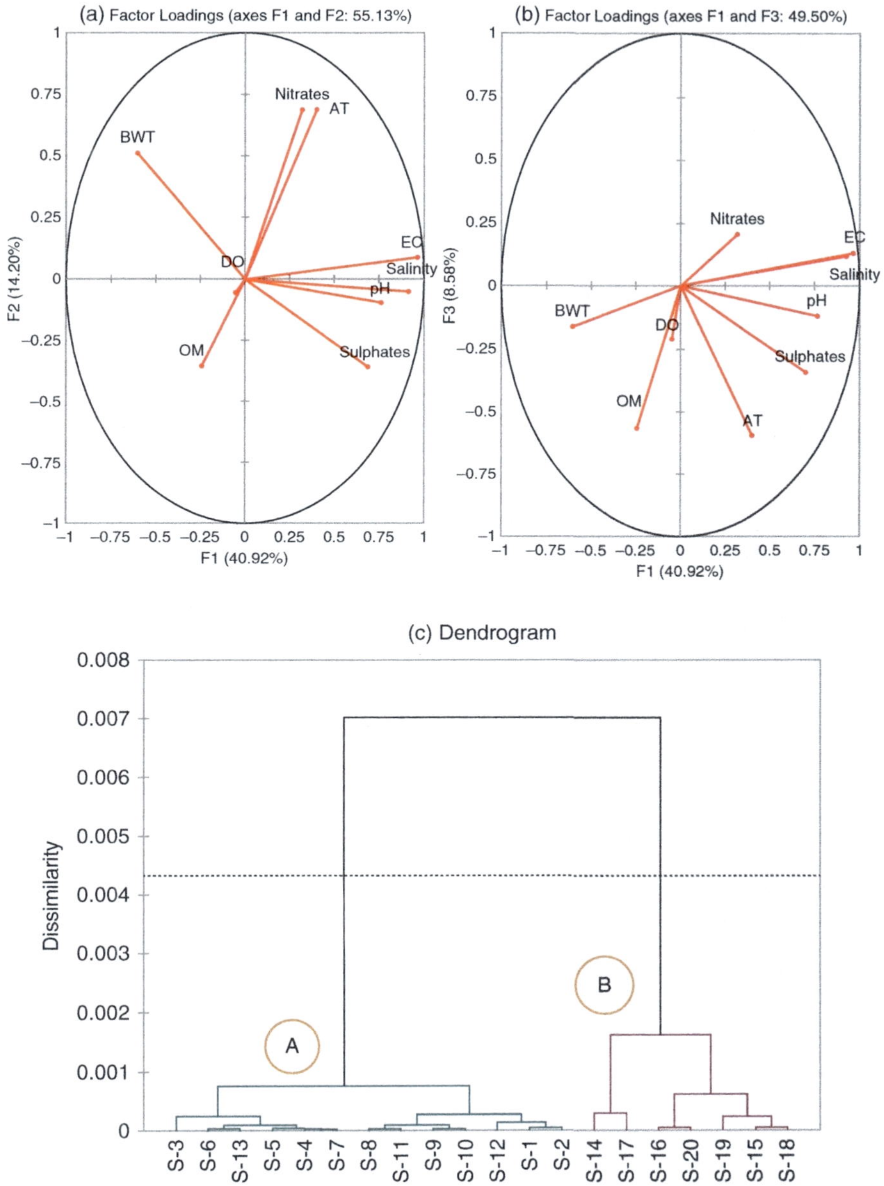

Fig. 5.5 (**a**) Factor loadings of physicochemical parameters (Factors 1 and 2) (**b**) Factor loadings of physicochemical parameters (Factors 1 and 3). (**c**) Dendogram of grouping of sampling stations of the study area

of water. The findings of this study concluded that there were significant variations in the physicochemical parameters in the estuary. The above examination of water also demonstrated that all of the research stations under study were fairly adequate, productive, and healthy for the aquatic ecosystem. The values of parameters are well within the recommended limit of APHA and WHO (World Health Organization), and therefore, can support the survival of aquatic organisms. As a result, assessing the water quality is a vital technique for monitoring the health of a water body. However, combining the evaluation of physicochemical parameters with different statistics provides a true picture of the water's quality. Constant monitoring and mitigating was required in order to safeguard the health of the water which enhances the production and growth of the Biosphere.

References

Abdus-Salam N, Adekola FA, Apata AO (2010) A physicochemical assessment of water quality of oil producing areas of Ilaje, Nigeria. Adv Nat Appl Sci 4(3):333–345

Ajayan A, AJIT KK (2016) On the seasonal changes in the surface water chemistry of Museum Lake, Thiruvananthapuram, Kerala, India. Pollution 2(2):103–114

APHA A (1998) Standard methods for the examination of water and wastewater. American Public Health Association Inc., Washington, DC

Damotharan P, Perumal NV, Arumugam M, Vijayalakshmi S, Balasubramanian T (2010) Seasonal variation of physico-chemical characteristics in Point Calimere coastal waters (south east coast of India). Middle-East J Sci Res 6(4):333–339

Gupta S, Mahato A, Roy P, Datta JK, Saha RN (2008) Geochemistry of groundwater, Burdwan District, West Bengal. India Environ Geol 53(6):1271–1282

Gupta N, Pandey P, Hussain J (2017) Effect of physicochemical and biological parameters on the quality of river water of Narmada, Madhya Pradesh, India. Water Sci 31(1):11–23

Jayaraju N, Sundara Raja Reddy BC, Reddy KR (2010) Impact of dredging on sediment of Krishnapatnam port, east coast of India: implications for marine biodiversity. J Environ Sci Eng 4(8):66–75

Jayaraju N, Sreenivasulu G, Reddy BSR, Lakshmanna B, Upendra B, Reddy AN (2021) Use of benthic foraminifera as a proxy for monitoring heavy metal pollution in the Swarnamukhi estuary, southeast coast of India. Environ Chem Ecotoxicol 3:249–260

Kaliyaperumal C (1992) Studies on the interrelationship between phytoplankton and zooplankton in the waterways of the Pitchavaram Mangroves (India). Doctoral dissertation, PhD thesis. Annamalai University, India, p 215

Karen B, Barid D (2001) Survey of heavy metals in the sediments of Swartkops River Estuary, Port Elizabeth South Africa. Water SA 27(4):461–466

Karleskint G, Turner RK, Small J (2009) Intertidal communities. In: Introduction to marine biology, 3rd edn. Cengage Learning, pp 356–411

Lloyd DS (1987) Turbidity as a water quality standard for salmonid habitats in Alaska. N Am J Fish Manag 7(1):34–45

Manjare SA, Vhanalakar SA, Muley DV (2010) Analysis of water quality using physicochemical parameters Tamdalge tank in Kolhapur district, Maharashtra. Int J Adv Biotechnol Res 1(2):115–119

PramodKumar M, Lakshmi Prasad T, Nagalakshmi K, Jayaraju N, Lakshmanna B (2020a) Concentrations of heavy metals as proxies of marine pollution along Nellore coast of South District, Andhra Pradesh. In: Heavy metals-their environmental impacts and mitigation. IntechOpen

PramodKumar M, Nagalakshmi K, Jayaraju N, Lakshmi Prasad T, Lakshmanna B (2020b) Deciphering water quality using WQI and GIS in Tummalapalle uranium mining area, Cuddapah Basin, India. Water Sci 34(1):65–74

Reddy BSR, Jayaraju N, Reddy KR, Reddy AN (2009) Pollution signatures of benthic foraminifear: a study from the pennar river estuary. India Res J Earth Sci 1(1):07–14

Satheeshkumar P, Khan AB (2012) Identification of mangrove water quality by multivariate statistical analysis methods in Pondicherry coast, India. Environ Monit Assess 184(6):3761–3774

Sharma MK, Kumar M (2020) Sulphate contamination in groundwater and its remediation: an overview. Environ Monit Assess 192(2):1–10

Sobha Rani S (2018) Seasonal changes in water quality parameters of Malidevi estuary (Nellore) East Coast of Andhra Pradesh, India. Int J Current Innov Adv Res 1(1):1–6

Sreenivasulu G, Jayaraju N, Sundara Raja Reddy BC, Lakshmi Prasad T (2015) J Coast Sci 2(2):34–39

Sundara Raja Reddy BC (2006) Estuarine pollution signatures on the Benthic foraminifera: a study from Nellore coast, East coast of India. Unpublished PhD thesis, S.V. University, Tirupathi, AP, India

Swain S, Sahu BK, Pattanaik S, Sahoo RK, Majhi A, Satapathy DR, Choudhury SB (2021) Anthropogenic influence on the physico-chemical parameters of Dhamra estuary and adjoining coastal water of the bay of Bengal. Mar Pollut Bull 162:111826

Velsamy G, Manoharan N, Ganesan S (2013) Analysis of physico-chemical variations in sea water samples Uppanar Estuary, Cuddalore, Tamilnadu, India. Int J Research in Biol Sci 3(2):80–83

Wang YB, Liu CW, Liao PY, Lee JJ (2014) Spatial pattern assessment of river water quality: implications of reducing the number of monitoring stations and chemical parameters. Environ Monit Assess 186(3):1781–1792

Yadav P, Yadav VK, Yadav AK, Khare PK (2013) Physico-chemical characteristics of a fresh water pond of Orai, UP, Central India. Octa J Biosci 1(2):177–184

Young JC, Clesceri LS, Kamhawy SM (2005) Changes in the biochemical oxygen demand procedure in the 21st edition of standard methods for the examination of water and wastewater. Water Environ Res 77:404–410

Chapter 6
Climatic Variability and Anthropogenic Forcing on Marine Ecosystems: Evidence from the Lakshadweep Archipelago

A. A. Fousiya, Javed N. Malik, and Supriyo Chakraborty

Abstract This study provides an overview of natural and anthropogenic impacts on marine ecosystems from the Lakshadweep Islands located in the Indian Ocean. The Lakshadweep Archipelago consists of low-lying landmasses highly impacted by climate change issues such as rising sea levels. The Archipelago comprises a group of 36 small islands, 10 of which are inhabited. The island ecosystem harbors rich biodiversity and supports a dense population, thus overexploiting the natural resources. While these islands have shown the impacts of climate change, the deleterious effects of direct human impacts including pollution from coastline developmental and tourism activities, over-exploitation, and unsustainable developments have likely compromised their adaptive capacity. The pristine coral reef ecosystems are getting vulnerable to severe climate change impacts that include rising sea levels, warming-induced coral bleaching, rainfall variability, and extreme weather events (cyclones). Climate change poses a severe threat to sea life, including fisheries, economies, and societies, particularly to the local community that relies heavily on natural resources. We have made a systematic effort to investigate the consequences of natural and anthropogenic forcing on these marine ecosystems, particularly the impacts of climate change.

Keywords Climate change impacts · Coral Islands · Anthropogenic impacts · Lakshadweep Archipelago

A. A. Fousiya (✉) · J. N. Malik
Department of Earth Sciences, Indian Institute of Technology, Kanpur, Uttar Pradesh, India

S. Chakraborty
Indian Institute of Tropical Meteorology, Ministry of Earth Sciences, Pune, Maharashtra, India

N. Jayaraju et al. (eds.), *Coasts, Estuaries and Lakes*,
https://doi.org/10.1007/978-3-031-21644-2_6

6.1 Introduction

Lakshadweep, the smallest Union Territory in India, is an archipelago of 36 islands comprising 12 atolls, 3 reefs, 5 submerged banks, and 10 inhabited islands. It has a total area of 32 km^2. The islands are low-lying, with lowest elevations of less than 1 m and maximum elevations of 6 m above mean sea level (MSL). A few islands that are larger in the area (such as Minicoy, Kavarati, and Agathi) are densely populated. The Archipelago also consists of several smaller but uninhabited islands. More than 80% of Lakshadweep is less than a meter above sea level. The Archipelago has a limited land area and experiences unpredictable weather extremes. These islands have become vulnerable to the threats associated with climate change, mainly the sea level rise, increased cyclone frequency, rising sea surface temperature, and ocean acidification. The islands face significant threats such as shrinking inland surface area due to coastline erosion, degradation of lagoon corals (associated with extreme events and anthropogenic pressure), and coastline local vegetation cover such as salt marshes, sea grasses, and mangroves. Hence, there is a significant impact on the land resources such as soil, groundwater, flora, and fauna along with the marine resources such as coral reef and lagoon fishing, especially in the case of Minicoy, Agathi, Kavarathi, Bangaram, and Suheli Islands. Such communities face extremely pressing challenges. The sea level rose by 1.3–1.7 mm each year over the twentieth century. Using satellite data, the rate of sea-level rise was calculated to be between 1.72 mm/year and 0.372 mm/year (1973–2010) (Sudha Rani et al. 2017; Chowdhury and Behera 2015). Sea-level rise for islands in the Indian Ocean region was determined to be trending at 1.22–0.3 mm/year between 1996 and 2010 (Palanisamy et al. 2014). Subtle salinity fluctuations are also brought on by warming and modifications in circulation patterns. Sea-level rise is a major problem in addition to climate considerations. Since the global mean sea level is anticipated to rise by 0.43–0.84 m by 2100 compared to 1986–2005, with large fluctuations at local and regional scales, depending on future warming scenarios (IPCC 2018).

6.2 Marine Archives

One of the best-known archives to study past regional changes in ocean surface conditions is obtained from the stable isotopic records of marine calcifiers such as massive coralline structures and clam shells (Pfeiffer and Dullo 2006; Lough and Barnes 2000). These marine aragonite-bearing organisms preserve high-resolution records with higher growth, thereby providing traces of ocean–atmosphere variations on sub-seasonal resolution ranging up to a few hundred years (Gagan et al. 2000; Mcculloch and Esat 2000). Since corals and clams have a high growth rate (up to about 1 cm/year), it is believed that they record the seasonal variation of climate (Chakraborty 2020). Due to their rapid development, the Lakshadweep corals' skeleton $\delta^{18}O$ records offer rich paleoclimatic data on exceptionally high-resolution time scales (Chakraborty and Ramesh 1993, 1998).

6.3 Resilience of Corals to Rising Sea Surface Temperature

It is a well-known fact that the climate of the equatorial Pacific Ocean plays a significant role in modulating the monsoon rainfall over India through the teleconnection process (Chowdary et al. 2021). According to Chakraborty et al. (2012), corals in the equatorial Pacific that record El Nino-induced warming are an excellent proxy for the variability of rainfall caused by the Indian summer monsoon. Similarly, the corals growing in the Lakshadweep region also hold promise in recording the past monsoon rainfall variability over India (Chakraborty 2020). Hence, we use the available coral records from the Lakshadweep region to study the surface ocean characteristics of the recent past. Due to vertical mixing caused by the SW monsoon, the Arabian Sea's summer surface water temperatures dropped by 3–4 °C (Chakraborty and Ramesh 1993; Luis and Kawamura 2004). This reduction in temperature is also evidenced by the isotopic ($\delta^{18}O$) records of corals from this region (Chakraborty and Ramesh 1992, 1993, 1997; Ahmad et al. 2011; Fousiya et al. 2016). However, long-term warming and/or cooling trends across the tropical seas are generally evident in coral $\delta^{18}O$ records.

Warm-water coral reefs are essential to tropical fisheries, other marine and human systems, and a wide variety of marine species. They are especially sensitive because if water temperatures rise beyond a threshold (1–2 °C above normal), they might die in large numbers. Since 1998, these circumstances have been present in a number of tropical oceans, causing extensive coral bleaching. It normally takes 10–15 years for a reef to recover following significant coral death due to bleaching. Sea-level rise, ocean acidification, coral degradation, and other climate change-related phenomena are already showing negative consequences on the marine environment of the Lakshadweep Archipelago (Fousiya et al. 2022a). Although certain coral species are more robust than others, and the effects of climate change differ depending on the locality, it seems inevitable that future climate change impacts will further degrade coral reefs. Other marine and coastal ecosystems may suffer major effects as a result, including the loss of coastal protection for numerous coral islands and low-lying regions as well as the high biodiversity that these reefs support.

The non-climatic impacts of human activity are also common in these islands, such as over-exploitation of natural resources and ecosystem-related marine pollution. Some marine animals, food webs, and ecosystems are more vulnerable than others, despite the fact that these stressors and their combined impacts are likely to affect practically all marine creatures. The effects on the human population living on these islands might be grave if adequate steps are not taken to slow down the impacts of climate change in the future. Hence, in this study, we provide an overview of the impact of climate and anthropogenic forcing on the coral reef ecosystems of the Lakshadweep Archipelago. Some of the datasets reported here were collected through field observations and the analysis of proxy records (Fousiya et al. 2016, 2022a).

6.4 Study Area

6.4.1 Geomorphology and Geology

The Lakshadweep Islands are situated between 225 and 450 km off Kerala's west coast, in the southeast corner of the Arabian Sea. In the Indian Ocean, the Lakshadweep Archipelago is home to a variety of coral reefs (Pillai 2001; Fig. 6.1). The geomorphology of the Lakshadweep Islands has been continuously affected by sea-level changes, due to which the coastal geomorphology of the Islands has been changing progressively. The Indian summer monsoon induced upwelling provides abundant nutrient that supports the islands' rich ecosystem. Low-lying coral islands formed mostly of calcareous sand and soils are often separated from the outer reef rim by a shallow lagoon on the western side of each island. The Lakshadweep Islands span a 32 km^2 geographical region and 28.5 km^2 land. The Lakshadweep has a 132 km long coastline. With 20,000 km^2 of territorial waters and almost 0.4 million km^2 of exclusive economic zone (EEZ), the lagoons have a total area of 4200 km^2.

6.4.2 Climate Variability in and Around the Lakshadweep Archipelago

The Lakshadweep experiences a typical tropical climate. The summer temperature ranges from 22–33 °C and winter temperatures from 20–32 °C. The area is strongly influenced by the south-westerly summer monsoon circulation (JJAS) and the north-easterly circulation in winter (Oct.–Dec.). The summer months receive the most rainfall (80%), while the winter months receive about 20%. The total annual rainfall is ~1510 mm. It should be noted that the southernmost island, Minicoy, which is closer to the equator, receives less rainfall during the summer monsoon compared to Amini and other islands, while the Minicoy acquires (24%) more northeast monsoon rain than Amini and the rest of the archipelago (17%). Since the moisture source for the summer and winter seasons are very different (Indian Ocean/Arabian Sea versus the Bay of Bengal), the isotopic composition of precipitation in this region shows distinct isotopic seasonality.

The Lakshadweep area is significantly impacted by the El Nino Southern Oscillation (ENSO) (Bokhari Friberg 2014; Chakraborty 2020). Ocean-atmosphere interactions (Webster et al. 1999) mostly control the climate of the Lakshadweep

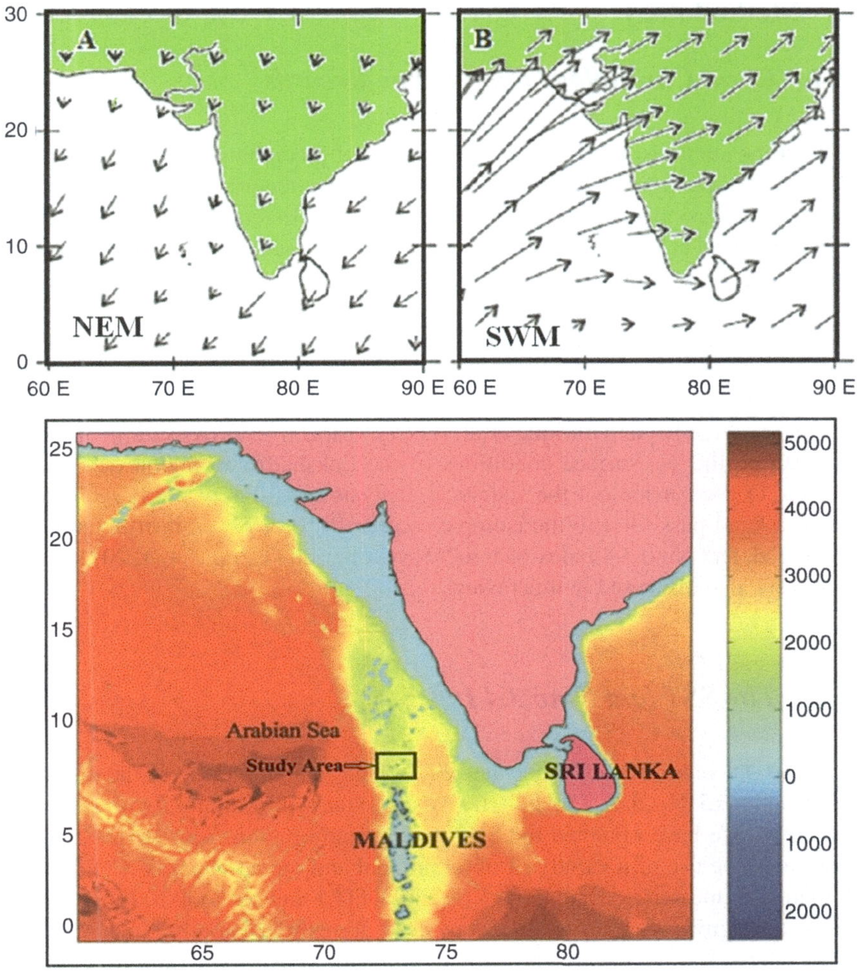

Fig. 6.1 The study area showing the Lakshadweep islands (lower panel). The upper panel shows the mean (**a**) NEM and (**b**) SWM wind pattern over the Lakshadweep Islands. The color coding shows bathymetry (m) of the Arabian Sea

Archipelago including tropical cyclone, storm surges, hurricanes, flooding and erosional processes. The annual mean sea surface temperature (SST) of the Lakshadweep region and its environs has gradually risen in recent years (Roxy et al. 2014). A 523-year-long tree-ring record used in a dendroclimatological study revealed a substantial correlation between the ENSO and the drought conditions in the Southern India (Borgaonkar et al. 2010).

6.5 Methodology

6.5.1 Coral Chronology and Isotope Records

We used the stable isotopic records of corals and precipitation isotopes from this region. A 24-year-long (1989–2013) oxygen isotopic record was generated from a *Porites sp.* The coral sample was collected from the Minicoy Island; the details are given in Fousiya et al. (2016). Briefly, there were around 24-year bands in the coral sample, and the average growth rate was 7.7 mm-y^1. For stable isotope ($\delta^{18}O$ and $\delta^{13}C$) studies, 100–200 µg of coral powder was removed along a vertical growth axis transect using low-powered hand-held drilling equipment. We have also analyzed boron isotopes (^{11}B) to study the surface ocean pH variability from the same shell samples. The details of the isotopic analysis ($\delta^{18}O$, $\delta^{13}C$) and (^{11}B) are given in Fousiya et al. (2016) and Tarique et al. (2021), respectively. To examine the effect of ENSO on the sea surface conditions of the Lakshadweep region (see Tarique et al. 2021), we carried out the statistical analysis on different proxy records. We applied a band-pass filter to the isotopic records ($\delta^{11}B$ & $\delta^{13}C$), instrumental SST, wind speed, and Nino 3.4 index over a 2–8-year period (Tarique et al. 2021). These records are also discussed in this review.

6.5.2 The SST and Nino 3.4 Index

The Nino 3.4 index representing the sea surface temperature over the equatorial Pacific region (5°N – 5°S, 150°E – 160°W) was sourced from http://www.cpc.ncep.noaa.gov.h. We have obtained the SST for the study region from HADISST. This report cited the use of a variety of observational and model datasets. COPEPOD's Interactive Time-series Explorer (COPEPODITE) online toolkit (https://www.st.nmfs.noaa.gov/copepod/toolkit/) was used to extract monthly mean Hadley Center Global Sea Ice and Sea Surface Temperature (HadISST), time series data of the study site for 0.5° × 0.5° box centered at 8.25°N, 73°E. Mean annual global SST conditions (Fig. 6.3) for the warmer years throughout the study period between 1969–70 (a), 1973–74 (b), 1977–78 (c), 1980–81 (d), and 1998–99 (e) were derived from online NOAA-CIRES Climate Diagnostics Centre, Boulder, Colorado (www.cdc.noaa.gov/). We also captured a few field photographs at (a) Agathi Island, (b) Bangaram tourism island, and (c) Suheli par uninhabited islands during the field visits to these islands.

6.6 Rainwater Sampling

We have also collected precipitation samples for isotopic analysis to understand the moisture dynamical processes over the Lakshadweep region (Fousiya et al. 2022b). We collected daily scale precipitation samples from the Minicoy Island; details are provided in Fousiya et al. (2022b)). The observed mean monthly surface temperature varies from 28.5–30.2 °C. The maximum temperature was observed during March and the lowest in August. Around ($n = 200$) rainwater samples were collected during the years 2015–2017: ($n = 63$) for 2015, ($n = 53$) for 2016, and ($n = 84$)-2017, respectively (from Jan to December). The rainwater samples were analyzed using an LGR (now ABB) Water and Water Vapor Analyzer (Model: TIWA-45-EP) at IITM, Pune. An internal laboratory water standard was used to check the precision and accuracy of the readings. The overall analytical precision obtained for $\delta^{18}O$ and (δD) was about 0.1‰ (<1‰). Daily precipitation information was sourced from TRMM 0.25 × 0.25 gridded data to observe the precipitation amount effect.

6.7 Result and Discussion

6.7.1 The Variability in Regional SST

The regional SST analysis using the Niño-3.4 ENSO index and SST over the Indo-Pacific region for the period 1965–2015 indicates that local SST over the sampling site is significantly correlated with the Niño-3.4 index.

We have observed an increase in regional SST patterns. The increased pattern associates well with the rising global SST pattern, especially for the following years: 1969, 1974, 1977, 1981, 1983, 1886, 1998, 2003, 2006, and 2010. Figure 6.2 depicts a significant variability in local SST. Since the coral $\delta^{18}O$ is inversely related to the temperature, we have an inverse correlation between the $\delta^{18}O$ and the SST (see Fig. 7, Fousiya et al. 2016). A drop of coral $\delta^{18}O$, up to −6.5‰, which is ca 1‰ lower relative to the mean during 1998, when the regional SST rose up to 31 °C. This observation also indicates an anomalous warming of the SST in 1998 (Fousiya et al. 2016), a year known for its extreme El Nino effect (Chowdary et al. 2021). The correlation between the Nino 3.4 index and the SST anomaly for this region shows a warming trend (0.3–1.5 °C) with an average of 0.9 °C for the last few decades. These variations can probably be due to local and regional atmosphere-ocean conditions. A gradual increasing trend in SST from 1977 to recent time is also observed for annual average SST. Local SST shows a significant correlation with the Nino 3.4 index $R^2 = 0.43$, with a more than 95% significance level. The annual mean SST of this region rose by 0.28 °C from 1985 to 2005. The western Indian Ocean's mean SSTs are currently at their greatest level since at least the 1950s and most certainly since the 1840s. According to Roxy et al. (2014), the western Indian Ocean warmed

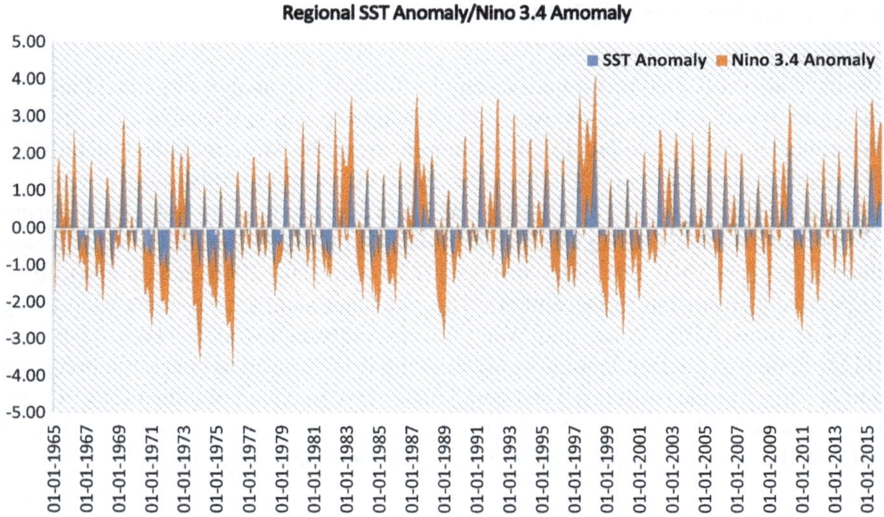

Fig. 6.2 Association between the equatorial Pacific climate and that of the study region. Red bars show the annual average NINO 3.4 SST anomaly. Blue bars represent regional (Lakshadweep) SST anomaly. The years which exceed 0.4 °C are associated with high summer-average SSTs, above the local summer average for the time interval studied

by 1.28 °C between 1901 and 2012. It is probable that the progressive increase in SST at the Lakshadweep region is imparting thermal stress on marine calcifiers. For example, the impact of thermal stresses was observed on the calcification process. An exceptionally low annual growth band in a *Tridacna Maxima* shell was found in the Minicoy Island area (Fousiya et al. 2022a). We have also observed anomalously low value of $\delta^{18}O$ in 2010 in the same clam sample. These evidences may be indicative of higher temperatures for longer periods of increased thermal stress, which could result in episodic coral bleaching.

Despite the upwelling impact on the Lakshadweep area, Savidge et al. (1990) found that the regional SST appears to be increasing. According to the findings, the temperatures may not have been caused by an upwelling directly but rather by an advective movement of water down the coast. Additionally, certain data indicate that during the past several decades, ENSO warm phases have occurred more frequently, which coincides with the more pronounced mean state of SST warming (Roxy et al. 2014). Our regional SST statistics provide evidence in favor of this claim.

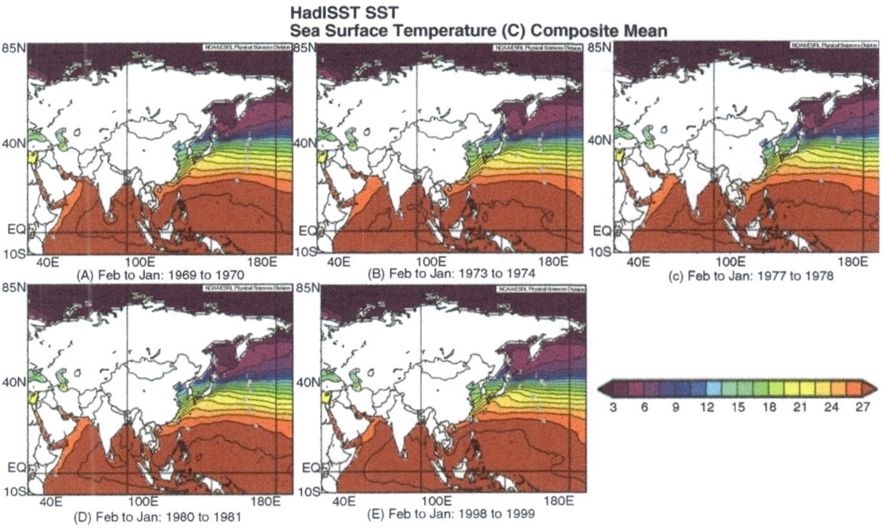

Fig. 6.3 The mean annual SST of the Indo-Pacific for the warmer years throughout the study period between 1969–1970 (**a**), 1973–1974 (**b**), 1977–1978 (**c**), 1980–1981 (**d**), and 1998–99 (**e**)

6.7.2 Analysis of Coralline $\delta^{11}B$ Records and Ocean pH Variability

A comparison of coral $\delta^{11}B$ (a proxy for surface ocean pH) with the Niño3.4 index showed a significant positive correlation (Tarique et al. 2021) for this region. Interestingly, on a seasonal scale, coral $\delta^{11}B$/pH had higher and lower values during each El Niño and La-Niña events, respectively (see Fig. 3 of Tarique et al. 2021). This shows the influence of ENSO on the physico-chemical condition of the sea surface of the reef environment. These findings are in line with earlier studies from the New Caledonia Pacific that suggested a similar effect of past El-Niño on $\delta^{11}B$/pH variability (Wu et al. 2018). Additionally, the teleconnections between local and global climatic processes have been investigated using coral records (Chakraborty et al. 2012, 2021). The regional SST is also influenced by the Pacific warm pool, and the coral $\delta^{18}O$ records have preserved the evidence for such warming events and thus help highlight the warm or extreme El-Nino events. This indicates the possibility of a local impact caused by vertical mixing and/or upwelling, which brings water masses with low $\delta^{13}C$ values to the surface and causes a general drop in the $\delta^{13}C$ values of the ocean's surface. This suggests that regions where deeper bottom water is upwelled or advected are where the yearly reduction in $\delta^{13}C$ amplitude is most significant. To assess the influence of ENSO and other instrumental parameters (SST and wind speed) on the $\delta^{11}B$ and $\delta^{13}C$ variability, 2–8-year band-pass filter was applied to these records (Fig. 6.4). The filtered $\delta^{11}B$ shows a strong association

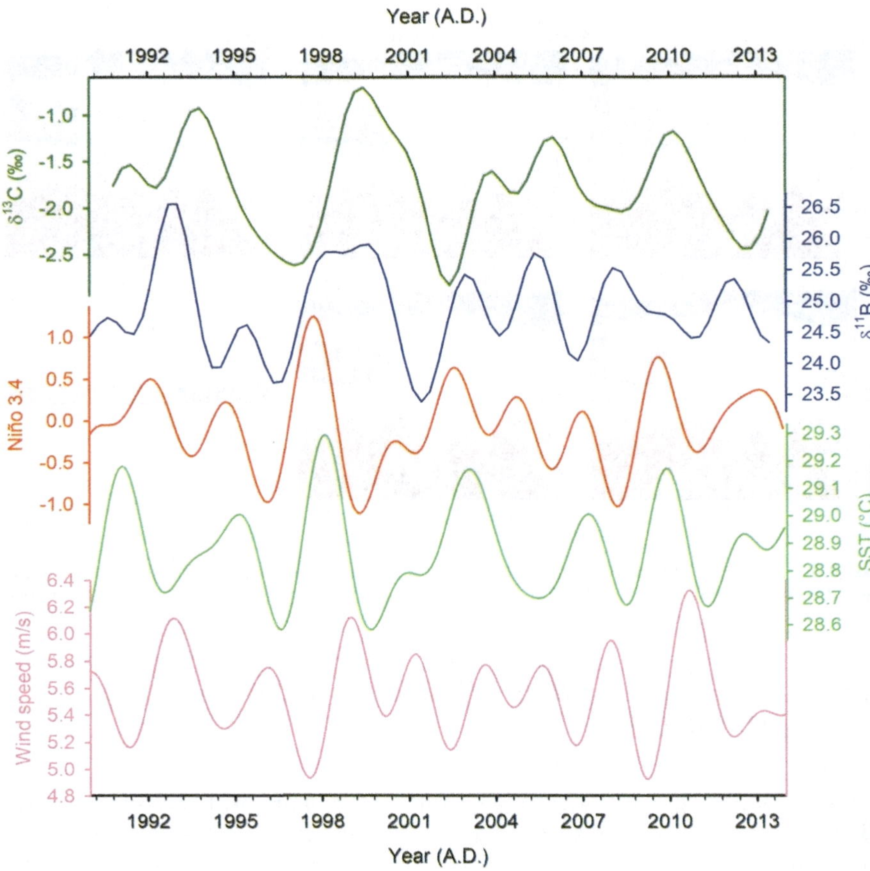

Fig. 6.4 The association of coralline isotopes ($\delta^{11}B$ and $\delta^{13}C$) and the ocean-atmosphere variabilities illustrated in the 2–8 year band pass filtered time series of various parameters

with the Nino 3.4 index and other ocean-atmosphere parameters such as SST and wind. This emphasizes ENSO's impact on the Lakshadweep region.

However, there are lead–lag relationships between different time series. Therefore, the fluctuations in our coral-based proxy records reflect a highly dependent nature with ocean climate oscillations following the pacing of ENSO (Torrence and Webster 1999). Instrumental observations (Sutton et al. 2019; Sutton et al. 2017) and proxy records (Shinjo et al. 2013; Wu et al. 2018) conclude that equatorial Pacific's seawater pH/pCO_2 change in inter-annual to decadal periods is dominantly controlled by ENSO occurrences. The high amplitude variations in $\delta^{11}B$/pH show a significant positive correlation with $\delta^{13}C$. Linear regression between $\delta^{11}B$ and $\delta^{13}C$ shows a significant positive correlation ($r = 0.40$, $n = 76$, $p < 0.01$) (see Figure 6 of Tarique et al. 2021), suggesting Dissolved Inorganic Carbon (DIC) fluctuations associated with productivity at Lakshadweep. In addition, pH variability at Lakshadweep is coupled with prominent SST, mixing, and upwelling variations modulated by ENSO. During the studied interval (1989–2013), pH variability due to local environmental factors and El-Niño teleconnection is large enough that it has masked the long-term ocean acidification due to the absorption of anthropogenic CO_2. A pH record from 1988 to 2013 was constructed using boron isotope research on corals from Lakshadweep in the Arabian Sea (Tarique et al. 2021). Using a multi-proxy approach (^{11}B [pH], ^{13}C [calcification processes], Sr/Ca, and Li/Mg [SST], Tarique et al. (2021) studied the impact of regional oceanographic processes in regulating surface ocean pH variability in the Arabian Sea. It has been noted that the Arabian Sea's pH fluctuation is mostly caused by differences in oceanic processes such as mixing, upwelling, and productivity (Wiggert et al. 2005). Since wind-driven upwelling and vertical mixing have a significant influence on the Lakshadweep area, they will affect the surface oceanographic processes. The robustness of these marine calcifiers, which grow under the strong impact of upwelling and significant fluctuation in physical and biological processes, has been revealed by coral-based pH reconstructions. The other possible threats to the coral ecosystems are untreated sewage and non-biodegradable solid waste, and oil slicks from automobile engine boats into the marine ecosystem that has a direct impact on the lagoon and coral reef system of these islands.

6.7.3 Erosion Associated with Sea-Level Rise

The formation of Lakshadweep Islands is connected to the volcanic activity of Reunion hotspot leaving a chain of islands in the Indian Ocean. These islands constantly change, erode, and grow as wind and waves alter in response to weather patterns and local geological factors. Some regions on the Northern side of Minicoy Island are completely eroded and made a pathway where a large boat can pass through after the major natural disaster caused by the Tsunami in 2004. Natural hazards such as extreme climatic, oceanographic, and geological events affect the physical, ecological, social, and economic landscapes of Lakshadweep islands and other coastal areas. Similarly, during cyclone Ockhi that occurred in December 2017, significant changes occurred, and several of the Island's landscapes were highly altered (Fousiya and Lone 2018). These short-term events have the potential to cause significant coastal erosion and have long-term consequences. Similarly, land erosion has been observed on almost every island of this archipelago occurring naturally, photographs of which are shown here from islands such as Agathi, Bangaram, and Suheli par (Fig. 6.5). Alterations in beach equilibrium conditions are caused by cyclic weather patterns over time scales ranging from months to decades. The process of adjustment occurs approximately on less daily basis, but it is usually only seen when substantial natural changes occur or when human disrupts the natural balance by developing coastal constructions within the active beach zone.

Fig. 6.5 Photographs of beach erosion representing (**a**) Agathi Island, (**b**) Bangaram tourism island and (**c**) Suheli par uninhabited island

6.7.4 Evidence of Cyclonic Activity on Rainwater Isotope from Minicoy Island

The Minicoy Island receives rainfall from the dual monsoon systems (the southwest (SWM)/Indian summer monsoon and northeast monsoon (NEM)/winter monsoon). We have also examined the behavior of rainwater isotopes from one of the Islands (Minicoy Island) of the Lakshadweep region from 2015 to 2017. During the observation period (2015–2017), the Minicoy Island received a significant amount of rainfall ~3743 mm in the SWM and 1453 mm during NEM. We observed seasonal variation in the precipitation isotope composition in Minicoy Island. The $\delta^{18}O$ of the SWM precipitation vary from −8.53 to 1.82‰ with an average of −1.13‰, respectively. The δD varies from −56.98 to 15.90 with an average value of −1.30. Similarly, during NEM, the maximum value of $\delta^{18}O$ is −4 to having an average difference of −6.39, and δD varies from −102.55 to 8.81 with a mean difference of −38.24, which is far more depleted than what is observed during SWM. Other than this general variation, the data reveals extremely negative $\delta^{18}O$ values below −10‰ during December 2015 and December 2017 associated with storm surges and severe cyclones such as Ockhi that occurred during the same time in December 2017. We have also collected the hourly based samples from this island during the cyclonic events of the Ockhi cyclone. It was observed that event-based isotope have a much-depleted value, which shows evidence of extreme cyclonic activity in these areas

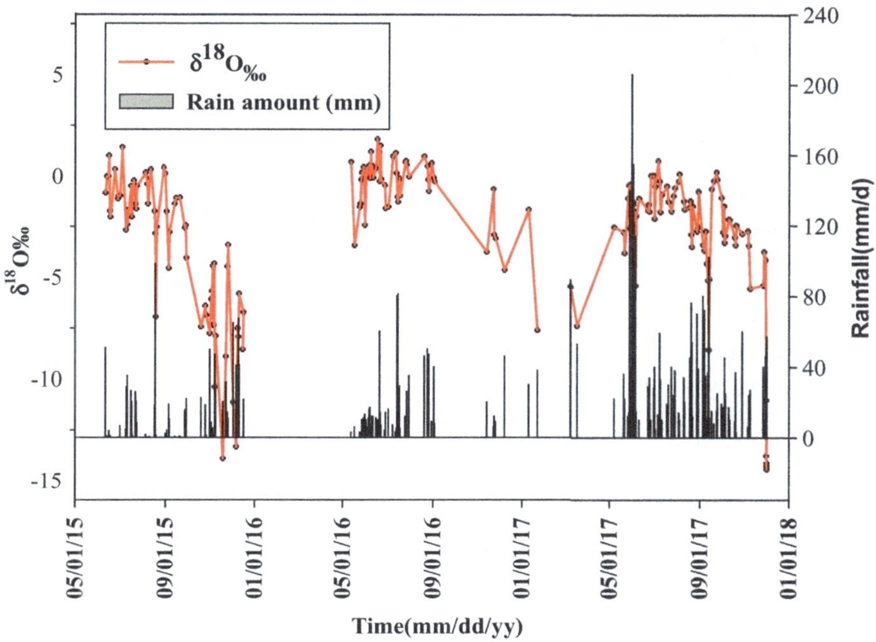

Fig. 6.6 The time series of rainfall (bar) and its oxygen isotopic values (line) for Minicoy Island during the observation period from 2015 to 2017

(Fousiya et al., 2022a; Fig. 6.6). During such cyclonic events, large-scale coral damage and loss of habitats occurred in this region. (Fousiya and Lone 2018). The assessment of amplitude, spatial scale, and duration of the isotope anomaly associated with tropical cyclones such as the one Ockhi confirms and supports the fact that strong, isotopically depleted rainfall from the tropical cyclone is likely to transmit a detectable isotope signal in a wide range of environmental proxies at a regional scale. Furthermore, we have observed depleted $\delta^{18}O$ values in rainwater isotopes associated with heavy rainfall events. The same isotopically depleted values are expected to be observed in the coral proxy as well and could provide some leads into anthropogenic-derived climate change in this region.

6.8 Summary and Conclusions

The study presents an overview of the climatic (natural) and historical (anthropogenic) induced variability from the Lakshadweep region with a focus on the Minicoy Island. We examined the coral isotope and rainwater isotope records and inferred the anthropogenic-induced short-term environmental changes from the Lakshadweep region.

In each of these proxy records, the environmental data given here illustrates basin-scale patterns on decadal-monthly periods. By using dominant and decreasing coral isotope records, the coral $\delta^{18}O$ records demonstrate event-based warming and/or freshening in the study area and concur on decadal timeframes with the NOAA extended reconstruction of sea surface temperature (ERSST). The signal in the western Arabian Sea is most likely being dominated by a seawater $\delta^{18}O$ freshening. According to regional oceanography or regional climatology, El-Nino variability was depicted by either reducing the values of $\delta^{18}O$ from the average or by modifying local seawater temperature or surface salinity. Through atmospheric teleconnections, ENSO influences seawater temperature and salinity beyond the Pacific. Given that the Indian Ocean is warming faster than the rest of the ocean waters and that model simulations predict that ENSO frequency and amplitude will likely increase in a warming environment, pH variability in the Indian Ocean is expected to increase. A rise in pH variability could be critical for the resilience and long-term sustenance of calcifying marine species in the Indian Ocean. The smallest union territory in the Arabian Sea is already facing challenges due to climate change, causing loss and damage to natural resources, as well as human displacement.

6.9 Recommendations

- Create awareness among the local community regarding climate change from the grass root level.
- Traditional knowledge must be incorporated into management processes, and scientific training must be provided to complement traditional knowledge in particular to Lakshadweep islands.
- Raising people's awareness of the value of marine ecosystems and the necessity of their preservation.
- Wise use of natural resources in order to maintain the natural sustainability.
- Conservation and protection of the coral reef system, coastline vegetation, Mangrove Ecosystem from further degradation.
- Creation of viable and sustainable coastal community choices and alternate means of subsistence to lessen the strain on reef resources.
- Step up ecological and socioeconomic research and monitoring to ensure that the right data and information are available to meet regional and national demands.
- Create methods for tracking information about coral reefs, such as monitoring data, and ensure managers and decision-makers can access them.
- Controlling invasive species and outright prohibiting the entrance of any invasive species to the lagoon region in particular.

Acknowledgment We thank the Dweep Panchayat of Minicoy Island for providing support and the necessary information.

References

Ahmad SM, Padmakumari VM, Raza W, Venkatesham K, Suseela G, Sagar N, Chamoli A, Rajan RS (2011) High-resolution carbon and oxygen isotope records from a scleractinian (Porites) coral of Lakshadweep Archipelago. Quat Int 238(1–2):107–114

Bokhari Friberg Y (2014) Oxygen isotopes in corals and their use as proxies for El Niño. Examensarbeten i Geologi vid Lunds universitet

Borgaonkar HP, Sikder AB, Ram S, Pant GB (2010) El Niño and related monsoon drought signals in 523-year-long ring width records of teak (Tectona grandis LF) trees from south India. Palaeogeogr Palaeoclimatol Palaeoecol 285(1–2):74–84

Chakraborty S (2020) Potential of reef-building corals to study the past Indian monsoon rainfall variability. Curr Sci 119(2):273–281. https://doi.org/10.18520/cs/v119/i2/273-281

Chakraborty S, Ramesh R (1992) Climatic significance of δ18O and δ13C variations in a banded coral (Porites) from Kavaratti, Lakshadweep Islands. In: Desai N (ed) Oceanography of the Indian Ocean. Oxford and IBH Publication (P) Ltd., New Delhi, pp 473–478

Chakraborty S, Ramesh R (1993) Monsoon-induced sea surface temperature changes recorded in Indian corals. Terra Nova 5(6):545–551

Chakraborty S, Ramesh R (1997) Environmental significance of carbon and oxygen isotope ratios of banded corals from Lakshadweep, India. Quat Int 37:55–65

Chakraborty S, Ramesh R (1998) Stable isotope variations in a coral (Favia speciosa) from the Gulf of Kutch during 1948–1989 AD: environmental implications. Proc Indian Acad Sci Earth Planet Sci 107(4):331–341

Chakraborty S, Goswami BN, Dutta K (2012) Pacific coral oxygen isotope and the tropospheric temperature gradient over Asian monsoon region: a tool to reconstruct past Indian summer monsoon rainfall. J Quat Sci 27(3):269–278. https://doi.org/10.1002/jqs.1541

Chakraborty S, Lone AM, Parekh A, Mohan PM (2021) Teleconnections between the Indian summer monsoon and climate variability: a proxy perspective. In: Chowdary J, Parekh A, Gnanaseelan C (eds) Indian summer monsoon variability. Elsevier, pp 131–154

Chowdary JS, Xie SP, Nanjundiah RS (2021) Drivers of the Indian summer monsoon climate variability. In: Chowdary J, Parekh A, Gnanaseelan C (eds) Indian summer monsoon variability. Elsevier, pp 1–28

Chowdhury P, Behera MR (2015) A study on regional sea level variation along the Indian coast. Procedia Eng 116:1078–1084

Fousiya AA, Lone AM (2018) Cyclone Ockhi and its impact over Minicoy Island, Lakshadweep, India. Curr Sci 115(5):819–820

Fousiya AA, Chakraborty S, Achyuthan H, Gandhi N, Sinha N, Datye A (2016) Stable isotopic investigation of Porites coral from the Minicoy Island. Indian J Geo-Mar Sci 45(11):1465–1470

Fousiya AA, Alberti M, Achyuthan H et al (2022a) Anomalous δ18O signal in a giant clam shell (Tridacna maxima) from the Lakshadweep Archipelago, India: signature of thermal stress during a coral bleaching event. Coral Reefs 41:1173–1185. https://doi.org/10.1007/s00338-022-02263-6

Fousiya AA, Arvind GH, Achyutan H, Chakraborty S, Chattopadhyay R, Datye A, Murkute C, Lone AM, Mohan PM, Yadava MG (2022b) Modulation of the precipitation isotopes by the dynamic and thermodynamic variables of the atmosphere in southern parts of India. Water Res. Res. 58:e2021WR030855

Gagan MK, Ayliffe LK, Warren Beck J, Cole JE, Druffel ERM, Dunbar RB, Schrag DP (2000) New views of tropical paleoclimates from corals. Qua Sci Rev 19(1–5):45–64

IPCC (2018) Global Warming of 1.5°C. An IPCC Special Report on theimpacts of global warming of 1.5°C above pre-industrial levelsand related global greenhouse gas emission pathways, in thecontext of strengthening the global response to the threat ofclimate change ed V Masson-Delmotte et al (Geneva,Switzerland: IPCC)

Lough JM, Barnes DJ (2000) Environmental controls on growth of the massive coral Porites. J Exp Mar Biol Ecol 245(2):225–243

Luis AJ, Kawamura H (2004) Air-sea interaction, coastal circulation and primary production in the eastern Arabian Sea: a review. J Oceanogr 60(2):205–218
McCulloch MT, Esat T (2000). The coral record of last interglacial sea levels and sea surface temperatures. Chem Geol 169(1–2):107–129
Palanisamy H, Cazenave A, Meyssignac B, Soudarin L, Wöppelmann G, Becker M (2014) Regional Sea level variability, total relative sea level rise and its impacts on islands and coastal zones of Indian Ocean over the last sixty years. Glob Planet Chang 116:54–67
Pfeiffer M, Dullo WC (2006) Monsoon-induced cooling of the western equatorial Indian Ocean as recorded in coral oxygen isotope records from the Seychelles covering the period of 1840–1994 AD. Quat Sci Rev 25(9–10):993–1009
Pillai VN (2001) Oceanographic aspects of Lakshadweep waters in relation to Skipjack tuna fisheries. Geol Surv India Spec Publ 56:125–128
Roxy MK, Ritika K, Terray P, Masson S (2014) The curious case of Indian Ocean warming. J Clim 27(22):8501–8509
Savidge G, Lennon J, Matthews AJ (1990) A shore-based survey of upwelling along the coast of Dhofar region, southern Oman. Cont Shelf Res 10(3):259–275
Shinjo R, Asami R, Huang KF, You CF, Iryu Y (2013) Ocean acidification trend in the tropical North Pacific since the mid-20th century reconstructed from a coral archive. Mar Geol 342:58–64
Sudha Rani NNV, Satyanarayana ANV, Bhaskaran PK (2017) Assessment of climatological trends of sea level over the Indian coast using artificial neural network and wavelet techniques. Pure Appl Geophys 174(4):1527–1546
Sutton AJ, Wanninkhof R, Sabine CL, Feely RA, Cronin MF, Weller RA (2017) Variability and trends in surface seawater pCO2 and CO2 flux in the Pacific Ocean. Geophys Res Lett 44(11):5627–5636
Tarique M, Rahaman W, Fousiya AA, Lathika N, Thamban M, Achyuthan H, Misra S (2021) Surface pH record (1990–2013) of the Arabian Sea from boron isotopes of Lakshadweep corals—trend, variability, and control. J Geophys Res Biogeosci 126:e2020JG006122
Torrence C, Webster PJ (1999) Interdecadal changes in the ENSO–monsoon system. J Clim 12(8):2679–2690
Understanding the role of nutrient limitation on plankton biomass over Arabian Sea via 1-D coupled biogeochemical model and bio-Argo observations
Webster PJ, Moore AM, Loschnigg JP, Leben RR (1999) Coupled ocean–atmosphere dynamics in the Indian Ocean during 1997–98. Nature 401(6751):356–360
Wiggert JD, Hood RR, Banse K, Kindle JC (2005) Monsoon-driven biogeochemical processes in the Arabian Sea. Prog Oceanogr 65(2–4):176–213
Wu HC, Dissard D, Douville E, Blamart D, Bordier L, Tribollet A, Le Cornec F, Pons-Branchu E, Dapoigny A, Lazareth CE (2018) Surface ocean pH variations since 1689 CE and recent ocean acidification in the tropical South Pacific. Nat Commun 9(1):1–13

Part III
Sediment Characteristics

Chapter 7
Geochemical Characterization of Suspended Sediments in the Nethravati Estuary, Southwest Coast of India: Insights to Redox Processes, Metal Sorption, and Pollution Aspect

G. P. Gurumurthy, Muguli Tripti, Keshava Balakrishna, Jean Riotte, Stephane Audry, and H. N. Udayashankar

Abstract Estuaries are the major conduits of material transfer from the continental river basins to the marine environments. In this study, suspended particulate matter (SPM) was collected from the estuarine region of Nethravati River, southwest India, over a one-year period on a seasonal basis to assess the metal fluxes to the eastern Arabian Sea. The focus was to understand the effect of secondary fluvial geochemical processes on metal chemistry as well as the role of Fe and Mn particles in heavy metal transport along a tropical micro-tidal estuary. The SPM content and its metal concentrations in the Nethravati estuary showed strong seasonal and spatial variability. The Fe–Mn oxyhydroxides formed the important carriers of metals in the estuarine region; however, it depended on the seasonality. The heavy metals displayed higher concentrations at a low Fe/Mn ratio in the estuary, which suggests a dominant role of Mn-oxyhydroxides as the metal carrier phase. Thus, the geochemi-

G. P. Gurumurthy (✉)
Birbal Sahni Institute of Palaeosciences, Lucknow, Uttar Pradesh, India
e-mail: gurumurthygp@bsip.res.in

M. Tripti
National Centre for Earth Science Studies, Ministry of Earth Sciences, Thiruvananthapuram, Kerala, India

K. Balakrishna · H. N. Udayashankar
Manipal Institute of Technology, Manipal Academy of Higher Education, Manipal, Karnataka, India

J. Riotte
Geosciences Environnement Toulouse, Université Paul Sabatier, Toulouse, France

Indo-French Cell for Water Sciences, Indian Institute of Science, Bangalore, Karnataka, India

S. Audry
Geosciences Environnement Toulouse, Université Paul Sabatier, Toulouse, France

N. Jayaraju et al. (eds.), *Coasts, Estuaries and Lakes*,
https://doi.org/10.1007/978-3-031-21644-2_7

cal assessment of SPM in the Nethravati estuary demonstrates redox cycling of metals coupled with adsorption–desorption of heavy metals onto the metal oxyhydroxides. Higher concentrations of heavy metals were observed in the estuarine bed sediments than in the suspended sediments. However, from the pollution point of view, the heavy metal concentrations were not enriched and were found to be within the limits of the National Oceanic and Atmospheric Administration (NOAA) guidelines.

Keywords Suspended particulate matter (SPM) · Biogeochemistry · Fe-Mn oxyhydroxides · Heavy metals · Tropical estuary · Arabian Sea

7.1 Introduction

River runoff forms the main pathway for the transfer of chemical elements and materials from continents to the open ocean. However, these materials undergo vigorous geochemical processes in the estuarine region. The geochemical processes affecting the elemental fluxes in the estuarine mixing zone include adsorption and desorption, colloidal flocculation, redox reactions, and complexation reactions (Sholkovitz 1978). As a result, an estuary can be an efficient zone of source and sink to sediment and associated chemical elements carried by the rivers. The source or sink pathway of an estuary is determined by the input from the external source, in situ processing in the water column (Turner 1996), relative mobility or degradation of the chemical element under varying redox conditions (Forstner et al. 1989; Guo et al. 1997), and postdepositional transformation in the sediments (Ridgway and Price 1987; Shaw et al. 1990). Estuarine sediments having high concentrations of organic matter and clay minerals facilitate the changes in redox conditions by chemical and microbial degradation of organic matter and mineral diagenesis (Froelich et al. 1979). Thus, the changes in the environmental and redox conditions in the estuary would affect the metal discharge fluxes.

The continental input of chemical elements to the open ocean is significantly affected by the sedimentation and biogeochemical processes occurring in the estuaries. These processes include the following: (i) particle mixing of continental and marine origin (Turner et al. 1994; Turner and Millward 2000); (ii) particle resuspension due to dredging/tidal action (Martino et al. 2002); (iii) phase redistribution of metals under varying redox, pH, and salinity gradient (Martino et al. 2002); (iv) flocculation and coagulation of colloidal material (Sholkovitz 1978); (v) biological uptake/regeneration (DeMaster et al. 1986; Windom et al. 1991) and organic matter production by phytoplankton; and (vi) disposal of domestic and industrial wastes (Allen and Rae 1986). The monitoring of estuarine chemistry (particulate and dissolved) as well as the impact of geochemical processes on the metal fluxes are important aspects of marine chemistry to establish the mass balance of chemical elements.

The particulate trace element geochemistry of estuaries is not well documented in the literature. Most studies reported either on dissolved or bed sediment chemistry data with scattered sampling, which does not take into account seasonal variation or sparse sampling points which does not consider changes in chemistry against the salinity (Biksham and Subramanian 1988; Ramesh et al. 1990; Ramanathan et al. 1993; Balachandran et al. 2005; Alagarsamy 2006). Limited studies have been reported on the estuarine geochemistry of suspended particulate matter (SPM) (Rahaman et al. 2010; Kessarkar et al. 2013; Suja et al. 2017; Mitra et al. 2020) over the Indian coast. With this background, the present study was taken up to achieve the following objectives: (i) to understand the seasonality of particulate metal chemistry along the salinity gradient in a humid tropical estuary, (ii) to study the effect of redox reactions and adsorption/desorption process on the estuarine particulate geochemistry, (iii) to elucidate the major carrier phase for heavy metals in the tropical river estuary, and (iv) to report the transition metal concentration and anthropogenic pollution in the Nethravati estuary located at critically polluted region (Mangalore) of the southwest coast of India as classified by the Ministry of Environment, Forest and Climate Change (MoEFCC), Government of India.

7.2 Materials and Methods

7.2.1 Study Area

In this study, the estuarine region of the Nethravati river has been monitored for heavy metal chemistry to understand its actual metal fluxes to the eastern Arabian Sea. The Nethravati estuary is located in the Karnataka coast, India. This estuary covers an area of 19.2 km^2. The Nethravati river joins the Arabian Sea at Mangalore, with the confluence of Gurupur river. The river length of Nethravati is about 147 km and that of Gurupur is about 87 km while their basins cover an area of about 3657 km^2 and 883 km^2, respectively. The mean annual water discharge of Nethravati and Gurupur rivers were 388 m^{3}s^{-1} and 84 m^{3}s^{-1}, respectively, while their annual sediment discharge was found to be 14×10^5 and 1×10^5 tons, respectively (data from Central Water Commission, Government of India). Thus, the Nethravati river discharged higher water and sediment to the eastern Arabian Sea in the Mangalore region. The Nethravati basin lies in the western edge of the Western Dharwar Craton and hosts the metamorphic transition of peninsular gneiss and southern granulite terrain in South India. The basin lithology is largely (83%) defined by migmatites and granodiorites, while the minor portion is composed of metabasalts, charnockites, laterites, and amphibolites. The detailed description of this river basin has been provided by Gurumurthy et al. (2014). The estuarine region of Nethravati river is very shallow and has almost reached the

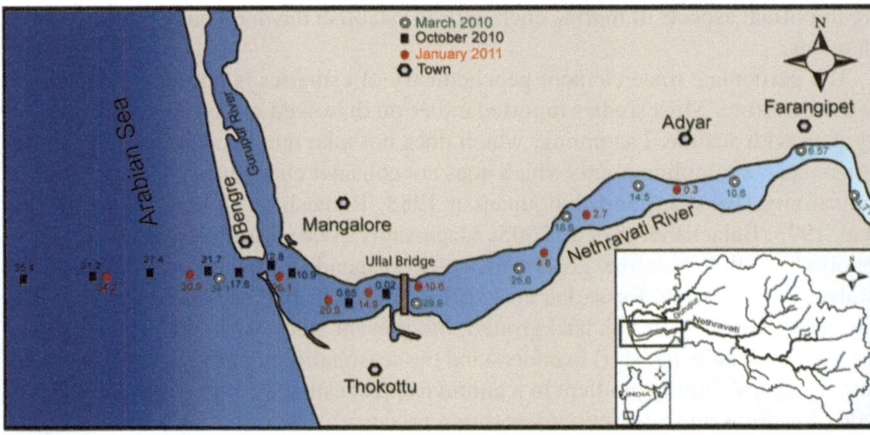

Fig. 7.1 Map of Nethravati-Gurupur estuary showing sampling locations during different sampling periods – premonsoon (March, 2010; empty circle), monsoon (October, 2010; black square), and postmonsoon (January, 2011; red circle). Corresponding salinity at each sampling site is presented

deltaic stage. The Nethravati estuary exhibits mixed type of diurnal tides (Reddy et al. 1979). The exposure of alluvial bed is observed during low tide. The sediments are mainly composed of sand (70 ± 7%) with minor proportion of silt (17 ± 6%), clay (12.63 ± 2%) and organic carbon (1.58 ± 0.3 mg kg^{-1}) (Reddy and Hariharan 1986; Sudhanandh et al. 2011). The freshwater flow from the river influenced the currents in the Nethravati estuary during the monsoon season while the tidal influence dominated during the non-monsoon season. However, the tidal influence is not seen beyond 15–20 km upstream of the Nethravati river mouth (Manjunatha and Harry 1994). About ~93% of the total river discharge is between June and October and only ~7% of the annual discharge is during the rest of the year (November-May). The river discharge during non-monsoonal months is negligible as the river water is being pumped from the Thumbe dam (located 5 km away from Farangipet in Nethravati river; Fig. 7.1) to Mangalore city to meet the domestic needs of about a million population.

7.2.2 Sampling Strategy and Analytical Techniques

This study was carried out to obtain the baseline geochemical database for suspended sediments on seasonal and spatial scales along the Nethravati estuary. The baseline data on the Nethravati estuary is important as it flows through Mangalore, which has been recently selected for development under the smart cities mission by the Government of India (https://smartcities.gov.in/node/193). The estuarine mixing transect samples were collected seasonally during premonsoon (March

2010), monsoon (October 2010) and postmonsoon (January 2011). The sampling was carried out during the low tide cycle with a mechanized boat along a salinity gradient. During the premonsoonal sampling, the saltwater ingress was observed till Farangipet, about 15 km upstream of the river mouth. The saltwater ingress reached only the Ullal bridge (~5 km inland) during the monsoonal and postmonsoonal seasons. The sample locations along with salinity during different seasons are given in Fig. 7.1. The pH, dissolved oxygen (DO; mg L^{-1}), and salinity (psu, practical salinity unit) were measured on-site through a portable multiparameter apparatus connected to their respective probes (HACH make). The salinity measurements in few samples were cross checked at laboratory using a salinity meter (Elico make) and were found to be in agreement. The water samples were collected in precleaned 1 L polypropylene bottles. The water samples were filtered through 0.22-μm pore size Nuclepore membrane filters (Whatman make) under a clean laminar flow bench (HEPA filtered). The first 100 ml of the filtration was systematically discarded to clean the membrane and filtration apparatus. Filtered solutions for trace elements measurements were acidified with double distilled HNO_3 to pH < 2 and stored in precleaned polypropylene bottles until analysis. The trace metals were measured through a quadrupole ICP-MS 7500ce (Agilent Technologies) after suitable dilution. Analytical measurement details are given in Gurumurthy et al. (2014).

In order to obtain the particulate material, another set of water samples was collected in a precleaned 5 L polythene container. These samples were filtered using 0.22 μm pore size cellulose acetate membrane filters (Sartorius make). The residues left over the membrane filters were washed with filtered water jet (same sample) and separated after centrifugation to recover the SPM. The particles were then dried at 50 °C in a hot air oven and weighed to obtain the SPM content. About 100 mg of SPM and bed sediment samples were digested using triacid mixtures (HNO_3:HCl:HF; 3:2:1 ratio) through a microwave digestion system (CEM make) in a clean room. The digestion was carried out at 180 °C for 30 minutes with a set pressure of 200 mm/Hg. The dried digested samples were later diluted by 3000–5000 times with nitric acid. The concentrations of major elements, Fe and Mn in the SPM were measured using ICP-AES while that of trace elements were measured using a quadrupole ICP-MS 7500ce (Agilent Technologies). An internal Re-In standard was used to control and calibrate the analytical drift. The repeated analysis of certified reference sediment, NIST SRM 1646a was carried out for measurement validation. The precision error in measurement was less than 10%, while the detection limit ranged between 1 ppb for V and <100 ppt for Cu, Ni, and Co.

Table 7.1 Physicochemical characteristics of water and metal concentrations data of particulate matter in the Nethravati River estuary

		Sal	T	pH	DO	SPM	Fe	Mn	Al	V	Cr	Co	Ni	Cu	Zn	Pb
Sample No.	DOS	psu	°C		mg L^{-1}	mg L^{-1}	wt%	wt%	wt%	mg L^{-1}	mg L^{-1}	mg L^{-1}	mg L^{-1}	mg L^{-1}	mg L^{-1}	mg L^{-1}
245	March 06, 2010	35.6	30.6	8.12	6.61	21.3	5.83	0.03	8.82	132.2	214.5	16.1	89.8	58.6	158.7	16.8
246		29.9	31.5	7.93	6.99	9.1	3.81	0.11	5.74	105.8	134.3	20.0	69.2	42.4	99.5	17.2
247		25.6	31.6	7.83	6.98	8.6	5.72	0.18	8.24	142.1	204.2	27.6	122.4	71.5	159.2	24.9
248		18.63	32.3	7.84	8.16	8	7.13	0.22	10.63	154.1	252.8	32.5	142.3	76.8	156.1	24.2
249		14.52	31.9	7.62	7.28	5.2	5.68	0.24	7.9	118.5	198.4	26.0	115.9	74.3	141.4	20.4
250		10.68	32.3	7.49	7.54	5.4	5.88	0.29	8.29	122.1	195.1	28.9	117.3	72.7	104.1	16.0
251		6.57	32.2	7.41	7.42	6.7	6.19	0.48	7.03	130.0	4000.8	33.0	128.1	78.1	154.4	391.3
252		4.71	32.2	7.41	7.46	4.6	5.8	0.55	8.14	117.9	179.4	33.2	136.7	101.7	267.3	20.4
341	October 09, 2010	0.02	26.8	7.27	7.44	21.9	9.2	0.06	13.63	201.4	293.5	22.8	148.9	106.3	206.6	22.5
342		0.62	27	7.38	7.22	19.4	9.37	0.07	14.69	210.8	338.0	25.3	165.6	88.6	126.2	21.7
343		10.88	26.8	7.6	7.15	17.3	6.03	0.04	9.08	147.1	218.1	17.4	99.3	57.4	83.6	14.7
344		12.78	28.3	8.09	7.16	14.5	6.41	0.05	9.77	162.0	227.1	18.6	107.1	59.1	91.8	15.3
345		17.64	28.6	8.29	7.21	10.8	1.2	0.35	0.62	27.6	2095.2	806.1	46.5	2439.7	401.9	1449.2
346		21.66	27.9	8.16	7.95	9.1	4.41	0.04	6.86	112.9	160.7	14.8	73.4	45.3	70.4	10.8
347		27.4	28.8	8.22	7.21	6.1	6.44	0.07	9.96	148.0	235.2	25.6	103.3	57.5	96.8	15.8
348		31.2	28.8	8.2	7.51	5.1	4.1	0.04	6.42	95.5	148.6	17.6	66.4	39.3	75.3	13.0
392	January 14, 2011	2.7	28.5	7.68	7.49	21.1	7.29	0.31	8.06	133.4	202.0	27.8	93.5	97.7	301.2	17.2
393		4.84	28.3	7.85	7.14	2.1	7.31	0.44	6.53	124.4	139.3	33.9	76.7	105.9	213.9	16.5
395		14.85	27.8	8.05	6.46	2.1	5.46	0.39	6.09	111.9	131.6	34.1	96.0	64.7	204.6	17.4
398		30.9	28	8.23	6.4	0.9	9.12	0.35	12.01	195.9	267.7	44.8	190.8	99.2	1039.7	48.3
Average			29.51	7.83	7.24	10.0	6.12	0.22	8.43	134.7	491.8	65.3	109.5	191.8	207.6	109.7
Std. dev.			2.05	0.33	0.43	6.9	1.93	0.17	3.03	40.5	928.8	174.5	36.0	529.5	213.2	326.1
UCC							3.53	0.08	8.15	97	92	17.3	47	28	67	17

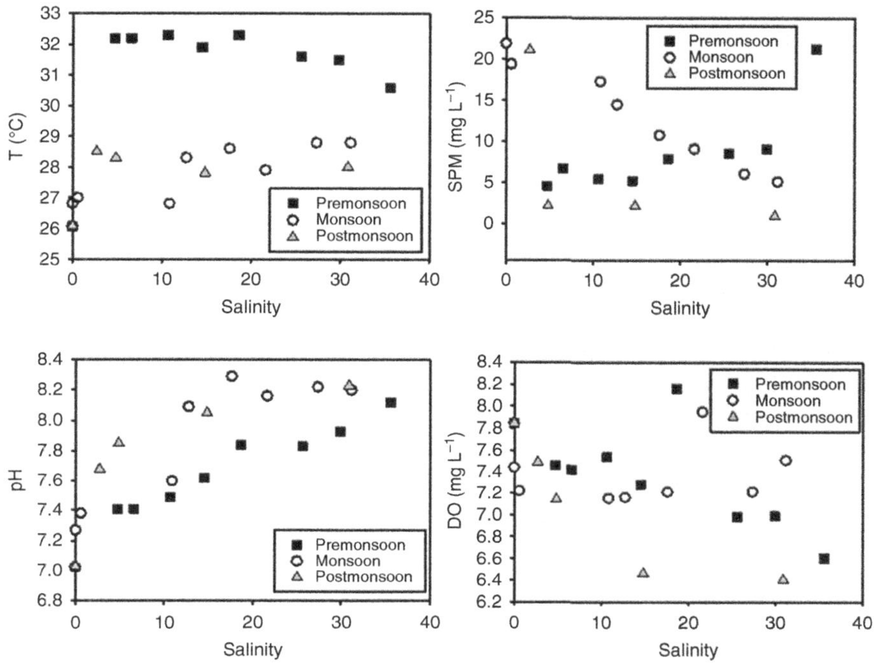

Fig. 7.2 Seasonal variations of T, pH, DO, and SPM along salinity gradient (in psu) in the Nethravati estuary

7.3 Results

7.3.1 General Physicochemical Characteristics of Water in the Nethravati Estuary

The measured physicochemical parameters of water and the SPM content in the Nethravati estuary are given in Table 7.1 and Fig. 7.2. The measured temperature of water in the Nethravati estuary for the study period was in the range of 26.8–32.3 °C (avg. 29.5 ± 2 °C). The temperature was higher during the premonsoon (avg. 31.4 ± 0.6 °C) and showed a decreasing trend along the salinity gradient. The temperature (avg. 27.9 ± 0.9 °C) was lesser during the monsoon and showed an increase with salinity. Lesser temperature (avg. 28.15 ± 0.31 °C) with negligible spatial variation was observed during the postmonsoon season. The pH of water was observed to be in the range of near neutral at the estuarine front to alkaline at the sea and increased linearly along the salinity gradient at all sampling periods. The estuarine water was well oxygenated and the concentration range of DO was 6.4–8.2 mg L^{-1} (avg. 7.2 ± 0.4 mg L^{-1}). The samples of mid-salinities (~20 psu) showed highly variable DO with a slightly lower value (6.5 mg L^{-1}) during post monsoon, while the maximum was observed during the monsoon and premonsoon sampling. The SPM concentration in the Nethravati river estuary was in the range of 4–21 mg L^{-1} (avg. 10 ± 7) for the sampling period which is similar to the previously reported SPM content (6.4–18 mg L^{-1}, Tripti et al. 2018) for Nethravati river during 2013 monsoon season. The SPM concentrations were higher during the monsoonal sampling compared to premonsoon and postmonsoon sampling (Fig. 7.2). The average SPM content measured during the study period (2010–2011) in the Nethravati estuary was lower than that of the Nethravati estuary during 2017 (19 mg L^{-1}; Fernandes et al. 2021), other tropical rivers (Hooghly = 20–3460 mg L^{-1} in eastern India, Mitra et al. 2020; Congo = 19 mg L^{-1}, Zambezi = 190 mg L^{-1} and Niger = 259 mg L^{-1} in Africa, Meybeck 2009; Amazon = 8–160 mg L^{-1}, Orinoco = 10–160 mg L^{-1} and Maroni = 7–36 mg L^{-1} in South America, Rousseau et al. 2019) and World River average (350 mg L^{-1}; Berner and Berner 1987). However, it was relatively higher than the SPM content of other smaller tropical rivers of the Indian west coast (nearby Swarna river = 9 ± 2.7 mg L^{-1}, Tripti et al. 2018 and several small rivers of Goa = 6–8 mg L^{-1}, Fernandes et al. 2017). The lower average SPM content of the Nethravati estuary reported in this study compared to earlier estimates is due to no sample collection during the peak monsoon season (June-September).

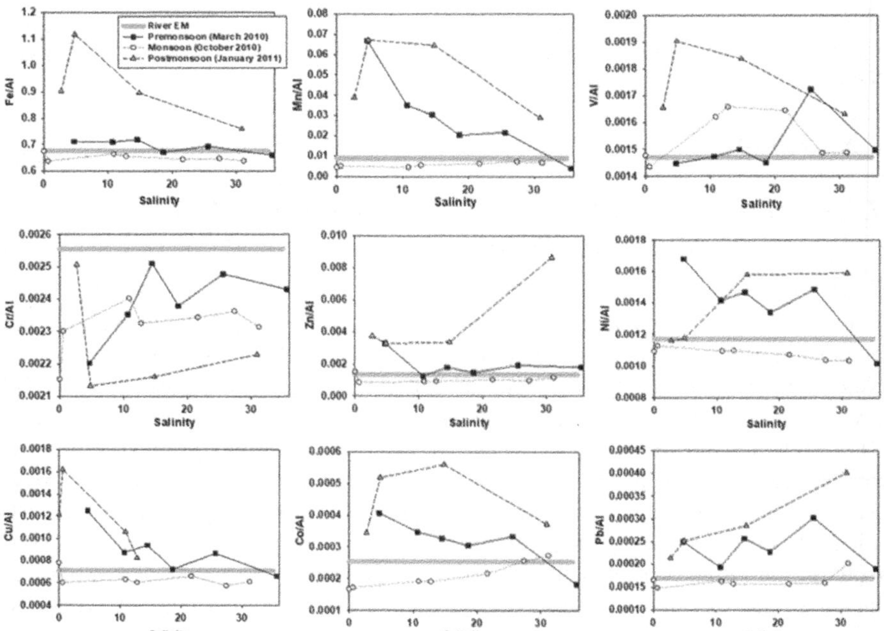

Fig. 7.3 Variation of metal/Al ratio in SPM along salinity gradient (in psu) in the Nethravati estuary. Note that the metal/Al ratio of river particulate matter (end member) is given in solid gray color line. The legend is common for all the figures

7.3.2 Particulate Elemental Concentrations in the Nethravati Estuary

The major element and trace metal concentrations measured in the particulate matter of the Nethravati estuary are given in Table 7.1 and Fig. 7.3. The concentrations of major elements like Na (avg. 3.9 ± 3.7 wt%), Ca (avg. 0.61 ± 0.56 wt%), and Mg (avg. 0.84 ± 0.59 wt%) in the estuarine particulates were higher compared to river particulate elemental concentrations. The mean concentrations of minor elements in the particulate matter were Fe (avg. 6.12 ± 1.9 wt %), Ti (avg. 0.72 ± 0.13 wt %), Al (avg. 8.43 ± 3.03 wt%), and Mn (avg. 0.22 ± 0.17 wt%). The obtained minor element concentrations were consistent with the previously reported values (Shankar and Manjunatha 1997), except for Mn which is higher in concentration by four times in this study. The concentrations of particulate heavy metals in the estuary were also found to be in good agreement with the earlier study and were as follows: V (avg. 135 ± 40 mg kg^{-1}), Cr (avg. 492 ± 928 mg kg^{-1}), Zn (avg. 208 ± 213 mg kg^{-1}), Ni (avg. 109 ± 35 mg kg^{-1}), Cu (avg. 192 ± 35 mg kg^{-1}), Co (avg. 65 ± 174 mg kg^{-1}), and Pb (avg. 110 ± 326 mg kg^{-1}).

7.3.3 Enrichment of Metals in SPM of the Nethravati Estuary

The enrichment factor for each metal was calculated using sample Al concentration as a tracer for terrigenous origin, following the equation by Eby (2004).

$$\mathrm{EF}(i) = \left(\frac{\left(\frac{x_{(i)}}{\mathrm{Al}} \right)_{\mathrm{sample}}}{\left(\frac{x_{(i)}}{\mathrm{Al}} \right)_{\mathrm{UCC}}} \right)$$

where EF is the enrichment factor and $x_{(i)}$ is the element under consideration. The EF value close to 1 attributes the source of metal "x" to geological strata, while EF > 1 indicates an anthropogenic source. However, it should be noted that EF > 1 does not necessarily mean anthropogenic influence (Tosiani et al. 2004). It could be due to the difference in the geological reference values which does not suit to the geological area under investigation (Viers et al. 2009). The enrichment factors for the SPM of the Nethravati estuary were obtained to understand the metal variability along the coastal stretch of Mangalore city (Fig. 7.4). The calculated enrichment factor for major elements (EF < 0.5) reflected depletion of Ca, Mg, Na (at lesser salinity) in the SPM relative to the upper continental crust (UCC) composition which suggested intense weathering in the catchment. The second group of elements (Mg, K, Fe, Mn, and Pb) had EF ranging from 0.5 to 1.5 in increasing order.

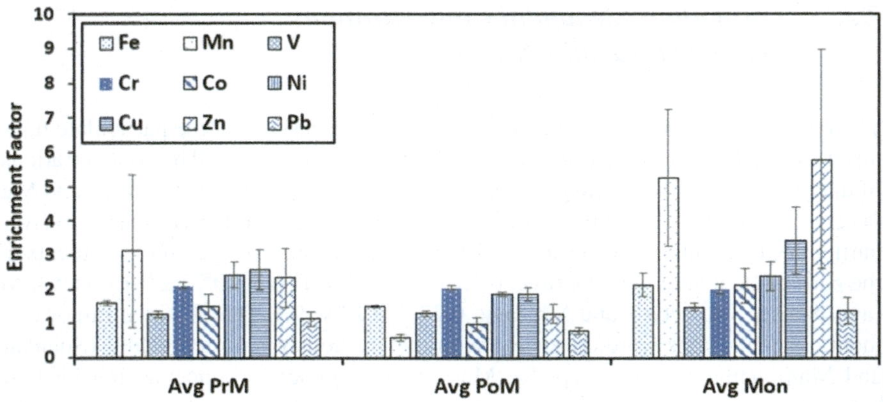

Fig. 7.4 Enrichment factor (EF) obtained for heavy metals measured in SPM of the Nethravati estuary during premonsoon (PrM), postmonsoon (PoM), and late monsoon (Mon) seasons. Avg represented average value and error bars reflected standard deviation in the obtained EF values

This group of elements was close to the UCC composition and suggested lesser elemental mobility during weathering as well as negligible anthropogenic contribution during their transport. The third group of metals which were enriched (EF > 1.5) relative to the upper continental crust included Fe, V, Cr, Co, Ni, Cu, and Zn. The heavy metals such as Cu, Ni, Zn, and Cr were showing EF > 2 suggesting an additional supply of these metals from other sources. In general, the particulate matter collected during the monsoonal months showed a high enrichment of heavy metals compared to non-monsoonal month particulate matter (Fig. 7.4). The three samples (246 and 251 belonging to premonsoon, and 345 belonging to monsoon) in the Nethravati estuary showed anomalous enrichment of heavy metals and could represent anthropogenic contamination. These samples have been further excluded while discussing geochemical processes.

7.4 Discussion

7.4.1 Seasonal Variability of Particulate Elemental Concentrations in the Tropical Estuary

The SPM concentration was observed to be decreasing toward the sea during the monsoon and postmonsoonal months, whereas it was increasing toward the sea during premonsoonal months (Fig. 7.2). The river discharge during the premonsoon was negligible, and the particulates in the estuary were produced either by the phytoplankton material or through resuspension of bed sediments. The higher concentrations of SPM during the premonsoon sampling toward the sea water end member could be due to wind-driven resuspension of the sediments forming an estuarine turbidity maximum (ETM) as observed in a recent study in the same estuary during 2017 (Fernandes et al. 2021) as well as other tropical estuaries in west coast of India (Kessarkar et al. 2009; Suja et al. 2017). The particulate matter in general is composed of four components like lithogenic material supplied from the rivers, biological material produced within the photic zones, authigenic precipitation of minerals through in situ geochemical processes, and contaminants originated from human activity. The geochemical study of dissolved and particulate elemental behavior in the Nethravati River suggested that the effect of secondary fluvial geochemical processes, redox reactions, and adsorption reactions had a significant effect on the particulate phase of the river water (Gurumurthy et al. 2014). Most of the lithogenic part can be traced in the water column using either unreactive Al or Ti as a tracer. In this study, the Al concentration has been used to trace the lithogenic components of the sedimentary phase. Any deviation from the end member ratio (metal/Al) is due to additional source contribution or metal cycling due to geochemical processes. Aluminum normalized ratios of metals (except Cr) in the Nethravati estuary were above the upper crustal values indicating their contribution from an additional source (Fig. 7.3). The phase redistribution of metals in the water column was due to geochemical processes or contaminants from the anthropogenic processes. The metal/Al ratio in the estuary decreases toward the sea at all seasons, except for Pb, Ni, Cr, and Zn during the postmonsoon. The enrichment of these metals at a salinity of 30.9 psu could be explained by the discharge of the Gurupur River, which received sewage effluents from the Mangalore City (Kavoor sewage treatment plant discharged domestic sewage). During the premonsoon, when the river runoff was low, the industries and the sewage treatment plant were not allowed to discharge effluents to the rivers.

On a spatial scale, during the monsoon, the metal/Al ratio showed lesser variability along the salinity gradient and the ratios were close to the riverine particulate average metal/Al ratio (Fig. 7.3). Most of the metals showed a lesser ratio compared to average riverine particulate metal/Al ratio, except for V which was having mid salinity increase in the estuary. During the post monsoon sampling, the estuarine particulate metal/Al ratios were observed to be higher than the average riverine particulate metal/Al ratio suggesting metal addition to the particulate phase in the estuary (Fig. 7.3). The metal/Al ratios of Fe, Mn, V, Cu, and Co were higher in the low salinity zones (lesser salinity) which then decreased toward the sea. The metal/Al ratios of other metals such as Pb, Zn, Ni, and Cr were increasing from freshwater end member to seawater. During the premonsoon sampling, the estuarine particulate metal/Al ratios were higher than the average riverine particulate metal/Al ratio suggesting metal addition to the particulate phase in the estuary (Fig. 7.3). The metal/Al profile along the salinity gradient showed a decrease in ratio with an increase in salinity and the ratio fell below the riverine particulate metal/Al ratio (except V, Zn and Pb) at maximum salinity zone (seawater).

7.4.2 *Geochemical Processes Controlling the Particulate Metal Abundance in Tropical Estuary*

7.4.2.1 Oxidation and Reduction of Metal Species

Iron and Mn form their respective insoluble metal oxyhydroxides under the oxidizing conditions of the aquatic system. The Fe-Mn oxyhydroxides are highly surface reactive in nature (Crerar et al. 1981). The dissolved oxygen concentration in the Nethravati estuary reflected a well-oxygenated condition, and as a result, it might have facilitated the formation of particulate Fe and Mn by forming insoluble Fe and Mn oxyhydroxides. In the Nethravati estuary, the measured dissolved Fe and Mn concentration, particularly at higher salinity, was below the instrument detection limit. The metal partitioning trends have been presented in Fig. 7.5. The partitioning of Fe and Mn between dissolved and particulate phases revealed that there was a phase redistribution of these metals along the salinity gradient in the Nethravati estuary. When there was a higher particulate fraction of metal, the dissolved fraction of metal was observed to be lesser and vice versa for the redox sensitive metals (Cr, Fe, Mn, and V) in the estuary (Fig. 7.5). The possible process which can explain the phase redistribution of these metals could be redox reaction at varying dissolved oxygen levels of the estuary. The effect of redox processes on the Fe and Mn concentration variability in the aquatic system is well understood and documented in the literature (Trefry and Presley 1982; Landing and Bruland 1987; Canfield 1989). Further, Fe and Mn oxyhydroxides are known to scavenge other surface reactive metals onto their surface and act as carriers of other metals in the aquatic system (Warren and Zimmerman 1994). However, the abundance and distribution of these metal hydroxides are dependent on the redox condition, salinity and residence time of metals in the estuarine environment. In the

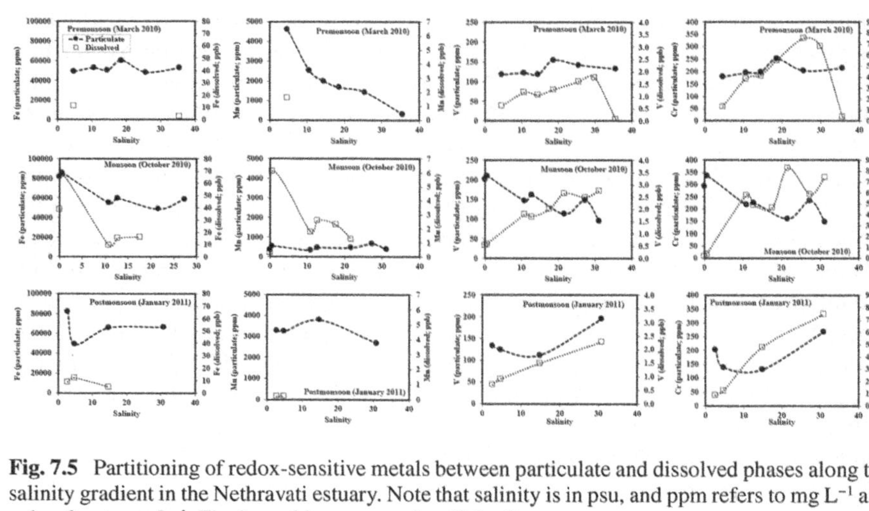

Fig. 7.5 Partitioning of redox-sensitive metals between particulate and dissolved phases along the salinity gradient in the Nethravati estuary. Note that salinity is in psu, and ppm refers to mg L^{-1} and ppb refers to μg L^{-1}. The legend is common for all the figures

Nethravati estuary, a significant part of Fe and Mn redistribution takes place in the low salinity region in all the sampling seasons. It is not clear whether the removal of Fe from the dissolved phase is by flocculation/coagulation or because of redox processes as few studies reported Fe flocculation in the low salinity regions of the estuary (Sholkovitz 1978). The abundance and distribution pattern of other redox-sensitive metals (Mn, V, and Cr) in the estuary supports the involvement of redox processes in the redistribution process. Moreover, the estimated distribution coefficients ($K_d = X_{SPM}/X_{dissolved}$, in L/kg, where X refers to metal concentration in SPM and dissolved phase; Turner 1996) for each metal in the Nethravati estuary also support higher phase redistribution along the salinity gradient and their average values for the study period exhibited the dominant order of $K_d(Fe) > K_d(Mn) > K_d(V) \approx K_d(Cr)$.

The oxidized species of Fe and Mn can be an important carrier of metals in the aquatic system. This necessitates the assessment of Fe and Mn abundance and their prevailing species in the estuary. In the Nethravati estuary, the Fe and Mn are being partitioned into particulate phase because of the prevailing oxidized conditions of the water column. The Fe/Mn ratio in the Nethravati estuary was plotted against the SPM concentration to evaluate their relative abundance in the estuarine particulates and their variability through time (Fig. 7.6a). The relationship between the particulate matter concentration and Fe/Mn ratio indicated that Fe and Mn oxyhydroxides make up a significant part of the particulate phase in the estuary. However, it should be noted that the Fe oxyhydroxide was dominant in the estuary during all the sampling seasons. This can be attributed to relatively higher and quicker oxidation of Fe than Mn in the estuary due to their thermodynamics (Froelich et al. 1979; Santos-Echeandia et al. 2009) and higher abundance of Fe in the estuarine water column. The Fe/Mn ratio in SPM of monsoon samples was about five times higher than that of premonsoon and postmonsoon samples having similar SPM content. This indicated that the abundance of Fe and Mn varied with season, and the organic and clay mineral fractions were abundant in the particulate phase during postmonsoon and premonsoon seasons. The plot of Fe/Mn against the salinity showed seasonality in the abundance of Fe and Mn oxyhydroxides (Fig. 7.6b). During the monsoon, the Fe/Mn ratio was observed to be

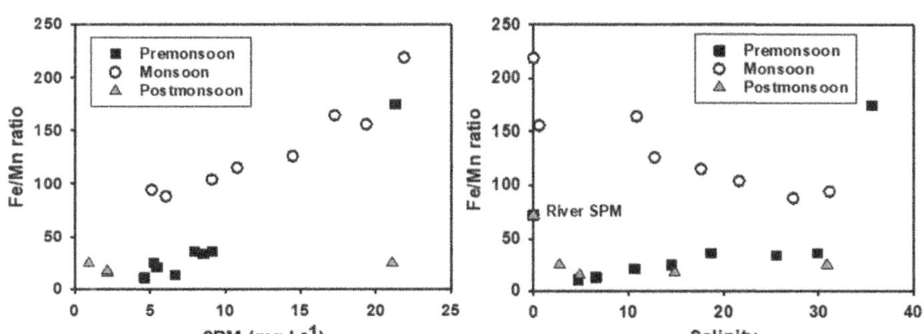

Fig. 7.6 Relationship of Fe/Mn ratio with (**a**) suspended particulate matter (SPM in mg L^{-1}) and (**b**) salinity (in psu) in the Nethravati estuary

decreasing with an increase in salinity whereas during postmonsoon and premonsoon, the Fe/Mn ratio was observed to be increasing with the salinity. The biological uptake of Fe might have been the possible reason for the observed depletion of Fe in the low salinity estuarine region during premonsoon and postmonsoon seasons. This was supported by the negligible river flow and the higher component of possible organic/clay mineral fraction in the SPM during premonsoon and postmonsoon. Further, the Fe/Mn ratio in the estuary varied significantly between the sampling periods.

7.4.2.2 Adsorption of Metals with Oxide Particles

The behavior and abundance of most metals in estuarine systems are dependent on the redox cycling of Fe and Mn, because of the scavenging capacity of oxyhydroxides (Sholkovitz and Copland 1982; Ingri 1985; Gunnars et al. 2002). The particulate heavy metals in the Nethravati estuary followed the trend of Fe and Mn (Fig. 7.7a–c). This suggested that the metal concentration in the estuary was

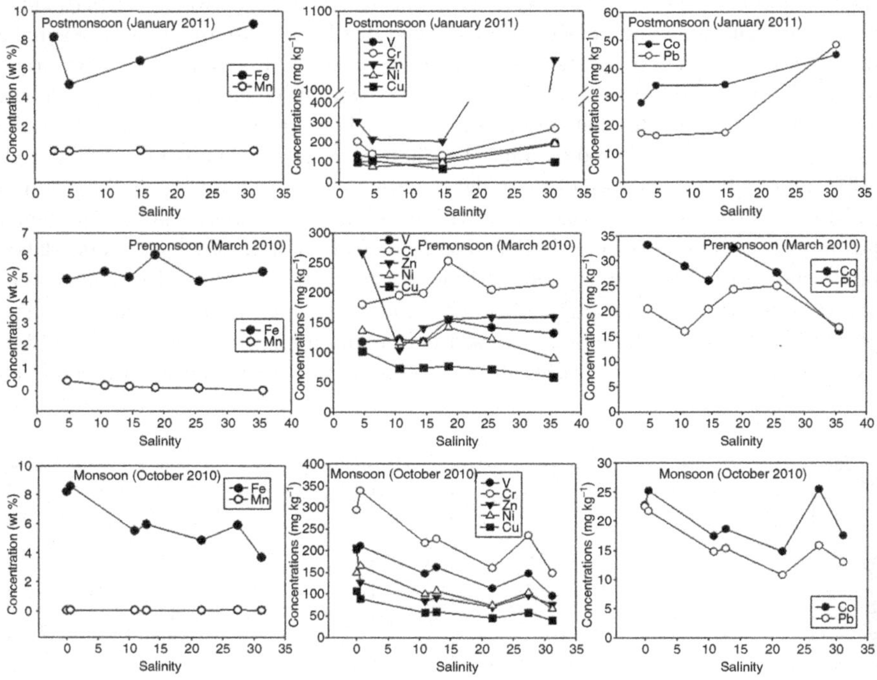

Fig. 7.7 (**a**) Variation in particulate minor elements and heavy metals along the salinity gradient (in psu) in the Nethravati estuary during January 2011 (postmonsoon sampling). (**b**) Variation of particulate minor elements and heavy metals along the salinity gradient (in psu) in the Nethravati estuary during March 2010 (premonsoon sampling). (**c**) Variation of particulate minor elements and heavy metals along the salinity gradient (in psu) in the Nethravati estuary during October 2010 (late monsoon sampling)

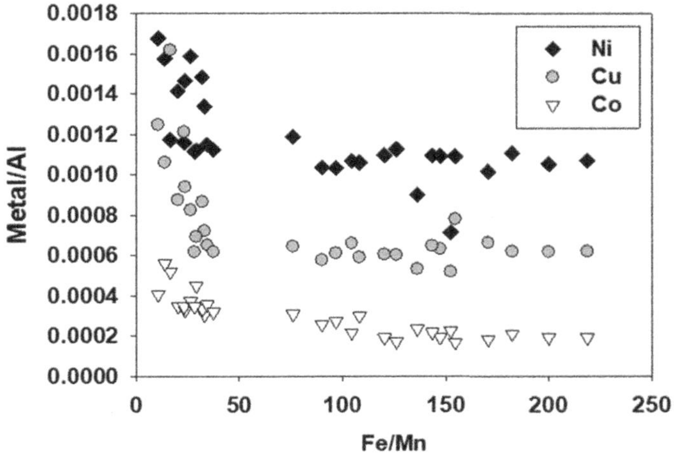

Fig. 7.8 Relationship between Al normalized metals (Ni, Co, and Cu) and the Fe/Mn ratio in SPM of the Nethravati estuary

controlled by the abundance and sorption process by oxyhydroxides. The concentration of metals in the estuary showed strong seasonal variation as well as spatial variation along the salinity gradient. During postmonsoon sampling, most of the metals assessed showed an increase in their concentration with an increase in salinity, i.e., higher concentrations of metals at higher salinity and vice versa (Fig. 7.7a). In the premonsoon sampling, the metal concentration showed mid salinity maxima and slightly lesser concentrations at marine end member salinity range (Fig. 7.7b). Similarly, during the monsoon sampling, the concentrations were lesser toward the marine end member salinity (Fig. 7.7c). In all the sampling seasons, the concentrations of heavy metals closely followed the trend similar to Fe and Mn in the estuary. A close look at the data indicated that the particulate metal concentrations generally mimicked the dissolved oxygen variation in the estuary, i.e., the higher the dissolved oxygen concentration, the higher the particulate Fe and Mn in the estuary. The plot of metal/Al ratios of surface reactive metals (Cu, Co, and Ni) against the Fe/Mn ratio in the estuary suggested the adsorption of these metals on to the oxyhydroxides of Fe and Mn in the estuary (Fig. 7.8). It was interesting to note that the higher concentrations of these metals were observed when the Fe/Mn ratio was lower in estuarine samples. This could be explained by the higher surface sites for the adsorption of metals given by Mn oxyhydroxides.

Table 7.2 Average metal concentrations in suspended particulate matter and bed sediment of tropical estuaries of India

Location	Fe	Mn	Co	Cu	Zn	Pb	Reference
	(wt%)	mg kg^{-1}					
Suspended Particulate Matter (SPM)							
Nethravati estuary	6.5 ± 1.5	0.2 ± 0.17%	75 ± 21	26.3 ± 8	206 ± 225	20 ± 8	This study
World River	5.8	0.17%	22.5	75.9	208	61.1	Viers et al. (2009)
Kali estuary	4.4	0.07 ± 1.6%	20.5	32	131	23	Suja et al. (2017)
Sal estuary	6.9–15.6	0.2–0.7%	24–106	57–108	156–540	25–125	Fernandes et al. (2019)
Zuari estuary	8.3	0.7%	33.1	101	281	61	Kessarkar et al. (2013)
Mandovi estuary	7.6	0.7%	36.9	101	245	30.6	Kessarkar et al. (2013)
Cauvery estuary	6.2	0.13%	100	60	500	40	Ramesh et al. (1990)[a]
Ganges estuary	9	0.34%	223	252	643	37	Ramesh et al. (1990)[a]
Hooghly estuary	6.2 ± 1.6	0.13 ± 0.2%		62 ± 13	178 ± 118	21 ± 10	Mitra et al. (2020)
Bed sediment							
Nethravati estuary	9.1	1.6%	35.2	79	103	26	This study
Sal estuary	0.2–7.38	184–1123	3.05–24	3–616	20–350	1.6–21	Fernandes et al. (2019)
Cochin estuary	0.266–7.4	45–921	0.4–30	1.3–147	3.4–4665	0.2–96	Salas et al. (2017)
Vellar estuary	3.2	756	37	109	97	82	Nethaji et al. (2017)
Mandovi estuary	2.2–49.7	1.6%	2.5–45	11–77	20–83	4.5–46	Alagarsamy (2006)
Narmada estuary	3.14	514	29	46	50	5	Biksham and Subramanian (1988)[b]
Tapti estuary	1.09	1300	36	126	118	5	Biksham and Subramanian (1988)[b]
Ganges estuary	3.1	553	36	26	71	29	Subramanian et al. (1988)[b]
Krishna estuary	4.23	1040	47	49	31	9	Biksham and Subramanian (1988)[b]
Godavari estuary	5.7	1070	47	82	54	11	Ramesh et al. (1990)[b]

(continued)

Table 7.2 (continued)

Location	Fe (wt%)	Mn mg kg^{-1}	Co	Cu	Zn	Pb	Reference
Cauvery estuary	1.76	319	64	12	26	10	Biksham and Subramanian (1988)[b]
National Oceanic and Atmospheric Administration (NOAA) Guidelines							
Effective Range-Low	–	–	–	34	150	46.7	Long et al. (1995)
Effective Range-Median	–	–	–	270	410	218	Long et al. (1995)

Note that Mn concentration is in mg kg^{-1} except mentioned in wt%. Average metal concentration in SPM of Nethravati estuary excludes the three samples as explained in text.
[a]Data taken from Alagarsamy and Zhang (2005)
[b]Data taken from Alagarsamy (2006)

7.4.3 *Heavy Metal Concentrations in Tropical Estuaries of India*

The heavy metal concentrations measured in SPM and bed sediments of the Nethravati estuary were compared with other Indian estuaries (Table 7.2). This provided insights into the heavy metal distribution status in the tropical estuaries and their actual flux to the northern Indian Ocean. The average values of metal concentration in SPM of the Nethravati estuary (Table 7.2) were obtained by excluding the three extreme samples (246, 251, and 345), as explained in Sect. 7.3.3. It was observed that the heavy metals reflected higher concentrations in the bed sediments of west coast estuaries (i.e., Mandovi, Tapi, Cochin, and Nethravati) than the east coast estuaries except the most recently reported Vellar estuary (Table 7.2). The heavy metal concentrations in the SPM of the Nethravati estuary were lower than that of its bed sediments. However, both suspended particulates and bed sediments were characterized by heavy metals which were within the concentration limits prescribed by the National Oceanic and Atmospheric Administration (NOAA) guidelines.

7.5 Conclusions

Seasonal samples of suspended sediments from the Nethravati estuary were assessed for the abundance and geochemical behavior of heavy metals under the tropical climate setting. The SPM content and its heavy metal concentrations showed strong seasonal and spatial variability in the Nethravati estuary. The geochemical variability of heavy metals in the estuarine region was driven by the redox cycling of metals (Fe, Mn, V, and Cr) coupled with the adsorption of heavy metals onto the metal oxyhydroxides. The Fe-Mn oxyhydroxides acted as a major carrier of heavy metals in the estuary. However, the relative abundance of Fe-Mn oxyhydroxides was dependent on the season and estuarine conditions. The Fe-oxyhydroxides were found to be abundant compared to Mn-oxyhydroxides in the Nethravati estuary. This study highlights the importance of hydrogenous particles in tropical estuaries for the transport of surface reactive metal contaminants. Although the heavy metal concentrations in estuarine bed sediments were higher than that in SPM, their concentrations were not enriched from the pollution point of view. The measured concentrations of heavy metals in the Nethravati estuary were within limits set by National Oceanic and Atmospheric Administration (NOAA) guidelines. Further, this study suggests that the dissolved metal assessment of water bodies for the pollution status needs to consider the phase transformation and adsorption of heavy metals which could obscure the actual contaminant signatures particularly in well-oxygenated tropical estuarine systems like the Nethravati estuary. Therefore, a detailed monitoring at seasonal and spatial scales will provide insights into the pollution status in the tropical estuaries.

Acknowledgments This work is funded by the Ministry of Environment and Forests (19/36/2006-RE), Government of India (PI: KB). GPG thanks the Director, BSIP and TM thanks the Director, NCESS, for the support and encouragement. TM acknowledges the Indian National Science Academy and the Department of Science and Technology, Government of India, for the INSPIRE Faculty fellowship. The authors thank the editors and two anonymous reviewers for their constructive comments and suggestions on the initially submitted version.

References

Alagarsamy R (2006) Distribution and seasonal variation of trace metals in surface sediments of the Mandovi estuary, west coast of India. Estuar Coast Shelf Sci 67:333–339

Alagarsamy R, Zhang J (2005) Comparative studies on trace metal geochemistry in Indian and Chinese rivers. Curr Sci India 89(2):299–309

Allen JRL, Rae JE (1986) Time sequence of metal pollution, Severn estuary, southwestern UK. Mar Pollut Bull 17(9):427–431

Balachandran KK, Laluraj CM, Nair M, Joseph T, Sheeba P, Venugopal P (2005) Heavy metal accumulation in a flow restricted tropical estuary. Estuar Coast Shelf Sci 65:361–370

Berner EK, Berner RA (1987) The global water cycle: geochemistry and environment. Prentice-Hall, Englewood Cliffs, New Jersey, ISBN: 0133571955/ 9780133571950

Biksham G, Subramanian V (1988) Elemental composition of Godavari sediments (central and southern Indian subcontinent). Chem Geol 70:275–286

Canfield DE (1989) Reactive iron in marine-sediments. Geochim Cosmochim Acta 53(3):619–632

Crerar DA, Means JL, Yuretich RF, Borcsik MP, Amster JL, Hastings DW, Knox GW, Lyon KE, Quiett RF (1981) Hydrogeochemistry of the New Jersey coastal plain, 2. Transport and deposition of iron, aluminum, dissolved organic matter, and selected trace elements in stream, ground, and estuary water. Chem Geol 33(1–4):23–44

DeMaster DJ, Kuehl SA, Nittrouer CA (1986) The effects of suspended sediments on geochemical processes near the mouth of the Amazon River: examination of biological silica uptake and the fate of particle-reactive elements. Cont Shelf Res 6:107–125

Eby GN (2004) Principles of environmental geochemistry, 1st edn. Brooks Cole, Kentucky, USA. ISBN: 0122290615/ 9780122290619

Fernandes LL, Purnachandra Rao V, Kessarkar PM, Suja S (2017) Estuarine turbidity maxima of six small rivers in Goa, central west coast of India. Hydrol Res 49(4):1234–1254

Fernandes LL, Kessarkar PM, Purnachandra Rao V, Suja S, Parthiban G, Kurian S (2019) Seasonal distribution of trace metals in suspended particulate and bottom sediments of four microtidal river estuaries, west coast of India. Hydrol Sci J 64(12):1519–1534

Fernandes LL, Kessarkar PM, Dhanayudapani I, Narvekar J, Suja S, Parab P (2021) Seasonal and tidal variations in suspended particulate matter dynamics of two microtidal rivers of Karnataka, central west coast of India. Arab J Geosci 14(8):1880

Forstner U, Ahlf W, Clamano W (1989) Studies on the transfer of heavy metals between sedimentary phases with a multi-chamber device: combined effects of salinity and redox variation. Mar Chem 28(1–3):145–158

Froelich PN, Klinkhmmer GP, Bender ML, Luedtke NA, Heath GR, Cullen D, Dauphin P, Hammond D, Hartman B, Maynard V (1979) Early oxidation of organic matter in pelagic sediments of the eastern equatorial Atlantic: suboxic diagenesis. Geochim Cosmochim Acta 43(7):1075–1090

Gunnars A, Blomqvist S, Johansoon P, Andersson C (2002) Formation of Fe (III) oxyhydroxide colloids in freshwater and brackish seawater, with incorporation of phosphate and calcium. Geochim Cosmochim Acta 66:745–758

Guo T, Delaune RD, Patrick WH Jr (1997) The influence of sediment redox chemistry on chemically active forms of arsenic, cadmium, chromium and zinc in estuarine sediment. Environ Int 23(3):305–316

Gurumurthy GP, Balakrishna K, Tripti M, Audry S, Riotte J, Braun JJ, Udaya Shankar HN (2014) Geochemical behaviour of dissolved trace elements in a monsoon-dominated tropical river basin, Southwestern India. Environ Sci Pollut Res 21(7):5098–5120

Ingri J (1985) Geochemistry of ferromanganese concretions and associated sediments in the Gulf of Bothnia. Ph.D. thesis. University of Luleå, Sweden

Kessarkar PM, Rao VP, Shynu R, Ahmad IM, Mehra P, Michael GS, Sundar D (2009) Wind-driven estuarine turbidity maxima in Mandovi estuary, Central west coast of India. J Earth Syst Sci 118(4):369–377

Kessarkar PM, Shynu R, Rao VP, Chong F, Narvekar T, Zhang J (2013) Geochemistry of the suspended sediment in the estuaries of the Mandovi and Zuari rivers, central west coast of India. Environ Monit Assess 185(5):4461–4480

Landing WM, Bruland KW (1987) The contrasting biogeochemistry of iron and manganese in the Pacific Ocean. Geochim Cosmochim Acta 51(1):29–43

Long ER, MacDonald DD, Smith SL, Calder FD (1995) Incidence of adverse biological effects within ranges of chemical concentrations in marine and estuarine sediments. Environ Manag 19(1):81–97

Manjunatha BR, Harry NA (1994) Geology of western coastal Karnataka. Geo Karnataka, Mysore Geological Department, Centenary volume, 109–116

Martino M, Turner A, Nimmo M, Millward GE (2002) Resuspension, reactivity and recycling of trace metals in the Mersey estuary, UK. Mar Chem 77(2–3):171–186

Meybeck M (2009) Africa. In: Likens GE (ed) Encyclopedia of inland waters. Academic Press, pp 295–305. ISBN 9780123706263

Mitra S, Sudarshan M, Jonathan MP, Sarkar SK, Thakur S (2020) Spatial and seasonal distribution of multielements in suspended particulate matter (SPM) in tidally dominated Hooghly river estuary and their ecotoxicological relevance. Environ Sci Pollut Res 27:12658–12672

Muller G (1969) Index of geoaccumulation in sediments of the Rhine river. GeoJournal 2:108–118

Nethaji S, Kalaivanan R, Viswam A, Jayaprakash M (2017) Geochemical assessment of heavy metals pollution in surface sediments of Vellar and Coleroon estuaries, southeast coast of India. Mar Pollut Bull 115(1–2):469–479

Rahaman W, Singh SK, Raghav S (2010) Dissolved Mo and U in rivers and estuaries of India. Implications to geochemistry of redox sensitive elements and their marine budgets. Chem Geol 278(3–4):160–172

Ramanathan AL, Vaithiyanathan P, Subramanian V, Das BK (1993) Geochemistry of the Cauvery estuary, East coast of India. Estuaries 16(3A):459–474

Ramesh R, Subramanian V, Van Grieken R (1990) Heavy metal distribution in sediments of Krishna river basin, India. Environ Geol 15:207–216

Reddy HR, Hariharan V (1986) Distribution of nutrients in the sediments of the Nethravati-Gurupur estuary, Mangalore. Indian J Fish 33:123–126
Reddy MPM, Hariharan V, Kurian NP (1979) Sediment movement and siltation in the navigational channel of old Mangalore Port. Proc Indian Acad Sci Earth Planet Sci 88(II):121–130
Ridgway IM, Price NB (1987) Geochemical association and post-depositional mobility of heavy metals in coastal sediments: Loch Etive, Scotland. Mar Chem 21(3):229–248
Rousseau TCC, Roddaz M, Moquet J-S, Delgado HH, Calves G, Bayon G (2019) Controls on the geochemistry of suspended sediments from large tropical South American rivers (Amazon, Orinoco and Maroni). Chem Geol 522:38–54
Salas PM, Sujatha CH, Ratheesh Kumar CS, Cheriyan E (2017) Heavy metal distribution and contamination status in the sedimentary environment of Cochin estuary. Mar Pollut Bull 119:191–203
Santos-Echeandia J, Prego R, Cobelo-Grarcía A, Millward GE (2009) Porewater geochemistry in a Galician Ria (NW Iberian Peninsula): implications for benthic fluxes of dissolved trace elements (Co, Cu, Ni, Pb, V, Zn). Mar Chem 117:77–81
Shankar R, Manjunatha BR (1997) Onshore transport of shelf sediments into the Nethravati-Gurupur estuary, west coast of India: geochemical evidence and implications. J Coast Res 13(2):331–340
Shaw TJ, Gieskes JM, Jahnke RA (1990) Early diagenesis in differing depositional environments: the response of transition metals in pore water. Geochim Cosmochim Acta 54(5):1233–1246
Sholkovitz ER (1978) The flocculation of dissolved Fe, Mn, Al, Cu, Ni, Co and Cd during estuarine mixing. Earth Planet Sci Lett 41(1):77–86
Sholkovitz ER, Copland D (1982) The chemistry of suspended matter in estuary water, a biologically productive lake with seasonally anoxic hypo-limnion. Geochim Cosmochim Acta 46:393–410
Subramanian V, Jha PK, Van Grieken R (1988) Heavy metals in the Ganges estuary. Mar Pollut Bull 19:290–293
Sudhanandh VS, Udayakumar P, Ouseph PP, Amaldev S, Narendra Babu K (2011) Dispersion and accumulation trend of heavy metals in coastal and estuarine sediments and its textural characteristics, a case study in India. J Hum Ecol 36(2):85–90
Suja S, Kessarkar PM, Fernandes LL, Kurian S, Tomer A (2017) Spatial and temporal distribution in suspended particulate matter of the Kali estuary, India. Estuar Coast Shelf Sci 196:10–21
Tosiani T, Loubet M, Viers J, Yanes C, Dupre B, Tapia J (2004) Major and trace elements in river borne materials from the Cuyuni Basin (southern Venezuela): evidence for organo-colloidal control on the dissolved load and elemental redistribution between the dissolved and suspended load. Chem Geol 211(3–4):305–334
Trefry JH, Presley BJ (1982) Manganese fluxes from Mississippi Delta sediments. Geochim Cosmochim Acta 46(10):1715–1726
Tripti M, Gurumurthy GP, Lambs L, Riotte J, Balakrishna K (2018) Water and organic carbon cycles in monsoon-driven humid tropics of the Western Ghats mountain belt, India: insights from stable isotope approach. J Geol Soc India 92:579–587
Turner A (1996) Trace-metal partitioning in estuaries: importance of salinity and particle concentration. Mar Chem 54(1–2):27–39
Turner A, Millward GE (2000) Particle dynamics and trace metal reactivity in estuarine plumes. Estuar Coast Shelf Sci 50(6):761–774
Turner A, Millward GE, Tyler AO (1994) The distribution and chemical composition of particles in a Macro-tidal estuary. Estuar Coast Shelf Sci 38(1):1–17
Viers J, Dupre B, Gailladet J (2009) Chemical composition of suspended sediments of World rivers: new insights from a new database. Sci Total Environ 407:853–868
Warren LA, Zimmerman AP (1994) Suspended particulate oxides and organic-matter interactions in trace-metal sorption reactions in a small urban river. Biogeochemistry 24(1):21–34
Windom H, Byrd J, Smith JR, Hungsperugs M, Dharmvanij S, Thumtrakul W, Yeats P (1991) Trace metal-nutrient relationships in estuaries. Mar Chem 32(2–4):177–194

Chapter 8
Geochemical Studies of Ilmenite from Bhimunipatnam to Konada Coastal Sands, East Coast of India, North Andhra Pradesh, India

K. Bangaku Naidu, M. Anji Reddy, K. S. N. Reddy, A. Lakshmi Venkatesh, and Ch. Ravi Sekhar

Abstract Titanium is an important industrial element and a key component of heavy mineral assemblages. Electron probe microanalysis (EPMA) was used to analyze the chemistry of ilmenites collected from Bhimunipatnam to Konada coastal sands, Andhra Pradesh, India's East Coast. TiO_2 (total Fe) from 28.79 to 55.33 wt.% (avg. 45.67 wt.%), FeO (total Fe) from 28.79 to 55.33 wt.% (avg. 45.67 wt.%), and Si, Mg, Mn, Ca, Al, V, Cr, Cu, and Zn are all present in trace amounts in detrital ilmenite. The composition of end member of ilmenites was calculated, revealing that they are made up of ilmenite—hematite derived from magma-sourced charnockites and ilmenite—geikielite, pyrophanite—hematite derived from the khondalite suite of the Eastern Ghats Granulite Rocks. They are characterized as ferrian ilmenites and have a Ti/(Ti + Fe) ratio of 0.35–0.64 (avg. 0.47), indicating that these ilmenite grains have undergone less modification. Ilmenite has an Mn/Mg ratio ranging from 0.04 to 113.23 (avg. 9.39), indicating that it is generated from metamorphic rocks. Backscattered electron (BSE) scans demonstrate their distinct angular to rounded shape, with some maintaining evidence of intergrowth, prior deformation, and rounded shape, all are influenced by the Gosthani and Champavathi rivers. In terms of environmental protection, the ilmenites (industrial specifications) in the current research area are better suitable for the Chlorate process, which extracts TiO_2 from ilmenite more efficiently than the Sulphate process.

K. Bangaku Naidu (✉)
Department of Geology, Sir. C. R. Reddy Degree College, Eluru, Andhra Pradesh, India

Center for Environment, Jawaharlal Nehru Technology University, Hyderabad, Telangana, India

M. Anji Reddy
Center for Environment, Jawaharlal Nehru Technology University, Hyderabad, Telangana, India

K. S. N. Reddy · A. Lakshmi Venkatesh · C. Ravi Sekhar
Department of Geology, Andhra University, Visakhapatnam, Andhra Pradesh, India

N. Jayaraju et al. (eds.), *Coasts, Estuaries and Lakes*,
https://doi.org/10.1007/978-3-031-21644-2_8

Keywords Ilmenite · Mineral chemistry · Provenance · Charnockite · Khondalite · EPMA

8.1 Introduction

Mineral beneficiation of deposits and metallurgical treatment, stratigraphic correlation, and provenance investigations all benefit from physical and chemical characterization of economic heavy minerals. Sediment composition is influenced by the initial source rock's mineralogical composition as well as numerous other variables like weathering, transportation, deposition, and diagenesis that change over time (Morton 1985; Morton and Halls Worth 1999). Because of their major, trace, and REE fingerprints, detrital heavy minerals such as rutile, garnet, ilmenite, zircon, and monazite are commonly employed for provenance characterization. The chemical composition varies depending on the paragenesis of the parent rocks (Hutton 1950; Buddington and Linsley 1964; Darby et al. 1985).

Many researchers attempted to study the geochemistry of ilmenite intergrowths, alteration, and other impurities will not only affect the overall grade of the ilmenite and Provides but also sensitive information on the alteration of ilmenite (Suresh Babu et al. 1994; Bhattacharyya et al. 1997; Ramakrishnan et al. 1997; Rao et al. 2002, 2005; Jagannadha Rao et al. 2005; Nair et al. 2006; Sundararajan et al. 2009b; Mohapatra et al. 2015). Leaching of elements (Fe) from the titanium mineral lattice by hydroxylation and environmental conditions of the deposit play role in alteration. Geochemistry of ilmenite by provenances in different rocks (Darby and Tsang 1987; Nayak and Mohapatra 1998) and coastal sands (Basu and Molinaroli 1989; Mitra et al. 1992; Ramadan et al. 2012; Rahman et al. 2014; Jayalakshmi et al. 2003; Hegde et al. 2006; Bhattacharyya et al. 2006; Dinesh et al. 2007; Bangaku Naidu et al. 2016; Ganapathi Rao et al. 2019) has been studied. To better understand chemical composition change and evaluate the range of physicochemical alterations with weathering, ilmenite from distinct beach placer deposits has been examined (Nair et al. 2009; Sundararajan et al. 2010; Nayak et al. 2012; Rao and Sengupta 2014; Acharya et al. 2015; Bangaku Naidu et al. 2017) Ilmenite's industrial applications (Gazquez et al. 2014; Bangaku Naidu et al. 2018).

This research uses EPMA to investigate the provenance and industrial applicability of ilmenite from Bhimunipatnam to the Konada coastal sands, East Coast of India, North Andhra Pradesh, India.

8.2 The Study Area's Geology and Location

The examined coastline length is between the Gosthani and Champavathi river mouths in the Visakhapatnam and Vizianagaram, Andhra Pradesh (17°52.005′ to 18°02.016′N latitude and 83°26.162′ to 83°36.545′ E longitude) and 20 km long, it has 65 N/12, O/5, and O/9 Top sheets that belong to the Survey of India (Fig. 8.1). These transient rivers make up the drainage system and have their origins in the Eastern Ghats Mountain ranges. These rivers discharge huge quantities of sand into the Bay of Bengal in Konada, Bhimunipatnam, and Chepala Uppada. Numerous geological and geomorphic characteristics were created in the study area by rivers, tiny creeks, and brisk seasonal winds. The dunes, which have a thickness of 12 m and run parallel to the coast, contain an area with a significant concentration of heavy minerals. The width of the coastal sand deposit is an average of 120 m running NE-SW over 1400 km of India's east coast is covered by the Eastern Ghats Granulite Belt. Major rock units in the Eastern Ghats include khondalites, charnockites (Divakara Rao 1984), anorthosites (Leelanandam 1990), and pyroxene granulites (Narayana et al. 1995).

8.3 Materials and Methods

8.3.1 Heavy Mineral Analysis

Forty-seven ilmenite grains were chosen from ten representative sediment samples and collected along ten perpendicular to the beach traverses with a 1 km interval, as shown on the map (Fig. 8.1). The samples were transported to the lab where they were dried and carefully combined, after coning and quartering, samples are reduced to 100 g. The samples were treated with dilute HCl to remove shell materials and H_2O_2 to remove the organic content after they have been properly washed and dried. The materials are subjected to textual analysis after they have been properly washed and dried. The sieved fractions, that is, coarse (+60 ASTM; 0.25 mm), medium (−60 to +120 ASTM; 0.25–0.125 mm), and fine (−120 to +230 ASTM; 0.125–0.062 mm) sands. These sieved samples were subjected to the mineral separation of lights and heaves by using bromoform.

8.3.2 Grain Picking

Ilmenites were identified in several samples using a binocular petrological microscope and selected for geochemical research based on their optical characteristics. From ten representative samples collected from the coastal sands of Bhimunipatnam and Konada, 47 ilmenite grains were chosen for chemical analysis.

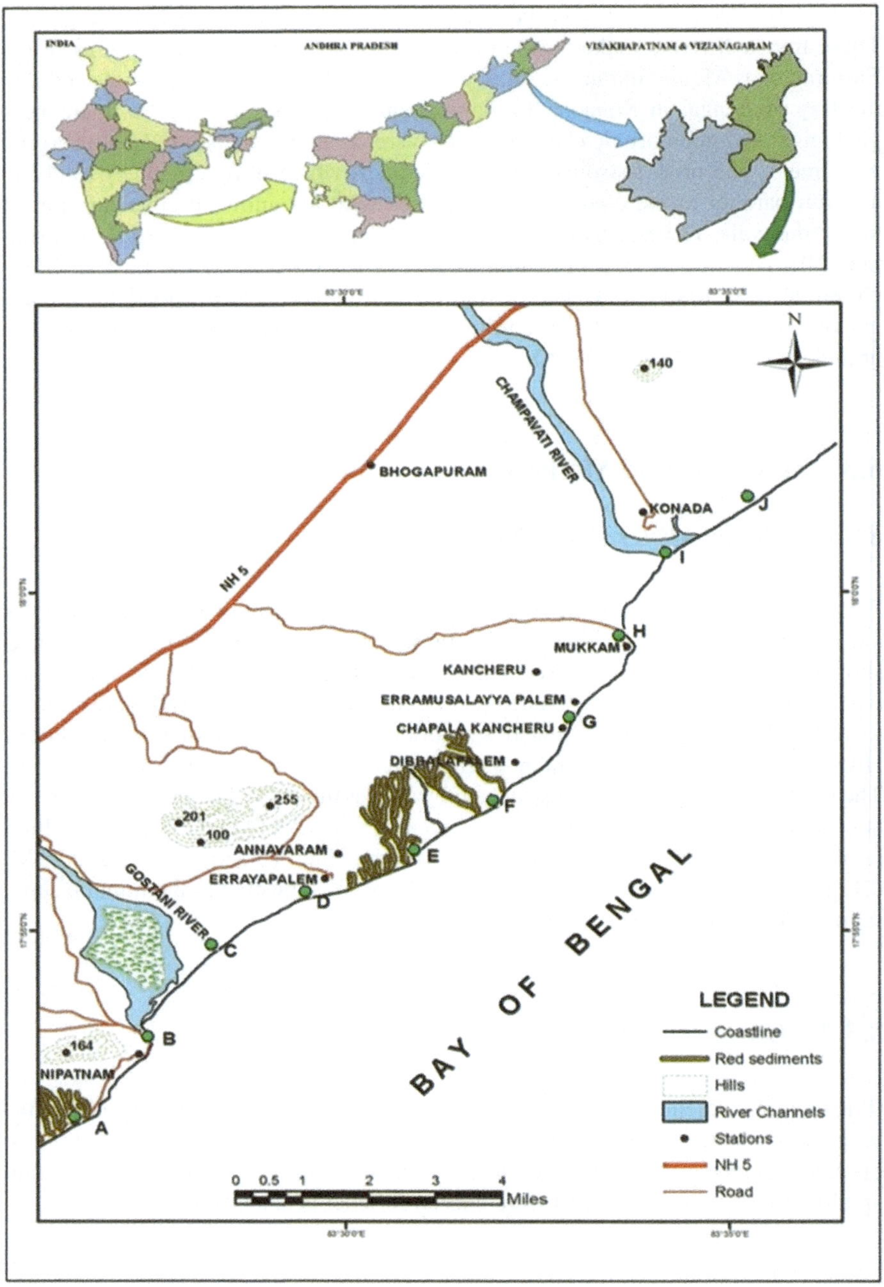

Fig. 8.1 Sample location map of the study area

8.3.3 Sample Preparation

Each sample's selected grains were mounted using epoxy resin on a glass slide of standard size for additional lapping and polishing. With the use of this technique, imaging can be done with both transmitted and reflected light microscopy. Grinding and sample removal are done to make sure the resin blocks' top and bottom are parallel. In order to create a smooth surface, lapping was done with fine abrasives, often silicon carbide. Samples were polished using a combination of very fine alumina slurry and fine SiC paper, with particle sizes ranging from 6 to 0.30 microns. The polishing and then removed from the polished sample by washing it in an ultrasonic cleaner with clean water. After drying in the air and being cleaned with a blow duster.

Before being tested with EPMA, polished ilmenite grains are carbon-coated to prevent change accumulation. Utilizing carbon evaporation in a vacuum, the coating is applied. The specimens' carbon coats are measured for thickness on a polished glass block.

8.3.4 Analytical Techniques

Chemical analysis of ilmenite is used by Electron Probe Micro Analyzer (CAMECA SX 100) in the Geological Survey of India, Hyderabad. Ilmenite grains were polished and sliced into 50 mm thin slices. Forty-seven ilmenite grains were exited by a beam current of 20 nA, the radius was kept at ~1 μm, and voltage of 15 kV. For calibration, the standard of natural minerals was used for most of the elements (Corundum-Al; Wollastonite-Ca; Almandine-Fe; Hematite-Fe; Diopside-Mg, Orthoclase for Si and K; Chromite-Cr; Albite-Na, Al; Ca; Rhodonite-Mn; TiO_2-Ti;). The other experimental settings are the same as those employed by Panday et al. in 2019. Table 8.1 provides ilmenite geochemical data and structure computed using two cations and three oxygens, as well as end-member compositions, from Bhimunipatnam to Konada coastal sands.

8.4 Results and Discussion

8.4.1 Distribution of Heavy Minerals

Total heavy minerals (THM) in beach sands range from 12.37 to 35.25 wt.% (avg. 21.41 wt.%), and their mineral composition is as follows, in decreasing order of weight percent abundance: ilmenite is 7.10, sillimanite is 6.09, zircon is 5.52, kyanite is 5.22, magnetite is 0.61, rutile is 4.33, leucoxene is 0.18, garnet is 0.25, monazite is 0.25, and other heavy minerals are 4.16.

Table 8.1 Ilmenite composition from Bhimunipatm–Konada coastal sands

Sam-ple. No	Grain No	SiO_2	TiO_2	Al_2O_3	FeO	MnO	MgO	CaO	Na_2O	K_2O	Cr_2O_3	P_2O_5	Si	Ti	Al	Fe	Mn	Mg	Ca	Na	K	Cr	P	Total	*Geikie-lite*	*Pyro-pha-nite*	*Esko-liate*	*Ilmen-ite*	*Hema-tite*	*Total*	*Mn/Mg*	*Fe + Ti*	*Ti/Fe + Ti*
A	1	0.01	49.83	0.02	48.38	0.25	0.03	0.04	0.01	0.01	0.03	0.00	0.00	0.96	0.00	1.03	0.01	0.00	0.00	0.00	0.00	0.00	0.00	2.00	*0.11*	*0.53*	*0.07*	*95.01*	*4.27*	*100*	*10.94*	*67.47*	*0.44*
	2	0.05	48.03	0.08	50.11	1.04	0.06	0.02	0.03	0.02	0.11	0.00	0.00	0.91	0.00	1.06	0.02	0.00	0.00	0.00	0.00	0.00	0.00	2.00	*0.23*	*2.23*	*0.21*	*88.45*	*8.89*	*100*	*22.30*	*67.74*	*0.42*
	3	0.03	47.14	0.14	48.89	0.19	0.06	0.02	0.05	0.00	0.02	0.01	0.00	0.92	0.00	1.06	0.00	0.00	0.00	0.00	0.00	0.00	0.00	2.00	*0.22*	*0.42*	*0.03*	*91.55*	*7.78*	*100*	*4.38*	*66.26*	*0.43*
	4	0.00	49.67	0.28	48.18	0.47	0.04	0.04	0.00	0.00	0.04	0.02	0.00	0.95	0.01	1.03	0.01	0.00	0.00	0.00	0.00	0.00	0.00	2.00	*0.14*	*1.01*	*0.07*	*93.94*	*4.83*	*100*	*15.85*	*67.22*	*0.44*
	5	0.03	47.51	0.00	50.37	0.29	1.07	0.00	0.03	0.01	0.14	0.03	0.00	0.89	0.00	1.05	0.01	0.04	0.00	0.00	0.00	0.00	0.00	2.00	*3.98*	*0.61*	*0.28*	*84.58*	*10.55*	*100*	*0.35*	*67.63*	*0.42*
B	6	0.01	50.02	0.15	47.47	0.25	0.04	0.04	0.04	0.01	0.04	0.03	0.00	0.97	0.00	1.02	0.01	0.00	0.00	0.00	0.00	0.00	0.00	2.00	*0.16*	*0.54*	*0.08*	*95.78*	*3.44*	*100*	*7.74*	*66.87*	*0.45*
	7	0.01	50.15	0.03	46.86	0.77	0.02	0.02	0.00	0.01	0.07	0.00	0.00	0.97	0.00	1.01	0.02	0.00	0.00	0.00	0.00	0.00	0.00	2.00	*0.08*	*1.67*	*0.14*	*95.19*	*2.92*	*100*	*46.78*	*66.48*	*0.45*
	8	0.02	49.51	0.01	47.35	0.59	0.04	0.01	0.05	0.03	0.09	0.03	0.00	0.96	0.00	1.02	0.01	0.00	0.00	0.00	0.00	0.00	0.00	2.00	*0.14*	*1.28*	*0.18*	*94.38*	*4.02*	*100*	*20.94*	*66.48*	*0.45*
	9	0.07	47.30	0.25	51.06	0.25	0.02	0.10	0.00	0.01	0.15	0.01	0.00	0.90	0.01	1.08	0.01	0.00	0.00	0.00	0.00	0.00	0.00	2.00	*0.08*	*0.53*	*0.30*	*89.04*	*10.05*	*100*	*14.36*	*68.03*	*0.42*
	10	0.10	49.40	0.32	46.24	0.52	0.14	0.08	0.02	0.03	0.21	0.04	0.00	0.96	0.01	1.00	0.01	0.01	0.00	0.00	0.00	0.00	0.00	2.00	*0.54*	*1.15*	*0.43*	*94.07*	*3.81*	*100*	*4.76*	*65.54*	*0.45*
C	11	0.07	50.00	0.01	45.95	0.76	0.06	0.13	0.03	0.00	0.05	0.00	0.00	0.98	0.00	1.00	0.02	0.00	0.00	0.00	0.00	0.00	0.00	2.00	*0.24*	*1.66*	*0.10*	*95.58*	*2.43*	*100*	*15.92*	*65.68*	*0.46*
	12	0.03	49.33	0.10	47.37	0.12	0.99	0.07	0.07	0.05	0.08	0.02	0.00	0.94	0.00	1.01	0.00	0.04	0.00	0.00	0.00	0.00	0.00	2.00	*3.73*	*0.26*	*0.15*	*90.11*	*5.74*	*100*	*0.16*	*66.38*	*0.45*
	13	0.02	46.92	0.02	51.68	0.65	0.03	0.00	0.15	0.01	0.04	0.04	0.00	0.89	0.00	1.09	0.01	0.00	0.00	0.01	0.00	0.00	0.00	2.00	*0.12*	*1.39*	*0.08*	*87.22*	*11.18*	*100*	*25.45*	*68.29*	*0.41*
	14	0.03	47.26	0.02	51.33	0.53	0.45	0.02	0.03	0.01	0.01	0.02	0.00	0.89	0.00	1.08	0.01	0.02	0.00	0.00	0.00	0.00	0.00	2.00	*1.68*	*1.13*	*0.03*	*86.34*	*10.82*	*100*	*1.52*	*68.22*	*0.42*
D	15	0.00	51.81	0.00	45.84	0.28	1.45	0.01	0.02	0.00	0.05	0.01	0.00	0.98	0.00	0.96	0.01	0.05	0.00	0.00	0.00	0.00	0.00	2.00	*5.42*	*0.58*	*0.11*	*91.57*	*2.33*	*100*	*0.24*	*66.68*	*0.47*
	16	0.02	48.96	0.02	48.83	1.06	0.01	0.04	0.05	0.03	0.01	0.03	0.00	0.94	0.00	1.04	0.02	0.00	0.00	0.00	0.00	0.00	0.00	2.00	*0.05*	*2.27*	*0.02*	*91.16*	*6.49*	*100*	*113.23*	*67.30*	*0.44*
	17	0.01	46.50	0.00	50.62	0.37	0.55	0.00	0.00	0.01	0.00	0.03	0.00	0.89	0.00	1.08	0.01	0.02	0.00	0.00	0.00	0.00	0.00	2.00	*2.09*	*0.79*	*0.00*	*86.28*	*10.83*	*100*	*0.86*	*67.21*	*0.41*
	18	0.03	49.36	0.02	48.36	0.02	0.01	0.00	0.02	0.01	0.04	0.01	0.00	0.96	0.00	1.04	0.00	0.00	0.00	0.00	0.00	0.00	0.00	2.00	*0.02*	*0.04*	*0.08*	*95.39*	*4.47*	*100*	*5.14*	*67.17*	*0.44*
E	19	0.12	48.61	0.10	46.14	0.62	1.03	0.02	0.03	0.00	0.00	0.07	0.00	0.94	0.00	1.00	0.01	0.04	0.00	0.00	0.00	0.00	0.00	2.00	*3.97*	*1.34*	*0.00*	*89.05*	*5.64*	*100*	*0.77*	*64.99*	*0.45*
	20	0.05	44.99	0.07	51.29	0.04	1.47	0.02	0.01	0.01	0.04	0.02	0.00	0.86	0.00	1.08	0.00	0.06	0.00	0.00	0.00	0.00	0.00	2.00	*5.53*	*0.09*	*0.08*	*79.82*	*14.48*	*100*	*0.04*	*66.83*	*0.40*
	21	0.02	51.03	0.03	46.91	0.62	1.30	0.01	0.02	0.01	0.04	0.02	0.00	0.96	0.00	0.98	0.01	0.05	0.00	0.00	0.00	0.00	0.00	2.00	*4.83*	*1.32*	*0.09*	*89.47*	*4.29*	*100*	*0.62*	*67.04*	*0.46*
	22	0.04	47.73	0.02	49.63	0.71	0.68	0.01	0.03	0.02	0.01	0.01	0.00	0.91	0.00	1.05	0.02	0.03	0.00	0.00	0.00	0.00	0.00	2.00	*2.56*	*1.51*	*0.02*	*86.60*	*9.31*	*100*	*1.34*	*67.18*	*0.43*
	23	0.03	49.55	0.01	48.46	0.26	0.71	0.00	0.00	0.02	0.07	0.03	0.00	0.94	0.00	1.02	0.01	0.03	0.00	0.00	0.00	0.00	0.00	2.00	*2.65*	*0.56*	*0.14*	*90.79*	*5.86*	*100*	*0.48*	*67.36*	*0.44*

Sample. No	Grain No	SiO_2	TiO_2	Al_2O_3	FeO	MnO	MgO	CaO	Na_2O	K_2O	Cr_2O_3	P_2O_5	Si	Ti	Al	Fe	Mn	Mg	Ca	Na	K	Cr	P	Total	*Geikielite*	*Pyrophanite*	*Eskoliate*	*Ilmenite*	*Hematite*	*Total*	*Mn/Mg*	*Fe + Ti*	*Ti/Fe + Ti*
F	24	0.19	58.37	0.18	39.35	0.34	1.54	0.03	0.03	0.00	0.09	0.00	0.00	1.10	0.01	0.82	0.01	0.06	0.00	0.00	0.00	0.00	0.00	2.00	*5.74*	*0.72*	*0.17*	*93.36*	*0.00*	*100*	*0.28*	*65.56*	*0.53*
	25	0.10	57.58	0.20	40.99	0.36	0.56	0.03	0.03	0.03	0.07	0.04	0.00	1.09	0.01	0.87	0.01	0.02	0.00	0.00	0.00	0.00	0.00	2.00	*2.10*	*0.77*	*0.15*	*96.98*	*0.00*	*100*	*0.83*	*66.37*	*0.52*
	26	0.03	57.70	0.01	39.98	0.09	1.26	0.01	0.05	0.01	0.09	0.01	0.00	1.10	0.00	0.85	0.00	0.05	0.00	0.00	0.00	0.00	0.00	2.00	*4.77*	*0.19*	*0.18*	*94.87*	*0.00*	*100*	*0.09*	*65.65*	*0.53*
	27	0.05	57.27	0.02	39.22	0.11	1.44	0.01	0.02	0.01	0.08	0.01	0.00	1.10	0.00	0.84	0.00	0.05	0.00	0.00	0.00	0.00	0.00	2.00	*5.48*	*0.24*	*0.16*	*94.11*	*0.00*	*100*	*0.10*	*64.81*	*0.53*
	28	1.65	57.37	0.10	39.27	0.04	0.10	0.00	0.00	0.01	0.15	0.57	0.04	1.11	0.00	0.84	0.00	0.00	0.00	0.00	0.00	0.00	0.01	2.00	*0.37*	*0.08*	*0.31*	*99.24*	*0.00*	*100*	*0.51*	*64.91*	*0.53*
G	29	0.01	50.50	0.00	48.12	0.26	0.05	0.03	0.02	0.01	0.07	0.00	0.00	0.97	0.00	1.02	0.01	0.00	0.00	0.00	0.00	0.00	0.00	2.00	*0.19*	*0.56*	*0.13*	*95.69*	*3.43*	*100*	*6.52*	*67.67*	*0.45*
	30	0.01	50.58	0.03	48.16	0.19	0.09	0.02	0.01	0.00	0.02	0.05	0.00	0.97	0.00	1.02	0.00	0.00	0.00	0.00	0.00	0.00	0.00	2.00	*0.34*	*0.41*	*0.04*	*95.88*	*3.33*	*100*	*2.73*	*67.75*	*0.45*
	31	0.08	56.66	0.21	36.95	0.35	0.50	0.03	0.02	0.01	0.08	0.01	0.00	1.14	0.01	0.82	0.01	0.02	0.00	0.00	0.00	0.00	0.00	2.00	*1.97*	*0.79*	*0.17*	*97.07*	*0.00*	*100*	*0.90*	*62.68*	*0.54*
	32	0.90	51.32	0.66	47.40	0.11	0.23	0.19	0.05	0.01	0.20	0.25	0.02	0.96	0.02	0.98	0.00	0.01	0.00	0.00	0.00	0.00	0.01	2.00	*0.83*	*0.23*	*0.39*	*94.10*	*4.44*	*100*	*0.64*	*67.60*	*0.46*
H	33	0.13	38.32	0.20	55.33	0.53	0.90	0.01	0.03	0.01	0.01	0.03	0.00	0.75	0.01	1.20	0.01	0.03	0.00	0.00	0.00	0.00	0.00	2.00	*3.46*	*1.15*	*0.02*	*69.92*	*25.45*	*100*	*0.75*	*65.97*	*0.35*
	34	0.43	67.63	0.53	28.79	0.25	0.97	0.01	0.01	0.02	0.10	0.14	0.01	1.31	0.02	0.62	0.01	0.04	0.00	0.00	0.00	0.00	0.00	2.00	*3.71*	*0.55*	*0.19*	*95.55*	*0.00*	*100*	*0.33*	*62.91*	*0.64*
	35	0.10	60.05	0.20	36.29	0.11	2.70	0.01	0.01	0.01	0.08	0.05	0.00	1.13	0.01	0.76	0.00	0.10	0.00	0.00	0.00	0.00	0.00	2.00	*10.06*	*0.24*	*0.16*	*89.54*	*0.00*	*100*	*0.05*	*64.20*	*0.56*
	36	0.38	66.96	0.41	30.58	0.28	0.76	0.06	0.08	0.02	0.11	0.07	0.01	1.28	0.01	0.65	0.01	0.03	0.00	0.00	0.00	0.00	0.00	2.00	*2.88*	*0.61*	*0.22*	*96.30*	*0.00*	*100*	*0.48*	*63.90*	*0.63*
I	37	0.85	56.32	0.87	39.03	0.01	0.01	0.01	0.04	0.02	0.03	0.03	0.02	1.10	0.03	0.85	0.00	0.00	0.00	0.00	0.00	0.00	0.00	2.00	*0.02*	*0.02*	*0.06*	*99.89*	*0.00*	*100*	*2.35*	*64.09*	*0.53*
	38	0.04	57.79	0.06	37.75	0.61	0.52	0.02	0.04	0.02	0.04	0.04	0.00	1.14	0.00	0.82	0.01	0.02	0.00	0.00	0.00	0.00	0.00	2.00	*2.03*	*1.36*	*0.09*	*96.53*	*0.00*	*100*	*1.51*	*63.98*	*0.54*
	39	0.04	49.84	0.01	49.04	0.52	0.11	0.01	0.01	0.03	0.02	0.03	0.00	0.95	0.00	1.04	0.01	0.00	0.00	0.00	0.00	0.00	0.00	2.00	*0.41*	*1.11*	*0.03*	*93.13*	*5.32*	*100*	*6.15*	*67.99*	*0.44*
	40	0.00	53.27	0.01	44.52	0.55	1.53	0.02	0.04	0.02	0.03	0.02	0.00	1.00	0.00	0.93	0.01	0.06	0.00	0.00	0.00	0.00	0.00	2.00	*5.70*	*1.15*	*0.06*	*93.02*	*0.06*	*100*	*0.46*	*66.53*	*0.48*
	41	0.03	50.88	0.02	48.14	0.28	0.63	0.01	0.04	0.01	0.02	0.03	0.00	0.96	0.00	1.01	0.01	0.02	0.00	0.00	0.00	0.00	0.00	2.00	*2.36*	*0.59*	*0.04*	*92.87*	*4.14*	*100*	*0.57*	*67.92*	*0.45*

(continued)

Table 8.1 (continued)

Sample. No	Grain No	SiO_2	TiO_2	Al_2O_3	FeO	MnO	MgO	CaO	Na_2O	K_2O	Cr_2O_3	P_2O_5	Si	Ti	Al	Fe	Mn	Mg	Ca	Na	K	Cr	P	Total	*Geikielite*	*Pyrophanite*	*Eskoliate*	*Ilmenite*	*Hematite*	*Total*	*Mn/Mg*	*Fe + Ti*	*Ti/Fe + Ti*
J	42	0.00	47.26	0.03	50.06	0.66	0.75	0.03	0.03	0.01	0.01	0.01	0.00	0.90	0.00	1.06	0.01	0.03	0.00	0.00	0.00	0.00	0.00	2.00	*2.81*	*1.40*	*0.01*	*85.52*	*10.25*	*100*	*1.12*	*67.23*	*0.42*
	43	0.01	51.16	0.00	47.06	0.20	0.55	0.01	0.02	0.02	0.03	0.01	0.00	0.98	0.00	1.00	0.00	0.02	0.00	0.00	0.00	0.00	0.00	2.00	*2.06*	*0.43*	*0.05*	*94.98*	*2.48*	*100*	*0.47*	*67.25*	*0.46*
	44	0.06	49.10	0.04	48.02	0.62	0.28	0.01	0.01	0.01	0.00	0.00	0.00	0.94	0.00	1.03	0.01	0.01	0.00	0.00	0.00	0.00	0.00	2.00	*1.07*	*1.33*	*0.00*	*92.08*	*5.51*	*100*	*2.82*	*66.76*	*0.44*
	45	0.04	49.36	0.29	49.35	0.03	0.01	0.03	0.03	0.00	0.58	0.02	0.00	0.94	0.01	1.04	0.00	0.00	0.00	0.00	0.00	0.01	0.00	2.00	*0.05*	*0.06*	*1.16*	*92.30*	*6.44*	*100*	*2.57*	*67.94*	*0.44*
	46	0.01	49.69	0.01	47.29	0.19	0.83	0.01	0.02	0.03	0.00	0.02	0.00	0.95	0.00	1.01	0.00	0.03	0.00	0.00	0.00	0.00	0.00	2.00	*3.16*	*0.40*	*0.00*	*96.43*	*0.00*	*100*	*0.29*	*66.54*	*0.45*
	47	0.08	51.69	0.01	45.71	0.11	0.92	0.00	0.00	0.01	0.05	0.02	0.00	0.99	0.00	0.97	0.00	0.03	0.00	0.00	0.00	0.00	0.00	2.00	*3.50*	*0.23*	*0.11*	*94.96*	*1.21*	*100*	*0.15*	*66.51*	*0.47*
Min.		*0.00*	*38.32*	*0.00*	*28.79*	*0.01*	*0.01*	*0.00*	*0.00*	*0.00*	*0.00*	*0.01*	*0.00*	*0.75*	*0.00*	*0.62*	*0.00*	*0.00*	*0.00*	*0.00*	*0.00*	*0.00*	*0.00*	*2.00*	*0.02*	*0.02*	*0.00*	*69.92*	*0.00*	*100*	*0.04*	*62.68*	*0.35*
Max.		*1.65*	*67.63*	*0.87*	*55.33*	*1.06*	*2.70*	*0.19*	*0.15*	*0.05*	*0.58*	*0.57*	*0.04*	*1.31*	*0.03*	*1.20*	*0.02*	*0.10*	*0.00*	*0.01*	*0.00*	*0.01*	*0.01*	*2.00*	*10.06*	*2.27*	*1.16*	*99.89*	*25.45*	*100*	*113.23*	*68.29*	*0.64*
Avg.		*0.16*	*51.49*	*0.14*	*45.67*	*0.38*	*0.62*	*0.03*	*0.03*	*0.01*	*0.08*	*0.05*	*0.00*	*0.99*	*0.00*	*0.97*	*0.01*	*0.02*	*0.00*	*0.00*	*0.00*	*0.00*	*0.00*	*2.00*	*2.32*	*0.81*	*0.16*	*91.74*	*5.14*	*100*	*9.39*	*66.40*	*0.47*

8.4.2 Ilmenite

Ilmenite is the most abundant ore mineral of titanium. This mineral has a solid solution with pyrophanite ($MnTiO_3$), geikielite ($MgTiO_3$), and hematite (Fe_2O_3). Ilmenite mineral has a large number of intergrowths of Ti–Fe oxides exsolutions, oxidation or hydrothermal process. Ilmenites contain low quantities of Ca., Zn, Cr, Al, Si, and Cu.

Ilmenite geochemistry is most important for the selection of processing methodology either chlorate route or sulfate route to produce titanium dioxide (TiO_2). The mineral end-member compositions of ilmenite are essential for ore dressing and mineral beneficiation processes. Ilmenite's economic value can be determined by how much of it has weathered or been altered. Ilmenite's economic worth is determined by the degree of weathering or change. The mineralogy and geochemical characteristics of ilmenite are very useful for understanding its provenance and industrial applicability.

The main sources of titanium dioxide pigment are the minerals ilmenite, leucoxene, and rutile which contain titanium oxide. Titanium dioxide (TiO_2) pigment is made from titanium slag and synthetic rutile, which are both produced by ilmenite. The three minerals that can carry the most titanium are ilmenite, leucoxene, and rutile, with ilmenite acting as the main source of titanium. Ilmenite is immediately heated in an electro-furnace to melt it into ferrotitanium alloys. Cutting tools can be made of titanium carbide, a durable and silent material. Ferro-Carbon-Titanium is an alloy used to create high-speed tools.

Ilmenites are distinguished from other rock types using the rhombohedra quaternary system diagram (Haggerty 1976; Nayak and Mohapatra 1998). The end-member compositions of Fe-Ti oxides from coastal sands were shown in a rhombohedral quaternary diagram ($FeTiO_3$–$MnTiO_3$–$MgTiO_3$–Fe_2O_3) in Fig. 8.2. Field 3 contains all of the samples, showing a basic rock suite, that is, khondalite and charnockite rocks.

Under varying temperature circumstances, the oxides in the TiO_2-FeO-Fe_2O_3 triangle. The tie line in Fig. 8.3 depicts the overall chemical results of the examined ilmenites. These figures were derived quantitatively from a typical examination of ilmenite grains from the research area's beach and dune sediments.

Ilmenite from the Bhimunipatnam–Konada coastal sands has an average TiO_2 value of 51.49 wt.%, whereas FeO (28.79–55.33 wt.%) has an average of 45.67 wt.% (Table 8.1). Cation leaching could explain the higher TiO_2 content, while exsolved hematite phases in ilmenite could explain the lower TiO_2 content (Jagannadha Rao et al. 2005). TiO_2 concentrations in ilmenites were used by early researchers to determine the type of source rocks. Igneous sources had TiO_2 less than 50 wt.%, whereas metamorphic sources had TiO_2 greater than 50 wt.% (Darby and Tsang 1987; Basu and Molinaroli 1989). There is more iron and less TiO_2 in ilmenite in feldspathic gneisses, whereas it has less iron and more TiO_2 in khondalites–charnockites (Divakara Rao and Murthy 1998). Higher-grade metamorphic rocks have an average TiO_2 of 51.49 wt.% in ilmenite grains.

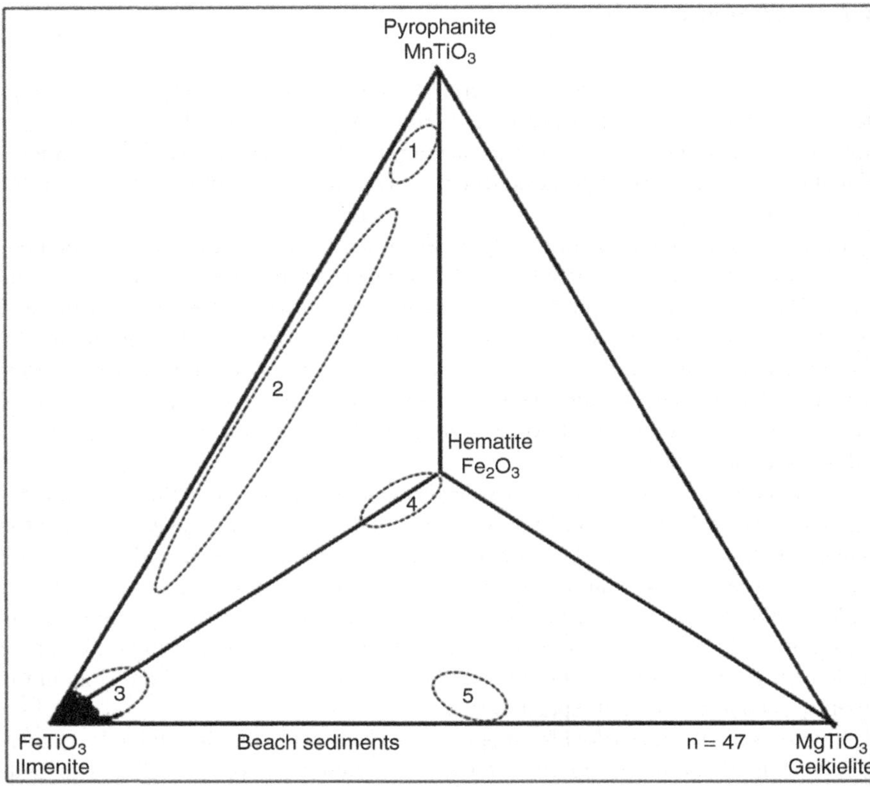

Fig. 8.2 Nayak and Mohapatra modified a quaternary diagram with proportions of pyrophanite, ilmenite, geikielite, and hematite as poles for ilmenite discrimination (1998). Mn-rich parametamorphites are found in field 1, intrusive acid suites, pegmatites, carbonatites, and extrusive acid suites are found in field 2, basic suites, such as amphibolites, granite gneisses, and basic igneous rocks are found in field 3, intrusive acid suites and anorthosite suites are found in field 4, and kimberlites are found in field 5

The FeO content of ilmenite from the research area's coastal sediments ranged from 28.79 to 55.33 wt.% (avg. 45.67 wt.%). The difference of FeO content in ilmenite could be linked to the degree of alteration, or leucoxenation, which results in SiO_2, Al_2O_3, Cr_2O_3, TiO_2, MgO, K_2O enrichment, and FeO and MnO loss. Hematite exsolved phases inside the ilmenite could account for the considerable range in FeO (Jagannadha Rao et al. 2005).

The MnO content of the ilmenite from 0.01 to 1.06 wt.% (avg. 0.38%) and MgO content ranges from 0.01 to 2.70 wt.% (avg. 0.62 wt.%) from the coastal sands of Bhimunipatnam–Konada. Ilmenite grains contain trace amounts of SiO_2 (0.01 wt.%) and Al_2O_3 (0.01 wt.%).

The elemental geochemistry of ilmenite is very useful for geochemical characterization. Elemental ratios such as Mn/Mg are widely used as a provenance

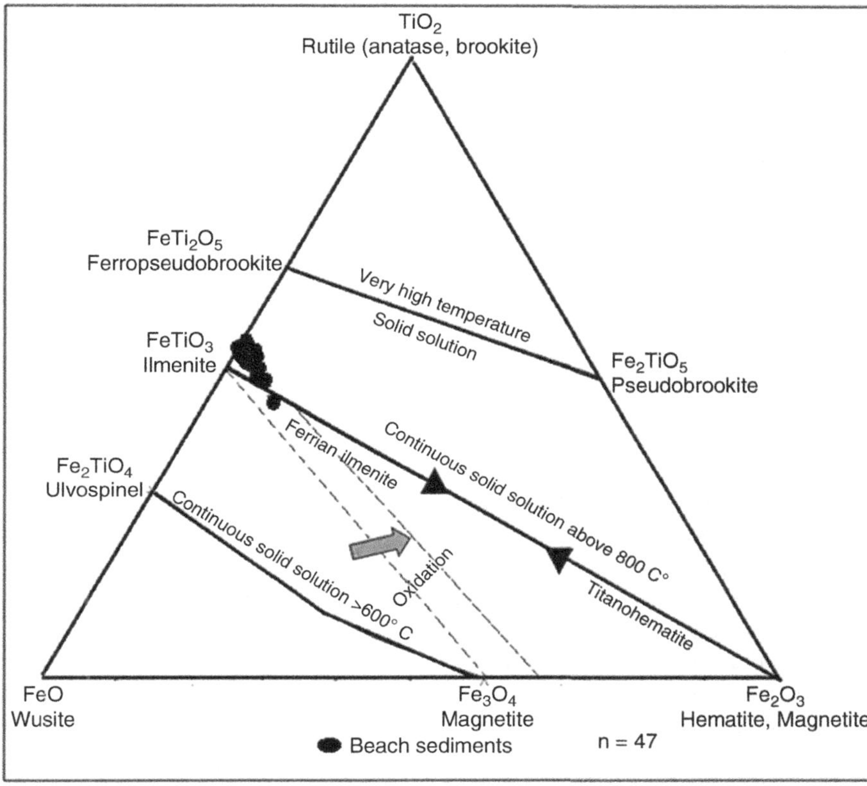

Fig. 8.3 TiO_2-FeO-Fe_2O_3 solid system diagram showing the composition and approximate equilibrium, tie lines (dashed lines) of analyzed ilmenite. TiO_2-FeO-Fe_2O_3 solid system diagram showing the composition and approximate equilibrium, tie lines (dashed lines) of analyzed ilmenite from the study area. (Modified after Buddington and Lindsley (1964), Broska et al. (2003))

indicator (Ganapathi Rao et al. 2019). The Mn/Mg ratios of ilmenite studied from Bhimunipatnam to Konada of coastal sands of this study vary from 0.05 to 113.23 (avg. 9.39) (Table 8.1). The present study area contains different rock types such as khondalites, calc-silicate rocks, quartzites, charnockites, basic granulites and granites of Archean to Precambrian age. These formations are highly migmatized and were termed as Eastern Migmatized zone (Ramam and Murthy 1997). The Mn/Mg ratios found in this study show that the majority of ilmenites come from metamorphic rocks (pyroxene granulites and khondalites), with a small percentage coming from basic charnockites and migmatites.

Ilmenite's weathering mechanisms have been described in terms of several elemental ratios. Based on the Ti/(Ti + Fe) ratios (Forst et al. 1983) divided the four stages of ilmenite alteration. These stages are described using the following terminology in a sequence of increasing stages of modification. The Ti/(Ti + Fe) ratio for four phases of alteration: leucoxene (>0.7), pseudorutile (0.60–0.70), hydrated ilmenite (0.50–0.60), and ferrian ilmenite (>0.5). Based on the chemistry of the key elements, this system has been adopted and used for the chemical characterization of ilmenite in this work.

Ti/(Ti + Fe) ratio of the coastal sands of Bhimunipatnam–Konada ranges from 0.35 to 0.64 (avg. 0.47). Seventy-five percent of ilmenite grains have Ti/ (Ti + Fe) ratio is <0.50 which indicates that they are ferrian ilmenite and the remaining 25% of samples range from 0.50 to 0.60 which indicates that they are hydrated ilmenite. The low Ti/(Ti + Fe) values and fresh ilmenite grains imply that these grains have undergone less alteration and are more recent in input to the research region, whereas hydrated ilmenites are more weathered than ferrian ilmenites.

The higher concentration range of TiO_2 content and MgO content <1.0% and Mn/Mg ratio ≥ 1 indicates that ilmenites are derived mainly from charnockites of calc-alkali magma. These ilmenites are formed from the khondalite suite of rocks with reduced TiO_2 content, MgO content >1.0%, and Mn/Mg 0.5%. In comparison to Chavara, Manavalakurichi, and Gopalpur dune ilmenites, the Ti/(Ti + Fe) values in the study region are low.

The scatter plot between Ti and (Ti/Ti + Fe) refers to the variation between different titanium mineral groups (Rahman et al. 2016) (Fig. 8.4). The studied ilmenite grains fall under the primary ilmenite field with less alteration in the narrow range of TiO_2, MgO content <0.50% and Mn/Mg ratio < 1 most of samples indicates that the ilmenites are derived from metamorphic rocks.

The alteration of ilmenite in Kerala deposits (Sundararajan et al. 2009a, b), Tamil Nadu deposits (Suresh Babu et al. 1994), Orissa deposits (Mohapatra et al. 2015), and multistage South Africa (Hugo and Cornell 1991). The alteration of ilmenites of

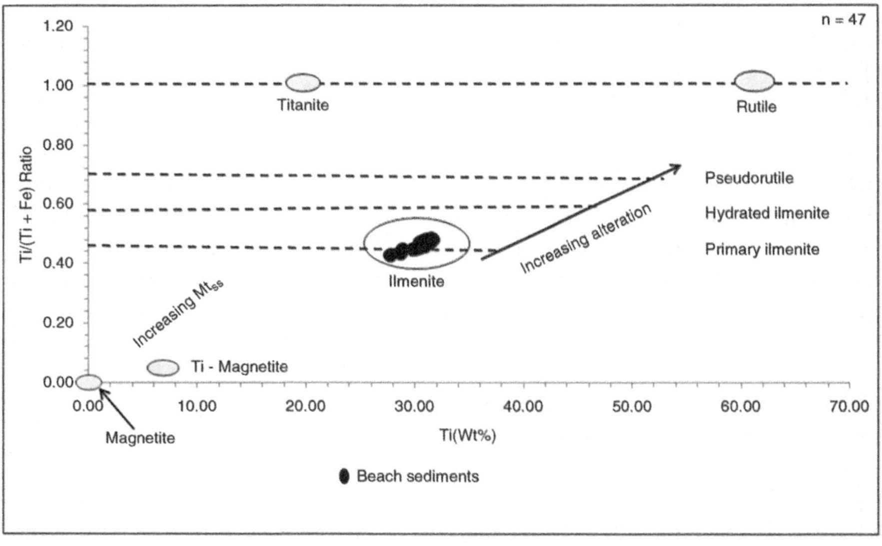

Fig. 8.4 Scatter plot of Ti (wt %) versus Ti/ (Ti + Fe) for Ti-rich components (Ilmenite) of the coastal sands

the study area slightly changes to that of Orissa placer deposits. The investigated area is under a subtropical environment. There is high ferrous iron and less TiO_2 ilmenite when compared to the west coast Manavalakurichi and Chavara deposits. It suggests that the present placer deposits are younger in age and have undergone the least weathering, because ilmenite contains considerable substitutions of Mn, Mg, V, and Cr (Deer et al. 1975).

Hematite is found in hematite laths, bands, and streaks in ilmenite, and they show polymodal distribution, fractures, and uneven patches within ilmenite. Ilmenite primary alteration pseudorutile with dark gray (1) and rimmed with secondary alteration product anatase (white) (2). Parallel bands of isolated ilmenite and/or pseudorutile surrounded by anatase (Fig. 8.5).

Partially altered ilmenite grains have a dark gray core or bands with anatase showing intense yellow internal reflection within and surrounding grains, according

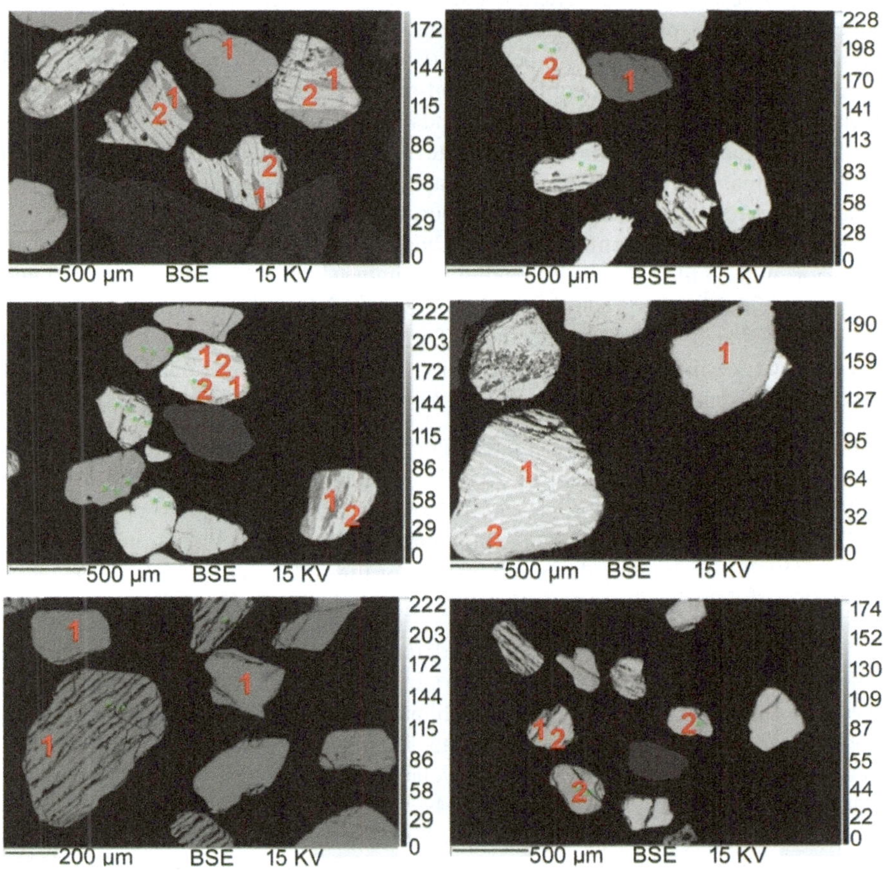

Fig. 8.5 Photomicrographs (reflected light) of polished sections of changed ilmenite grains displaying different phases of alteration

to photomicrographs of polished sections obtained with reflected light (Fig. 8.5). During the intermediate stage of alteration anatase, the dark gray phase is pseudorutile, which appears in bands inside parallel sheets of original ilmenite/pseudorutile (Fig. 8.5). Isolated ilmenite/pseudorutile residues surrounded by anatase are common (Fig. 8.5). In the last stages of transformation, anatase completely replaces ilmenite (Fig. 8.5).

The difference in ilmenite content could be attributed to the origin and level of alteration of the parent rocks. Ilmenite and exsolved hematite can occur at lower temperatures and are miscible with Fe_2O_3 and $FeTiO_3$ in metamorphic rocks of the Eastern Ghats of Rocks (7–13 kbar, 900–1100 °C). Exsolved hematite has intergrown with some ilmenite grains and vice versa. Ilmenite and hematite lamellae have bimodal thickness distributions, which may indicate that they crystallized and exsolved at distinct periods from the $FeTiO_3$ to Fe_2O_3 solid solution (Ramdohr 1969; Ahmed et al. 1992; Acharya et al. 1999).

There have been few geochemical studies of ilmenite placer deposits in India, especially coastal placer deposits in Bhimunipatnam–Konada (the current study region). Ilmenite (69–99 wt.%), hematite (0–25 wt.%), geikielite (0–10 wt.%), and pyrophanite (0–1 wt.%), according to end-member compositions of Fe–Ti oxides from Bhimunipatnam–Konada coastal sands (Table 8.1).

The chemical composition of ilmenite is shown in Table 8.2. Ilmenite from Bhimunipatnam–Konada (BK) in Andhra Pradesh has a superior chemical purity when compared to ilmenite from other sources.

In terms of TiO_2, ilmenites from the United States, Australia, and Malaysia are superior, with Quilon (Q grade) having the highest TiO_2 concentration (60%) and Manavalakurichi (MK grade) having the lowest (55%). When compared to that of the United States, Australia, and India (Q, MK, and TN), the TiO_2 concentration (avg. 51 wt.%) in the current study region, Bhimunipatnam–Konada (BK), is of lower quality, but it is superior to the OR Grade.

Industrial specifications TiO_2 greater than 50%, SiO_2 less than 1.5%, Al_2O_3 less than 1.0%, MnO_2.1.0%, Cr_2O_3 0.1%, P_2O_5 0.05% (sulfate route), P_2O_5 0.1%, CaO 0.2%, MgO 1.0% (chlorate route), U + Th1000 ppm, preferred 500 ppm, ore size 100 microns >50% and 40 microns >90% are the industrial specifications for ilmenite. Table 8.2 lists the ilmenite requirements for the sulfate and chlorate procedures for TiO_2 extraction.

The sulfate method is no longer considered since it generates gaseous hydrogen sulfide and sulfurdioxide during initial heating, causing air pollution. As a result, in terms of environmental protection, the chlorate method is preferable to the sulfate approach for extracting TiO_2 from ilmenite. Ilmenite is appropriate for chlorate process in the current study region, based on chemical parameters, and U + Th concentration must be investigated, as it has not been determined in the current study.

Titanium, the 'lightweight champion' of metals, has a variety of uses. Titanium's high strength and corrosion resistance make it a significant and strategic material. Owing to its small weight, it has been utilized in the production of surgical equipment as well as in the chemical, electrical, aerospace, and aviation industries. Owing

Table 8.2 Ilmenite quality from various deposits around the world

	United States	Australia					India					Present study area
Oxides (wt %)	Florida	Australia1	Australia2	Norway-Rock	Malaysia	South Africa	Q	MK	OR	TN	AP	BK (AP)
TiO_2	63.5	54–56	61	43.8	52	49.5	60	55	50.2	51–54	48–52	*51.49*
Fe_2O_3	27–31	18–24	32.5	12.7	6.5	25	25.5	18.9	12.8	11.0–16.0	7.0–17.4	*45.67*
FeO	4.7	16–22	3.6	34.3	34.3	22.5	9.7	20.9	34.1	28–33	30.9–37.0	
Al_2O_3	0.73–1.38	0.5–1.0	1.2	1.2	1.3	0.7	1.1	0.8	0.6	0.5–0.55	0.30–0.77	*0.13*
SiO_2	0.06–0.79	0.4–0.7	0.85	1.9	0.7	0.6	0.9	0.9	0.8	0.44–0.90	0.40–0.45	*0.16*
ZrO_2	0.04–0.19	0.05–0.15	0.25	0.01	0.13	NA	0.4	0.06	0.01	NA	NA	……
MnO	0.86–1.63	1.4–1.7	1.06	0.33	26	1.2	0.4	0.4	0.55	0.3–0.37	0.38–1.14	*0.38*
Cr_2O_3	0.1–0.12	0.03–0.04	0.11	0.02	0.02	0.2	0.13	0.08	0.05	0.04–0.06	0.04–0.07	*0.08*
V_2O_5	0.13–0.16	0.12–0.22	0.18	0.2	0.04	0.3	0.15	0.22	0.24	0.21–0.25	0.20–0.22	……
MgO	0.17–0.31	0.01–0.10	0.23	3.2	0.13	0.6	0.6	1	0.6	0.6–0.67	0.48–0.82	*0.62*
CaO	0.04–0.15	0.01–0.10	0.02	0.5	0.13	0.1	0.2	0.2	0.2	0.02–0.03	0.04–0.13	*0.03*
P_2O_5	0.12–0.18	0.02–0.04	0.14	0.04	0.1	NA	0.2	0.12	0.03	0.02	0.02–0.57	*0.05*
U + Th (ppm)	69–162	40–140	160	<2	NA	<20	150	225	50–60	39–78	NA	*NA*

Q Chavara, *MK* Manavalakurichi, *OR* Orissa, *TN* Tamil Nadu, *AP* Andhra Pradesh, *BK* Bhimunipatnam–Konada to its outstanding qualities of low specific gravity, considerable hiding power, high refractory index, non-toxicity, opacity, and titanium dioxide (TiO_2) is a white pigment used in welding rod coating, ceramics, chemicals, textiles, cosmetics, paints, plastics, paper, rubber, fabric, and pharmaceuticals.

8.5 Conclusions

1. Ilmenite–hematite is from charnockite and ilmenite–geikielite–pyrophanite–hematite is from khondalite suite of rocks.
2. The higher concentration range of TiO_2 content and MgO content <1.0% and Mn/Mg ratio are ≥1 indicating that some ilmenite grains are derived mainly from charnockites of calc-alkali magma.
3. The lower concentration range of TiO_2 content, MgO content >1.0%, and Mn/Mg <0.5% suggest that the khondalite suite of rocks is also the main contribution of ilmenites in the study area.
4. The Ti/(Ti + Fe) ratio ranges from 0.35 to 0.64 (avg. 0.47), indicating that these ilmenite grains have undergone less alteration and have a fresh look. The subangular to subrounded appearance indicates that the ilmenites contributed to the red sediments recently and over a short distance.
5. The TiO_2 content of ilmenite from Bhimunipatnam–Konada (BK) is lower than that of the United States of America, Australia, and Q, MK, TN (India), but higher than that of Chhatrapur, Orissa, India (avg. 51.49 wt.%) (avg. 50.2%).
6. In this investigation, the chlorate process extracts TiO_2 from ilmenite better than the sulphate method in terms of environmental protection.

Acknowledgments We are grateful to the authorities of the Centre for Environment, Jawaharlal Nehru Technological University Hyderabad, and the Head, Department of Geology, Andhra University Visakhapatnam. The first author is also thankful to the authorities of Sir C. R. Reddy College, Eluru, for their constant encouragement and EPMA in the Geological Survey of India, Hyderabad, to carry out the geochemical analysis. The first author thanks the University Grant Commission (UGC) for financial assistance to Dr. D. S. Kothari Post Doctoral Fellow is acknowledged.

References

Acharya BC, Das SK, Muralidhar J (1999) Mineralogy, mineral chemistry and magnetic behaviour of ilmenite from Chhatrapur Coast, Orissa. Indian J Earth Sci 26:45–51

Acharya BC, Nayak B, Das SK (2015) Mineralogy and mineral chemistry of placer deposit around Jhatiapodar, Odisha. J Geol Soc India 86:137–147

Ahmed S, Pal T, Mitra S (1992) Ilmenites from Cox's Bazar beach sands, Bangladesh: their intergrowths. J Geol Soc India 40:29–41

Bangaku Naidu K, Reddy KSN, RaviSekhar C, GanapatiRao P, Murali Krishna KN (2016) REE geochemistry of monazites from coastal sands between Bhimunipatnam and Konada, East Coast of India, Andhra Pradesh. Curr Sci 110(8):1550–1559

Bangaku Naidu K, Reddy KSN, RaviSekhar C, GanapatiRao P, Murali Krishna KN (2017) Heavy mineral studies of recent coastal sands between Gosthani and Champavathi river confluences, East Coast of India, Andhra Pradesh. J Indian Geol Congr 9(2):45–56

Bangaku Naidu K, Reddy KSN, Anji Reddy M (2018) Industrial suitability of recent coastal sands of ilmenite, rutile, garnet and monazites of Gosthani and Champavathi River confluences, Andhra Pradesh, East Coast of India. Int J Sci Res Sci Eng Tech 4(1):167–178

Basu A, Molinaroli E (1989) Provenance characteristics of detrital opaque Fe–Ti oxide minerals. J Sediment Petrol 59:922–934

Bhattacharyya S, Sengupta R, Chakraborty M (1997) Elemental chemistry of ilmenite – an indicator of provenance. J Geol Soc India 50(6):787–789

Bhattacharyya P, Bhattacharyya KK, Kumar V, Nayak B, Mohandas PN, Vidyadhar A, Swamy N, Chattoraj US, Maulik SC (2006) Studies on medium grade ilmenite-depleted beach placer deposit of Chavakkad-Ponnani Coast, Kerala. In: Proceedings of the international seminar on mineral processing technology held in Chennai, India, pp 544–552

Broska I, Uher P, Ondrejka M (2003) Geochemical and mineralogical characterization of the Fe-Ti oxide paragenesis in the magmatic and hydrothermal systems. Slovak Academy of Sciences. Web page: geol.sav.sk

Buddington AF, Linsley DH (1964) Iron-titanium oxides minerals and synthetic equivalents. J Pet 5:310–357

Darby DA, Tsang YW (1987) Variation in ilmenite element composition within and among drainage basins: implications for provenance. J Sediment Petrol 57(5):831–838

Darby DA, Tsang YW, Council EA III (1985) Detrital ilmenite composition: implications for coastal sand sources and dispersal pathways. Geol Soc Am 17(7):559

Deer WA, Howie RA, Zussman J (1975) In rock forming minerals, vol 5. Longman, London, pp 339–346

Dinesh AC, Nambiar AR, Unnikrishnan E, Jayaprakash C, Venkateswara Rao C (2007) Mn and Mg and Mn/Mg ratio in detrital ilmenite – an indicator of provenance. J Appl Geochem 9(1):112–119

Divakara Rao V (1984) Khondalites from the Eastern Ghats granulite belt, geochemistry and origin. Geophy Res Bull 21(3):233–242

Divakara Rao V, Murthy NN (1998) Thermobarometry and mineral chemistry of granulite facies rocks from Eastern Ghats India. Gondw Res 1(2):267–274

Frost MT, Grey IE, Harrow Field IR, Mason K (1983) The dependence of alumina and silica contents on the extent of the alteration of weathered ilmenites of their provenance. Sed Geol 77:235–247

Ganapathi Rao P, Reddy KSN, Ravi Sekhar C, Bangaku Naidu K, Murali Krishna KN (2019) Mineral chemistry of ilmenite from red sediments, Bhimunipatnam Coast, Andhra Pradesh, East Coast of India. J Geol Soc India 93:101–108

Gazquez MJ, Bolivar JP, Garcia-Tenorio R, Vaca F (2014) A review of production cycle of titanium dioxide pigment. Mater Sci Appl 5:441–458

Haggerty SE (1976) Opaque mineral oxides in terrestrial igneous rocks. In: Rumble D III (ed) Oxide minerals, vol 3. Mining. Soc. America (Rev. Mineral.), pp 101–130

Hegde VS, Shalini G, GosaviKanchanagouri D (2006) Provenance of heavy minerals with special reference to ilmenite of the Honnavar beach, central west coast of India. Curr Sci 91(5):644–648

Hugo VE, Cornell DH (1991) Altered ilmenites in Holocene dunes from Zululand, South Africa, Petrographic studies for multistage alteration. S Afr J Geol 94:365–378

Hutton CO (1950) Studies of heavy detrital minerals. Geol Soc Am Bull Econ Geol 61:635–716

Jagannadha Rao M, Venkata Ramana J, Venugopal R, Chandra Rao M (2005) Geochemistry and ore mineralogy of ilmenite from beach placers of the Visakhapatnam – Bhimunipatnam deposit, Andhra Pradesh. J Geol Soc India 66:147–150

Jayalakshmi K, Nair KM, Hisao K, Santhosh M (2003) Mineralogical and geochemical variations as indicators of provenance in the heavy mineral deposits of Ambalapuzha beach sands, SW coast of India. J Geosci 46(10):157–168
Leelanandam C (1990) The anorthosite complexes and Proterozoic mobile belt of Peninsular India: a review. Dev Precambrian Geol 8:409–436
Mitra S, Ahmed SS, Moon HS (1992) Mineralogy and chemistry of the opaques of Cox's Bazaar (Bangladesh) beach sands and the oxygen fugacity of their provenance. Sediment Geol 77:235–247
Mohapatra S, Behera P, Das SK (2015) Heavy mineral potentiality and alteration studies for ilmenite in Astaranga beach sands, district Puri, Odisha, India. J Geosci Environ Prot 3(1):31–37
Morton AC (1985) Heavy minerals in provenance studies. In: Zuffa GG (ed) Provenance of arenites. Reidel, Dordrecht, pp 249–277
Morton AC, Halls Worth CR (1999) Processes controlling the composition of detrital heavy mineral assemblages in sandstones. Sediment Geol 124:3–29
Nair AG, Suresh Babu DS, Vivekanandan KL, Vlach SRF (2006) Differential alteration of ilmenite in a tropical beach placer, Southern India: microscopic and electron probe evidences. Resour Geol 56(1):75–81
Nair AG, Suresh Babu DS, Damodaran KT, Shankar R, Prabhu CN (2009) Weathering of ilmenite from Chavara deposit and its comparison with Manavalakurichi placer ilmenite, South Western India. J Asia Earth Sci 34:115–122
Narayana BL, Rama Rao P, Reddy GLN, Divakara Rao V (1995) Geochemistry and origin of megacrystic charnockites and granites from Eastern Ghat Granulite Belt. In: Proceedings of the symposium on India and Antarctica during the Precambrian and granulite and crustal processes in East Gondwana, Andhra University, Visakhaptanam, vol 1–3, p 36
Nayak BR, Mohapatra BK (1998) Two morphologies of pyrophanite in Mn-rich assemblages, Gangpur Group, India. Mineral Mag 62(6):847–856
Nayak B, Mohanty S, Bhattacharyya P (2012) Heavy minerals and the characters of ilmenite in the beach placer sands of Chavakkad-Ponnani, Kerala coast. J Geol Soc India 79:259–266
Pandey M, Dinesh P, Devsamridhi A, Chalapathi Rao NV, Pant NC (2019) Analytical protocol for U-Th-Pb chemical dating of monazite using CAMECA SXFive EPMA installed at the mantle petrology laboratory, Department of Geology, Banaras Hindu University, Varanasi, India. J Geol Soc India 93:46–50
Rahman MA, Pownceby MI, Haque N, Bruckard WJ, Zaman MN (2014) Characterization of titanium rich heavy mineral concentrates from the Brahmaputra River basin, Bangladesh. Appl Earth Sci 123(4):222–233
Rahman MA, Pownceby MI, Haque N, Bruckard WJ, Zaman MN (2016) Valuable heavy minerals from the Brahmaputra River sands of Northern Bangladesh. Appl Earth Sci (Trans Inst Min Metall B) 125(3):174–188
Ramadan F, Zalamah A, Saad AM (2012) Provenance of heavy minerals in some occurrences of placer deposits along tip of Egyptian Mediterranean coastal plain. J Appl Sci Res 8(11):5322–5332
Ramakrishnan C, Mani R, Suresh Babu DS (1997) Ilmenite from the Chavara deposit, India: a critical evaluation. Mineral Mag 61:233–242
Ramam PK, Murthy VN (1997) Geology of Andhra Pradesh. Geol. Soc. India. publ., Bangalore, 245 pp
Ramdohr PF (1969) The ore minerals and their intergrowths. Pergamon Press, Oxford/London, 1174 pp
Rao DS, Sengupta D (2014) Electron microscopic studies of ilmenite from the Chhatrapur coast, Odisha, India, and their implications in processing. J Geochem 2014(4):1–8
Rao DS, Murthy GVS, Rao KV, Das D, Chintalapudi SN (2002) Alteration characteristics of beach placer ilmenite from the Chhatrapur coast, Orissa, India. J Appl Geochem 4:47–59
Rao DS, Vijayakumar TV, Prabhakar S, BhaskarRaju G, Ghosh TK (2005) Alteration characteristics of ilmenites from South India. J Miner Mater Charact Eng 4(1):47–59

Sundararajan M, Bhat KH, Babu N, Janaki MEK, Mohan Das PN (2009a) Characterization studies on ilmenite of Ullal and Suratkal along Karnataka coastline, west coast of India. J Miner Mater Charact Eng 8(6):479–493
Sundararajan M, Bhat KH, Velusamy S, Babu N, Janaki MEK, Sasibhooshanan S, Mohan Das PN (2009b) Characterization of ilmenite from Kerala Coastline, India: implications in the production of synthetic rutile. J Miner Mater Charact Eng 8(6):427–438
Sundararajan M, Bhat KH, Velusamy S (2010) Investigation on mineralogical and chemical characterization of ilmenite deposits of Northern Kerala coast, India. World Appl Sci J 9(3):333–337
Suresh Babu DS, Thomas KA, Mohan Das PN, Damodaran AD (1994) Alteration of ilmenite in the Manavalakurichi deposit, India. Clay Clay Miner 42(5):567–571

Chapter 9
Study of Beach Sand from Harihareshwar, Shrivardhan, and Diveagar Beach of Raigad District, Maharashtra, India

Dnyaneshwar Wayal, Animesh Mishra, and Prasanna Lavhale

Abstract In this study, sand samples from three beaches were analyzed to know the mineral content. Two samples (high tide boundary and low tide boundary) from each beach were selected to study the minimum variation in composition. This study reveals that plagioclase feldspars [andesine $(Ca,Na)Al_{1\text{-}2}Si_{3\text{-}2}O_8$, bytownite $(Ca,Na)Al(Al,Si)Si_2O_8$] occur almost in all localities except andesine at Diveagar and bytownite at high tide boundary of Harihareshwar and low tide boundary of Shrivardhan. Augite of the pyroxene group [$(Ca,Na)(Mg,Fe,Al)(Si,Al)_2O_6$] arise in all locations except the low tide boundary of Shrivardhan. Zirconium carbonate mineral sabinaite [$Na_4Zr_2TiO_4(CO_3)_4$] appears at low tide settings of all three beaches. Albite is abundant in the low tide region of Diveagar beach, and calcite in a minor amount at Shrivardhan beach and the high tide boundary of Diveagar. Hematite is barely found.

Keywords Beach sand · XRD · Feldspars · Pyroxenes · Sabinaite · West coast · Raigad

9.1 Introduction

Generally, minerals are derived from the essential and accessory minerals present in the source rock and deposited in the basin as unconsolidated sediments. These concentrated unconsolidated sediments are then forced to rework by various agents such as wind, waves, and tides in the marginal marine environment. Mineral

D. Wayal (✉)
P. G. Department of Geology, Government Institute of Science, Aurangabad, Maharashtra, India

A. Mishra · P. Lavhale
Deogiri College, Aurangabad, Maharashtra, India

N. Jayaraju et al. (eds.), *Coasts, Estuaries and Lakes*,
https://doi.org/10.1007/978-3-031-21644-2_9

assemblages in the sand vary from region to region depending on several factors such as source rocks in the provenance, prevailing climatic conditions, agents and mechanism of transport, and hydraulic conditions during deposition (Borreswar 1957). Pyroxenes and feldspars are frequently occurring minerals in the studied samples, where calcite is common cementing minerals. Heavy minerals are volumetrically insignificant in samples. We, therefore, seek methods to somehow separate heavy minerals from the bulk of the sand. The obvious way to do that is to use heavy liquids with a density greater than quartz (2.65 g/cm^3) but lighter than other minerals. Several liquids with slightly different densities have been used. That's the reason why there is a range instead of a fixed value. The liquid used to separate heavy minerals from the rest is usually bromoform (2.89 g/cm^3). Unfortunately, heavy liquids having the least properties are not common (Barsdate 1962). The mineral content of sand is useful for studying the provenance of sand or consolidated material. "Provenance" is a term geologist use to the place where the sand grain broke out of its parent rock and transported as sediment. How provenance can be studied? We analyze the minerals and make sure of their composition. Unfortunately, as compositional variations occur in the mineral suits due to so many factors; detailed studies into the provenance often give debatable results. These factors are mostly weathering, burial diagenesis, hydrodynamic sorting, and mechanical abrasion during transport. Heavy mineral placer deposits containing variable proportions of ilmenite, magnetite, garnet, zircon, rutile, monazite, tourmaline, etc., are reported from beaches along the Indian coastline (Siddiquie et al. 1982; Rajamanickam 1983; Gujar et al. 1988, 2004, 2010; Wagle et al. 1989; Gujar 1995; Angusamy et al. 2005). On the west coast, beach placers have been reported in Kerala (Prabhakara 1968) and Ratnagiri (Mane and Gawade 1974). The heavy mineral concentration is called a placer. Some minerals in the sand and their properties are given in Table 9.1 (https://www.sandatlas.org/heavy-minerals).

9.2 Study Area

The samples were collected from the west coast of the Raigad district, Maharashtra, India. This area falls under a humid tropical climate, so chemical weathering is a dominant process. The average rainfall is 3000 mm/year. Samples were collected from the three beaches, namely, Harihareshwar beach (HB), Shrivardhan beach (SB), and Diveagar beach (DB) (Fig. 9.1).

- HB (17°59′43″ N, 73°01′10″ E) is pristine, arcuate, and about 3 km long with prominent wave directions E, NE, and SE. It has very fine sand. The Savitri river meets the Arabic ocean just near the HB and forms the estuary. The terrigenous load is carried by the river deposits at the estuary near Harihareshwar town. So the beach sand is influenced by the continental sediments in a tiny amount at the sampling site. The sea at HB is almost always turbulent. The coast is rocky, and the force of the waves is considerable.

Table 9.1 Minerals in the sand and their properties (https://www.sandatlas.org/heavy-minerals)

Minerals	Density	Stability in weathering	Stability in digenesis	Provenance
Anatase	3.8–3.9	High	High	Felsic igneous rocks, hydrothermal veins, altered product of titanite or ilmenite
Andalusite	3.1–3.1	High	Low	Metamorphic rocks
Amphibole	3.0–3.5	Low	Low	Igneous and metamorphic rocks
Apatite	3.1–3.3	Low	High	Igneous and metamorphic rocks
Cassiterite	6.9–7.0	High	Moderate	Felsic plutonic rocks, hydrothermal deposits
Chloritoid	3.5–3.8	Moderate	Moderate	Metamorphic rocks
Chromite	4.4–5.0	High	High	Mafic and ultramafic igneous rocks
Clinopyroxene	2.9–3.5	Low	Low	Igneous and metamorphic rocks
Corundum	3.9–4.0	High	High	Silica-poor igneous rocks, pelitic metamorphic rocks, hornfels, metamorphosed carbonates, mafic igneous rocks
Epidote	3.1–3.5	Low	Low	Mostly metamorphic rocks, less in igneous rocks
Garnet	3.5–4.3	Moderate	Moderate	Mostly metamorphic but igneous also
Ilmenite	4.7–4.7	Moderate	Moderate	Igneous and metamorphic rocks
Kyanite	3.5–3.6	High	Moderate	Metamorphic rocks, rarely in igneous rocks
Magnetite	5.1–5.2	High	High	Igneous and metamorphic rocks, hydrothermal veins
Monazite	5.0–5.3	High	High	Igneous and metamorphic rocks
Olivine	3.2–4.3	Low	Low	Mostly mafic and ultramafic igneous rocks
Orthopyroxene	3.2–3.9	Low	Low	Mafic and ultramafic igneous rocks, high grade metamorphic rocks
Pumpellyite	3.1–3.2	High	Moderate	Metamorphic rocks
Rutile	4.2–5.5	High	High	Igneous and metamorphic rocks
Sillimanite	3.2–3.2	High	Low	Metamorphic rocks, sometimes granite
Staurolite	3.7–3.8	High	Moderate	Metamorphic rocks
Titanite	3.4–3.5	Moderate	Moderate	Igneous and metamorphic rocks
Topaz	3.4–3.5	High	High	Felsic igneous rocks, metamorphic rocks
Tourmaline	3.0–3.1	Low	High	Granitic pegmatites, some metamorphic rocks
Xenotime	4.2–5.1	High	High	Igneous and metamorphic rocks
Zircon	4.6–4.7	High	High	Igneous and metamorphic rocks

- SB (18°02′34″ N, 73°00′30″ E) is straight in outline and about 4–5 km long. It has fine to moderate sand with prominent wave directions E, NE. The beach sand was carried by high-velocity wind away from the beach, creating the wavy structure along the shoreline leads to the extension of the beach.

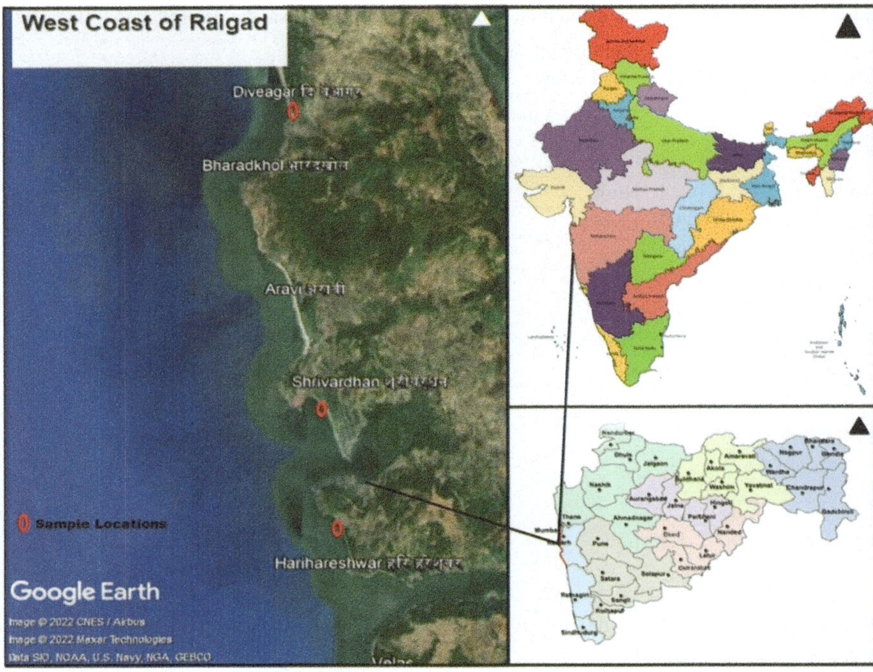

Fig. 9.1 Generalized map of the study area

- DB (18°10′12″ N, 72°59′04″ E) is arcuate and has a 6 km long stretch with white sand and clear water. There is extreme erosion on cliffs during the high tide period. The eroded material contributes to the influence in the sand by eroded particles. DB has moderate to coarse sand with prominent wave directions E, NE, SE.

9.3 Physiography of the Area

The southern west part of the district is covered by beaches. Murud and Shrivardhan tehsils cover with small pockets of reserved forest and limited forest cover and many creeks along the coast. Pen, Uran, and Tala tehsils have mangroves, marshes, and tidal flats.

The Raigad district has three physiographic divisions.

(i) About 20% of the district is consumed by the coastal zone in the west.
(ii) The central zone covers about 1/3rd of the district, consisting of fertile land in the low-lying area.
(iii) The hilly zone in the eastern part is highly uneven in altitude and covered with forest. This hill range is characterized by ruggedness and uneven topography,

with a Crestline of peaks and saddles forming the eastern horizon. Ulhas, Panvel, and Patalganga are the three main rivers in the northern part. Kundalika River is the main river in the central part, whereas in the southern part Savitri River is the main river.

9.4 Geology of the Study Area

Geological and geophysical studies indicate 100–80 Ma existence of the west coast as the Indian subcontinent broke away from Madagascar. Further, the west coast of Raigad appeared as an abrupt cliff some 1000 m in elevation. Basalt is the predominant rock found in the hills, reaching a thickness of 3 km. Residual laterite and bauxite ores are also found in the southern hills (Cheepurupalli et al. 2012). Approximately 70–66 mya, at the close of the Mesozoic Era, the basaltic magma pours out of the long and narrow fissures. So, the basaltic and lateritic lithology is common in the study area. These lava flows spread out in the form of horizontal sheets. The thin soil covers of variable thickness occur in the plains and valleys. The horizontal deposition and lateral extent with almost incredible uniformity in their composition and appearance is a key feature of these flows.

9.5 Materials and Methods

Thirty sand samples (*c.* 1 Kg each) were collected from the beach. The samples were dried in a hot air oven at 150 degrees to remove the moisture. From the dried samples, to ensure uniformity, coning and quartering are applied to the 500 gm sample to avoid the analytical error. Two samples (high tide boundary and low tide boundary) from each beach were selected to study the minimum variation in composition. The samples were crushed by using a mortar and pestle for X-ray diffraction (XRD) analysis.

9.5.1 Stability of Minerals

Minerals have different degrees of stability in various conditions. Minerals with low stability generally do not appear in the sediments for a longer duration of time, depending upon the relative stabilities (Fig. 9.2).

Resistance of Minerals to Weathering		
	Primary Minerals	Secondary Minerals
most weathering resistant	zircon	anatase
↑	rutile	
	tourmaline	gibbsite
	ilmenite	
	garnet	hematite
	quartz	
	epidote	goethite
	titanite	kaolinite
	muscovite	
	K-feldspar	clay minerals
	plagioclase	
	hornblende	calcite
	chlorite	
	augite	gypsum
↓	biotite	
	serpentine	pyrite
least weathering resistant	volcanic glass	halite
	apatite	
	olivine	other salts

Fig. 9.2 List of minerals and their resistance to weathering. (Modified after Birkeland 1999)

9.6 Results and Discussion

9.6.1 Harihareshwar Beach (HB) (Fig. 9.3a)

Hb1 (high tide boundary)
Hb1 has an abundant amount of plagioclase feldspars andesine and bytownite. The amount of Ca–Mg pyroxenes diopside-enstatite solution and augite is comparatively less. The composition of sand indicates the influence of the Savitri river, which flows the NW direction from the Deccan basaltic region. The fewer number of pyroxenes is due to their less stability in turbid water (Pettijohn 1941).

Hb2 (low tide boundary)
A considerable amount of plagioclase feldspars andesine and bytownite are present in Hb2 as on Hb1. The Hb2 sand sample has a minor amount of a zirconium carbonate mineral sabinaite and a very less amount of iron oxide mineral hematite (Fe_2O_3). The sabinaite content indicates that the source is vugs of carbonatite or sodalite. Sabinaite is commonly present in the silico-carbonatite in small amygdule-like

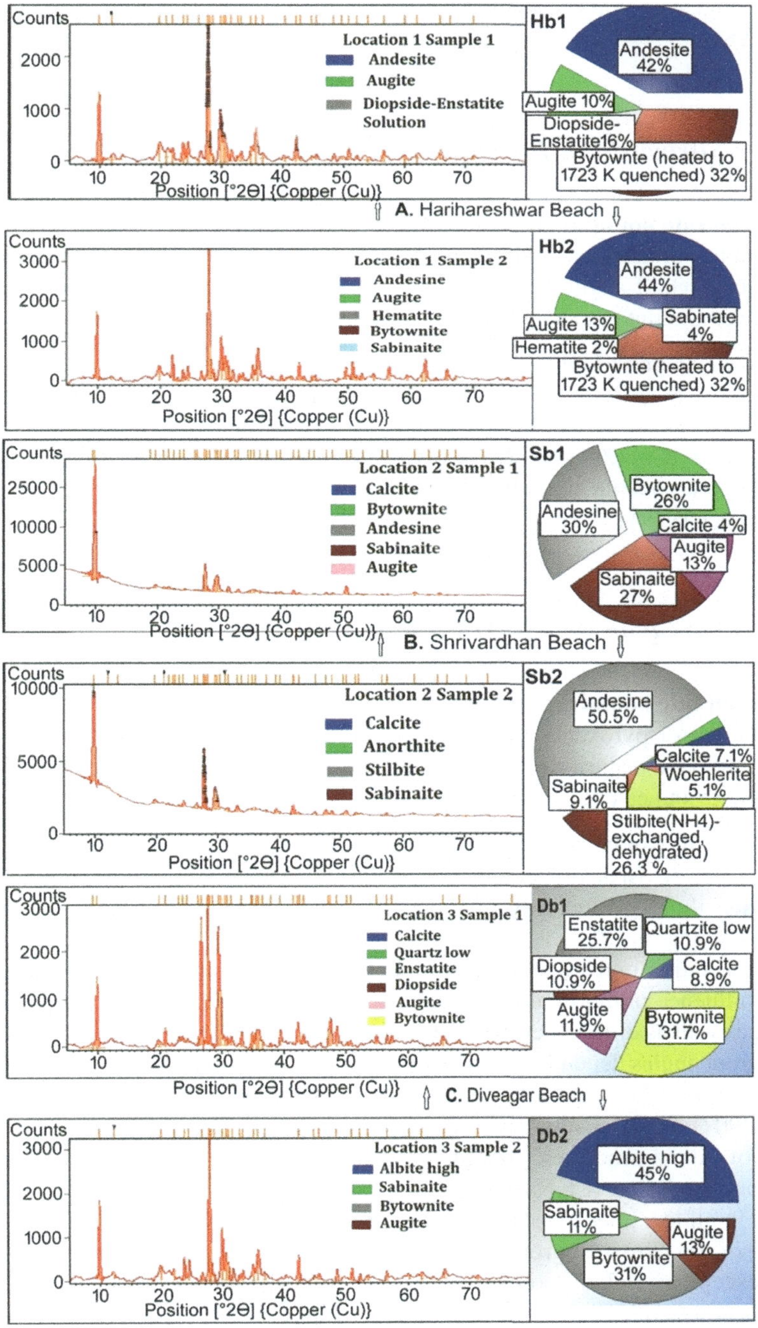

Fig. 9.3 XRD analysis reports of beach sand (**a**) Quantification of minerals from Harihareshwar beach (Hb1 and Hb2). (**b**) Quantification of minerals from Shrivardhan beach (Sb1 and Sb2). (**c**) Quantification of minerals from Diveagar beach (Db1 and Db2)

pockets and as coatings in cavities lined with calcite, dawsonite, and quartz (Jambor et al. 1980).

9.6.2 Shrivardhan Beach (SB) (Fig. 9.3b)

Sb1 (high tide boundary)
The amount of andesine and bytownite is less in Sb1 than Hb1 and Hb2, whereas the percentage of augite and sabinaite is maximizing in Sb1 as compared to Hb2. The fewer amount of calcite ($CaCO_3$) indicates the carbonate content of the source.

Sb2 (low tide boundary)
The amount of andesine is rapidly increasing in low-tide regions by up to 50.5%. The pyroxenes are absent in this sample. A considerable amount of sabinaite and wohlerite [$NaCa_2$ $(Zr,Nb)Si_2O_7(O,OH,F)_2$] is present in fewer amounts indicating the silico-carbonatite or ijolite-pyroxenite or nepheline-syenite source (Mariano and Roeder 1989).

9.6.3 Diveagar Beach (DB) (Fig. 9.3c)

Db1 (high tide boundary)
The amount of pyroxenes (diopside, enstatite, and augite) is about 50% in Db1, indicating poor maturity of sand. Bytownite is present in a considerable amount. Low quartz and calcite occur in a minor amount.

Db2 (low tide boundary)
Sodium-rich feldspar albite and calcic feldspar bytownite occur in huge amounts indicating the good maturity of sand. Sabinaite and augite are also present in a minor amount.

The variation in the mineral distribution pattern in studied samples is attributed to the influence of morphological features and coastline configuration (Gujar et al. 2001), as well as differences in the energy conditions of the waves striking the coast at different angles depending upon their local configurations. Conversely, depletion in heavy mineral content may result in severe diagenetic dissolution (Garzanti and Andò 2007). Alternatively, textural differences may be due to the degree of dominance of fluvial processes (Ibrahim et al. 1992). In the present study, plagioclase feldspars [andesine $(Ca,Na)Al_{1\text{-}2}Si_{3\text{-}2}O_8$, bytownite $(Ca,Na)Al(Al,Si)Si_3O_8$] occur almost in all localities except andesine at Diveagar and bytownite in Hb1 and Sb2. Augite of the pyroxene group [$(Ca,Na)(Mg,Fe,Al,Ti)(Si,Al)_2O_6$] arises in all locations except in Sb2. Zirconium carbonate mineral sabinaite [$Na_4Zr_2TiO_4(CO_3)_4$] appears at low tide settings in Hb2, Sb2, and Db2. Albite is abundant in Db2, and calcite in a minor amount in Sb1, Sb2, and Db1. Dehydrated stilbite is abundant,

Table 9.2 Distribution of minerals in each sample

Samples	Compound name
HB/Hb1	Andesine, augite, diopside-enstatite
HB/Hb2	Andesine, augite, hematite, bytownite, sabinaite
SB/Sb1	Andesine, augite, calcite, bytownite
SB/Sb2	Dehydrated stilbite, andesine, calcite, sabinaite, wohlerite, calcite
DB/Db1	Diopside, enstatite, augite, bytownite, quartz low, calcite
DB/Db2	Albite, sabinaite, augite, bytownite

while wohlerite in fewer amount occur in Sb2. A minor amount of low quartz occurs in Db1 (Table 9.2).

9.7 Conclusion

The studied samples have sand, silt, and silty clay with variable proportions of quartz, feldspar, calcareous materials, and a few heavy minerals. XRD analysis of washed and dried samples has been carried out to know the abundance of minerals. Almost all beaches contain feldspars and pyroxenes, which are the basic composition of Deccan basalt. Bytownite and andesine of the feldspar group are aggregated plenty. Diopside and augite of the pyroxene family occur in considerable amounts. The rare Zr-carbonate mineral sabinaite is found in a significant amount associated with calcite indicating the presence of silico-carbonatite or spilite source near the deposition. Hematite is barely found in the samples. Based on the XRD results of sand samples of these three beaches, it is inferred that the source rock of the unconsolidated sediment may be basalt (Deccan trap basalt), silico-carbonatite, and spilite of the Arabian ocean.

Acknowledgments Authors are thankful to Prof. Ashok Tejankar, Principal, Deogiri College, Aurangabad, for help to carried out field work and to provide lab facility for sample analysis. We are thankful to Dr. Mohan Sonar and Dr. Ranjana Gawande for help in the analysis and improvement in manuscript writing. We are grateful to Dr. Pranaya Diwate for editing the text. We are grateful Sophisticated Analytical Instrument Facility (SAIF), IIT Bombay, for XRD analysis. The authors are thankful to anonymous reviewers for their useful comments on initial version of the manuscript.

References

Angusamy NJ, Sahayam DJ, Gandhi SM, Rajamanickam GV (2005) Coastal placer deposits of central Tamil Nadu, India. Mar Georesour Geotechnol 23:137–174

Barsdate RJ (1962) Rapid heavy mineral separation. J Sediment Petrol 32:608

Borreswar RC (1957) Beach erosion and concentration of heavy minerals. J Sediment Petrol 27:143–147

Cheepurupalli NR, Radha BA, Reddy KS, Dhanamjayarao EN, Dayal AM (2012) Heavy mineral distribution studies in different micro-environments of Bhimunipatnam coast, Andhra Pradesh, India. Int J Sci Res Publ 2(5)

Garzanti E, Andò S (2007) Heavy mineral concentration in modern sands: implications for provenance interpretation. Dev Sedimentol 58:517–545

Gujar AR (1995) Morphogenetic controls on the distribution of the littoral placers along central west coast of India. In: Proceedings of the recent researches on geology along west coast of India, pp 171–180

Gujar AR, Nagendernath B, Banerjee R (1988) Marine minerals: the Indian perspective. Mar Min 7:317–350

Gujar AR, Rajamanickam GV, Wagle BG (2001) Shoreline configurations control on the concentration of nearshore heavy minerals: a case study from Konkan-Maharashtra, central west coast of India

Gujar AR, Mislankar PG, Ambre NV (2004) Coastal placers of Konkan Maharashtra-their distribution, mineralogy and geomorphic environment of formation. India J Geomorphol 9:123–130

Gujar AR, Ambre NV, Mislankar PG, Iyer SD (2010) Ilmenite, magnetite and chromite beach placers from South Maharashtra, central west coast of India. Resour Geol 60(1):71–86

Ibrahim N, Muhammad MJ, Zaidi SMS (1992) Textural and compositional variations in beach sands along Karachi coast

Jambor JL, Sturman BD, Weatherly GC (1980) Sabinaite, a new anhydrous zirconium-bearing carbonate mineral from Montreal Island, Quebec. Can Mineral 18:25–29

Mane RB, Gawade MK (1974) Reports on the prospecting of ilmenite beach sands of Ratnagiri district. Unpublished report (145)

Mariano AN, Roeder PL (1989) Wöhlerite: chemical composition, cathodoluminescence and environment of crystallization. Can Mineral 27:709–720

Pettijohn FJ (1941) Persistence of heavy minerals and geologic age. J Geol 49(6):610–625

Prabhakara RG (1968) Sediments of the near-shore region off Neendakara-Kayamkulam coast and the Ashtamudi and Vatta estuaries, Kerala. India Bull Nat Sci India 30:513–551

Rajamanickam GV (1983) Geological Investigation of offshore heavy mineral placers of Konkan Coast, Maharashtra. Unpublished PhD thesis. Indian School of Mines, Dhanbad, 247p

Siddiquie HN, Rajamanickam GV, Gujar AR, Ramana MV (1982) Geological and geophysical exploration for offshore ilmenite placers off the Konkan coast, Maharashtra, India. In: Proceedings of the 14th offshore technology conference, Houston, Texas, pp 749–755

Wagle BG, Gujar AR, Mislankar PG (1989) Impact of coastal features on beach placers—a case study using remote sensing. In: Proceedings of the 21st offshore technology conference, Houston, Texas, pp 229–233

Chapter 10
Impact of Seasonal Sediment Dynamics on Beach Morphology: A Case Study from the Govindampalli–Durgarajupatnam Coast, East Coast of India

M. Pramod Kumar, B. Praveena, T. Lakshmi Prasad, K. Nagalakshmi, N. Jayaraju, B. Lakshmanna, and T. Siva Prathap

Abstract The current investigation was carried out in four consecutive seasons, along the Govindampalli to Durgarajupatnam (GP-DP) coast. The current project work was chosen, in the backdrop, of the Indian government's plan to set up a major port at Durgarajupatnam, in order to reduce the cargo traffic at Krishnapatnam port on the East coast of India. Furthermore, the investigators believe that the information collected through this study will serve as baseline data for the stakeholders and policymakers when designing the port. An extensive investigation of beach morphodynamics studies with respect to seasons has been carried out by making transects along the coast. Each length of the GP-DP coast was divided into three segments and with two traverses chosen for each segment (a total of six profiles). The data (sediment column levels along three microenvironments in each transect) obtained from the four consecutive seasons (December 2015 to June 2017) were compared and used to calibrate the sediment accretion or erosion levels at each point of measurement which was then used to decipher the accretion and erosion zones, sedimentation trends and cyclic patterns in the study area. Sweep zones and the changes in the beach slopes have been identified.

Keywords Beach · Profile studies · Beach morphology · Sedimentation trends · Sweep zones · Beach slope

---

M. Pramod Kumar · B. Praveena · T. Lakshmi Prasad · T. Siva Prathap
Department of Earth Sciences, Yogi Vemana University, Kadapa, Andhra Pradesh, India

K. Nagalakshmi (✉)
Govt. Degree College, Anantapur, Andhra Pradesh, India

N. Jayaraju
Department of Geology, Yogi Vemana University, Kadapa, Andhra Pradesh, India

B. Lakshmanna
DEEE, Indian Institute of Chemical Technology, Hyderabad, Telangana, India

N. Jayaraju et al. (eds.), *Coasts, Estuaries and Lakes*,
https://doi.org/10.1007/978-3-031-21644-2_10

10.1 Introduction

The dynamic activity of beaches is determined by profile investigations. Longshore sand movement, wave characteristics, and rhythmic topography are the different components that cause profile changes from deposition to erosion (Hesp 2006; Mishra et al. 2001). The nature of shoreline changes and mechanism of coastal accretion/erosion as well as their causes by possible controlling factors such as offshore bathymetry, changes in wind direction, waves, climate changes, and sediment supply (Pye and Neal 1994) The accreted and eroded beaches assume convex and concave profiles, respectively (Díez et al. 2017; Lakshmi Prasad et al. 2021).

Beach profile studies have been used to determine the beach slope, width and volume as well as to understand how beach morphology has changed through time (Wilson et al. 2014). A few researchers have developed two- and three-dimensional morphological stages to understand the sequentiality of a beach transition from depositional to erosional and erosional to depositional phases (Tiwari et al. 2019). Beach material and beach slope influence the morphodynamic changes of a beach. The volume of sand transport and beach gradient have an inverse relationship, which has been demonstrated graphically demonstrated (Dora et al. 2014). When the gradient was lower, greater the beach width, sediment volume of the surf zone, and spacing of the shore normal and shore parallel morphologies (Masselink et al. 2010). While the breaker wave power provides the energy to move a beach through various beach stages, the beach gradient and beach material determine the horizontal and lateral scales of the morphodynamics (Otto et al. 2021).

Steeper coarse-grained beaches are more reflective. These are in favor of lower reflective beach stages with ridge and runnel, welded bar, and incipient bar. The lower gradient fine sand beaches are more dissipative and favor higher reflective stages with outer and multiple bars, wide deep channels between the beach, and parallel bar and rip currents. The low to moderate gradient medium sand beaches potentially encompass the entire spectrum of beach stages (Aragonés et al. 2016).

Many researchers (Abanades et al. 2014; Yin et al. 2019; Jackson and Nordstrom 2020; Bayle et al. 2021; De Schipper et al. 2021) conducted beach morphological studies to establish the seasonal impacts and factors controlling sedimentation trends as well as to demarcate the erosion and accretion zones. It is imperative to establish an erosion and accretion zone that will serve as baseline data for policymakers and executive bodies when building ports and taking precautionary measures to prevent erosion in the vicinity of coastal zones. During the Southwest and Northeast monsoons, researchers (Saravanan and Chandrasekar 2010; Nagalakshmi et al. 2018; Dhanalakshmi et al. 2019; Noujas et al. 2019; Waghmare et al. 2020) studied the prevailing conditions of wind patterns, waves, and currents systems along the east and west coast of India.

The main objective of the present investigation is to sediment dynamics and to study the impact of seasonal sediment on beach morphology. The investigation area, namely, Govindampalli–Durgarajupatnam coast was chosen in light of that the government of India's plan to build a major port at Durgarajupatnam to reduce the cargo traffic at the Krishnapatanam port on the East coast of India. Against this backdrop, the present study was carried out to determine the seasonal impact on beach morphology, besides deciphering the erosion and accretion zones and sweep zones in the studied area.

10.2 Study Area

The Govindampalli to Durgarajupatnam coast (GP-DP coast), is geographically located in the southeastern part of the Nellore district on the East Coast of India. It lies on 14°0′10″–14°02′30″ N latitudes and 80°08′20″–80°19′00″ E longitudes. The Geological Survey of India has classified it as toposheet No. 66 B3, 66 B4 & 66 C1& C5 on a scale of 1:50,000 (Fig. 10.1). The total stretch of the study area, that is, the GP-DP coast is approximately 7 km. The Swarnamukhi river, having a drainage area of 3225 sq. km is the primary source of sediments. This is one of the important, independent, and ephemeral rivers in south India. Swarnamukhi River is about 130 km long. This eventually drains into the Bay of Bengal at Govindampalli village, Nellore coast. There are two tidal openings or creeks to the south of the Swarnamukhi River conjunction. Obviously, during stormy and cyclonic conditions, these tidal inlets are prone to periodic openings and closures. The study area has a subtropical climate with temperatures ranging from 22.5 to 25 °C (minimum) and 30 to 32 °C (maximum) with a mean annual rainfall of 1070 mm.

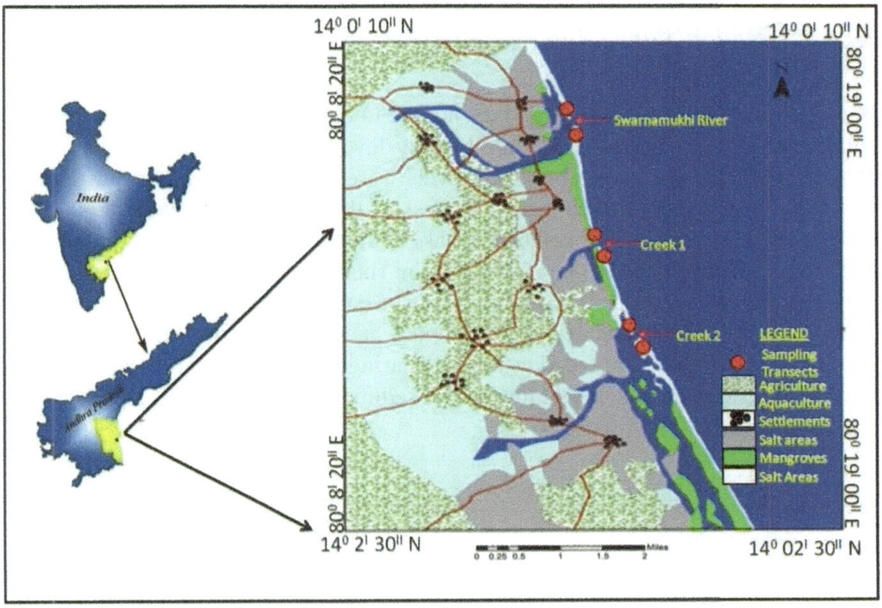

Fig. 10.1 Location map of the study area with sampling stations

10.2.1 Geology and Geomorphology

The hinterland geology consists of Quaternary alluvium, laterites, and Triassic sandstones. The Archean basement gneisses, amphibolites, and migmatized politic schists are observed and above these rocks, recent alluvial sediments are deposited (Nagalakshmi et al. 2018). The investigation area more or less covers geomorphological features, namely, sandbars, mudflats, fluviatile deposits, marine deposits, salt-affected lands, aquaculture ponds, mangroves, and other geomorphological characteristics.

10.3 Methodology

The study area is divided into three segments, which are Govindampalli, Tupilipalem, and Durgarajupatnam. Two stations were chosen in each segment to carry out beach dynamic studies and named Govindampalli North (GP-N), Govindampalli South (GP-S), Tupilipalem North (TP-N), Tupilipalem (TP-S), Durgarajupatnam North (DP-N), and Durgarajupatnam (DP-S). The stations/profiles were named North and South with reference to the openings of the Swarnamukhi River at GP and the other two tidal openings at TP and DP. The beach morphological changes were investigated in different seasons over a period of 2 years from December 2015 to June 2017.

10.3.1 Beach Surveying Method

Beach level changes were recorded during the lowest low tide using the surveyor's Dumpy level, 3 m long Cross staff, Dumpy level stand, and Global Positioning System (GPS). The changes in beach levels were recorded for every 2 m interval along the profile section. The same traverses of the beach profile were maintained every season by using GPS positions. Base stations were chosen from unaffected and identifiable regions and a reference point (benchmark) was fixed at each transect on a foredune, ensuring that it was undisturbed even by abnormal storm waves and could be easily relocated and occupied at subsequent profile surveys. The benchmark height was recorded as a back bearing reading that can be used to calibrate beach slope angles, determination of sedimentation trends, sweep zones and accretion, and erosion trends.

10.4 Results and Discussion

10.4.1 Temporal Variations in Sedimentation Trends from December 2015 to June 2017 Seasons

Each traverse is classified into three microenvironments to better understand changes in the sediment column and sedimentation trends: dune, backshore, and foreshore.

10.4.2 Govindampalli Segment

10.4.2.1 Govindampalli-North (GP-N) Profile

GP-N profile was first surveyed in December 2015. The total width of the profile was 56 m; the foreshore width was 26 m. From the dune to the backshore the beach was concave in shape and the foreshore slope was moderately steep. Seasonal changes in beach levels were shown (Fig. 10.2a). Seasonal changes in erosion/accretion of sediment column at profile transects are given (Table 10.1).

From December 2015 to June 2016, the dune environment was under deposition with the sediment column ranging from 2 to 85 cm (av. 45 cm). A high rate of deposition has occurred in the backshore region and with the accreted sediment column ranging from 90 to 140 cm (av. 114 cm). The deposition was noticed in the foreshore region with the sediment column ranging from −8 to 40 cm (av. 20 cm).

Between June 2016 and December 2016 GP-N, the dune and the backshore exhibited erosion with column ranges from −35 to 6 cm (av. −17 cm), and −36 to −21 cm (av −27 cm), respectively, whereas foreshore regions accreted with sediment column ranges from 77 to 213 cm (av. 131 cm), respectively. Between December 2016 and June 2017 all the micro-environments, namely, dune, backshore, and foreshore were eroded, with sediment columns ranging from −13 to −11 cm (av. −12 cm), −59 to −10 cm (av. −39 cm), and foreshore −213 to −23 cm (av. −80 cm), respectively.

During the investigation, there was no cyclic pattern in any of the three microenvironments. However, cumulative accretion/erosion data reveals that the accretion of sediments occurred in the GP-N region.

10.4.2.2 Govindampalli-South (GP-S)

GP-S recorded a beach width of 56 m and a foreshore width was 20 m when profiled in December 2015. The shape from the dune to the backshore region was concave and the slope of the beach was moderately steep in the foreshore region. The changes in sediment levels along with the profile during the period of the study were shown

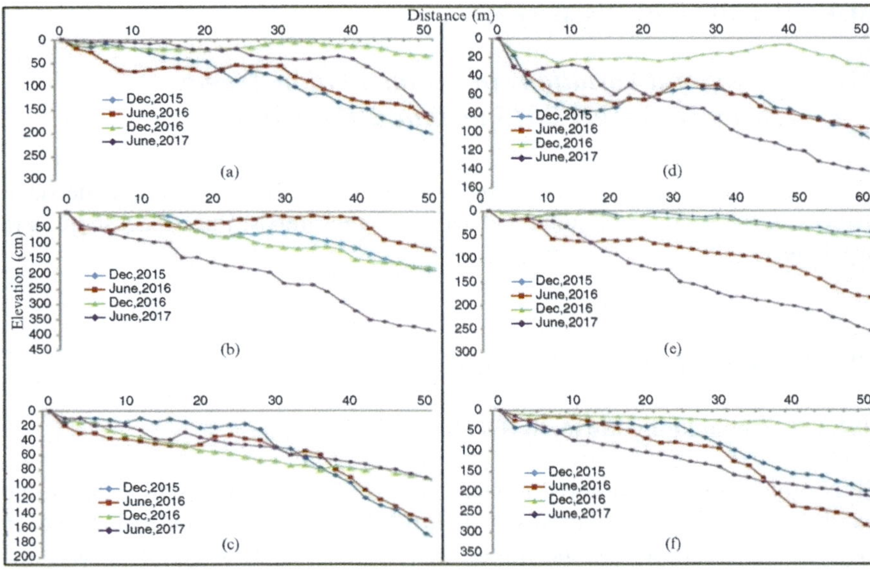

Fig. 10.2 Seasonal sand levels (cm) from GP to DP coast (distance in meters). (**a**) GP-N, (**b**) GP-S, (**c**) TP-N, (**d**) TP-S, (**e**) DP-N, (**f**) DP-S

(Fig. 10.2b). Seasonal variations in deposition/erosion of sediment column at each point of measurement of the transect were shown (Table 10.1).

The backshore and foreshore environments were accreted between December 2015 and June 2016. The accreted sediment column measures between −23 and 46 cm (av. 17 cm) and −7 and 48 cm (av. 15 cm), respectively. The sediment column in the dune habitats ranges from −46 to −14 cm (av. −29 cm) indicating erosion.

From June 2016 to December 2016, the backshore and foreshore environments showed erosion, with sediment columns ranging from −67 to 25 cm (av. −37 cm) and −68 to 27 cm (av. −13 cm), respectively. The dune environment shows deposition as the deposited sediment column ranges from 16 to 45 cm (av. 29 cm).

The dune, backshore, and foreshore environments significantly eroded during the period between December 2016 and June 2017 with eroded sediment column ranges from −76 to −20 cm (av. −52 cm), −100 to −48 cm (av. −74 cm), and −200 to −125 cm (av. −171 cm), respectively.

No cyclic patterns of erosion and deposition were recorded in all three environs during the period of study. However, the cumulative result exhibits that the erosion of sediments at GP-S is due to northeast shoreline currents drifting the sediments toward the north region, that is, GP-N.

Table 10.1 GP-N and GP-S sediment columns eroded or accreted between two consecutive measurements at each point on the profile (with reference to December 2015 data)

GP-N seasonal sediment column Differences in elevation						GP-S seasonal sediment column Differences in elevation					
Environment	Distance (m)	June 2016	December 2016	June 2017	(+) or (−)	Environment	Distance (m)	June 2016	December 2016	June 2017	(+) or (−)
Dune	2	2	6	−13	−5	Dune	2	−14	16	−20	−18
	4	8	−5	−13	−10		4	−20	20	−36	−36
	6	53	−16	−12	25		6	−22	20	−54	−56
	8	77	−35	−11	31		8	−46	45	−73	−74
	10	85	−35	−12	38		10	−43	43	−76	−76

(continued)

Table 10.1 (continued)

GP-N seasonal sediment column Differences in elevation						GP-S seasonal sediment column Differences in elevation					
Environment	Distance (m)	June 2016	December 2016	June 2017	(+) or (−)	Environment	Distance (m)	June 2016	December 2016	June 2017	(+) or (−)
Backshore	12	90	−31	−10	49	Backshore	12	−23	25	−66	−64
	14	95	−25	−26	44		14	−16	−2	−48	−66
	16	98	−24	−35	39		16	11	−32	−75	−96
	18	107	−27	−43	37		18	24	−22	−65	−63
	20	118	−36	−42	40		20	46	−46	−64	−64
	22	129	−29	−44	56		22	45	−45	−71	−71
	24	140	−21	−39	80		24	24	−32	−80	−88
	26	124	−27	−51	46		26	25	−52	−67	−94
	28	127	−22	−59	46		28	6	−50	−66	−110
Foreshore	30	22	77	−46	53		30	10	−59	−95	−144
	32	19	89	−37	71		32	21	−67	−97	−143
	34	25	99	−37	87		34	26	−57	−100	−131
	36	6	112	−30	88	Foreshore	36	42	−60	−125	−143
	38	15	118	−23	110		38	48	−68	−147	−167
	40	13	129	−28	114		40	26	−63	−146	−183
	42	10	135	−44	101		42	12	−34	−172	−194
	44	29	131	−56	104		44	−7	−2	−174	−183
	46	37	123	−66	94		46	−4	4	−183	−183
	48	40	127	−88	79		48	1	4	−176	−171
	50	30	145	−122	53		50	−3	13	−183	−173
	52	20	163	−143	40		52	7	22	−186	−157
	54	19	180	−180	19		54	21	16	−191	−154
	56	−8	213	−213	−8		56	19	27	−200	−154

10.4.3 Tupilipalem Segment

Tupilipalem beach was studied for seasonal beach profiles from December 2015 to June. A small tidal creek (creek 1) joins the Bay of Bengal between the north and south beaches.

10.4.3.1 Tupilipalem Beach-North (TP-N)

In December 2015, the width of the TP-N was 54 m wide and the foreshore width was 20 m. Seasonal beach level changes in the sediment column at each point of observation along the transverse are shown (Fig. 10.2c and Table 10.2).

The total beach was under deposition during the period from December 2015 to June 2016, with deposited sediment column ranging from −4 to 17 cm (av. 10 cm), for dune, it ranges from 5 to 112 cm (50 cm) for the backshore and from −3 to 12 (av. 3 cm) for the foreshore.

Between June 2016 and December 2016, an insignificant amount of sediments was eroded at the dune, ranging from −11 to −2 cm (av. −7 cm). But, in the backshore and foreshore environments, it varies from −2 to 32 cm (av. 16 cm) and 145 to 287 cm (av. 212 cm), respectively.

Between December 2016 and June 2017, the sediment columns in the dune and foreshore environments deposited sand levels ranging from −36 to 20 cm (av. 0.2 cm) and −1 to 21 cm (av. 12 cm), respectively. The backshore region shows erosion of the sediment column ranging from −102 to 4 cm (av.−64 cm). No cyclicity of erosion and deposition in dune and backshore regions but deposition occurred through all seasons in the foreshore region. The net results from December 2015 to June 2017, indicate deposition in the dune, backshore, and foreshore regions.

10.4.3.2 Tupilipalem-South (TP-S)

The changes in sand levels during four seasons and season to season, deposition/erosion of sediment column at each point of measurement of the profile transects is shown (Fig. 10.2d and Table 10.2).

The TP-S exhibits a significant amount of deposition in all microenvironments during the period from December 2015 to June 2016 in all microenvironments. The deposited sediment column ranges from 19 to 55 cm (av. 43 cm) at the dune, 35 to 49 cm (av. 41 cm) at the backshore, and 26 to 66 cm (av. 39 cm) at the foreshore.

The dune, backshore, and foreshore environments show prominent accretion between June 2016 and December 2016 with sediment columns ranging from −8 to 29 cm (av. 11 cm), 9 to 35 cm (av. 23 cm), and 10 to 73 cm (av. 40 cm), respectively.

The dune shows insignificant erosion from December 2016 to June 2017, with eroded sediment column ranges from −19 to −3 cm (av. −11 cm). Backshore and foreshore environments show high erosional conditions with sediment columns

Table 10.2 TP-N and S sediment columns (cm) eroded or accreted between two consecutive measurements at each point on the profile (with reference to December 2015 data)

TP-N seasonal sediment column Differences in elevation						TP-S seasonal sediment column Differences in elevation					
Environment	Distance (m)	June 2016	December 2016	June 2017	(+) or (−)	Environment	Distance (m)	June 2016	December 2016	June 2017	(+) or (−)
Dune	2	−4	−6	−4	−14	Dune	2	19	−8	−6	5
	4	13	−11	11	13		4	44	−2	−3	39
	6	12	−10	20	22		6	55	11	−9	57
	8	17	−7	10	20		8	46	29	−19	56
	10	14	−2	−36	−24		10	51	25	−17	59

TP-N seasonal sediment column					
Differences in elevation					
Environment	Distance (m)	June 2016	December 2016	June 2017	(+) or (−)
Backshore	12	43	−2	−45	−4
	14	51	0	−62	−11
	16	50	1	−67	−16
	18	56	2	−60	−2
	20	15	12	−74	−47
	22	5	25	−80	−50
	24	6	29	−87	−52
	26	12	29	−93	−52
	28	57	32	−100	−11
	30	91	22	−102	11
	32	104	18	2	124
	34	112	23	4	139
Foreshore	36	10	145	−1	154
	38	0	159	8	167
	40	−1	173	8	180
	42	3	193	8	204
	44	0	202	17	219
	46	−3	219	11	227
	48	−1	234	14	247
	50	11	244	16	271
	52	12	257	21	290
	54	−2	287	19	304

TP-S seasonal sediment column					
Differences in elevation					
Environment	Distance (m)	June 2016	December 2016	June 2017	(+) or (−)
Backshore	12	49	30	−14	65
	14	48	30	−95	−17
	16	40	35	−105	−30
	18	35	29	−94	−30
	20	35	31	−105	−39
	22	36	27	−113	−50
	24	42	15	−114	−57
	26	44	9	−119	−66
	28	39	11	−115	−65
	30	40	9	−125	−76
Foreshore	32	35	19	−137	−83
	34	37	17	−141	−87
	36	26	25	−141	−90
	38	31	29	−142	−82
	40	32	30	−149	−87
	42	34	40	−156	−82
	44	33	48	−172	−91
	46	39	52	−177	−86
	48	36	64	−189	−89
	50	43	67	−192	−82
	52	51	73	−200	−76
	54	66	10	−142	−66

ranging from −125 to −14 cm (av. −100 cm) and −200 to −137 cm (av. −162 cm), respectively.

The succession of erosion and deposition was not observed in all three microenvironments, but by the end of the seasonal cycle erosion of sediments occurred at TP-S.

10.4.4 Durgarajupatnam Segment

At the time of profiling in December 2015, the beach was 52 m wide. Sand level sand season-to-season changes in deposition/erosion of sediment column at each point of measurement on the profile are recorded (Fig. 10.2e and Table 10.3).

10.4.4.1 Durgarajupatnam Beach-North (DP-N)

Between December 2015 and June 2016, the backshore region of DP-N had a significant amount of deposition with the sediment column ranging from −11 to 194 cm (av. 77 cm). The dune and foreshore regions showed erosion with sediment columns ranging from −18 to 5 cm (av. −8 cm) and −81 to −4 cm (av. −53 cm), respectively.

Profile studies reveal that the dune and backshore environments were eroded between June 2016 and December 2016. The eroded sediment column ranges from −11 to −4 cm (av. −7 cm) and −94 to 180 cm (av. −38 cm), respectively. Acute deposition with 132–270 cm (av. 218 cm) can be found along the foreshore region. From December 2016 to June 2017, the dune, backshore, and foreshore regions were under erosion with sediment column size ranging from −68 to −10 cm (av. −37 cm), −195 to −68 cm (av. −104 cm), and −174 to −143 cm (av. −160 cm), respectively.

The backshore does not show erosion and deposition cycle, and the dune environment shows erosion during all three seasons, although the net result is the erosion of sediments. The foreshore regions, on the other hand, exhibit cyclicity (Table 10.4).

10.4.4.2 Durgarajupatnam Beach-South (DP-S)

Seasonal beach levels in DP-S and season-to-season changes in deposition/erosion of sediment column at each point of measurement on the profile are shown (Fig. 10.2f and Table 10.3).

The total beach of DP-S shows erosion from December 2015 to June 2016. For dune, backshore, and foreshore, the eroded sediment column ranged from −81 to

Table 10.3 DP-N and DP-S eroded or accreted between two consecutive measurements at each point on the profile (with reference to December 2015 data)

DP-N seasonal sediment column Differences in elevation						DP-S seasonal sediment column Differences in elevation					
Environment	Distance (m)	June 2016	December 2016	June 2017	(+) or (−)	Environment	Distance (m)	June 2016	December 2016	June 2017	(+) or (−)
Dune	2	−1	−10	−10	−21	Dune	2	9	−12	3	0
	4	5	−11	−24	−30		4	−31	−12	2	−41
	6	−18	−4	−36	−58		6	7	−15	2	−6
	8	−17	−5	−48	−70		8	12	−32	11	−9
	10	−9	−5	−68	−82		10	−81	67	−38	−52

(continued)

Table 10.3 (continued)

DP-N seasonal sediment column Differences in elevation						DP-S seasonal sediment column Differences in elevation					
Environment	Distance (m)	June 2016	December 2016	June 2017	(+) or (−)	Environment	Distance (m)	June 2016	December 2016	June 2017	(+) or (−)
Backshore	12	8	−16	−68	−76	Backshore	12	−85	68	−49	−66
	14	21	−25	−75	−79		14	−89	70	−65	−84
	16	30	−35	−80	−85		16	−94	73	−81	−102
	18	37	−44	−88	−95		18	−89	66	−99	−122
	20	46	−63	−92	−109		20	−83	59	−97	−121
	22	67	−73	−98	−104		22	−103	64	−121	−160
	24	128	−74	−103	−49		24	−99	59	−125	−165
	26	153	−82	−111	−40		26	−103	67	−131	−167
	28	173	−88	−115	−30		28	−107	68	−130	−169
	30	194	−94	−121	−21		30	−117	72	−154	−199
	32	−11	180	−195	−26	Foreshore	32	−123	74	−157	−206
Foreshore	34	−4	132	−143	−15		34	−131	80	−164	−215
	36	−19	164	−156	−11		36	−130	86	−179	−223
	38	−46	200	−155	−1		38	−134	85	−184	−233
	40	−64	220	−149	7		40	−150	81	−178	−247
	42	−66	232	−162	4		42	−150	80	−181	−251
	44	−67	231	−161	3		44	−162	84	−180	−258
	46	−61	236	−162	13		46	−180	94	−185	−271
	48	−59	237	−167	11		48	−186	93	−183	−276
	50	−67	259	−168	24		50	−201	105	−189	−285
	52	−81	270	−174	15		52	−211	111	−187	−287

Table 10.4 Erosion and accretion phases and Beach slope variability of Govindampalli to Durgarajupatnam coast during the study period

GP-N	Phase	Season	Beach slope	Average size of sediment column accreted or eroded (cm)		
		December 2015 (NE)	3.96°	Dune	Backshore	Foreshore
	Erosion (−)/ Accretion (+)	June 2016 (SW)	3.71°	45	114	20
		December 2016 (NE)	2.56°	−17	−27	131
		June 2017 (SW)	3.38°	−12	−39	−80
		+ or − (Net)		16	48	71
GP-S	**Phase**	**Season**		**Average size of sediment column accreted or eroded (cm)**		
		December 2015 (NE)	2.68°	Dune	Backshore	Foreshore
	Erosion (−)/ Accretion (+)	June 2016 (SW)	2.42°	−29	17	15
		December 2016 (NE)	3.16°	29	−37	−13
		June 2017 (SW)	4.50°	−52	−74	−171
		+ or − (Net)		−52	−94	−169
TP-N	**Phase**	**Season**		**Average size of sediment column accreted or eroded (cm)**		
		December 2015 (NE)	2.56°	Dune	Backshore	Foreshore
	Erosion (−)/ Accretion (+)	June 2016 (SW)	2.26°	10	50	3
		December 2016 (NE)	2.09°	−7	16	212
		June 2017 (SW)	3.36°	0.2	−64	12
		+ or − (Net)		3	2	227
TP-S	**Phase**	**Season**		**Average size of sediment column accreted or eroded (cm)**		
		December 2015 (NE)	3.46°	Dune	Backshore	Foreshore
	Erosion (−)/ Accretion (+)	June 2016 (SW)	2.59°	43	41	39
		December 2016 (NE)	2.39°	11	23	40
		June 2017 (SW)	4.82°	−11	−100	−162
		+ or − (Net)		43	−36	−83

(continued)

Table 10.4 (continued)

DP-N	Phase	Season		Average size of sediment column accreted or eroded (cm)		
		December 2015 (NE)	2.19°	Dune	Backshore	Foreshore
	Erosion (−)/ Accretion (+)	June 2016 (SW)	2.08°	−8	77	−53
		December 2016 (NE)	1.78°	−7	−38	218
		June 2017 (SW)	3.99°	−37	−104	−160
		+ or – (Net)		−52	−65	9
DP-S	**Phase**	**Season**		**Average size of sediment column accreted or eroded (cm)**		
		December 2015 (NE)	2.68°	Dune	Backshore	Foreshore
	Erosion (−)/ Accretion (+)	June 2016 (SW)	3.72°	−17	−97	−160
		December 2016 (NE)	1.99°	−1	67	89
		June 2017 (SW)	4.20°	−4	−105	−179
		+ or – (Net)		−22	−135	−250

NE Northeast, *SW* Southwest

12 cm (av. −17 cm), −117 to −83 cm (av. −97 cm), and −211 to −123 cm (av. −160 cm), respectively.

Between June 2016 and December 2016, the dune showed significant erosion with sediment columns ranging from −32 to 67 cm (av. −1 cm), backshore and foreshore shows significant deposition with sediment columns between 59 and 73 cm (av. 67 cm) and 74 and 111 cm (av. 89 cm), respectively.

The entire beach was under erosion in all three microenvironments with sediment columns ranging from −38 to 11 cm (av. −4 cm); −154 to −49 cm (av. −105 cm), and −189 to −157 cm (av. −179 cm), respectively. Between December 2016 and June 2017, dune erosion was minimal, with sediment columns ranging from −32 to 67 cm (av. −0.8 cm), while backshore and foreshore deposition was significant, with sediment columns ranging from 59 to 73 cm (av. 70 cm) and 74 to 111 cm (av. 89 cm) correspondingly.

The total beach of DP-S, during the seasonal cycle from December 2015 to June 2017 was under erosion (Table 10.4). The eroded and creek input sediments together transported toward DP-N beach due to northeast (NE) alongshore currents and further transported toward TP and GP beaches.

10.4.5 Sweep Zones

Beaches have considerable temporal variability in third-dimension mobility throughout time. The term "sweep zone" expresses the limits of this mobility over a long time. The line joining the highest points on all the profiles indicates the height above which the beach is unlikely to extend. Similarly, the line joining the lowest points indicates the level below which the material is unlikely to be eroded. The zone between the upper and lower curves is called the "sweep zone" and is defined as the vertical envelope within which movement of beach material may take place by wave action (Tiwari et al. 2019). The sweep zone is referred to as the profile envelope of temporal variation (Noraisyah et al. 2014). The beach mobility is proportional to the sweep zone, occupied by the profiles over prolonged periods.

The sweep zone indicates the maximum and minimum possible conditions of the beach profile during a short period of time. The sweep zones of the six profile stations from the GP-DP coast are shown (Fig. 10.3). The sweep zones exhibit that the subaerial beach vertical mobility is minimum in all the dune environments. In the backshore environment, vertical mobility is significant in GP-N and TP-N beaches and the mobility is high in GP-S, TP-S, DP-N, and DP-S beaches. Vertical mobility is very high under the foreshore environments of all regions. At the lower end, the upper and lower sweep zone profiles of the subaerial portion remain widely separated at all stations. This suggests that vertical mobility is significant in the surf zone.

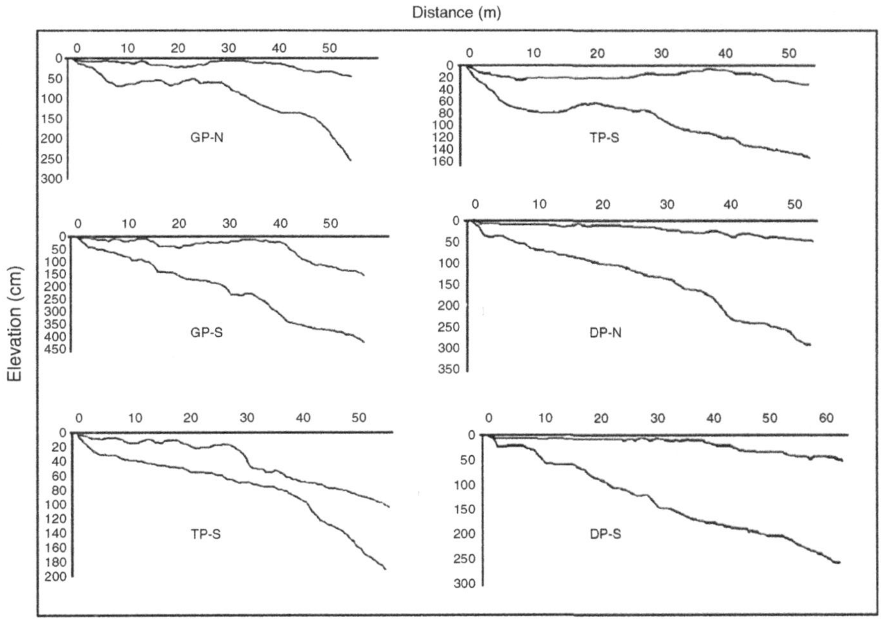

Fig. 10.3 Sweep zones from GP to DP coast

10.4.6 Beach Slopes

The average beach slope values divide the beaches into three categories flat (with a slope <2°), moderate beaches (with a slope >2° to <3.5°), and steep beaches (with >3.5°) (Lakshmi Prasad et al. 2021). The beach slope angles at six profiles show decreasing trend during the first three seasons and then an abruptly increase in slope angles in June 2017. This clearly signifies that this area is under erosion and the beach slope variability of six stations through the four seasonal cycles is recorded (Table 10.4).

The slope of GP-N beach varies from 2.56° to 3.96°. The beach slope was very high (3.96°) in December 2015, gradually declining to 3.71°, further decreasing to 2.56° in December 2016, but abruptly increased to 3.38° in June 2017. The slope in the GP-S region ranges from 2.42° to 4.5° (Table 10.4). The entire beach slope was moderately steep in the first three seasons, and a sharply increased to 4.5° in the last season is observed. The slope angles depend on erosion/accretion processes that occurred during monsoon periods.

The slope angle on the TP-N and S beaches is from 2.09° to 3.36° and 2.39° to 4.82°, respectively. The entire TP-N and TP-S were moderately steep in the first three seasons and a sharp increase in slope angle in June 2017 (3.36° at TP-N and 4.82° at TP-S). The slope of the DP-N beach ranges from 1.78° to 3.99° (Table 10.4). The beach was moderately steep during the first two seasons and nearly flat in December 2016 (1.78°) and very steep slope in June 2017 (3.99°).

The slope in DP-S ranges from 1.99° to 4.2°. Initially, the beach was moderately steep (2.68°) in December 2015 and shows a very steep slope (3.72°) in June 2016. Then followed by a low angle (1.99°) in Dec 2016 and then abruptly increased (4.2°) in June 2017. The slope decreased in the northeast (NE) monsoon season (December 2016) and increased in the southwest (SW) monsoon season (June 2017).

10.4.7 Inferences

The following is the comprehensive outcome of individual transects and segment-by-segment. In the study area, except for the GP-S and DP-S profiles, the majority of the traverses show no cyclic nature of deposition or erosion of sediments during the study period. GP-S exhibits erosion irrespective of the monsoon seasons, whereas DP-S shows the cyclic nature of erosion, deposition, and erosion in June 2016 (SW), June 2016 (NE), and June (2017), respectively. The remaining traverses GP-N, TP-N, TP-S, and DP-N exhibit deposition from June 2016 to December 2016, and significant erosion during June 2017. The non-cyclic nature of erosion and accretion of sediment in the study area during the period of study is attributed or tentatively interpreted to the prevalent hypothetical cell circulation systems and associated rip zones of convergence and divergence of waves energies, currents, shoreline topography, tidal cycles and presence of different water masses, etc. (Lakshmi Prasad et al. 2021).

The net results of the erosion and accretion phases across the study period unravel that in the GP segment, GP-S beach is undergoing erosion while the GP-N region is undergoing accretion. Evidently, the same trend is shown in the TP segment, that is, the TP-S region befalls in erosion and accretion at the TP-N region. It is imperative to note that erosion is rampant in the DP segment. The presence of the Swarnamukhi River at GP (sediment input during monsoon times) and the presence of a tidal creek at TP, as well as dominant, prevailing, northeast long currents, contribute to eroding in the southern regions and accretion in the northern regions (Sreenivasulu et al. 2017; Nagalakshmi et al. 2018).

Last season in June 2017, a very severe cyclonic storm name Vardah cause sediment erosion in all segments and an abrupt increase in beach slopes (Vardah. 2017).

10.5 Conclusions

Profile studies were used to track changes in beach morphology throughout the GP-DP coast, from December 2015 to June 2017 (four consecutive seasons; two southwest and two Northeast Monsoon seasons). A total of six profiles (three microenvironments in each profile) were considered. The obtained data reveals that the majority of the microenvironments along the six traverses do not show cyclic erosion and accretion patterns. The existence of the non-cyclic nature of erosion and deposition is attributed or tentatively interpreted to the widespread hypothetical cell circulation systems and associated rip zones of convergence and divergence of wave energies, currents, shoreline topography, and tidal cycles. However, it is envisioned to compare erosion or accretion calculations at each point of measurement on all profiles from season to season. The GP-S traverse exhibits erosion throughout the season as well as the cyclic nature of erosion, deposition, and erosion pattern recorded along the DP-S transect. The remaining profiles, namely, GP-N, TP-N, TP-S, and DP-N, show significant deposition during June 2016 (SW) and December 2016 (NE) then followed by acute erosion in June 2017. By the end of the seasonal cycle, sediment erosion is dominant on the south beaches such as GP-S, and TP-S while sediment deposition occurs on the north beach, that is, GP-N and TP-N. Anomalous erosional condition prevails at the DP-N and S irrespective of the seasons. This is a proposed site for port construction by the Government of India. Erosion at the southern beaches and deposition at the northern beaches is ascribed to the dominant, prevailing, Northeast long currents that aid the sediments to erode in the southern regions and accretion in the Northern regions. In all the regions vertical mobility is very high in the beach environments. The slope angle of all beaches abruptly increased at the end of the seasonal cycle. In order to understand the sediment dynamics of the study area, more research is needed. This will help to suggest whether the area under investigation is an ideal site for the construction of a major port as proposed by the government of India.

References

Abanades J, Greaves D, Iglesias G (2014) Wave farm impact on the beach profile: a case study. Coast Eng 86:36–44

Aragonés L, Serra JC, Villacampa Y, Saval JM, Tinoco H (2016) New methodology for describing the equilibrium beach profile applied to the Valencia's beaches. Geomorphology 259:1–11

Bayle PM, Beuzen T, Blenkinsopp CE, Baldock TE, Turner IL (2021) A new approach for scaling beach profile evolution and sediment transport rates in distorted laboratory models. Coast Eng 163:103794

De Schipper MA, Ludka BC, Raubenheimer B, Luijendijk AP, Schlacher T (2021) Beach nourishment has complex implications for the future of sandy shores. Nat Rev Earth Environ 2(1):70–84

Dhanalakshmi S, Kankara RS, ChenthamilSelvan S (2019) Impact assessment of sea level rise over coastal landforms: a case study of Cuddalore coast, south-east coast of India. Environ Earth Sci 78(16):1–14

Díez J, Cánovas V, Uriarte A, Medina R (2017) Characterization of the dry beach profile: a morphological approach. J Coast Res 33(6):1292–1304

Dora GU, Kumar VS, Vinayaraj P, Philip CS, Johnson G (2014) Quantitative estimation of sediment erosion and accretion processes in a micro-tidal coast. Int J Sediment Res 29(2):218–231

Hesp PA (2006) Sand beach ridges: definitions and re-definition. J Coast Res:72–75

Jackson NL, Nordstrom KF (2020) Trends in research on beaches and dunes on sandy shores, 1969–2019. Geomorphology 366:106737

Lakshmi Prasad T, Satyanarayana Reddy K, Asadi SS (2021) Evalution of beach morphology. Lulu Publications

Masselink G, Russell P, Blenkinsopp C, Turner I (2010) Swash zone sediment transport, step dynamics and morphological response on a gravel beach. Mar Geol 274(1–4):50–68

Mishra P, Mohanty PK, Murty ASN, Sugimoto T (2001) Beach profile studies near an artificial open-coast port along south Orissa, east coast of India. J Coast Res:164–171

Nagalakshmi K, Kumar MP, Prasad TL, Jayaraju N, Lakshmanna M, Sreenivasulu G (2018) A study on textural parameters of beach sands along some parts of the Nellore coast, east coast of India: implications to depositional environment. J Ind Geophys Union 22(5):558–567

Noraisyah S, Parham PR, Mohd-Lokman H, Rosnan Y, Nor-Antonina A, Rokiah S (2014) Sandy beach morphodynamic classifications index based on wright and short at selected beaches along Terengganu, Malaysia. In: Geological Society of Malaysia National Geoscience. Conference

Noujas V, Kankara RS, Selvan SC (2019) Shoreline management plan for embayed beaches: a case study at Vengurla, west coast of India. Ocean Coast Manag 170:51–59

Otto P, Piter A, Gijsman R (2021) Statistical analysis of beach profiles–a spatiotemporal functional approach. Coast Eng 170:103999

Pye K, Neal A (1994) Coastal dune erosion at Formby Point, north Merseyside, England: causes and mechanisms. Mar Geol 119(1–2):39–56

Saravanan S, Chandrasekar N (2010) Potential littoral sediment transport along the coast of South Eastern Coast of India. Earth Sci Res J 14(2):153–160

Sreenivasulu G, Jayaraju N, Reddy BSR, Prasad TL, Lakshmanna B, Nagalakshmi K (2017) Coastal morphodynamics of Tupilipalem coast, Andhra Pradesh, southeast coast of India. Curr Sci:823–829

Tiwari M, Rathod TD, Ajmal PY, Bhangare RC, Sahu SK (2019) Distribution and characterization of microplastics in beach sand from three different Indian coastal environments. Mar Pollut Bull 140:262–273

Vardah (2017) Cyclones Vardah and ARB 02 – Wikipedia

Waghmare SM, Hanamgond PT, Mitra D, Koti BK, Shinde PS (2020) Application of remote sensing and GIS techniques to study sediment movement along Harwada Beach, Uttar Kannada, West Coast of India. J Coast Res 36(6):1121–1129

Wilson JJ, Sujitha SB, Shruti VC, Shaeema ZA, PrasannaKumar S, Chandrasekar N (2014) Seasonal variability of beach characteristics between Candoliam and Colva coast, Goa, India

Yin K, Xu S, Huang W, Li R, Xiao H (2019) Modeling beach profile changes by typhoon impacts at Xiamen coast. Nat Hazards 95(3):783–804

Chapter 11
Heavy Minerals Studies of Coastal Sands from Bavanapadu to Kalingapatnam, Andhra Pradesh, East Coast of India

A. Lakshmi Venkatesh, K. S. N. Reddy, K. Bangaku Naidu, Ch. Aruna, N. Ankita Varma, and K. Sandeep Kumar

Abstract To better understand the heavy mineral assemblage, fraction-wise distribution, and concentration, 40 surface sediment samples were analyzed from Bavanapadu to Kalingapatnam, Andhra Pradesh. Ilmenite, sillimanite, garnet, rutile, kyanite, zircon, monazite, and other heavy minerals make up the heavy mineral assemblage in coastal sediments from Bavanapadu to Kalingapatnam. Total heavy minerals (THM) range from 13.10 to 27.98 wt% (avg. 20.54 wt%) in all sediment samples, and their assemblage is not uniform in berm (27.98 wt%), dune (23.26 wt%), backshore (19.21 wt%), and foreshore (19.21 wt%) (13.10 wt%). The total heavy mineral (THM) weight percentage is lower in coarse fractions than in fine fractions. The mineral assemblages of transparent heavy minerals (ilmenite, sillimanite, garnet, rutile, zircon, leucoxene, kyanite, and monazite) found in the Eastern Ghats Mobile Belt (EGMB) rocks are suggestive of their derivation from a heterogeneous provenance containing high-grade metamorphic rocks such as charnockite and khondalite, which characterize the Precambrian age. The heavy mineral assemblage in various sand units (i.e., foreshore, berm, backshore, and dune) indicates that the sediments are derived from the Eastern Ghat suite of rocks' khondalite and charnockite groups.

Keywords Foreshore · Backshore · Berm and dune · Heavy minerals · Khondalites · Charnockites

---

A. Lakshmi Venkatesh (✉) · K. S. N. Reddy · C. Aruna · N. Ankita Varma · K. Sandeep Kumar
Department of Geology, Andhra University, Visakhapatnam, Andhra Pradesh, India

K. Bangaku Naidu
Department of Geology, Sir.C. R. Reddy Degree College, Eluru, Andhra Pradesh, India

N. Jayaraju et al. (eds.), *Coasts, Estuaries and Lakes*,
https://doi.org/10.1007/978-3-031-21644-2_11

11.1 Introduction

The heavy mineral assemblage, their distribution, concentration, process, and depositional circumstances, as well as their significance for understanding the provenance of sediments between Bavanapadu and Kalingapatnam coast, were studied in the Srikakulam, which is next to the analyzed area. Bavanapadu, located between Kalingapatnam and Baruva, is one of the primary greenfield ports proposed by the Government of Andhra Pradesh. The northeast monsoon brings rain to this coastal zone. Visakhapatnam lies to the south, the Bay of Bengal is to the east, and Brahmapur is to the north. Charnockites, khondalite, and gneisses make up the majority of the bedrock. From the Pleistocene to the Holocene, the beach terrace and the majority of the bedrock along Andhra Pradesh's south-eastern coast were developed (Sahayam 2010).

Because heavy mineral suites were deposited by numerous geological processes such as wind, water, glaciers, and gravity, heavy mineral investigations are conducted to understand sediment provenance. These are more resistant to physical and chemical weathering, revealing valuable information about the mineralogical composition of source places.

Heavy mineral studies along the east coast of India were studied by Mahadevan and Sriramadas (1948, 1954), Mahadevan and Sathapathi (1948), Mahadevan and Rao Nageswara (1950), Sriramadas (1951), Borreswara Rao and La Fond (1958), Mahadevan et al. (1958), Sastry et al. (1981, 1987), Ramamohana Rao et al. (1982), Dhanunjaya Rao et al. (1989), Deva Varma et al. (1989), Rao et al. (1993, 2001), Sreenivasa Rao et al. (1995), Rajasekhara Reddy et al. (1998, 2001), Mohan and Rajamanickam (2000), Angusamy and Rajamanickam (2000), Ravi et al. (2001), Chandrasekhar et al. (2005), Dhana Raju (2006), Reddy et al. (2007, 2012), Acharya et al. (2009), Ramasamy and Karikalan (2010), Anil et al. (2011), Cheepurupalli et al. (2012), Vinoth Kumar and Asaithambi (2013), Joevivek and Chandrasekar (2014), Suresh Gandhi and Raja (2014), Mohapatra et al. (2015), Murali Krishna et al. (2016), Bangaku naidu et al. (2017), Mohammad et al. (2020), and Ganapathi Rao et al. (2020).

Heavy mineral assemblages in current coastal sands from Bavanapadu to Kalingapatnam are discussed, as well as their concentration, distribution, and provenance

11.2 Study Area

The current research region is located between 18°20′ and 18°35′ N latitudes and 84°5′ and 84°30′ E longitudes, with a 42-kilometer shoreline as indicated in Fig. 11.1. Srikakulam is the district headquarters of Srikakulam and is close to the study area. The climate of Srikakulam is mildly humid. The average yearly temperature is 19.3–35.7 °C, with the maximum in May and the lowest in January.

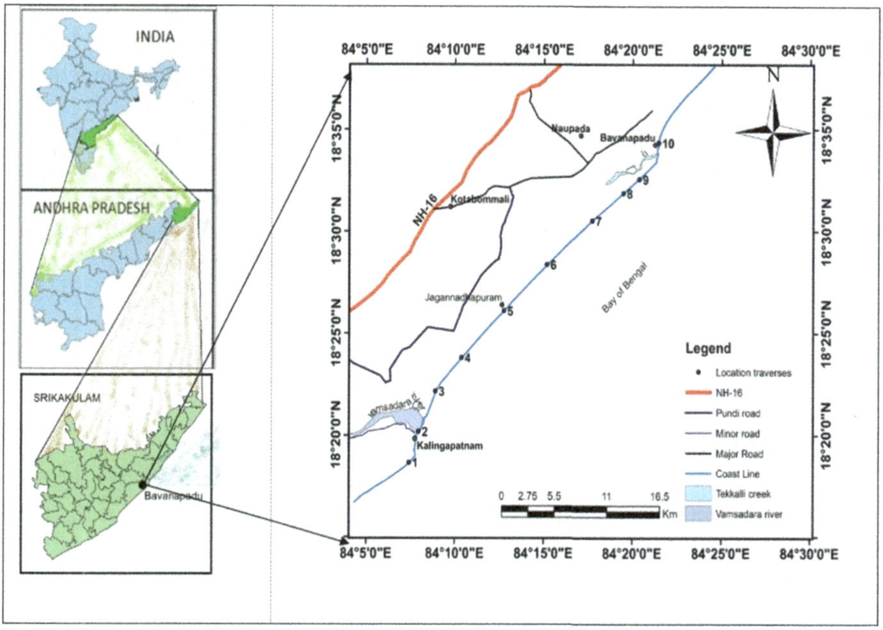

Fig. 11.1 Sample location map of the study area

Furthermore, the current research region, which is located between the Bavanapadu and Kalingapatnam beaches, has a humid climate, which means that it is hot and humid with monsoon-controlled seasonal rainfall.

11.3 Methodology

Forty sediment samples were taken by entering a PVC tube (4 inches diameter) to a depth of 30 cm from the surface (10 in dune, 10 in backshore, 10 in berm, and 10 in foreshore). All of the samples (40) were taken from places with the lowest low-tide (LT) and highest high-tide (HT). To remove salts from the sediment samples, they are repeatedly rinsed with distilled water and then dried. After drying, a sub-sample weighing about 100 grammes was prepared using the Jones riffle splitter; the samples were treated with (1:10) dilute hydrochloric acid (HCl) and (1:8) hydrogen peroxide (H_2O_2), respectively, to remove carbonate, shell content and organic debris, and then dried. This method usually takes 12 hours to provide good results.

Different sizes of sieves were used to separate the +230 mesh materials, resulting in three fractions (+60, +120, and +230). These sub-samples were utilized to separate light and heavy minerals (sink-float method) using bromoform (Sp. Gr 2.89, g/cm^3) in accordance with standard procedure. The lighter and heavier fractions were utilized to quantify the weight percentage of heavy minerals in three size fractions after these samples were dried in a hot air oven. The weight % of heavy minerals is computed using the magnetic minerals extracted by bar magnetic. Canada balsam consists of heavy minerals put on glass slides. Heavy mineral determination requires a polished strewn slide.

The relative abundance of heavy mineral grains is calculated using a typical petrological microscope's point-counting approach. After counting 250–300 grains on each dispersed slide, the number percentage (percent) can be computed. Tables 11.1, 11.2, 11.3, and 11.4 show the distribution of heavy mineral weight percentages in the +60, +120, and +230 fractions, as well as the whole sample.

11.4 Results and Discussion

11.4.1 Distribution of Heavy Minerals Percentages

The total heavy mineral weight percentage in the coarse fraction (+60) of the foreshore is 1.38 (Table 11.1), with 42% garnet, 36% Ilmenite, 14% sillimanite, 3% rutile, and 3% other heavy minerals such as hypersthenes. THM wt. percent in the medium fraction (+120) is 7.52 (Table 11.2), with individual heavy minerals accounting for 36% ilmenite, 31% sillimanite, 29% garnet, 1% rutile, 0.7% monazite, 0.2% kyanite, and 2% miscellaneous heavy minerals. THM wt percent in the fine fraction (+230) is 3.53 (Table 11.3), with 50% ilmenite, 39% sillimanite, 12% garnet, 3% kyanite, 1% rutile, 1% monazite, 1% zircon, and 1% other heavy minerals. THM weight percent in the shoreline is 13.10. The THM contains 42% ilmenite, 31% sillimanite, 28% garnet, 1% kyanite, 1% monazite, and 1% other heavy minerals (Table 11.4). The heavy mineral association in shoreline coastal sediments shows a decreasing abundance of ilmenite, garnet, sillimanite, and rutile, modest concentration increases of zircon, kyanite, and monazite (less than 1%), and other heavy minerals amphibole, leucoxene, and epidote.

The average weight percentage of Heavy minerals in the berm zone ranges from 3.15 to 80.35 (avg. 27.98%) in the research region. The heavy mineral weight percentage ranges from 0.24 to 11.20 (avg. 3.66%) in the +60 fraction, 1.24 to 58.57 (avg. 18.68%) in the +120 fraction, and 0.75 to 10.57 (avg. 5.47%) in the +230 fraction (Table 11.4). The average total heavy mineral weight percent in the coarse fraction (+60) is 3.66 (Table 11.1).

There are 39% garnets, 30% ilmenites, 23% sillimanite, 4% rutile, 0.8% zircons, 0.6% kyanite, and 2% other heavy minerals in the THM distribution pattern of minerals. The overall heavy mineral weight percentage in the medium portion (+120) is 18.68 (Table 11.2). Ilmenite is 45%, sillimanite is 27%, garnets are 25%, rutile is 2%, monazite is 0.9%, and other heavy minerals are 2%. The total heavy minerals weight percent in the finer fraction (+230) is 5.47 (Table 11.3). Ilmenite is 53%, sillimanite is 20%, garnet is 13%, rutile is 3%, kyanite is 3%, monazite is 2%, and zircon is 0.5%. Other heavy minerals are 3%. THM weight % is 27.98 in a berm

Table 11.1 Distribution of heavy minerals (wt%) of coastal sediments from Bavanapadu to Kalingapatnam beach in +60 fractions

Sample no	Sil	Gar	Kya	Ilm	Mon	Rut	Zir	Oth	Mag	THM%
Foreshore										
1	0.11	0.06	0.01	0.02	0.01	0.08	0.00	0.00	0.00	0.30
2	0.08	0.05	0.01	0.14	0.00	0.06	0.00	0.00	0.00	0.34
3	0.09	0.09	0.01	0.09	0.01	0.03	0.00	0.00	0.00	0.31
4	0.99	4.28	0.00	3.04	0.00	0.26	0.00	0.33	0.00	8.90
5	0.18	0.42	0.00	0.60	0.00	0.03	0.00	0.06	0.00	1.28
6	0.10	0.15	0.00	0.24	0.00	0.01	0.00	0.02	0.00	0.52
7	0.26	0.58	0.00	0.54	0.00	0.01	0.00	0.03	0.00	1.42
8	0.07	0.09	0.00	0.17	0.00	0.00	0.00	0.01	0.00	0.34
9	0.05	0.04	0.00	0.12	0.00	0.01	0.00	0.02	0.00	0.23
10	0.03	0.03	0.00	0.07	0.00	0.00	0.00	0.01	0.00	0.13
Min	0.03	0.03	0.01	0.02	0.01	0.00	0.00	0.00	0.00	0.13
Max	0.99	4.28	0.01	3.04	0.01	0.26	0.00	0.33	0.00	8.90
Avg.	0.20	0.58	0.00	0.50	0.00	0.05	0.00	0.05	0.00	1.38
Berm										
1	3.01	2.00	0.00	1.54	0.49	0.53	0.00	0.00	0.00	7.57
2	0.27	0.36	0.00	0.35	0.00	0.19	0.00	0.00	0.00	1.17
3	0.05	0.29	0.00	0.14	0.00	0.11	0.00	0.00	0.00	0.58
4	1.65	6.57	0.00	2.53	0.00	0.22	0.24	0.00	0.00	11.20
5	1.40	1.71	0.05	1.82	0.00	0.21	0.00	0.18	0.00	5.38
6	1.33	2.88	0.00	2.92	0.00	0.04	0.04	0.35	0.00	7.56
7	0.77	1.19	0.09	0.98	0.00	0.00	0.00	0.11	0.00	3.14
8	0.07	0.05	0.00	0.10	0.00	0.02	0.00	0.01	0.00	0.24
9	0.53	0.27	0.03	0.89	0.04	0.30	0.00	0.00	0.00	2.06
10	0.16	0.30	0.00	0.64	0.01	0.00	0.01	0.04	0.00	1.16
Min	0.05	0.05	0.00	0.10	0.00	0.00	0.00	0.00	0.00	0.24
Max	3.01	6.57	0.09	2.92	0.49	0.53	0.24	0.35	0.00	11.20
Avg.	0.84	1.42	0.02	1.09	0.05	0.15	0.03	0.06	0.00	3.66
Backshore										
1	0.45	0.50	0.00	0.14	0.04	0.00	0.00	0.00	0.00	1.12
2	1.05	3.49	0.00	0.80	0.00	0.69	0.00	0.34	0.00	6.38
3	0.03	0.05	0.00	0.14	0.00	0.00	0.00	0.00	0.00	0.22
4	2.14	7.00	0.56	2.77	0.00	0.00	0.00	0.00	0.00	12.47
5	0.17	0.37	0.01	0.44	0.00	0.04	0.00	0.13	0.00	1.16
6	0.45	2.07	0.00	1.02	0.00	0.06	0.00	0.08	0.00	3.68
7	1.25	1.28	0.00	1.33	0.00	0.08	0.00	0.11	0.00	4.04
8	0.29	0.26	0.00	0.19	0.01	0.03	0.00	0.05	0.00	0.83
9	0.70	0.16	0.00	1.25	0.00	0.00	0.00	0.06	0.00	2.17
10	0.02	0.02	0.00	0.02	0.00	0.01	0.00	0.00	0.00	0.07
Min	0.02	0.02	0.00	0.02	0.00	0.00	0.00	0.00	0.00	0.07
Max	2.14	7.00	0.56	2.77	0.04	0.69	0.00	0.34	0.00	12.47

(continued)

Table 11.1 (continued)

Sample no	Sil	Gar	Kya	Ilm	Mon	Rut	Zir	Oth	Mag	THM%
Avg.	0.73	1.85	0.09	0.91	0.01	0.13	0.00	0.09	0.00	3.72
Dune										
1	0.12	0.13	0.00	0.04	0.01	0.00	0.00	0.00	0.00	0.30
2	0.19	0.71	0.00	0.47	0.19	0.00	0.00	0.00	0.00	1.56
3	0.13	0.60	0.00	0.20	0.00	0.00	0.00	0.00	0.00	0.93
4	0.11	0.43	0.01	0.08	0.00	0.00	0.00	0.00	0.00	0.63
5	0.02	0.04	0.00	0.04	0.00	0.00	0.00	0.01	0.00	0.11
6	0.08	0.14	0.00	0.08	0.00	0.00	0.00	0.01	0.00	0.31
7	0.13	0.12	0.00	0.13	0.00	0.00	0.00	0.02	0.00	0.40
8	0.14	0.03	0.00	0.23	0.00	0.00	0.00	0.00	0.00	0.40
9	0.32	0.43	0.00	0.50	0.00	0.42	0.00	0.12	0.00	1.79
10	0.03	0.07	0.00	0.05	0.01	0.04	0.00	0.00	0.00	0.19
Min	0.02	0.03	0.00	0.04	0.00	0.00	0.00	0.00	0.00	0.11
Max	0.32	0.71	0.01	0.50	0.19	0.42	0.00	0.12	0.00	1.79
Avg.	0.13	0.29	0.00	0.20	0.03	0.07	0.00	0.02	0.00	0.71

Sil sillimanite, *Gar* garnet, *Kya* kyanite, *Ilm* ilmenite, *Mon* monazite, *Rut* Rutile, *Zir* zircon, *Oth* other heavy minerals, *THM%* total heavy mineral %, *Min* minimum, *Max* maximum, *Avg.* average

environment. It comprises 45% ilmenite, 25% garnet, 25% sillimanite, 2% rutile, 1% monazite, zircon, mangitite, and kyanite, and 1% monazite, zircon, mangitite, and kyanite are 1% each and 2% other heavy minerals (Table 11.4). The garnet, ilmenite, sillimanite, and rutile decrease in their concentration with the increase in the course to fine fractions. Whereas zircon, monazite other heavy minerals (epidote, amphibole, and staurolite) increase (up to 2%) in the berm zone.

In the backshore region of the research area, the weight percentage of heavy minerals ranges from 5.44 to 44.66 (avg. 19.21%) (Table 11.4). The heavy mineral weight percentage ranges from 0.07 to 12.47 (avg. 3.72%) in the +60 fraction (Table 11.1), 2.43 to 31.26 (avg. 11.03%) in the +120 fraction (Table 11.2), and 0.84 to 12.86 (avg. 5.24%) in the +230 fraction (Table 11.3). The average wt. percent of THM in the coarser fraction (+60) is 3.72 (Table 11.1). This comprises 50% garnet, 24% ilmenite, 20% sillimanite, 4% rutile, 2% kyanite, and 3% other heavy minerals. The average THM wt. percent in the middle portion (+120) is 11.03 (Table 11.2). Sillimanite makes up 33%, ilmenite 31%, garnet 31%, zircon, and rutile 2% each, and other heavy minerals 2%. THM weight percent is 5.24 in the finer fraction (+230) (Table 11.3). Ilmenite makes up 45%, sillimanite 34%, garnet 17%, rutile and kyanite 1% each, zircon and monazite 1% each, and other heavy minerals 3%. THM wt.% in the backshore environment is 19.21, with 34% ilmenite, 30% sillimanite, 31% garnet, 2% rutile, and 1% to 2% additional heavy minerals such as zircon, kyanite, and monazites (Table 11.4). The heavy mineral composition in Backshore coastal sediments shows a rising abundance of ilmenite, followed by sillimanite, garnet, and rutile, with small concentration increases of monazite, zircon, and kyanite, as well as other heavy minerals including amphibole, staurolite, and epidote (up to 2%).

Table 11.2 Distribution of heavy minerals (wt%) of coastal sediments from Bavanapadu to Kalingapatnam coast in −60 to +120 fraction

Sample no	Sil	Gar	Kya	Ilm	Mon	Rut	Zir	Oth	Mag	THM %
Foreshore										
1	0.09	0.01	0.01	1.09	0.00	0.01	0.00	0.00	0.01	1.20
2	4.65	1.09	0.06	0.47	0.33	0.27	0.00	0.00	0.00	6.87
3	0.09	0.09	0.00	0.23	0.02	0.03	0.00	0.00	0.01	0.46
4	8.36	11.44	0.00	13.58	0.00	0.00	0.00	0.00	0.00	33.38
5	3.55	1.14	0.00	1.81	0.00	0.17	0.00	0.35	0.00	7.03
6	0.34	0.31	0.00	0.84	0.08	0.10	0.00	0.00	0.00	1.68
7	0.77	0.26	0.00	0.79	0.00	0.04	0.00	0.11	0.00	1.97
8	0.97	0.23	0.00	0.98	0.00	0.08	0.00	0.15	0.00	2.41
9	1.46	0.35	0.06	0.41	0.00	0.14	0.00	0.04	0.01	2.46
10	0.09	0.04	0.00	0.04	0.00	0.03	0.00	0.00	0.01	0.21
Min	0.09	0.01	0.00	0.04	0.00	0.00	0.00	0.00	0.00	0.21
Max	8.36	11.44	0.06	13.58	0.33	0.27	0.00	0.35	0.01	33.38
Avg.	2.40	2.20	0.02	2.74	0.06	0.10	0.00	0.08	0.00	7.52
Berm										
1	7.19	2.10	0.00	1.98	0.00	0.75	0.00	0.47	0.00	12.50
2	3.48	0.68	0.12	1.41	0.17	0.00	0.00	0.00	0.10	5.85
3	4.53	2.62	0.04	1.83	0.00	0.17	0.00	0.00	0.01	9.19
4	9.21	16.21	0.00	31.74	0.75	0.00	0.66	0.00	0.05	58.57
5	10.47	7.59	0.00	10.21	0.31	0.25	0.00	1.04	0.04	29.87
6	6.20	4.71	0.00	8.82	0.00	0.12	0.00	0.48	0.03	20.33
7	2.93	2.40	0.00	5.24	0.00	0.19	0.00	0.43	0.01	11.20
8	4.55	1.98	0.04	4.74	0.00	0.48	0.00	0.60	0.13	12.39
9	0.99	0.71	0.00	1.52	0.00	0.00	0.00	0.00	0.01	3.22
10	0.28	0.31	0.01	0.56	0.00	0.04	0.00	0.05	0.01	1.24
Min	0.28	0.31	0.00	0.56	0.00	0.00	0.00	0.00	0.00	1.24
Max	10.47	16.21	0.12	31.74	0.75	0.75	0.66	1.04	0.13	58.57
Avg.	5.05	4.65	0.03	8.36	0.17	0.23	0.11	0.34	0.04	18.68
Backshore										
1	1.42	1.48	0.06	0.70	0.08	0.27	0.00	0.00	0.00	4.00
2	5.07	5.05	0.00	8.73	0.00	0.00	0.00	0.69	0.02	19.54
3	3.32	1.11	0.04	0.87	0.11	0.09	0.00	0.00	0.02	5.54
4	7.91	11.85	0.00	9.76	0.00	1.00	0.73	0.00	0.01	31.26
5	2.44	1.97	0.00	2.28	0.00	0.14	0.00	0.35	0.00	7.18
6	3.81	2.39	0.05	2.76	0.00	0.21	0.00	0.26	0.01	9.48
7	3.64	2.41	0.00	3.13	0.00	0.12	0.00	0.35	0.01	9.65
8	2.98	1.03	0.07	1.59	0.11	0.17	0.00	0.00	0.04	5.95
9	1.66	0.25	0.13	0.29	0.00	0.05	0.05	0.00	0.01	2.43
10	2.08	0.72	0.00	0.64	0.05	0.08	0.05	0.00	0.02	3.62
Min	1.42	0.25	0.00	0.29	0.00	0.00	0.00	0.00	0.00	2.43
Max	7.91	11.85	0.13	9.76	0.11	1.00	0.73	0.69	0.04	31.26

(continued)

Table 11.2 (continued)

Sample no	Sil	Gar	Kya	Ilm	Mon	Rut	Zir	Oth	Mag	THM %
Avg.	3.64	3.36	0.04	3.40	0.04	0.26	0.13	0.20	0.02	11.03
Dune										
1	8.34	5.78	0.35	3.45	0.09	0.28	0.00	0.00	0.00	18.29
2	2.82	1.83	0.00	3.36	0.05	0.04	0.00	0.45	0.00	8.56
3	9.02	6.92	0.23	8.59	0.00	0.26	0.00	0.00	0.00	25.02
4	9.08	5.52	0.00	8.58	0.00	0.13	0.83	0.00	0.01	24.13
5	7.31	2.01	0.00	2.65	0.00	0.00	0.00	0.47	0.00	12.45
6	10.90	5.61	0.00	7.54	0.00	0.65	0.00	0.85	0.01	25.54
7	3.08	1.26	0.00	2.18	0.00	0.09	0.00	0.18	0.03	6.80
8	4.82	1.64	0.00	1.80	0.10	0.16	0.00	0.00	0.01	8.53
9	5.47	1.54	0.13	1.25	0.57	0.31	0.00	0.38	0.01	9.64
10	1.10	0.20	0.02	0.18	0.03	0.07	0.00	0.02	0.00	1.62
Min	1.10	0.20	0.00	0.18	0.00	0.00	0.00	0.00	0.00	1.62
Max	10.90	6.92	0.35	8.59	0.57	0.65	0.83	0.85	0.03	25.54
Avg.	6.16	3.29	0.09	4.03	0.12	0.22	0.14	0.27	0.01	13.98

Sil sillimanite, *Gar* garnet, *Kya* kyanite, *Ilm* ilmenite, *Mon* monazite, *Rut* Rutile, *Zir* zircon, *Oth* other heavy minerals, *THM%* total heavy mineral %, *Min* minimum, *Max* maximum, *Avg.* average

In the dune environment of the research area, the weight percentage of heavy minerals ranges from 7.56 to 38.45 (avg. 23.26%) (Table 11.4). The weight percentage ranges from 0.11 to 1.79 (avg. 0.71%), 1.62 to 25.54 (avg. 13.98%), and 3.27 to 16.29 (avg. 8.86%) in the +60 fraction (Table 11.4). The average total heavy mineral (THM) weight percentage in the coarse fraction (+60) is 0.71 (Table 11.1), with 40% garnet, 29% ilmenite, 10% rutile, 19% sillimanite, 4% monazite, and 3% other heavy minerals. The average wt. percent of THM in the medium fraction (+120) is 13.98 (Table 11.2), consisting of 44% sillimanite, 29% ilmenite, 24% garnets, 2% rutile, 1% monazite, 1% zircon, and 2% other heavy minerals. In the fine fraction (+230) the average wt.% of THM is 8.86 (Table 11.3). in which 44% ilmenite 34% sillimanite, 20% garnets, 2% kyanite, 2% rutile, zircon, and monazite are <1% each, and 2% other heavy minerals. The total amount of heavy minerals in a dune environment is 23.26%, with sillimanite accounting for 39%, ilmenite for 36%, garnets for 22%, rutile for 2%, zircons for 1%, and other heavy minerals accounting for 2% (Table 11.4). The ilmenite and sillimanite fractions rise in the dune environment, while garnet percentages decrease.

In the research region, the proportion of total heavy minerals (THM) ranges from 1.70 to 49.90 (avg. 13.10%) (Table 11.4). Heavy mineral percentages range from 0.13 to 8.90 (avg. 1.38%) in the +60 fraction, 0.21 to 33.38 (avg. 7.52%) in the +120 fraction, and 0.57 to 7.62 (avg. 3.53%) in the +230 fraction (Table 11.4).

Compare the number of minerals in all four microenvironments (dune, backshore, berm, and foreshore). The garnet is the most prevalent mineral in Berm settings, but ilmenite is the most abundant mineral on the shoreline, backshore, and dune. Khondalite and charnockite are heavy minerals found in Eastern Ghats Mobile Belt (EGMB) rocks.

Table 11.3 Distribution of heavy minerals (wt%) of coastal sediments from Bavanapadu to Kalingapatnam coast in −120 to +230 fraction

Sample no	Sil	Gar	Kya	Ilm	Mon	Rut	Zir	Oth	Mag	THM%
Foreshore										
1	1.80	0.20	0.00	1.15	0.15	0.08	0.09	0.00	0.02	3.46
2	0.98	0.62	0.09	0.81	0.06	0.05	0.00	0.00	0.07	2.62
3	0.63	0.15	0.07	0.22	0.03	0.01	0.01	0.00	0.04	1.11
4	0.53	0.76	0.08	6.06	0.16	0.03	0.00	0.00	0.07	7.62
5	1.09	0.67	0.00	3.00	0.00	0.08	0.00	0.14	0.09	4.98
6	0.52	0.19	0.00	0.46	0.01	0.03	0.00	0.03	0.09	1.24
7	0.23	0.12	0.00	0.21	0.00	0.00	0.00	0.01	0.00	0.57
8	2.17	0.69	0.08	2.07	0.00	0.06	0.00	0.19	0.01	5.25
9	3.86	0.33	0.50	1.17	0.08	0.00	0.07	0.00	0.08	6.00
10	0.84	0.25	0.00	0.07	0.04	0.06	0.00	0.10	0.10	1.35
Min	0.23	0.12	0.00	0.07	0.00	0.00	0.00	0.00	0.00	0.57
Max	3.86	0.76	0.50	6.06	0.16	0.08	0.09	0.19	0.10	7.62
Avg.	1.40	0.41	0.11	1.78	0.06	0.04	0.02	0.06	0.06	3.53
Berm										
1	0.56	0.60	0.01	1.21	0.00	0.00	0.00	0.06	0.19	2.44
2	0.49	0.37	0.04	0.80	0.05	0.00	0.00	0.00	0.14	1.75
3	0.40	0.20	0.04	0.46	0.01	0.00	0.01	0.00	0.20	1.12
4	0.57	0.51	0.09	9.04	0.37	0.00	0.00	0.00	0.18	10.57
5	1.26	0.87	0.00	3.52	0.29	0.29	0.02	0.24	0.19	6.49
6	0.80	0.41	0.00	1.63	0.06	0.14	0.03	0.10	0.11	3.17
7	0.50	0.33	0.00	1.45	0.00	0.05	0.00	0.03	0.20	2.36
8	6.97	2.18	0.05	6.28	0.00	0.81	0.00	0.81	0.17	17.09
9	0.69	0.30	0.17	0.91	0.00	0.00	0.01	0.00	0.20	2.08
10	0.30	0.13	0.04	0.26	0.01	0.00	0.01	0.00	0.25	0.75
Min	0.30	0.13	0.00	0.26	0.00	0.00	0.00	0.00	0.11	0.75
Max	6.97	2.18	0.17	9.04	0.37	0.81	0.03	0.81	0.25	17.09
Avg.	1.65	0.68	0.05	2.91	0.10	0.18	0.01	0.17	0.18	5.47
Backshore										
1	0.61	0.85	0.02	2.57	0.00	0.00	0.00	0.08	0.17	4.13
2	1.12	1.59	0.08	4.77	0.00	0.00	0.00	0.00	0.21	7.56
3	1.83	0.54	0.12	0.72	0.00	0.00	0.02	0.00	0.18	3.23
4	0.07	0.13	0.01	0.65	0.07	0.00	0.00	0.00	0.14	0.93
5	1.17	0.40	0.00	2.25	0.07	0.05	0.01	0.19	0.10	4.15
6	1.01	1.04	0.00	1.71	0.09	0.06	0.03	0.11	0.13	4.05
7	0.66	0.81	0.00	1.66	0.06	0.25	0.00	0.08	0.25	3.53
8	1.70	0.85	0.43	4.84	0.08	0.00	0.04	0.00	0.12	7.94
9	0.30	0.11	0.00	0.34	0.02	0.03	0.00	0.04	0.13	0.84
10	6.38	2.12	0.00	3.66	0.00	0.26	0.00	0.44	0.18	12.86
Min	0.07	0.11	0.00	0.34	0.00	0.00	0.00	0.00	0.10	0.84
Max	6.38	2.12	0.43	4.84	0.09	0.26	0.04	0.44	0.25	12.86

(continued)

Table 11.3 (continued)

Sample no	Sil	Gar	Kya	Ilm	Mon	Rut	Zir	Oth	Mag	THM%
Avg.	1.78	0.89	0.09	2.36	0.04	0.08	0.01	0.12	0.16	5.24
Dune										
1	7.31	5.07	0.31	3.03	0.07	0.24	0.00	0.25	0.03	16.29
2	3.90	1.73	0.00	2.74	0.00	0.00	0.00	0.14	0.09	8.50
3	1.77	2.02	0.57	8.14	0.00	0.00	0.00	0.00	0.10	12.50
4	1.70	1.11	0.21	8.11	0.25	0.00	0.00	0.00	0.10	11.38
5	2.12	0.79	0.00	1.95	0.08	0.00	0.05	0.20	0.06	5.20
6	2.97	1.53	0.00	5.30	0.11	0.35	0.09	0.29	0.28	10.64
7	1.17	0.68	0.00	1.33	0.00	0.03	0.00	0.07	0.19	3.27
8	2.70	1.05	0.00	3.45	0.11	0.18	0.10	0.27	0.20	7.86
9	1.81	0.61	0.06	2.31	0.09	0.24	0.00	0.22	0.16	5.33
10	2.37	1.05	0.30	1.82	0.04	0.00	0.19	0.00	0.23	5.77
Min	1.17	0.61	0.00	1.33	0.00	0.00	0.00	0.00	0.03	3.27
Max	7.31	5.07	0.57	8.14	0.25	0.35	0.19	0.29	0.28	16.29
Avg.	3.03	1.78	0.17	3.97	0.08	0.12	0.05	0.14	0.15	8.86

Sil sillimanite, *Gar* garnet, *Kya* kyanite, *Ilm* ilmenite, *Mon* monazite, *Rut* Rutile, *Zir* zircon, *Oth* other heavy minerals, *THM%* total heavy mineral %, *Min* minimum, *Max* maximum, *Avg.* average

11.4.2 Comparative Abundance of Heavy Minerals

The surface variation graphs (Fig. 11.2) show the relative abundance of heavy minerals in three size fractions (+60, +120, and +230) as well as the average total heavy mineral (THM) weight percentages at various study locations.

The average heavy mineral concentration is higher (8.30%) at traverse 4 and lower (0.39%) at traverse 10. In the course (+60) portion, the average heavy mineral concentration is higher (8.30%) at traverse 4 and lower (0.39%) at traverse 10. In the medium fraction, average heavy mineral concentration is higher (36.84%) at traverse 4 and lower (1.67%) at traverse 10, while in the fine (+230) fraction, heavy mineral concentration is higher (9.53%) at traverse 8 and lower (2.43%) at traverse 10 in the study area. In this case, the medium (+120) fraction of the research region has a higher heavy mineral concentration than the course (+60) and fine fractions (230).

11.4.3 Along Shore Variation of Individual Minerals

The line graphs of individual heavy minerals, such as ilmenite, garnet, sillimanite, kyanite, rutile, monazite, zircon, magnetite, and other heavy minerals, are shown in Fig.11.3.

Ilmenite concentration is higher at 4 traverses and lower at 10 traverses. In the 4 traverses, there was a high concentration of sillimanite, while in the 10 traverses,

Table 11.4 Heavy minerals (wt%) distribution in different microenvironmental units between Bavanapadu to Kalingapatnam coast

Sample no	Sil	Gar	Kya	Ilm	Mon	Rut	Zir	Oth	Mag	THM%
Foreshore										
1	2.00	0.27	0.02	1.26	0.16	0.17	0.09	0.00	0.03	4.96
2	5.71	1.76	0.16	1.42	0.39	0.38	0.00	0.00	0.07	9.83
3	0.81	0.33	0.08	0.54	0.06	0.07	0.01	0.00	0.05	1.88
4	9.88	16.48	0.08	22.68	0.16	0.29	0.00	0.33	0.07	49.90
5	4.82	2.23	0.00	5.41	0.00	0.28	0.00	0.55	0.09	13.29
6	0.96	0.65	0.00	1.54	0.09	0.14	0.00	0.05	0.09	3.44
7	1.26	0.96	0.00	1.54	0.00	0.05	0.00	0.15	0.00	3.96
8	3.21	1.01	0.08	3.22	0.00	0.14	0.00	0.35	0.01	8.00
9	5.37	0.72	0.56	1.70	0.08	0.15	0.07	0.06	0.09	8.69
10	0.96	0.32	0.00	0.18	0.04	0.09	0.00	0.11	0.11	1.69
Min	0.81	0.27	0.00	0.18	0.00	0.05	0.00	0.00	0.00	0.91
Max	9.88	16.48	0.56	22.68	0.39	0.38	0.09	0.55	0.11	49.90
Avg.	3.81	3.46	0.13	5.20	0.11	0.18	0.02	0.18	0.06	13.10
Berm										
1	10.76	4.70	0.02	4.73	0.49	1.28	0.00	0.53	0.19	22.51
2	4.24	1.41	0.10	2.56	0.22	0.19	0.00	0.00	0.24	8.77
3	4.98	3.11	0.04	2.43	0.01	0.28	0.01	0.00	0.21	10.89
4	11.43	23.29	0.09	43.31	1.12	0.22	0.90	0.00	0.23	80.34
5	13.13	10.17	0.05	15.55	0.60	0.75	0.02	1.46	0.23	41.74
6	8.33	8.00	0.00	13.37	0.06	0.30	0.07	0.93	0.14	31.06
7	4.20	3.92	0.09	7.67	0.00	0.24	0.00	0.57	0.21	16.70
8	11.59	4.21	0.05	11.12	0.00	1.31	0.00	1.42	0.30	29.72
9	2.21	1.28	0.26	3.32	0.04	0.30	0.01	0.00	0.21	7.36
10	0.74	0.74	0.04	1.46	0.02	0.04	0.02	0.09	0.26	3.15
Min	0.74	0.74	0.00	1.46	0.00	0.04	0.00	0.00	0.14	3.15
Max	13.13	23.29	0.26	43.31	1.12	1.31	0.90	1.46	0.30	80.34
Avg.	7.12	7.07	0.08	12.52	0.31	0.52	0.16	0.54	0.22	27.98
Backshore										
1	2.48	2.83	0.08	3.41	0.12	0.27	0.00	0.08	0.17	9.25
2	7.24	10.13	0.08	14.30	0.00	0.69	0.00	1.03	0.23	33.48
3	5.18	1.70	0.16	1.73	0.11	0.09	0.02	0.00	0.20	8.99
4	10.12	18.98	0.57	13.18	0.07	1.00	0.73	0.00	0.15	44.66
5	3.78	2.74	0.01	4.97	0.07	0.23	0.01	0.67	0.10	12.49
6	5.27	5.50	0.05	5.49	0.09	0.33	0.03	0.45	0.14	17.21
7	5.55	4.50	0.00	6.12	0.06	0.45	0.00	0.54	0.26	17.22
8	4.97	2.14	0.50	6.62	0.20	0.20	0.04	0.05	0.16	14.72
9	2.66	0.52	0.13	1.88	0.02	0.08	0.05	0.10	0.14	5.44
10	8.48	2.86	0.00	4.32	0.05	0.35	0.05	0.44	0.20	16.55
Min	2.48	0.52	0.00	1.73	0.00	0.08	0.00	0.00	0.10	5.44
Max	10.12	18.98	0.57	14.30	0.20	1.00	0.73	1.03	0.26	44.66

(continued)

Table 11.4 (continued)

Sample no	Sil	Gar	Kya	Ilm	Mon	Rut	Zir	Oth	Mag	THM%
Avg.	5.69	5.95	0.18	6.50	0.08	0.40	0.14	0.37	0.18	19.21
Dune										
1	15.77	10.98	0.66	6.52	0.17	0.52	0.00	0.25	0.03	34.88
2	6.91	4.27	0.00	6.57	0.24	0.04	0.00	0.59	0.09	18.62
3	10.92	9.54	0.80	16.93	0.00	0.26	0.00	0.00	0.10	38.45
4	10.89	7.06	0.22	16.77	0.25	0.13	0.83	0.00	0.11	36.14
5	9.45	2.84	0.00	4.64	0.08	0.00	0.05	0.68	0.06	17.76
6	13.95	7.28	0.00	12.92	0.11	1.00	0.09	1.15	0.29	36.49
7	4.38	2.06	0.00	3.64	0.00	0.12	0.00	0.27	0.22	10.47
8	7.66	2.72	0.00	5.48	0.21	0.34	0.10	0.27	0.21	16.79
9	7.60	2.58	0.19	4.06	0.66	0.97	0.00	0.72	0.17	16.76
10	3.50	1.32	0.32	2.05	0.08	0.11	0.19	0.02	0.23	7.58
Min	3.50	1.32	0.00	2.05	0.00	0.00	0.00	0.00	0.03	7.58
Max	15.77	10.98	0.80	16.93	0.66	1.00	0.83	1.15	0.29	38.45
Avg.	9.19	5.25	0.25	8.21	0.21	0.37	0.17	0.43	0.15	23.26

Sil sillimanite, *Gar* garnet, *Kya* kyanite, *Ilm* ilmenite, *Mon* monazite, *Rut* Rutile, *Zir* zircon, *Oth* other heavy minerals, *THM%* total heavy mineral %, *Min* minimum, *Max* maximum, *Avg.* average

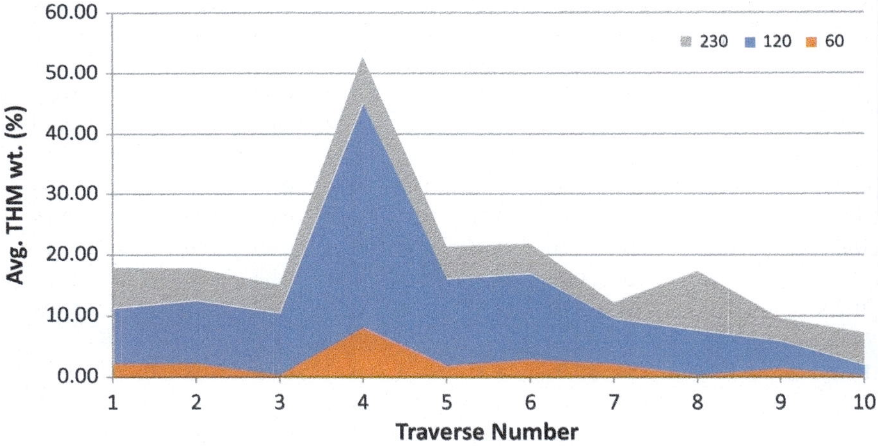

Fig. 11.2 Fraction-wise variation of average total heavy minerals (THM) in different traverses in the study area

there was a low concentration. Garnet has a high concentration in 4 traverses and a low concentration in 9 and 10. One traverse had a high concentration of rutile while ten traverses had a low concentration. Other heavy minerals with a high concentration in the 5 traverses but a low concentration in the 4 traverses (i.e. amphibole, staurolite, and epidote). The study area contains less than 2% of monazite, kyanite, and magnetite. The ilmenite, sillimanite, and garnet are more abundant in the study

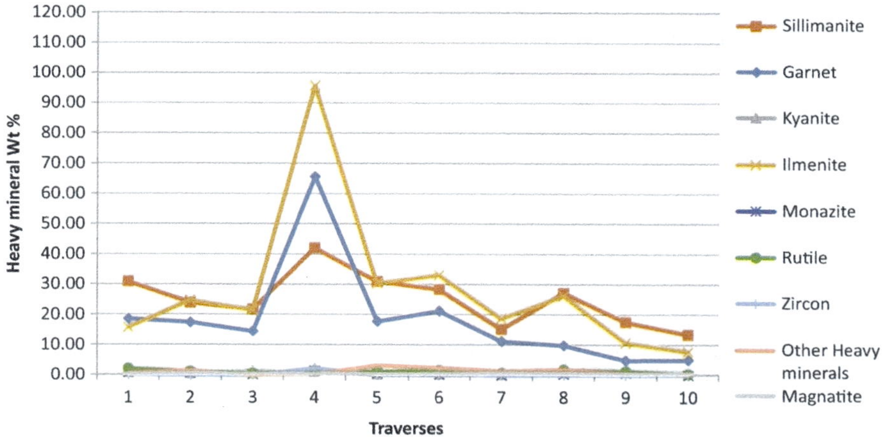

Fig. 11.3 Alongshore variation of individual minerals in Bavanapadu to Kalingapatnam coastal sediments

area. There is no systematic increase or decrease of any heavy minerals in the region. This indicates that the source of heavy minerals is not a point source Bangaku Naidu et al. (2017).

11.5 Conclusion

In the coastal sediments, the total heavy minerals (THM) wt. percent ranges from 13.10 to 27.98 (avg. 20.54 wt%). Berm (27.98%) has the highest Heavy Mineral content, followed by dune (23.26%), backshore (19.21%), and foreshore (19.21%) (13.10%).

Ilmenite, sillimanite, and garnet are found in greater abundance in the study region. The concentration of heavies in the medium fraction is higher than the course fraction and fine fraction in all sand units (i.e., foreshore, berm, backshore, and dune). In the research area, there is no systematic rise or reduction in any heavy mineral. This suggests that the heavy mineral source is not a single point. In particular size fractions of surficial sediments, heavy minerals are preferentially enriched. Garnet and sillimanite tend to concentrate in coarser fractions, whereas opaque minerals (such as ilmenite) concentrate in finer fractions. Beach placer minerals in my research area contain garnet, sillimanite, ilmenite, rutile, zircon, leucoxene, monazite, and other heavy minerals. The majority of the sediments come from the Eastern Ghats Mobile Belt's khondalite and charnockite group of rocks, based on their heavy mineral content.

Acknowledgments The authors are grateful to the Head of the Department of Geology, Andhra University, for providing the required facilities and for their consistent encouragement in carrying out our research. Mr. A Lakshmi Venkatesh would like to express his gratitude to the Department of Science and Technology (DST), Government of India, New Delhi, for funding this research project through the INSPIRE Fellowship program in the form of an SRF, and the authors would also like to express their gratitude to Dr. G. Seenuvasulu, Editor, Springer Nature, for his valuable suggestions and assistance in publishing this work.

References

Acharya BC, Nayak BK, Kumar Das S (2009) Heavy mineral placer sand deposits of Kontiagarh area, Ganjam district, Orissa, India. Resour Geol 59(4):388–399

Angusamy N, Rajamanickam GV (2000) Distribution of heavy minerals along the beach from Mandapam to Kanyakumari, Tamilnadu. J Geol Soc India 56:199–211

Anil C, Chandrasekar N, Rajamanickam GV, Gujar AR (2011) An evaluation of major placer minerals along the Valinokkam-Tuticorin coast, Southern Tamil Nadu, India. Int J Earth Sci Eng 4(6):1031–1035

Bangaku Naidu K, Reddy KSN, Ravi Sekhar C, Ganapati Rao P, Murali Krishna KN (2017) Heavy mineral studies of recent coastal sands between Gosthani and Champavathi river confluences, East Coast of India, Andhra Pradesh. J Indian Geol Cong 9(2):45–56

Borreswara Rao C, La Fond EC (1958) Study of the deposition of heavy mineral sands at the confluences of some rivers along the east coast of India. Andhra Uni Mem Oceanogr 2:48–60

Chandrasekhar N, Anil C, Paul DK, Rajamanickam GV, Loveson VJ (2005) Geospatial application in the study of beach placers along the coast of Gulf of Mannar, India. Geocarto Int 20(2):69–74

Cheepurupalli NR, Anu Radha B, Reddy KSN, Dhanamjaya Rao EN, Dayal AM (2012) Heavy mineral distribution studies in different microenvironments of Bhimunipatnam coast, Andhra Pradesh, India. Int J Sci Res Publ 2:1–10

Deva Varma D, Rao AT, Rao KSR (1989) Heavy mineral characteristics of Vamsadhara River along the east coast of India. Indian J Earth Sci 16(34):166–172

Dhana Raju R (2006) Delta region of the east coast of India: a potential target for exploration of heavy minerals. J Geol Soc India 67:669–674

Dhanunjaya Rao D, Krishnaiah Setty B, Rami Naidu CH (1989) Heavy mineral content and textural characteristics of coastal sands in the Krishna – Godavari, Gosthani – Champavathi, and Penna river deltas of Andhra Pradesh, India: a comparative study. Explor Res At Miner 2:147–155

Ganapathi Rao P et al (2020) Geochemical studies of monazites from coastal sands of Kandivalasa–Dibbalapalem coast, Andhra Pradesh. J Earth Syst Sci 129:143

Joevivek V, Chandrasekar N (2014) Seasonal impact on beach morphology and the status of heavy mineral deposition central Tamil Nadu coast, India. J Earth Syst Sci 123(1):135–149

Mahadevan C, Rao Nageswara B (1950) Black sand concentrates of the Vizagapatam Coast. Curr Sci 19:48–49

Mahadevan C, Sathapathi N (1948) Home of monazite. Curt Sci 17:297

Mahadevan C, Sriramadas AS (1948) Monazites in the beach sands of Vizagapatam District. Proc Indian Acad Sci 27A:275–227

Mahadevan C, Sriramadas A (1954) Effective of high waves in the formation of coastal black sand deposits. Andhra Univ Mem Oceanogr 1:51–56

Mahadevan C, Narayan Das GR, Nagaraja Rao N (1958) Prospecting and evaluation of beach placers along the coastal belt of India. In: Proceedings of the 2nd international conference on the peaceful uses of atomic energy, vol 2, Geneva, pp 103–106

Mohammad A et al (2020) Textural characteristics, heavy mineral distribution and grain-microtextures of recent sediment in the coastal area between the Sarada and Gosthani rivers, east coast of India. Int J Sediment Res 35(5):484–503

Mohan PM, Rajamanickam GV (2000) Buried placer mineral deposits along the East coast between Chennai and Pondicherry. J Geol Soc India 56:1–13
Mohapatra S, Behera P, Das SK (2015) Heavy mineral potentiality and alteration studies for ilmenite in Astaranga beach sands, district Puri, Odisha, India. J Geosci Environ Prot 3:31e37
Murali Krishna KN et al (2016) Heavy mineral studies on late quaternary red sediments of Bhimunipatnam, Andhra Pradesh, East Coast of India. J Geol Soc India 88:637–647
Rajasekhara Reddy D, Malathi V, Reddy KSN, Varma DD (1998) Heavy minerals in different environments of beaches between Pudimadaka and Pentakota, East coast of India. J India Acad Geol Sci 41(2):47–54
Rajasekhara Reddy D, Siva Sankar Prasad V, Malathi V, Reddy KSN, Varma DD (2001) Economic potential of the heavy minerals of the beaches between Baruva and Bhavanapadu, Andhra Pradesh. J Geol Soc India 57:443–449
Ramamohana Rao T, Shanmukha Rao C, Sanyasi Rao K (1982) Textural analysis and mineralogy of the black sand deposits of Visakhapatnam – Bhimunipatnam coast, Andhra Pradesh, India. J Geol Soc India 23:284–289
Ramasamy P, Karikalan R (2010) Distribution and percentage of heavy minerals in coastal geomorphological landforms in Palk Strait, southeast coast of India. Middle-East J Sci Res 5(1):49–53
Rao AT, Rao PN, Deva Varma D, Purnachandra Rao K (1993) Coastal sediments along Visakhapatnam-Bhimunipatnam region, Andhra Pradesh, India. J Indian Assoc Sediment 12:1–9
Rao RG, Sahoo P, Panda NK (2001) Heavy mineral sand deposit of Orissa. Explor Res At Miner 13:23–52
Ravi GS, Gajapathi Rao R, Yugandhara Rao A (2001) Coastal heavy mineral sand deposits of Andhra Pradesh. Special issue on beach and inland heavy mineral sand deposits of India. Explor Res At Miner 13:53–85
Reddy KSN, Lakshmi Prasad T, Babu Rao N (2007) Relationship of heavy mineral redistribution in different micro environments to seasonal changes of beach processes in an embayed beach of Yarada to Gangavaram, North Coastal Andhra Pradesh. J Geol Soc India 70:963e974
Reddy KSN, Deva Varma D, Dhanamjayarao EN, Lakshmi Prasad T (2012) Distribution of heavy minerals in Nizampatnam- Lankavanidibba coastal sands, Andhra Pradesh, east coast of India. J Geol Soc India 79:411e418
Sahayam D (2010) Distribution of arsenic and mercury in subtropical coastal beachrock, Gulf of Mannar, India. J Earth Syst Sci 119:129–135
Sastry AVR, Swamy ASR, Rao P (1981) Distribution of sands along Visakhapatnam-Bhimunipatnam beach. Indian J Mar Sci 1:369–370
Sastry AVR, Swamy ASR, Vasudev K (1987) Heavy minerals of beach sands along Visakhapatnam-Bhimunipatnam, east coast of India. Indian J Mar Sci 16:39–42
Sreenivasa Rao P, Satyanarayana G, Swamy ASR (1995) Heavy minerals of modern and relict sediments of the Nizampatnam bay, East Coast of India. Indian J Mar Sci 24:166–170
Sriramadas A (1951) The black sand concentrates of the Vizagapatam Beach. Q J Geol Min Metal Soc India 23:169–180
Suresh Gandhi M, Raja M (2014) Heavy mineral distribution and geochemical studies of coastal sediments between Basant Nagar and Marakkanam, Tamil Nadu, India. J Radiat Res Appl Sci 7:256–268
Vinoth Kumar KC, Asaithambi T (2013) Seasonal variation of mineral assemblage along the west coast of Kanya Kumari district. Indian J Adv Chem Sci 2(1):71–77

Chapter 12
Mineral Chemistry of Ilmenites as a Source Indicator for Coastal Sediments Between Vamsadhara and Nagavali River Mouth, North Coastal, Andhra Pradesh

Ch. Ravi Sekhar, K. S. N. Reddy, K. N. Murali Krishna, K. Bangaku Naidu, P. Ganapathi Rao, K. Veera Krishna, and A. Lakshmi Venkatesh

Abstract Ilmenites come from coastal sediments between the mouths of the Vamsadhara and Nagavali rivers, north coastal Andhra Pradesh, was studied using ilmenite end-member components. Ilmenite mineral chemistry has been studied from various environments to understand provenance by electron microprobe analyzer (EPMA). This study reveals that the ilmenite with the end-member components of Fe-Ti oxides is mainly ilmenite and has minor proportions of hematite, geikielite, and pyrophanite. The end-member compositions of Fe-Ti oxides and manganese/magnesium ratio indicate all the ilmenites of beach, dune, and estuarine environments are from the pyroxene granulites, khondalites, basic charnockites, and migmatites. Ilmenites are ferrian types and Ti/ (Ti + Fe) ratio is <0.5 indicating these are recently contributed to placer deposits. Ilmenites are mainly concentrated in fine fraction (+230) 51.50%. Ilmenite contains average TiO_2 content is 52% with a low concentration of trace elements.

Keywords Ilmenite · Mineral chemistry · Provenance · Charnockite · Khondalite · EPMA

C. Ravi Sekhar (✉) · K. S. N. Reddy · K. Veera Krishna · A. Lakshmi Venkatesh
Department of Geology, Andhra University, Visakhapatnam, Andhra Pradesh, India

K. N. Murali Krishna
Department of Civil Engineering, Sasi Institute of Technology & Engineering, Tadepalligudem, Andhra Pradesh, India

K. Bangaku Naidu
Department of Geology, Sir. C. R. Reddy College, Eluru, Andhra Pradesh, India

P. Ganapathi Rao
Department of Geology, M.R. (A) College, Vizianagaram, Andhra Pradesh, India

N. Jayaraju et al. (eds.), *Coasts, Estuaries and Lakes*,
https://doi.org/10.1007/978-3-031-21644-2_12

12.1 Introduction

Ilmenite ($FeTiO_3$) is a titanium-rich ore mineral. It has a solid solution with pyrophanite ($MgTiO_3$), geikielite ($MnTiO_3$), and hematite (Fe_2O_3). Generally, under microscope, ilmenite exhibits a significant amount of intergrowths with other Ti-Fe oxides due to exsolutions, hydrothermal, or oxidation processes. There are also trace amounts of Si, Al, Ca, Cr, Cu, and Zn in ilmenites. The mineral chemical characteristics of ilmenite are very useful to understand the provenance of ilmenite-bearing coastal sediments and also helpful in the selection of adaption methods of beneficiation for better recovery and extraction process.

Ilmenite geochemistry is most important for the selection of processing methodology either the chlorite route or sulfite route to produce titanium dioxide (TiO_2). The mineral end-member compositions of ilmenite are essential for ore dressing and mineral beneficiation processes. Ilmenite's economic value can be determined by how altered or weathered it is. Ilmenite deposits in India have not received much attention in terms of mineral chemistry studies, and the coastal placer deposit at Nagavali-Vamsadhara (the subject of the current study) in particular. Rao and Sengupta (2014), Laxmi et al. (2014), Mohapatra et al. (2015), Acharya et al. (2015), and Ganapati Rao et al. (2019) are a few studies that looked at the ilmenite geochemistry of India's east coast.

12.2 Study Area and Regional Geology

The study area is a coastal stretch that extends to 33 km from the Nagavali to the Vamsadhara river mouths, Srikakulam district, North coastal Andhra Pradesh, India (Fig.12.1). The geographical coordinates of the study area are between 18°12.410′N and 18°21.995′N latitude, between 83°55.432′E and 84°08.730′E longitude. The examined coastal region is a portion of a sedimentary basin next to the Granulite Belt of the Eastern Ghats in the middle (EGGB). The three main types of rocks are as follows: (i) charnockite group of rocks; (ii) basic granulites formed from tholeiitic magma; and (iii) khondalite group of rocks (Ramam and Murthy 1997; Yugandhara Rao et al. 2001). The general trend of the rock formation is NW-SE following the major lineament trend (Fig. 12.2).

12.3 Methodology

12.3.1 Sample Collection and Preparation

In the research region, 38 sediment samples from coastal sediment were collected over 17 traverses parallel to the coast, with 2 km between traverses (Fig. 12.1). A polyvinyl chloride pipe with a 3-inch diameter and 30-cm length was used at each

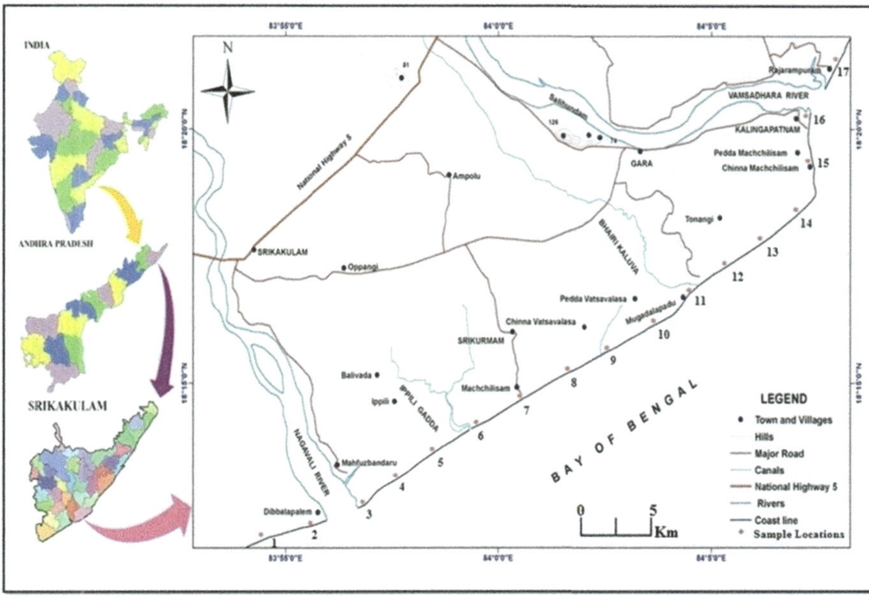

Fig. 12.1 The study area and sample location map

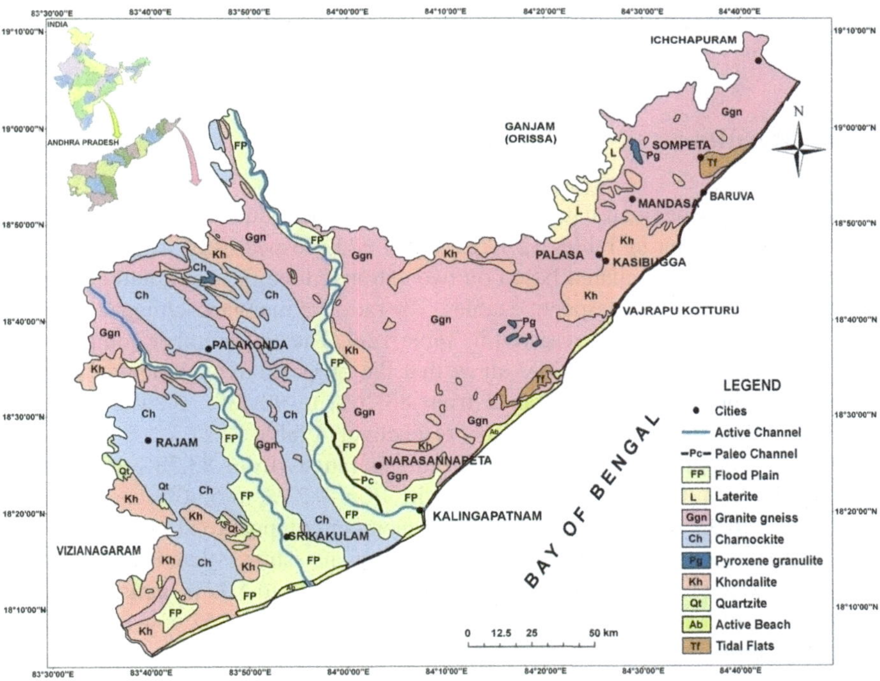

Fig. 12.2 Map of the Geology of Srikakulam district

station to collect sediment samples, penetrating the sediment layers to a depth of 10 cm. By using the coning and quartering procedure, the sediment samples were decreased, and a representative amount was taken for sediment treatment to separate the heavy minerals. A total of 38 ilmenite grains were selected for mineral chemistry analysis after being differentiated from other grains in each sample using a binocular petrological microscope.

Every sample's chosen ilmenite mineral grains were put on an epoxy resin slide of standard size for additional coating, polishing, and grinding to make sure the top and bottom of the resin blocks were parallel to one another. Using a fine silicon carbide abrasive, lapping is used to create a smooth surface (class 600). Samples were polished using very fine alumina slurry with a size range of 6–0.30 microns and fine silicon carbide (SiC) paper. To get rid of polishing grit and other surface impurities, polished samples are subsequently rinsed in clean water in an ultrasonic cleaner. After that, the sample was cleaned using a fan and allowed to air dry. To obtain good electron conductivity and interaction, the samples are coated with a coating of carbon.

An electron probe microanalyzer (EPMA) was used to analyze the mineral grains of ilmenite (CAMECA SX-100). An electron beam with a 15 kV acceleration voltage and a 20 nA beam current stimulated the polished surfaces of 38 ilmenite granules. We maintained a ~1 μm beam radius. Most of the element calibrations were performed using natural mineral standards (almandine-Fe: rhodonite-Mn; TiO_2-Ti; diopside-Mg, Ca; albite-Na, Al; chromite-Cr; hematite-Fe; wollastonite-Ca; corundum-Al; and orthoclase for Si and K).

12.4 Results

The detailed geochemical data of ilmenites from beach, dune, and estuarine environments were given in Tables 12.1, 12.2, and 12.3, respectively. The structure of ilmenites has been calculated based on two cations, three oxygens, and end members are also given in the respective tables. The ratios of manganese/magnesium and Ti/(Ti + Fe) of ilmenites were given in Table 12.4. The end-member compositions of beach sediments' Fe-Ti oxides show that the proportions of ilmenite, pyrophanite, geikielite, and hematite range from 83.26% to 98.53%, 0.29% to 1.25%, 0.20% to 4.94%, and 0.25% to 11.41%, respectively (Table 12.1).

Ilmenite component in a dune environment ranges from 83.78% to 97.67%, pyrophanite from 0.22% to 4.21%, geikielite from 0.23% to 5.90%, and hematite from 0.59% to 9.91%, according to the end-member composition (Table 12.2). The end-member composition of estuarine environments shows that ilmenite varies from 91.17% to 96.89%, pyrophanite varies from 0.31% to 5.77%, geikielite varies from 1.55% to 7.02%, and hematite varies from 0.69% to 2.52%, with eskolaite being negligible in all environments (Table 12.3).

Ilmenites originating from several groups of rocks are distinguished using a rhombohedral quaternary system (Haggerty 1976; Nayak and Mohapatra 1998; Nayak et al.

Table 12.1 Chemical composition of ilmenites of coastal sediment from beach environment between Vamsadhara and Nagavali river mouths

Traverse no.	4		8		14		30		32				
Grain no.	1	2	3	4	5	6	7	8	9	10	Min.	Max.	Avg.
SiO_2	0.030	0.020	0.000	0.050	0.070	0.070	0.020	0.010	0.080	0.030	0.080	0.080	0.040
TiO_2	52.640	45.940	52.140	52.370	53.020	50.190	51.300	52.880	51.840	52.420	53.020	53.020	51.470
Al_2O_3	0.000	0.040	0.010	0.020	0.000	0.050	0.060	0.000	0.000	0.000	0.060	0.060	0.020
Cr_2O_3	0.020	0.230	0.150	0.030	0.030	0.090	0.070	0.070	0.000	0.080	0.230	0.230	0.080
FeO	46.620	49.410	44.610	45.620	44.210	46.300	46.170	44.810	45.480	46.010	49.410	49.410	45.920
MnO	0.590	0.320	0.520	0.190	0.140	0.360	0.250	0.170	0.190	0.340	0.590	0.590	0.310
MgO	0.110	1.090	1.130	0.980	1.310	0.940	0.050	1.320	0.320	0.290	1.320	1.320	0.750
CaO	0.050	0.000	0.010	0.060	0.000	0.000	0.000	0.000	0.000	0.000	0.060	0.060	0.010
Na_2O	0.030	0.000	0.000	0.010	0.030	0.010	0.000	0.000	0.000	0.010	0.030	0.030	0.010
K_2O	0.000	0.000	0.010	0.000	0.000	0.000	0.000	0.020	0.000	0.000	0.020	0.020	0.000
	100	97.05	98.57	99.33	98.80	98.01	97.91	99.27	97.91	99.18	100	100	99
On the basis of three oxygen													
Si	0.000	0.000	0.000	0.000	0.000	0.000	0.000	0.000	0.000	0.000	0.000	0.000	0.000
Ti	1.000	0.890	1.000	0.990	1.010	0.960	0.990	1.000	1.000	1.000	1.010	1.010	0.980
Al	0.000	0.000	0.000	0.000	0.000	0.000	0.000	0.000	0.000	0.000	0.000	0.000	0.000
Cr	0.000	0.000	0.000	0.000	0.000	0.000	0.000	0.000	0.000	0.000	0.000	0.000	0.000
Fe	0.980	1.060	0.950	0.960	0.940	0.990	0.990	0.940	0.980	0.980	1.060	1.060	0.980
Mn	0.010	0.010	0.010	0.000	0.000	0.010	0.010	0.000	0.000	0.010	0.010	0.010	0.010
Mg	0.000	0.040	0.040	0.040	0.050	0.040	0.000	0.050	0.010	0.010	0.050	0.050	0.030
Ca	0.000	0.000	0.000	0.000	0.000	0.000	0.000	0.000	0.000	0.000	0.000	0.000	0.000
Na	0.000	0.000	0.000	0.000	0.000	0.000	0.000	0.000	0.000	0.000	0.000	0.000	0.000
K	0.000	0.000	0.000	0.000	0.000	0.000	0.000	0.000	0.000	0.000	0.000	0.000	0.000
	2.000	2.000	2.000	2.000	2.000	2.000	2.000	2.000	2.000	2.000	2.000	2.000	2.000

(continued)

Table 12.1 (continued)

Traverse no.	4		8		14		30		32				
Grain no.	1	2	3	4	5	6	7	8	9	10	Min.	Max.	Avg.
End-member compositions													
Geikielite	0.400	4.150	4.260	3.680	4.930	3.580	0.200	4.940	1.220	1.090	4.940	4.940	2.850
Pyrophanite	1.250	0.700	1.110	0.400	0.290	0.770	0.550	0.360	0.410	0.740	1.250	1.250	0.660
Eskolaite	0.040	0.470	0.300	0.050	0.050	0.190	0.140	0.130	0.000	0.160	0.470	0.470	0.150
Ilmenite	98.070	83.260	93.900	95.180	95.620	91.780	98.530	94.570	98.370	98.010	98.530	98.530	94.730
Hematite	0.250	11.410	0.440	0.690	0.000	3.680	0.580	0.000	0.000	0.000	11.410	11.410	1.710
	100	100	100	100	100	100	100	100	100	100	100	100	100

Table 12.2 Chemical composition of ilmenites from coastal sediment of dune environment between Vamsadhara and Nagavali river mouths

Traverse no.	3		4		6		8		14		26		28		32				
Grain no.	11	12	13	14	15	16	17	18	19	20	21	22	23	24	25	26	Min.	Max.	Avg.
SiO_2	0.020	0.030	0.010	0.050	0.040	0.020	0.030	0.000	0.180	0.150	0.180	0.080	0.010	0.050	0.040	0.000	0.000	0.180	0.060
TiO_2	51.330	47.070	51.980	52.130	50.740	50.890	51.810	50.850	58.620	48.930	56.670	51.180	53.310	49.720	52.140	52.300	47.070	58.620	51.850
Al_2O_3	0.010	0.040	0.000	0.000	0.020	0.000	0.030	0.020	0.080	0.070	0.240	0.070	0.000	0.030	0.000	0.000	0.000	0.240	0.040
Cr_2O_3	0.000	0.100	0.090	0.100	0.000	0.030	0.130	0.000	0.030	0.010	0.050	0.000	0.000	0.040	0.020	0.030	0.000	0.130	0.040
FeO	45.170	48.580	45.590	45.470	47.190	46.110	44.120	45.930	32.970	46.600	33.340	45.150	43.000	46.720	45.650	44.090	32.970	48.580	44.110
MnO	0.130	0.100	0.550	0.580	0.750	0.200	0.510	0.240	0.870	0.550	1.810	0.460	0.270	0.250	0.250	0.150	0.100	1.810	0.480
MgO	1.200	1.560	0.630	0.690	0.100	0.470	1.550	0.670	0.060	0.810	0.110	0.490	1.450	0.810	0.830	1.440	0.060	1.560	0.800
CaO	0.020	0.000	0.000	0.020	0.020	0.000	0.010	0.010	0.050	0.000	0.010	0.050	0.000	0.010	0.000	0.000	0.000	0.050	0.010
Na_2O	0.010	0.010	0.000	0.000	0.030	0.000	0.020	0.000	0.010	0.000	0.020	0.010	0.020	0.000	0.000	0.000	0.000	0.030	0.010
K_2O	0.000	0.000	0.000	0.000	0.000	0.000	0.000	0.000	0.000	0.000	0.000	0.000	0.020	0.000	0.000	0.010	0.000	0.020	0.000
	97.880	97.480	98.840	99.040	98.880	97.730	98.200	97.720	92.850	97.140	92.420	97.490	98.070	97.610	98.940	98.020	92.420	99.040	97.390
On the basis of three oxygens																			
Si	0.000	0.000	0.000	0.000	0.000	0.000	0.000	0.000	0.000	0.000	0.000	0.000	0.000	0.000	0.000	0.000	0.000	0.000	0.000
Ti	0.990	0.900	0.990	0.990	0.970	0.980	0.990	0.980	1.210	0.950	1.170	0.990	1.020	0.960	0.990	1.000	0.900	1.210	1.010
Al	0.000	0.000	0.000	0.000	0.000	0.000	0.000	0.000	0.000	0.000	0.010	0.000	0.000	0.000	0.000	0.000	0.000	0.010	0.000
Cr	0.000	0.000	0.000	0.000	0.000	0.000	0.000	0.000	0.000	0.000	0.000	0.000	0.000	0.000	0.000	0.000	0.000	0.000	0.000
Fe	0.960	1.030	0.970	0.960	1.000	0.990	0.940	0.990	0.760	1.000	0.770	0.970	0.920	1.000	0.970	0.940	0.760	1.030	0.950
Mn	0.000	0.000	0.010	0.010	0.020	0.000	0.010	0.010	0.020	0.010	0.040	0.010	0.010	0.010	0.010	0.000	0.000	0.040	0.010
Mg	0.050	0.060	0.020	0.030	0.000	0.020	0.060	0.030	0.000	0.030	0.000	0.020	0.050	0.030	0.030	0.050	0.000	0.060	0.030
Ca	0.000	0.000	0.000	0.000	0.000	0.000	0.000	0.000	0.000	0.000	0.000	0.000	0.000	0.000	0.000	0.000	0.000	0.000	0.000
Na	0.000	0.000	0.000	0.000	0.000	0.000	0.000	0.000	0.000	0.000	0.000	0.000	0.000	0.000	0.000	0.000	0.000	0.000	0.000
K	0.000	0.000	0.000	0.000	0.000	0.000	0.000	0.000	0.000	0.000	0.000	0.000	0.000	0.000	0.000	0.000	0.000	0.000	0.000
	2.000	2.000	2.000	2.000	2.000	2.000	2.000	2.000	2.000	2.000	2.000	2.000	2.000	2.000	2.000	2.000	2.000	2.000	2.000

(continued)

Table 12.2 (continued)

Traverse no.	3		4		6		8		14		26		28		32				
Grain no.	11	12	13	14	15	16	17	18	19	20	21	22	23	24	25	26	Min.	Max.	Avg.
End-member compositions																			
Geikielite	4.550	5.900	2.380	2.590	0.390	1.810	5.860	2.550	0.230	3.120	0.440	1.870	5.490	3.090	3.140	5.460	0.230	5.900	3.050
Pyrophanite	0.290	0.220	1.180	1.250	1.620	0.420	1.090	0.520	2.030	1.210	4.210	1.010	0.580	0.530	0.540	0.330	0.220	4.210	1.060
Eskolaite	0.010	0.190	0.170	0.210	0.000	0.060	0.270	0.000	0.070	0.020	0.110	0.000	0.000	0.070	0.040	0.050	0.000	0.270	0.080
Ilmenite	93.730	83.780	95.620	95.340	95.170	96.160	91.660	95.140	97.670	90.380	95.250	96.310	93.930	92.200	95.690	94.160	83.780	97.670	93.890
Hematite	1.430	9.910	0.650	0.610	2.830	1.540	1.110	1.790	0.000	5.270	0.000	0.800	0.000	4.110	0.590	0.000	0.000	9.910	1.920
	100	100	100	100	100	100	100	100	100	100	100	100	100	100	100	100	100	100	100

Table 12.3 Chemical composition of ilmenites from estuarine sediment of Vamsadhara and Nagavali rivers

Traverse no.	NR 1		NR 2		NR 3		VR 1		VR 2		VR 3				
Grain no.	27	28	29	30	31	32	33	34	35	36	37	38	Min.	Max.	Avg.
SiO_2	0.030	0.050	0.030	0.060	0.030	0.050	0.050	0.020	0.060	0.000	0.000	0.060	0.000	0.060	0.040
TiO_2	52.620	51.600	52.750	52.520	52.300	52.270	53.560	51.310	52.110	52.980	51.830	53.850	51.310	53.850	52.480
Al_2O_3	0.010	0.000	0.030	0.000	0.020	0.020	0.000	0.000	0.000	0.040	0.000	0.000	0.000	0.040	0.010
Cr_2O_3	0.020	0.030	0.040	0.070	0.120	0.030	0.000	0.080	0.070	0.060	0.040	0.250	0.000	0.250	0.070
FeO	44.720	46.040	43.810	45.950	46.050	46.840	44.930	47.360	46.400	43.330	45.880	43.760	43.330	47.360	45.420
MnO	0.160	0.280	0.500	0.670	0.320	0.220	0.150	0.290	0.620	2.720	0.260	0.290	0.150	2.720	0.540
MgO	0.840	0.410	1.280	0.650	0.950	0.490	1.550	0.440	0.780	0.780	0.660	1.900	0.410	1.900	0.890
CaO	0.000	0.080	0.030	0.000	0.010	0.010	0.010	0.000	0.000	0.000	0.000	0.040	0.000	0.080	0.020
Na_2O	0.020	0.020	0.000	0.000	0.010	0.000	0.020	0.010	0.030	0.000	0.000	0.030	0.000	0.030	0.010
K_2O	0.020	0.010	0.020	0.000	0.010	0.000	0.000	0.000	0.010	0.000	0.000	0.000	0.000	0.020	0.010
	98.44	98.51	98.48	99.91	99.82	99.93	100	99.51	100	99.92	98.68	100	98	100	99
On the basis of three oxygens															
Si	0.000	0.000	0.000	0.000	0.000	0.000	0.000	0.000	0.000	0.000	0.000	0.000	0.000	0.000	0.000
Ti	1.010	0.990	1.010	0.990	0.990	0.990	1.000	0.970	0.980	1.000	0.990	1.010	0.970	1.010	0.990
Al	0.000	0.000	0.000	0.000	0.000	0.000	0.000	0.000	0.000	0.000	0.000	0.000	0.000	0.000	0.000
Cr	0.000	0.000	0.000	0.000	0.000	0.000	0.000	0.000	0.000	0.000	0.000	0.000	0.000	0.000	0.000
Fe	0.950	0.980	0.930	0.970	0.970	0.990	0.930	1.000	0.970	0.910	0.980	0.910	0.910	1.000	0.960
Mn	0.000	0.010	0.010	0.010	0.010	0.000	0.000	0.010	0.010	0.060	0.010	0.010	0.000	0.060	0.010
Mg	0.030	0.020	0.050	0.020	0.040	0.020	0.060	0.020	0.030	0.030	0.030	0.070	0.020	0.070	0.040
Ca	0.000	0.000	0.000	0.000	0.000	0.000	0.000	0.000	0.000	0.000	0.000	0.000	0.000	0.000	0.000
Na	0.000	0.000	0.000	0.000	0.000	0.000	0.000	0.000	0.000	0.000	0.000	0.000	0.000	0.000	0.000
K	0.000	0.000	0.000	0.000	0.000	0.000	0.000	0.000	0.000	0.000	0.000	0.000	0.000	0.000	0.000
	2.000	2.000	2.000	2.000	2.000	2.000	2.000	2.000	2.000	2.000	2.000	2.000	2.000	2.000	2.000

(continued)

Table 12.3 (continued)

Traverse no.	NR 1		NR 2		NR 3		VR 1		VR 2		VR 3				
Grain no.	27	28	29	30	31	32	33	34	35	36	37	38	Min.	Max.	Avg.
End-member compositions															
Geikielite	3.200	1.550	4.830	2.420	3.570	1.840	5.740	1.660	2.910	2.930	2.510	7.020	1.550	7.020	3.350
Pyrophanite	0.350	0.600	1.060	1.420	0.690	0.470	0.310	0.610	1.320	5.770	0.570	0.600	0.310	5.770	1.150
Eskolaite	0.050	0.060	0.090	0.130	0.230	0.060	0.010	0.170	0.150	0.120	0.090	0.490	0.010	0.490	0.140
Ilmenite	96.410	96.890	94.020	95.340	94.210	96.520	93.940	95.040	93.770	91.170	96.040	91.880	91.170	96.890	94.600
Hematite	0.000	0.910	0.000	0.690	1.310	1.100	0.000	2.520	1.850	0.000	0.800	0.000	0.000	2.520	0.770
	100	100	100	100	100	100	100	100	100	100	100	100	100	100	100

Table 12.4 Ratios of Mn/Mg and Ti/(Ti + Fe) of ilmenite from sediments between Vamsadhara and Nagavali river mouths

Traverse no.	Grain no.	Mn/Mg	Ti/(Ti + Fe)
Beach environment			
4	1	7.100	0.470
	2	0.380	0.420
8	3	0.590	0.470
	4	0.250	0.470
14	5	0.130	0.480
	6	0.490	0.460
30	7	6.250	0.460
	8	0.160	0.480
32	9	0.770	0.470
	10	1.520	0.470
	Min.	0.130	0.420
	Max.	7.100	0.480
	Avg.	1.650	0.460
Dune environment			
3	11	0.140	0.470
	12	0.080	0.430
4	13	1.120	0.470
	14	1.090	0.470
6	15	9.350	0.450
	16	0.530	0.460
8	17	0.420	0.480
	18	0.460	0.460
14	19	20.040	0.580
	20	0.870	0.450
26	21	21.690	0.570
	22	1.220	0.470
28	23	0.240	0.490
	24	0.390	0.450
32	25	0.390	0.470
	26	0.140	0.480
	Min.	0.080	0.430
	Max.	21.690	0.580
	Avg.	3.640	0.480
Estuarine environment			
NR 1	27	0.240	0.480
	28	0.870	0.460
NR 2	29	0.500	0.480
	30	1.330	0.470
NR 3	31	0.430	0.470
	32	0.580	0.460

(continued)

Table 12.4 (continued)

Traverse no.	Grain no.	Mn/Mg	Ti/(Ti + Fe)
VR 1	33	0.120	0.480
	34	0.830	0.460
VR 2	35	1.020	0.460
	36	4.450	0.490
VR 3	37	0.510	0.470
	38	0.190	0.490
	Min.	0.120	0.460
	Max.	4.450	0.490
	Avg.	0.920	0.470

2012) (Fig. 12.3). The five fields in this picture are as follows: (1) parametamorphites (Mn-rich); (2) intrusive (pegmatites and carbonatites); (3) basic suits (amphibolites and granite gneisses); (4) intrusive acid and anorthosite suites; and (5) kimberlites. All ilmenites fall in field 3 of the basic suite of rocks, which is the ilmenite field. The end-member compositions of Fe-Ti oxides from beach, dune, and estuarine environments were plotted in the rhombohedral quarternary ($FeTiO_3$-$MnTiO_3$-$MgTiO_3$-Fe_2O_3) diagrams (Haggerty 1976; Nayak and Mohapatra 1998).

Ilmenites formed from beach sediment had an average TiO_2 content of 52.48%, ranging from 51.31% to 53.85% for dune sediments, 47.07% to 58.62% for beach sediments, and 51.31% to 53.85% for estuarine sediments (Tables 12.1, 12.2, and 12.3). which is comparatively higher or lower than the hypothetical ilmenite (Deer et al. 1992) Higher titanium dioxide (TiO_2) concentration could result from the other cations in ilmenite leaching, while lower titanium dioxide (TiO_2) content could be the existence of exsolved phases of hematite (Jagannadha Rao et al. 2005). Detrital ilmenites from metamorphic sources have a narrow range of TiO_2 content, with a mean of roughly 47% TiO_2 (Basu and Molinaroli 1989).

FeO content varies from 32.97% to 48.58% (avg. 44.11%) in dune environments, from 44.21% to 49.41% (range 44.21–49.41%) in beach environments, and from 43.33% to 47.36% (range 43.33–47.36%) in estuarine environments (avg. 45.42%). The significant proportion of FeO in ilmenite may be due to exsolved stages of hematite. The MnO percentage of the ilmenite varies from 0.14% to 0.59% (avg. 0.31%) of sediments in beach environments, from 0.10% to 1.81% (avg. 0.48%) in dune environments, and from 0.15% to 2.72% in estuarine environments (avg. 0.54%). The MgO level varies from 0.05% to 1.32% (avg. 0.75%) in beach environments, from 0.06% to 1.56% (avg. 0.80%) in dune environments, and from 0.41% to 1.90% in estuarine environments (avg. 0.89%). In all conditions, ilmenite grains contain traces of SiO_2 (0.04%) and Al_2O_3 (0.02%) (Tables 12.1, 12.2, and 12.3).

The elemental geochemistry of ilmenite is very useful for geochemical characterization and elemental ratios such as manganese/magnesium are widely used as a provenance indicator. The manganese/magnesium ratio of ilmenites from beach sediments ranges from 0.31 to 7.10 (avg.1.65). More than 75% of the ilmenite samples show Mn/Mg ratio is ≤ 1. In the ilmenites of dune sediments, manganese/

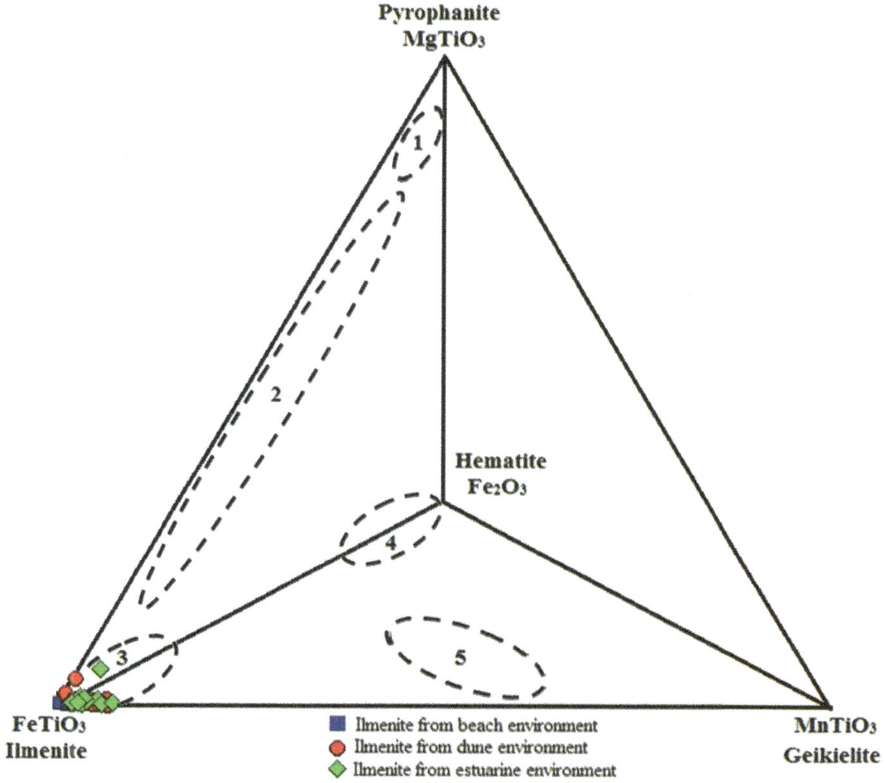

Fig. 12.3 Rhombohedral quaternary plot with a proportion of ilmenite, pyrophanite, hematite, and geikielite as poles (Haggerty 1976; Nayak and Mohapatra 1998). Ilmenites from beach, dune, and estuarine environment of the study area

magnesium ratio varies from 0.08 to 21.69 (avg. 3.64). More than 62% of ilmenite samples show manganese/magnesium ratio is ≤1 and in the ilmenite of estuarine sediments, manganese/magnesium ratio ranges from 0.12 to 4.45 (avg. 0.92) in the ilmenite of estuarine sediments (Table 12.4). More than 75% of ilmenites of estuarine environment also show a manganese/magnesium ratio of ≤1.

Several investigations on ilmenite of dunal sands Southwest coast of India and the Tamil Nadu coast indicate manganese/magnesium ratio is ≤1. The manganese/magnesium ratio of ilmenites from dune sands and source rocks of the southwest coast of India (Dinesh et al. 2007) is ≤1, and they suggested that the source rock for the ilmenites are mainly khondalite gneisses, and charnockites in the hinterland which was earlier reported by Aswathanaryana et al. (1964), Mallik et al. (1987), and Unnikrishnan (1988).

The manganese/magnesium ratio is ≤1, according to several studies on ilmenite of dunal sands on the Tamil Nadu and Southwest Indian coasts. Dinesh et al. (2007) suggested that the source rock for ilmenites is primarily khondalite gneisses and charnockites in the hinterland, which were previously reported by Aswathanaryana et al. (1964), Mallik et al. (1987), and Unnikrishnan (1988). Bhattacharyya et al. (1997) noticed that the high Manganese/Magnesium ratio > 9 for ilmenites from the Chhatrapur deposit corroborates the provenance of charnockites, migmatites, and granulites of Eastern Ghats as suggested earlier (Sengupta et al. 1990). The ratio of 2.56 for dune ilmenites of Visakhapatnam indicates various sources of basaltic and metasedimentary, rocks (Bhattacharyya et al. 1997). Ilmenites of Bhimunipatnam-Visakhapatnam coastal sands Manganese/Magnesium ratio ranges from 0.39 to 5.16 (Jagannadha Rao et al. 2005). The Manganese/Magnesium ratio of ilmenites from sapphirine granulite, charnockites, and khondalite of Eastern Ghat Group of rocks, Visakhapatnam is ≤1 (Kamineni and Rao 1988). Ilmenites of Chhatrapur (Acharya and Das 2001; Rao et al. 2005) and Ekakula dune sands (Acharya and Das 2001) of Orissa show a manganese/magnesium ratio of ≤1. The Tamil Nadu coastal ilmenites are mainly derived from metamorphic rocks. The manganese/magnesium ratios of ilmenites range from 1.69 to 3.59 in southeastern Bangladesh (Ahmed and Islam 2001) indicate that they are derived from plutonic rocks.

The present study area contains different suit of rock types such as khondalites, calc-silicate rocks, quartzites, charnockites, basic granulites, and granites of Archean to Precambrian age. These formations are highly migmatized and were termed as Eastern migmatized zone (Ramam and Murthy 1997). The manganese/magnesium ratios of the present work show that most of the ilmenites were derived from metamorphic rocks (pyroxene granulites and khondalites), the minor portion was derived from basic charnockites and migmatites.

Ilmenite's weathering mechanism has been proposed to be described by the ratio of Ti/(Ti + Fe). As the weathering mechanism progresses, the terms for the various stages are as follows: (a) Ti/(Ti + Fe) of ferrian ilmenite (0.50), (b) hydrated ilmenite (0.50–0.60), (c) pseudo rutile (0.60–0.70), and (d) leucoxene (>0.70) (Frost et al. 1983). In this work, ilmenite's chemical characterization and phases of modification were determined using the aforementioned classification.

The Ti/(Ti + Fe) ratio of the sediments from the beach environment ranges from 0.42 to 0.48 (avg. 0.46) (Table 12.4). These ilmenite grains have Ti/(Ti + Fe) ratio is <0.50 which indicates that they are ferrian ilmenite. In a dune environment, the Ti/(Ti + Fe) ratio of the sediments varies from 0.43 to 0.58 (avg.0.48) (Table 12.4) except for three samples all other ilmenite grains have Ti/(Ti + Fe) ratio is <0.50 which indicates that they are ferrian ilmenite. In an estuarine environment, the Ti/(Ti + Fe) ratio of the sediments average is 0.47and ranges from 0.46 to 0.49 (Table 12.4); these ilmenite grains have Ti/(Ti + Fe) ratio is <0.50 which indicates that they are ferrian ilmenite. The lower values of Ti/(Ti + Fe) and fresh grains of ilmenite indicate these grains have undergone less alteration.

Ilmenite's degree of alteration is determined by the deposit's geological history and weathering conditions (Hugo and Cornell 1991; Suresh Babu et al. 1994). Indian placer deposits have suffered varying degrees of change, with Kerala deposits showing the most alteration, Tamil Nadu deposits showing mild alteration, and Orissa deposits showing the least alteration (Suresh Babu et al. 1994). The present study area ilmenites have less alteration like ilmenites of Orissa placer deposits. The investigated area is under a subtropical environment. The high ferrous iron and less TiO_2 ilmenite compared to the west coast Manavalakurichi and Chavara deposits. It suggests that the present placer deposits are younger in age and have undergone the least weathering.

12.5 Conclusions

1. The end-member components of Fe-Ti oxides are mainly ilmenite and minor proportions of hematite, geikielite, and pyrophanite. The end-member compositions of Fe-Ti oxides and Manganese/Magnesium ratio indicate all the ilmenites of beach, dune, and estuarine environments are from the pyroxene granulites, khondalites, basic charnockites, and migmatites.
2. Ilmenites are ferrian types and $Ti/(Ti + Fe)$ ratio is <0.5 indicating these are recently contributed to placer deposits.
3. Ilmenites are mainly concentrated in fine fraction (+230) 51.50%. Ilmenite contains average TiO_2 content is 52% with a low concentration of trace elements.

Acknowledgments I am grateful to the Lord Jesus Christ for giving me the opportunity to write this post and for His bountiful grace and blessings. The corresponding author highly appreciates the financial support provided by the Basic Scientific Research (BSR), UGC, New Delhi Program. The HOD of the Geology Department at Andhra University is thanked by the corresponding author for expanding the lab facilities.

References

Acharya BC, Das SK (2001) Beach placer deposits of Orissa. In: Rajamanickam GV (ed) Hand book of placer mineral deposits. New Academic Publishers, pp 130–137

Acharya BC, Nayak BK, Das SK (2015) Mineralogy and mineral chemistry of placer deposit around Jhatipodar, Odisha. J Geol Soc India 86:137–147

Ahmed SS, Islam MDB (2001) Economic minerals in the beach sands of south eastern Bangladesh. In: Rajamanickam GV (ed) Hand book of placer mineral deposits. New Academic Publishers, Delhi, pp 280–294

Aswathanaryana U, Subramaniam AP, Balakrishna S (1964) Origin of the heavy mineral sand deposits of the southwest coast of India. In: Advancing frontiers in geology and geophysics Indian Geophysical Union Hyderabad, pp. 481–489

Basu A, Molinaroli E (1989) Provenance characteristics of detrital opaque Fe-Ti oxide minerals. J Sediment Petrol 59(6):922–934

Bhattacharyya S, Sengupta R, Chakraborty M (1997) Elemental chemistry of ilmenite – an indicator of provenance. J Geol Soc India 50(6):129–140

Deer WA, Howie RA, Zussman J (1992) An introduction to rock forming minerals. Logman Group Ltd, Harlow, p 712

Dinesh AC, Nambiar AR, Unikrishnan E, Jayaprakash C, Venkateswara Rao C (2007) Mn and Mg and Mn/Mg ratio in detrital ilmenite – an indicator of provenance. J Appl Geochem 9(1):112–119

Frost MT, Grey IE, Harrowfield IR, Mason K (1983) The dependence of alumina and silica contents on the extent of alteration of weathered ilmenites from Western Australia. Mineral Mag 47:201–208

Ganapati Rao P, Reddy KSN, Ravi Sekhar C, Bangaku Naidu K, Murali Krishna KN, Reddy GVR (2019) Provenance studies of Ilmenite from Red Sediments, Bhimunipatnam Coast, East Coast of India. J Geol Soc India 93:101–108

Haggerty SE (1976) Opaque mineral oxides in terrestrial igneous rocks. In: Rumble D III (ed) Oxide minerals, vol 3. Mineralogical Society of America (Rev. Mineral.), pp 101–300

Hugo VE, Cornell DH (1991) Altered ilmenites in Holocene dunes from Zululand, South Africa, Petrographic Studies for multistage alteration. S Afr J Geol 94:365–378

Jagannadha Rao M, Ramana JV, Venugopal R, Rao MC (2005) Geochemistry and ore mineralogy of ilmenites from beach placers of the Visakhapatnam –Bhimunipatnam deposit, Andhra Pradesh. J Geol Soc India 66:147–149

Kamineni DC, Rao AT (1988) Sapphirine – granulites from the Kakanuru area, Eastern Ghats, India. Am Mineral 73:692–700

Laxmi T, Bhima Rao R, Mukherjee PS (2014) Recovery of badlands ilmenite for preparation of titanium rich slag using thermal plasma. Revista INNOVER 1(2):1–9

Mallik TK, Vasudevan V, Verghese A, Machado T (1987) The black sand placer deposits of Kerala beach, Southwest, India. Mar Geol 77:129–150

Mohapatra S, Behera P, Das SK (2015) Heavy mineral potentiality and alteration studies for ilmenite in Astaranga beach sands, district Puri, Odisha, India. J Geosci Environ Prot 3:31–37

Nayak BR, Mohapatra BK (1998) Two morphologies of pyrophanite in Mn-rich assemblages, Gangpur Group. India Mineral Mag 62(6):847–856

Nayak B, Mohanty S, Bhattacharyya P (2012) Heavy minerals and the characters of ilmenite in the beach placer sands of Chavakkad-Ponnani, Kerala coast, India. J Geol Soc India 79:259–266

Ramam PK, Murthy VN (1997) Geology of Andhra Pradesh. Geological Society of India Publication, pp 81–82

Rao DS, Sengupta D (2014) Electron microscopic studies of ilmenite from the Chhatrapur coast, Odisha, India, and their implications in processing. J Geochem:1–8

Rao DS, Vijaya Kumar TV, Subba Rao S, Prabhakar S, Bhaskar Raju G (2005) Alteration characteristics of the Manavalakurichi beach placer ilmenite, Tamilnadu. J Appl Geochem 7(2):195–200

Sengupta R, Bhattacharyya S, Rana RS, Mitra SK, Jain VK (1990) Preliminary studies of offshore heavy mineral placers, Gopalpur-Chhatrapur coast, Orissa. Indian J Geol 62(1):27–37

Suresh Babu DS, Thomas KA, Mohan Das PN, Damodaran AD (1994) Alteration of ilmenite in the Manavalakurichi deposit, India. Clays Clay Miner 42(5):567–571

Unnikrishnan VP (1988) Texture, mineralogy and provenance of the beach sands of south Kerala. Ph.D. Thesis (Unpublished). Cochin University of Science and Technology, Cochin, p 120

Yugandhara Rao A, Ravi GS, Gajapathi Rao R, Krishnan S, Azam Ali M, Banerjee DC (2001) Srikurmam–a major heavy mineral beach and dune sand deposit of the east coast of India. In: Hand book of placer mineral deposits. New Academic Publishers, Delhi, pp 53–63

Chapter 13
Major and Trace Elements in the Sediments of the Gollumutta Paya Estuary of the Krishna River, East Coast of India

K. Veera Krishna, G. Swathi, Ch. Ravi Sekhar, G. Veeraswamy, P. Krishna Kumari, R. Demudu Naidu, T. Sankar Rao, and V. Asha

Abstract To study the spatial distribution of geochemical elements of the Gollumutta Paya estuary of the Krishna River, a total of 12 sediment samples were collected in April and December 2016. The sediment samples were analyzed to estimate the concentrations of the elements (major and trace) using Inductively Coupled Plasma-Mass Spectrometry (ICP-MS). Three major elements, namely, Fe, Mn, and Al, and 18 trace elements, namely, Li, Be, V, Cr, Co, Ni, Cu, Zn, As, Rs, Sr, Ag, Cd, Cs, Ba, Pb, Ti, and U have been identified in the Krishna River estuarine sediments.

Keywords Major elements · ICP-MS · Estuary · Sorption · Provenance · Paleoenvironments

---

K. Veera Krishna (✉) · C. Ravi Sekhar · P. Krishna Kumari · R. Demudu Naidu · T. Sankar Rao · V. Asha
Department of Geology, Andhra University, Visakhapatnam, Andhra Pradesh, India

G. Swathi
Department of Geosciences, Dr. B. R. Ambedkar University, Srikakulam, Andhra Pradesh, India

G. Veeraswamy
Department of Civil Engineering, Malla Reddy Institute of Technology, Hyderabad, Telangana, India

N. Jayaraju et al. (eds.), *Coasts, Estuaries and Lakes*,
https://doi.org/10.1007/978-3-031-21644-2_13

13.1 Introduction

The geochemical study of sediments is a powerful tool to identify the Earth's surface processes since the sediments record their effects on provenance, generation, transport, deposition, and environmental conditions (Maroof Azam and Tripathi 2016). A significant advancement in the technology of analytical geochemistry-enabled analysis of sediment samples and to publish valuable papers during the last 10 years. The geochemical composition of sediments has been utilized to calculate weathering indices, namely, the chemical index of alteration (CIA). The CIA has been applied widely to investigate the weathering conditions of the provenance of the sediments and for understanding paleoclimate operating at the time of the deposition of sediments. As sedimentary records store the original lithologic properties, document changes in climatic conditions existed at the time of the deposition of sediments. Hence, they are widely used in the reconstruction of paleoenvironments (Shilling et al. 2020). Recent research has found links between certain chemical constituents and environmental changes. Many processes, including weathering, erosion, transit, and sediment deposition, are shown to influence chemical elemental abundance (Ngeuutchoua et al. 2019; Skurzynski et al. 2020). One of the most often used approaches for determining the sedimentary environment and origin is the quantitative analysis elements in a given sample (Lei et al. 2017; George et al. 2019; Van de Velde et al. 2020). The Gollumutta Paya estuary sediments were analyzed to show the concentrations and to reconstruct the paleoclimate, paleoenvironment, and pollution consequences. With a view to understand the prevailing physicochemical, and biological conditions in controlling the fixation and migration of the chemical constituents of the sediments, the major, and trace the elemental study of the sediments of the Gollumutta Paya estuary have been taken up.

It has been observed that Na, K, and Mg are related to the geology of the provenances whereas Fe, Ni, Mn, Cr, Cu, Co, Pb, and Zn are potential pollutants. The distribution of trace elements among the various phases under natural conditions is necessary to monitor pollution. Several studies have reported on elemental partitioning between geochemical phases. Toxic trace elements include Cu, Cd, Pb, Cr, As, and Hg, which are required for living things at specific quantities but are toxic to live species at higher concentrations. These elements are normally not hazardous in their metallic state, but they become toxic when present in fine powder form. Their combinations with other elements are usually harmful to living creatures. Copper, chromium, and lead are the most common contaminants in water resources. Soils that have excessively high amounts of heavy metals are said to be polluted and cannot be used for agricultural crop cultivation. The rise in heavy metal concentrations could be ascribed to human activity, agricultural production, and mining. All these processes increase heavy metal concentrations in soil solutions, which eventually end up in the tissues of various plant and animal species, as well as human bodies, and heavy metals can accumulate to hazardous levels in the environment.

13.2 The Study Area

The Krishna River rises near Mahabaleswar in the Western Ghats in the Maharastra State in Central India. It is one of the longest rivers (1500 km) in India that empties into the Bay at Hamsala Divi in Andhra Pradesh on the east coast. The area covers 3700 km^2 along Andhra (Lat. 15°42′–16°31′ N and Long. 80°30′–81°15′ E). The Krishna River passes through the Eastern Ghats at Vijayawada and flows through an open deltaic area (96 km) (Fig.13.1). The Krishna River basin has an area of 260,000 Km^2.

13.3 Geology of the Krishna River Basin

The geochemistry of the sediments of the estuary depends on the lithology of the source area. Here the Krishna River basin is the main source area, and therefore, the geology of the above basin is to be discussed. The geology of the Krishna River basin is shown Fig. 13.2. The major geological formations traversed by the river from the source area consist of the Deccan Traps ranging in age from Upper Cretaceous to Eocene. At some places, differentiated rocks are generally fine-grained, non-porphyritic, doleritic or basaltic with abundant labradorite and pigeonite. Minerals of hydrothermal stages, such as zeolites, calcite, chalcedony, and its varieties, are commonly observed either as amygdales or as crystals radiating from geodes. Upon weathering, the trap-rocks usually give rise either to laterite or to bauxite, and the soil derived from the weathering is called "black cotton soil." The Krishna River flows down from the Deccan Plateau through crystalline rocks composed primarily of Dharwars, different gneisses, and granites, including charnockites. The Dharwars are gneisses and granites with regional metamorphism, including charnockites. The Dharwars are igneous and sedimentary rocks that have been regionally metamorphosed and are formed of phyllites, schists, quartzites, amphibolites, and granites. Quartzites, limestones, and shales make up the Cuddapah formations. In its last stages of travel, the Krishna River passes through khondalites, which give birth to brick-red soil after extensive weathering. The kaolinite mineral dominates the clay mineral in the brick-red soil (Nagelschmidt et al. 1940).

13.4 Methodology

A total of 12 sediment samples, 6 each during April and December 2016, were collected from the study area (Fig. 13.1). These sediment samples were analyzed to estimate the concentrations of the elements (major and trace) using the ICP-MS inductively coupled plasma-mass spectrometry. For the analysis of the geochemical elements, a quantity of 5 mg of powdered sediment from each sample was taken

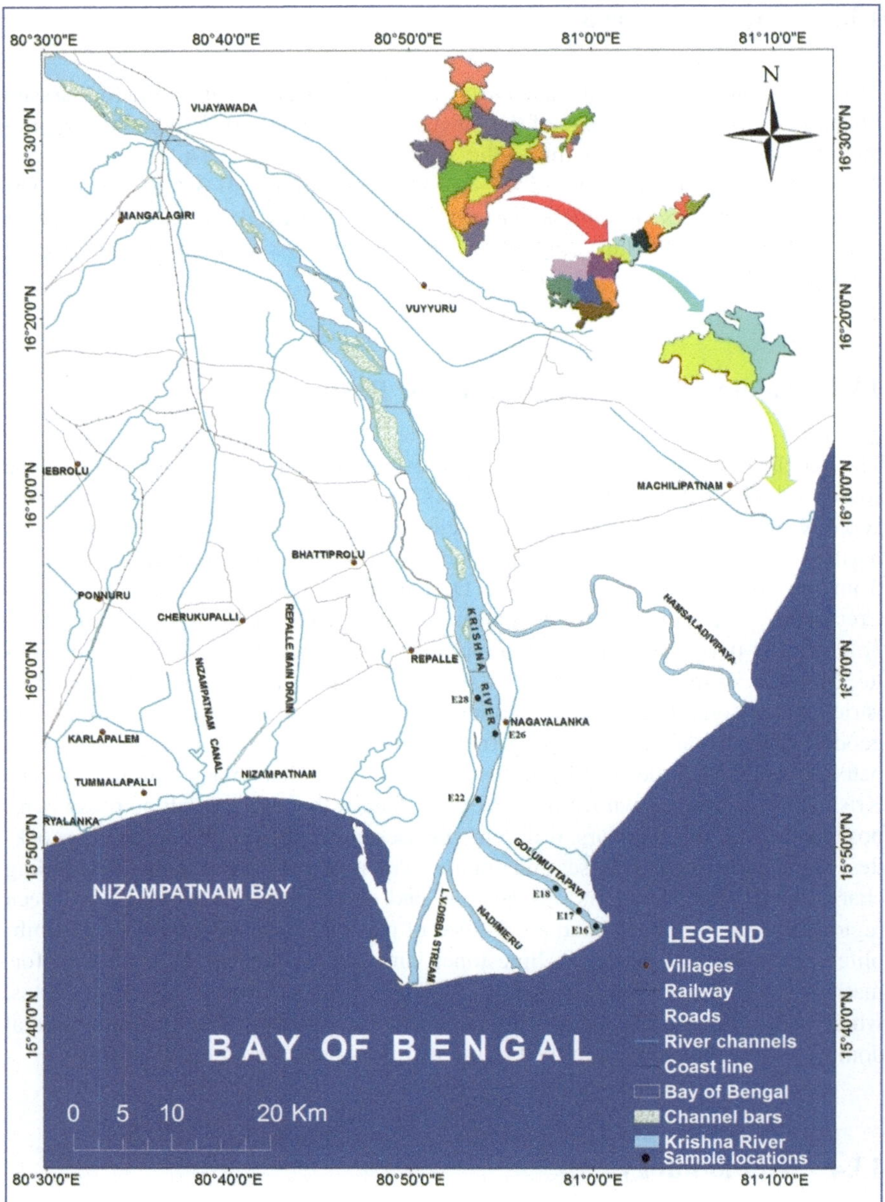

Fig. 13.1 Map of the study area and sample locations

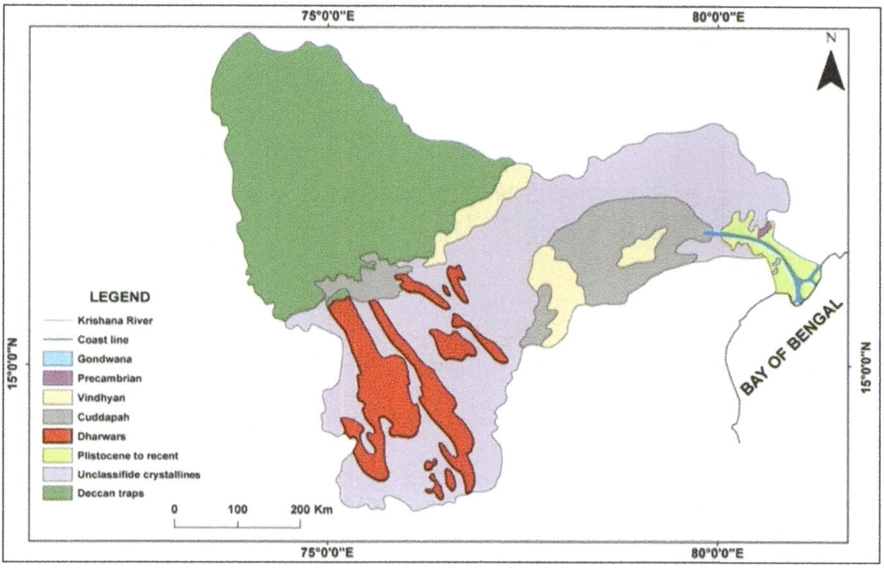

Fig. 13.2 Geology of the drainage basin of Krishna River (Source: Geological Survey of India)

into a Teflon beaker and 10.0 ml of the acid mixture (HH:HNO_3: $HCLO_4$: 7:3:1) was added to it. This sample was swirled until completely moist and kept overnight for the digestion of the sediment. Then, it was heated on a hotplate at 150 °C. After 1 hour, the lid of the beaker was removed, and the sample was allowed to dry to form a crystalline paste. Again, 5 ml of the acid mixture was added to the crystalline paste and allowed to dry. An amount of 20 ml of 1:1 HNO_3 was added and heated at 70 °C to dissolve the precipitate, 5 ml of 1 PPM ISTD was added to it. To make up the volume, 250 ml of 3D water was added. This sample was stored in a 60 ml HDPE sample bottle and labeled with code and sample number. In ICP-MS, the source was mounted horizontally, and the sample was introduced into the plasma by a standard ICP nebulization process. Thus, the concentration of the elements was determined. The concentration of elements in the sediments of the Gollumutta Paya estuary is recorded (Tables 13.1 and 13.2).

Table 13.1 Concentration of major and minor elements in the sediments of the Gollumutta Paya estuary in April 2016

Elements	E-16	E-17	E-18	E-22	E-26	E-28	Min	Max	Avg.
Li	305.08	313.01	403.01	361.01	364.00	236.04	236.04	403.01	330.36
Be	4.01	4.02	3.01	6.08	6.01	4.11	3.01	6.08	4.54
Al	41640.03	55083.12	170406.10	50431.28	68150.06	21399.13	21399.13	170406.10	67851.62
V	97.04	206.18	98.00	102.11	1.02	191.01	1.02	206.18	115.89
Cr	92.04	87.13	103.01	99.72	103.03	49.13	49.13	103.03	89.01
Mn	385.71	410.11	519.18	65.03	46.01	998.07	46.01	998.07	404.02
Fe	7109.07	4105.10	728.11	9805.12	10201.98	29948.18	728.11	29948.18	10316.26
Co	4.01	9.18	7.06	9.01	1.05	18.06	1.05	18.06	8.06
Ni	9.96	7.23	9.63	10.01	34.01	12.41	7.23	34.01	13.88
Cu	20.02	6.15	19.12	12.03	98.03	67.15	6.15	98.03	37.08
Zn	21.04	27.11	5.12	119.02	120.07	73.41	5.12	120.07	60.96
As	30.04	28.07	41.02	1.08	17.07	24.09	1.08	41.02	23.56
Rb	7.04	1.06	19.11	9.02	1.02	38.98	1.02	38.98	12.70
Sr	17.01	107.01	51.08	61.02	124.05	1028.91	17.01	1028.91	231.51
Ag	98.01	110.19	117.01	118.01	117.02	108.09	98.01	118.01	111.39
Cd	9.09	4.07	8.27	12.01	11.03	10.42	4.07	12.01	9.15
Cs	34.02	31.04	29.00	31.12	38.03	38.14	29.00	38.14	33.56
Ba	204.02	210.15	209.16	260.15	338.01	255.81	204.02	338.01	246.22
Pb	62.08	91.02	97.17	0.15	0.03	0.08	0.03	97.17	41.75
Ti	0.02	0.01	0.02	0.11	0.01	0.07	0.01	0.11	0.04
U	0.02	0.04	0.01	0.26	0.02	0.01	0.01	0.26	0.06

Table 13.2 Concentration of major and minor elements in the sediments of the Gollumutta Paya estuary, December 2016

Elements	E-16	E-17	E-18	E-22	E-26	E-28	Min	Max	Avg.
Li	555.48	563.02	553.50	560.67	564.03	536.44	536.44	564.03	555.52
Be	7.11	7.20	6.82	7.01	7.45	6.43	6.43	7.45	7.00
Al	64647.63	85683.92	370486.90	70437.98	98256.96	41399.74	41399.74	370486.90	121818.90
V	143.84	306.88	126.26	142.55	2.42	297.06	2.42	306.88	169.83
Cr	115.94	100.54	158.24	124.78	173.14	59.33	59.33	173.14	121.99
Mn	785.57	700.44	859.78	565.53	66.61	1148.77	66.61	1148.77	687.78
Fe	9139.70	7905.30	928.27	10602.25	30231.98	39948.98	928.27	39948.98	16459.41
Co	16.41	15.98	14.16	10.11	2.35	25.76	2.35	25.76	14.13
Ni	12.96	19.23	10.63	19.21	44.11	18.81	10.63	44.11	20.83
Cu	31.58	38.85	26.42	17.13	113.35	82.95	17.13	113.35	51.71
Zn	138.64	33.20	6.42	189.12	250.03	83.47	6.42	250.03	116.81
As	41.08	40.27	55.59	3.68	20.27	34.90	3.68	55.59	32.63
Rb	13.74	3.62	29.17	18.03	2.03	44.68	2.03	44.68	18.54
Sr	21.02	192.65	71.69	80.82	144.76	2027.47	21.02	2027.47	423.07
Ag	146.71	147.89	147.16	148.59	147.52	148.02	146.71	148.59	147.65
Cd	15.89	16.47	16.27	16.54	16.30	15.45	15.45	16.54	16.15
Cs	47.07	47.95	46.00	47.45	48.08	44.87	44.87	48.08	46.90
Ba	364.82	354.65	409.36	360.67	531.65	355.92	354.65	531.65	396.18
Pb	82.79	110.58	101.91	107.64	0.03	0.23	0.03	110.58	67.19
Ti	0.14	0.04	0.17	0.11	93.88	72.17	0.04	93.88	27.75
U	0.42	0.34	0.34	0.26	0.43	6.09	0.26	6.09	1.31

13.5 Results and Discussion

13.5.1 *Major and Trace Elements*

In the present study, 3 major elements, namely, Fe, Mn, and Al, and 18 minor elements, namely, Li, Be, V, Cr, Co, Ni, Cu, Zn, As, Rb, Sr, Ag, Cd, Cs, Ba, Pb, Ti, and U were identified from the Gollumutta Paya estuary in April and December 2016. The values are shown in Fig. 13.3.

13.5.2 *Major Elements*

13.5.2.1 Iron (Fe)

The concentration of iron in the sediments of the Gollumutta Paya estuary ranges from 728.10 to 29,948.18 ppm in April and from 928.265 to 39,948.98 ppm in December (Fig. 13.3; Tables 13.1 and 13.2). The iron concentration in these sediments is thought to be derived from the source rocks which are mainly the Eastern Ghats, Dharwars, and Deccan Traps. The Deccan basalts contribute more iron than the Eastern Ghats. The Eastern Ghats sources yield more ilmenite than magnetite whereas the Dharwars source more magnetite than ilmenite (Nagamalleswara Rao and Kshira Sagar 1992). It is thought that iron is derived from the weathering of all the above continental rocks. The common heavy minerals of the area consisting of magnetites, ilmenites, and ferromagnesian minerals including pyroxenes, amphiboles, and garnets contribute to iron in a sand-sized fraction of the sediments. Ferrous minerals such as augites, feldspars, hornblendes, magnetites, and ilmenites are the sources of iron (Periakali et al. 2000; Malathi 2013).

13.5.2.2 Manganese (Mn)

The concentration of manganese in the sediments of the Gollumutta Paya estuary ranges from 46.01 to 998.07 ppm in April and from 66.61 to 1148.77 ppm in December, (Fig. 13.3; Tables 13.1 and 13.2). The manganese concentration in these sediments is thought to be derived from the rocks of the Eastern Ghats which consist of khondalites, kodurites, gondites, leptynites, and charnockites, and these rock formations at places are enriched in manganese ores. Apparently, basalts of the Deccan Traps provide more manganese to the present study areas. The weathering of manganese-bearing garnets and opaques (ilmenite and magnetite) might be the sources for manganese concentration in the bulk sediments.

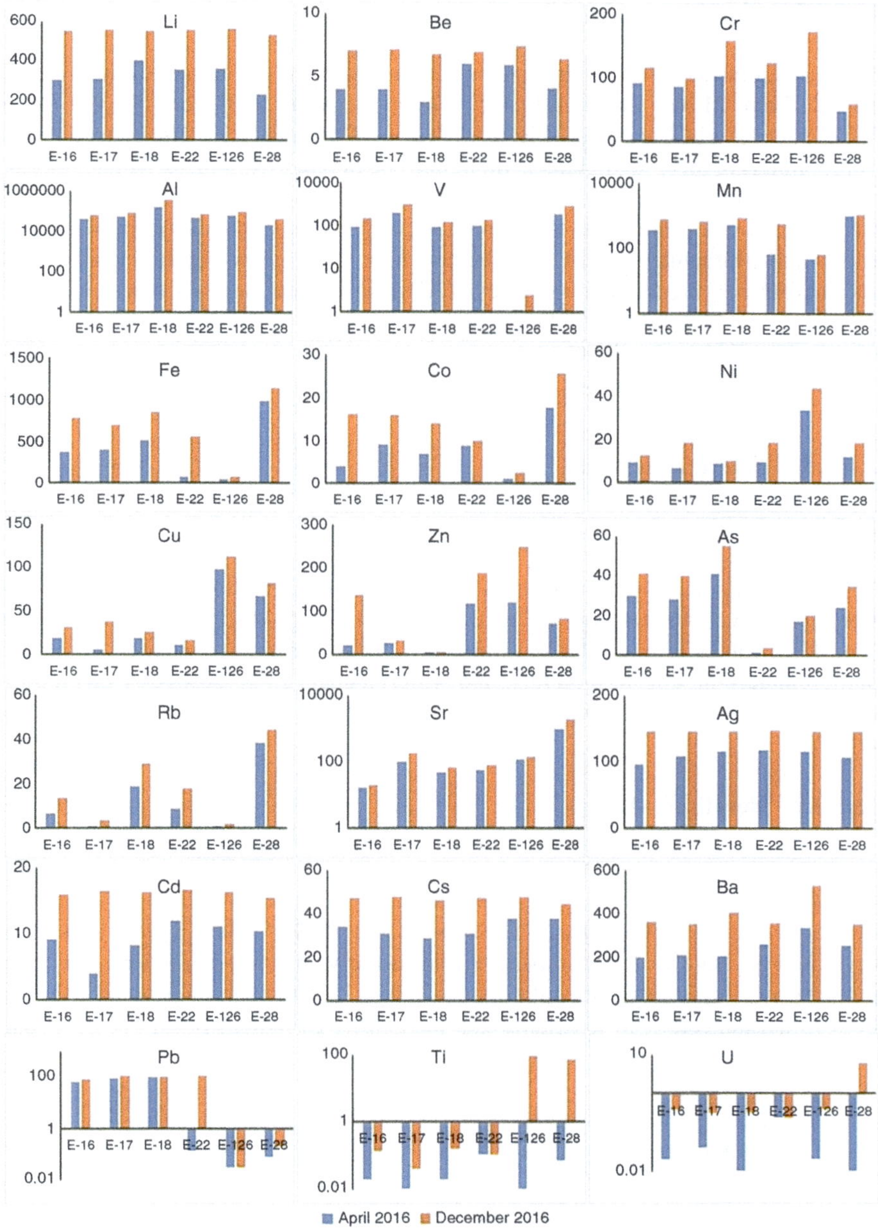

Fig. 13.3 Bar diagram of concentrations of the major and trace elements in the Gollumutta Paya estuary

13.5.2.3 Aluminum (Al)

The concentration of aluminum in the sediments of the Gollumutta Paya estuary ranges from 21,399.13 to 1,70,406.1 ppm in April and from 41,399.74 to 3,70,486.9 ppm in December (Fig. 13.3; Tables 13.1 and 13.2). The concentration of aluminum in these sediments is thought to be derived from the detrital aluminosilicates from the continents. It has been used as an index of the inorganic or argillaceous fraction of sediments. Aluminum occurs in feldspar and heavy minerals like garnet and sillimanite derived from the host rock as mentioned earlier. Clay minerals and feldspars are the major sources of Al in marine sediments. Studies on chemical extraction of sediments indicate that Al is largely located in the clay minerals, where it is held in crystal lattices.

13.5.3 Trace Elements

13.5.3.1 Lithium (Li)

The concentration of lithium in the sediments of the Gollumutta Paya estuary ranges from 236.035 to 403.01 ppm in April and 536.435 to 564.025 ppm in December (Fig. 13.3; Tables 13.1 and 13.2). Lithium is a lithophile metallic element occurring predominately in silicate minerals, and widely occurs as an accessory element in K-feldspars, biotite, and amphiboles.

13.5.3.2 Beryllium (Be)

The concentration of beryllium in the sediments of the Gollumutta Paya estuary ranges from 3.011 to 6.08 ppm in April and 6.425 to 7.445 ppm in December (Fig. 13.3; Tables 13.1 and 13.2). The small size of the Be^{+2} ion (35 ppm) influences beryllium to behave as incompatibly during early magmatic fractionation. This property and the unstable behavior of beryllium at high temperatures promote beryllium to get concentrated in granite and pegmatite. Beryllium occurs in beryl and chrysoberyl.

13.5.3.3 Vanadium (V)

The concentration of iron in the sediments of the Gollumutta Paya estuary ranges from 1.015 to 206.18 ppm in April and 2.415 to 306.88 ppm in December (Fig. 13.3; Tables 13.1 and 13.2). Vanadium occurs in magnetite, pyroxenes, amphiboles, and biotite. According to Rankama and Sahama (1950), vanadium is strongly enriched in titaniferous iron ores which are the early separated oxides. It is largely found in magnetite if the sediments are sourced from igneous rocks.

13.5.3.4 Chromium (Cr)

The concentration of chromium in the sediments of the Gollumutta Paya estuary ranges from 49.125 to 103.031 ppm in April and 59.325 to 173.135 ppm in December (Fig. 13.3; Tables 13.1 and 13.2). In silicate minerals, Cr^{3+} exhibits diadochus replacement with Fe^{3+} and Al^{3+} (Rankama and Sahama 1950). Chromium probably is carried in lattice positions with degraded clays. Both the Deccan basalts and the Eastern Ghats seem to provide the source of Cr. Some of the Cr is carried in the lattice position of clay minerals both montmorillonite and illite, but most Cr is fixed in oxide cations. High chromium concentrations are usually found to occur in sedimentary basins. This is because, these basins receive a higher input of ferromagnesian minerals such as micas, clays, especially illite, chromite, and chromium spinel.

13.5.3.5 Cobalt (Co)

The cobalt concentration in the sediments of the Gollumutta Paya estuary ranges from 1.05 to 18.055 ppm in April and from 2.35 to 25.755 ppm in December (Fig. 13.3; Tables 13.1 and 13.2). Cobalt occurs in divalent and trivalent states in rocks. Cobalt migrates more readily than a nickel. Its geochemistry is the same as that of nickel. Cobalt tends to be in solution, carried for long distances from continental sources, and is deposited in the hydrolysate sediments. Fe and Mn oxides occur as grain coatings in the sediments and are common carriers of Ni and Co. Cobalt is associated as an accessory element with amphiboles, garnets, micas, olivine, pyroxenes, and sphalerite. It may also associate with the iron sulfides like pyrite, arsenopyrite, and pyrrhotite, and with oxide minerals such as magnetite.

13.5.3.6 Nickel (Ni)

The concentration of nickel in the sediments of the Gollumutta Paya estuary ranges from 7.23 to 34.01 ppm in April and from 10.63 to 44.11 ppm in December (Fig. 13.3; Tables 13.1 and 13.2). Nickel occurs in a divalent state in rocks. Ni tends to be in solution, carried for long distances from continental weathering products, and is deposited in the hydrolysate sediments. It occurs in clay either by adsorption or exchange mechanism (Madhusudhana Rao 1983). High nickel concentration in most montmorillonite relative to other clay minerals is due to the fact that nickel strongly favors octahedral layers of the montmorillonite lattice (Cronan et al. 1982). Nickel occurs as fragments of ferromagnesian silicates and clays.

13.5.3.7 Copper (Cu)

The copper concentration in the sediments of the Gollumutta Paya estuary ranges from 6.145 to 98.03 ppm in April and from 17.13 to 113.35 ppm in December (Fig. 13.3; Tables 13.1 and 13.2). The association of Cu with the hydrate of Fe and Mn oxides could be stronger, especially where the source is more diffuse. Copper distribution is governed by sedimentary structures, namely, grain size and angularity. It may also be fixed in clay minerals by adsorption and cation exchange. It is introduced into the sediments from detrital heavy minerals and clay minerals. It might be fixed in the lattices of ferromagnesian silicate minerals such as garnets, pyroxenes, and amphiboles derived from the continental rocks as mentioned earlier. It is also widely distributed in small amounts such as ilmenite (Jagannadha Rao et al. 2005), magnetite, rutile, zircon, and sillimanite. It is observed that copper is sourced from ilmenite, magnetite, rutile, and zircon.

13.5.3.8 Zinc (Zn)

The zinc values of the Gollumutta Paya estuary range from 5.118 to 120.07 ppm in April and from 6.415 to 250.03 ppm in December (Fig. 13.3; Tables 13.1 and 13.2). The zinc concentration of these sediments is thought to be derived from the earlier-mentioned continental rocks. Zinc occurs in several minerals such as sphalerite, smithsonite, and zincite. The quantity of ferromagnesian silicates, detrital oxides, and clay particles predominantly control the distribution of zinc in sedimentary rocks. Magnetite is the primary zinc carrier in mafic rocks, whereas biotite is the most important mineral in granite. It is widely dispersed in the form of trace amounts in a range of other minerals such as ilmenite, magnetite, zircon, sillimanite, and muscovite.

13.5.3.9 Arsenic (As)

The concentration of arsenic in the sediments of the Gollumutta Paya estuary ranges from 1.08 to 41.01 ppm in April and from 6.415 to 250.03 ppm in December (Fig. 13.3; Tables 13.1 and 13.2). The arsenic concentration of these sediments is thought to be derived from the rocks of the Deccan Traps, Dharwars, and Eastern Ghats. Arsenic is incorporated into a limited amount in the principal rock-forming silicate minerals by replacing Fe^{+3} or Al^{+3} with arsenic. Arsenic is abundant in sedimentary rocks in clays, Fe and Mn oxides, sulfides, and phosphates.

13.5.3.10 Rubidium (Rb)

The rubidium concentration in the sediments of the Gollumutta Paya estuary ranges from 1.021 to 38.98 ppm in April and from 2.025 to 44.68 ppm in December (Tables 13.1 and 13.2). The rubidium concentration of these sediments is thought to be derived from the continental rocks as mentioned earlier.

13.5.3.11 Strontium (Sr)

The concentration of strontium in the sediments of the Gollumutta Paya estuary ranges from 17.011 to 1028.91 ppm in April and 21.015 to 2027.47 ppm in December (Fig. 13.3; Tables 13.1 and 13.2). The strontium concentration of the sediments is thought to be derived from the rocks of the Eastern Ghats, Deccan Traps, and Dharwars.

13.5.3.12 Silver (Ag)

The concentration of silver in the sediments of the Gollumutta Paya estuary ranges from 98.005 to 118.01 ppm in April and from 146.705 to 148.59 ppm in December (Fig. 13.3; Tables 13.1 and 13.2). The silver concentration of these sediments is thought to be derived from the Eastern Ghats, Dharwars, and the Deccan Traps. Silver concentrations in sediments are typically governed by pH, the occurrence of organic matter, and the source material. Under low acidity circumstances, Ag^+ can replace K^+ in clay minerals. Because of this, as well as preferred sorption, silver is more concentrated in clay than in silt or sand. The silver content is mostly determined by pH and E^h conditions, as well as the concentration of halide ions, which is particularly relevant in coastal environments. It is more mobile in acidic settings and highly oxidative in situations associated with high ionic strength solutions where soluble complexes with anions, particularly chloride, can form.

13.5.3.13 Cadmium (Cd)

The cadmium concentration in the sediments of the Gollumutta Paya estuary ranges from 4.07 to 12.01 ppm in April and from 15.44 to 16.53 ppm in December (Fig. 13.3; Tables 13.1 and 13.2). The cadmium concentration of the sediments is thought to be derived from the continental rocks of the Eastern Ghats, Dharwars, and the Deccan Traps. Cadmium is a common substitution for Cu, Hg, Pb, and Zn in sulfide minerals, particularly sphalerite, and to a lesser extent in other Zn minerals such as smithsonite. It can also be found in trace levels in silicate minerals like biotite and amphiboles. Cadmium is commonly found in ilmenite, magnetite, monazite, rutile, and zircon.

13.5.3.14 Cesium (Cs)

The concentration of cesium in the sediments of the Gollumutta Paya estuary ranges from 29.00 to 38.135 ppm in April and from 44.86 to 48.08 ppm in December (Fig. 13.3; Tables 13.1 and 13.2). The cesium concentration of these sediments is thought to be derived from the rocks of the Eastern Ghats. It can, however, make a limited substitution for K in micas and K-feldspars, which contain most of the cesium in the Earth's lithosphere.

13.5.3.15 Barium (Ba)

The value of barium in the sediments of the Gollumutta Paya estuary ranges from 204.02 to 338.01 ppm in April and 354.65 to 531.65 ppm in December (Fig. 13.3; Tables 13.1 and 13.2). Barium might be derived from the Eastern Ghat rocks. It is commonly found in K-feldspars and micas due to the substitution of K^+ with Ba^{2+}, which has a comparable ionic radius. K-feldspars have higher barium contents than phyllosilicates. The Ba^{++} ion also serves as a Ca^{++} substitution in plagioclase and pyroxenes, and the silicate minerals apatite and calcite. This can be absorbed is ascribed to the concentration of Ba in sedimentary rocks.

13.5.3.16 Lead (Pb)

The concentration of lead in the sediments of the Gollumutta Paya estuary ranges from 0.03 to 97.17 ppm in April and from 0.03 to 110.575 ppm in December (Fig. 13.3; Tables 13.1 and 13.2). Lead occurs in several minerals such as galena, anglesite, and cerussite. It is also distributed in small amounts in other several minerals such as K-feldspars, plagioclase, micas, zircon, and magnetite. Lead primarily occurs in the structure of potassium feldspars, micas of magmatic and metamorphic rocks. Still, some lead may be transported by clay minerals and ferric iron oxides. Lead is associated with fine clays.

13.5.3.17 Titanium (Ti)

The concentration of titanium in the sediments of the Gollumutta Paya estuary ranges from 0.005 to 0.11 ppm in April and from 0.035 to 93.88 ppm in December (Fig. 13.3; Tables 13.1 and 13.2). Generally, Ti exists in insoluble, heavy minerals, namely, anatase, ilmenite, rutile, and sphene. In clay sediments, titanium is present as TiO_2 minerals. In the present study, titanium is thought to be derived from the Eastern Ghats, Deccan Traps, and Dharwars.

13.5.3.18 Uranium (U)

The concentration of uranium in the sediments of the Gollumutta Paya estuary ranges from 0.005 to 0.255 ppm in April and from 0.255 to 6.085 ppm in December (Fig. 13.3; Tables 13.1 and 13.2). Uraniumoccurs in monazite, uranite, brannerite, and carnotite. Uranium may be associated with minerals such as ilmenite, magnetite, monazite, rutile, and zircon.

13.6 Conclusions

A total of 12 samples were collected from the sediments of the Gollumutta Paya estuary in April and December 2016 to study the distribution of the geochemical elements. Three major elements, namely, Fe, Mn, and Al, and 18 minor elements, namely, Li, Be, V, Cr, Co, Ni, Cu, Zn, As, Rs, Sr, Ag, Cd, Cs, Ba, Pb, Ti, and U were identified from the sediments of the estuary. Based on the distribution of all the above elements, the nature of the source rocks, the transport mechanism, and the depositional environment of the sediments could be predicted which will help as tools in the delineation of the paleo-depositional environments of the sediments. These studies are also useful to establish the provenance and diagenesis. Another important area of applications of the geochemical elements, especially heavy metals, is that they are among the most persistent pollutants that occurs in water, sediments, and organisms indicating the pollution sources. This helps to take precautions against the pollutants to save the environment by estimating the pollution levels of waters and sediments not only in the study area but also in other areas which are similar in characteristic features of the source rocks as well as the transporting mechanism, environmental, and climatic conditions. These observations are helpful to take remedial measures to save the environment from pollution and to give advice to the people to take necessary measures to save themselves, and for biodiversity conservation.

Acknowledgments We, the authors of this chapter, are grateful to Prof. R. Kaladhar, a retired Professor of the Department of Geology, Andhra University, Visakhapatnam, for his help in drafting the manuscript of this chapter. We are thankful to the technicians of the Institute of Advanced Studies in Bay of Bengal for their help in the ICP-MS analysis of the sediment samples for the geochemical elements. We are also thankful to Dr. Ch. Ravi Sekhar for preparing the necessary diagrams for this chapter.

References

Cronan DS, Glasby GP, Moorby SA, Thomson J, Knedler KE, Mc Dougall JC (1982) A submarine hydrothermal manganese deposit from the S.W. Pacific Island arc. Nature 298:456–458

George CF, Macdonald DIM, Spagnolo M (2019) Deltaic sedimentary environments in the Niger Delta, Nigeria. J African Earth Sci 160:103592
Jagannadha Rao M, Venkata Ramana J, Venu Gopal R, Chandra Rao M (2005) Geochemistry and ore-minerology of ilmenite from beach placers of the Visakhapatnam-Bhimunipatnam deposit, Andhra Pradesh. J Geol Soc India 66(2):147
Lei KY, Liu CY, Zhang L, Wu BL, Cun XN, Sun L (2017) Element geochemical characteristics of the Jurassic mudstones in the Northern Ordos Basin: implications for tracing sediment sources and paleoenvironment restoration. Acta Sedimentol Sin 35(3):621–636
Madhusudhana Rao Ch (1983) Geochemistry of the sediments of the central western continental shelf of India. Ph.D. Thesis. Andhra University, Waltair (Unpublished)
Malathi V (2013) Geochemical studies of clays from coastal red sediments from Yerrapalem to Chepalakancheru, Vizianagaram District, Andhra Pradesh
Maroof Azam M, Tripathi JK (2016) Recent contributions in the field of sediment geochemistry. Proc Indian Ntl Sci Acad 82(3)
Nagamalleswara Rao B, Kshira Sagar TVSR (1992) A study on radiation environment along parts of central east coast of India. Indian J Environ Prot 12:646–646
Nagelschmidt G, Desi AD, Muir A (1940) The minerals in the clay fractions of a black cotton soil and a red earth from Hyderabad, Deccan State, India. J Agric Sci 30:645–647
Ngeuutchoua G, Bessa AZE, Eyong JT, Zandjio DD, Djaoro HB, Nfada LT (2019) Geochemistry of Cretaceous fine-grained silisiclastic rocks from Upper Mundeck and Logbadjeck Formations, Douala sub-basin, SW Cameroon: implications for weathering intensity, provenance, palaeoclimate, redox conditions & tectonic setting. J Afr Earth Sci 152:215–236
Periakali P, Eswaramoorthi S, Subramanian S, Jaisankar P (2000) Geochemistry of Pichavaram mangrove sediments, southeast coast of India. J Geol Soc India 55(4):387–394
Rankama K, Sahama G (1950) Geochemistry. University of Chicago Press, Chicago, 912 p
Shilling AM, Colcord DE, Karty J, Hansen A, Freeman KH, Njau JK, Stanistreet IG, Stollhofen H, Schick KD, Toth N, Brassell SC (2020) Biogeochemical evidence from OGCP Core 2A sediments for environmental changes proceeding deposition of Tuff IB and climatic transition in Upper Bed I of the Olduvai Basin. Palaeogeogr Palaeoclimatol Palaeoecol 555:109824
Skurzynski J, Jary Z, Kenis P, Kubik R, Moska P, Raczyk J, Seul C (2020) Geochemistry and mineralogy of the late Pleistocene loess-palaeosol sequence in Zlota (near Sandomierz, Poland): Implications for weathering, sedimentary recycling, and provenance. Geoderma 375:114459
Van de Velde SJ, Hylen A, Kononets M, Maezocchi U, Leermaker M, Choumiline K, Per Hall OJ, Meysman FJR (2020) Elevated sedimentary removal of Fe, Mn, trace elements following a transient oxygenation event in the Eastern Gotland Basin, central Baltic Sea. Geochim Cosmochim Acta 271:16–32

Part IV
Biodiversity/Bio-indicators/Ecological Studies

Chapter 14
Assessment of Trace Metal Contamination in *Saccostrea cucullata* (Born, 1778) from the Coast of South Andaman Island, India

S. Chetan, Abhijeet Purkayastha, and S. Venu

Abstract The Andaman and Nicobar group of Islands sustains a very rich diversity of terrestrial as well as aquatic organisms. Bivalves harbour in the majority of the coastal habitats and *Saccostrea cucullata* was the dominant bivalve species in all the study sites. One-time sampling was carried out in eight different study sites to understand the population abundance through the quadrat method. The samples were collected and brought to the laboratory to determine the concentration of trace metal elements. The number of individual bivalves was observed to be high in Corbyn's Cove and Haddo, medium in Brookshabad beach, Rangachang, Burmanallah and South Point and low in the North Bay and Kodiyaghat. The concentration of manganese exceeded 0.6 μg/g in Corbyn Cove samples and is the highest observed in this study, whereas, manganese concentration in South Point and North Bay exceeded 0.2 μg/g. Iron concentration in North Bay was maximum with a value of 6 μg/g, followed by Haddo with approximately 4 μg/g, south point with a concentration of 3 μg/g and other stations showed concentrations less than 2 μg/g. Copper showed the highest concentrations in Kodiyaghat with 1.2 μg/g. Zinc concentration was constant with values between 2.5 and 3.1 μg/g. The BCF

S. Chetan
Department of Ocean Studies and Marine Biology, Pondicherry University, Port Blair, Andaman and Nicobar Islands, India

Department of Zoology, St. Joseph's College, Bengaluru, Karnataka, India

A. Purkayastha
Department of Ocean Studies and Marine Biology, Pondicherry University, Port Blair, Andaman and Nicobar Islands, India

Biological Oceanography Division, National Institute of Oceanography, Dona Paula, Goa, India

S. Venu (✉)
Department of Ocean Studies and Marine Biology, Pondicherry University, Port Blair, Andaman and Nicobar Islands, India
e-mail: svenu.omb@pondiuni.ac.in

N. Jayaraju et al. (eds.), *Coasts, Estuaries and Lakes*,
https://doi.org/10.1007/978-3-031-21644-2_14

values for iron and copper could not be ascertained because their concentrations in seawater in all the study sites were undetectable. Apart from the BCF value, the highest bioconcentration value was obtained from Brookshabad beach (2.6) and the lowest value in Kodiyaghat (3.6). Zinc showed the highest bioconcentration value of 212.3 in Kodiyaghat and the lowest value of 38.7 in South Point. Analysis of variance (ANOVA) test showed significant variation (p-value < 0.05) among the study sites with respect to both water and tissues, and zinc showed the maximum correlation. From the study, it is quite clear that the obtained trace element values fall within the standard values, thereby indicating the study site is less polluted.

Keywords Trace elements · Andaman · ANOVA · Bioconcentration Factor · Oyster

14.1 Introduction

Trace metals are metals that are present in low concentrations in the environment. For living organisms, minimal quantities of these metals are required for everyday metabolism, hence the name trace metals. However, in higher quantities, these trace metals are proven toxic to living beings and determine the health of the ecosystem (Abhilash et al. 2013). Trace metals are found in negligible quantities in marine environments. The problem arises when these trace metals are accumulated in biological tissues due to filter feeding. Filter feeding is aquatic feeding in which the animal takes in water and separates plankton and organic matter that is consumed as food. Filter feeders have specialised mouthparts for this function. The trace metals accumulated in this process may end up in different pathways due to their bioavailability and lead to the formation of new chemical species (Bordin et al. 1992). Trace metals are released into the marine environments, primarily due to industries and factories, agriculture and mining, and the secondary sources are transportation, urban effluents and municipal garbage.

Earlier studies on metals in aquatic environments were carried out in late 1960 in Canada. Pequegnat et al. (1969) carried out studies of zinc requirements on marine organisms. Another study was done by Hobden (1970) on iron metabolism in freshwater mussels. Phillips et al. (1982) studied the toxicological significance of trace metals in the seafood of Hong Kong. Szefer and Szefer (1990) worked on metals in molluscs and associated bottom sediments of the Baltic Sea. In the Netherlands, Bordin et al. (1992) worked on trace metals in the bivalve *Macoma balthica* in 1992. Bou-olayan et al. (1995) worked on the metal accumulation in *Pinctada radiata* in Kuwait. Usero et al. (1997) of Spain studied trace metal in *Ruditapes decussatus* and *Ruditapes philippinarum* in Boening (1999) evaluated bivalves as biomonitors of heavy metals. Campanella et al. (1999) studied the trace metals in seagrass and molluscs of the Mediterranean. Conti and Cecchetti (2003) also used algae and molluscs in their trace metal biomonitoring study. Usero et al. (2005) of Spain also worked on heavy metal concentrations in molluscs. Tapia et al. (1990) studied trace metals in bivalves from the Pacific coast of Chile. Bodin et al. (2013) assessed the trace metals in mangrove ecosystems from Senegal, West Africa using molluscs. Dabwan and Taufiq (2016) of Malaysia used bivalves as bioindicators for heavy metal detection. Li et al. (2017) re-described the taxonomic classification of oyster species in Myanmar.

The earliest study in India was by Rao and Rao (1971) on the distribution of trace elements in the Bay of Bengal. Bhosle and Matondkar (1978) studied the variation in trace metals in two populations of *Perna viridis* from Goa. Lakshmanan and Nambisan (1983) studied the seasonal variations in trace metal content in bivalve molluscs, *Villorita cyprinoides*, *Meretrix casta* and *Pernaviridis*. Mitra and Choudhury (1993) worked on the trace metals in macrobenthic molluscs of the Hooghly estuary. Biswas et al. (2013) worked on the accumulation of trace metals in *Saccostrea cucullata* of West Bengal. In Andaman and Nicobar Islands, Sarma et al. (2013) worked on the impact of coastal pollution on the biological, biochemical and nutritional status of edible oysters in Phoenix Bay and North Wandoor. Abhilash et al. (2013) worked on trace metal bioconcentration in *Saccostrea cucullata* in Andaman waters. Seetharaman et al. (2015) worked on the coastal pollution impacts and the microbial and mineral profile of *Crassostrea rivularis* in Andaman waters. The present study was conducted to understand the abundance of *Saccostrea cucullata,* an efficient filter feeder in the South Andaman region, with an objective to determine the amount of accumulated trace element and their respective bioconcentration factor (BCF) to understand the levels of pollution in this region.

14.2 Materials and Methods

14.2.1 Study Site

The study sites cover the eastern part of South Andaman Island, which opens to the Andaman Sea. The entire stretch has certain regions with rocky coasts where the oysters are predominantly found. There are eight regions where *Saccostrea cucullata* was dominant and the sampling was undertaken in these areas namely North Bay (11°42′19.4″N; 92°44′32.9″E), Haddo (11°40′50.6″N; 92°43′33.3″E), South Point (11°39′27.0″N; 92°45′14.8″E), Carbyn's Cove (11°38′31.9″N; 92°44′47.5″E), Brookshabad beach (11°37′48.8″N; 92°45′02.8″E), Rangachang (11°34′31.0″N; 92°44′16.0″E), Burmanallah (11°33′39.1″N; 92°43′52.7″E) and Kodiyaghat (11°31′49.9″N; 92°43′25.4″E) as shown in Fig. 14.1.

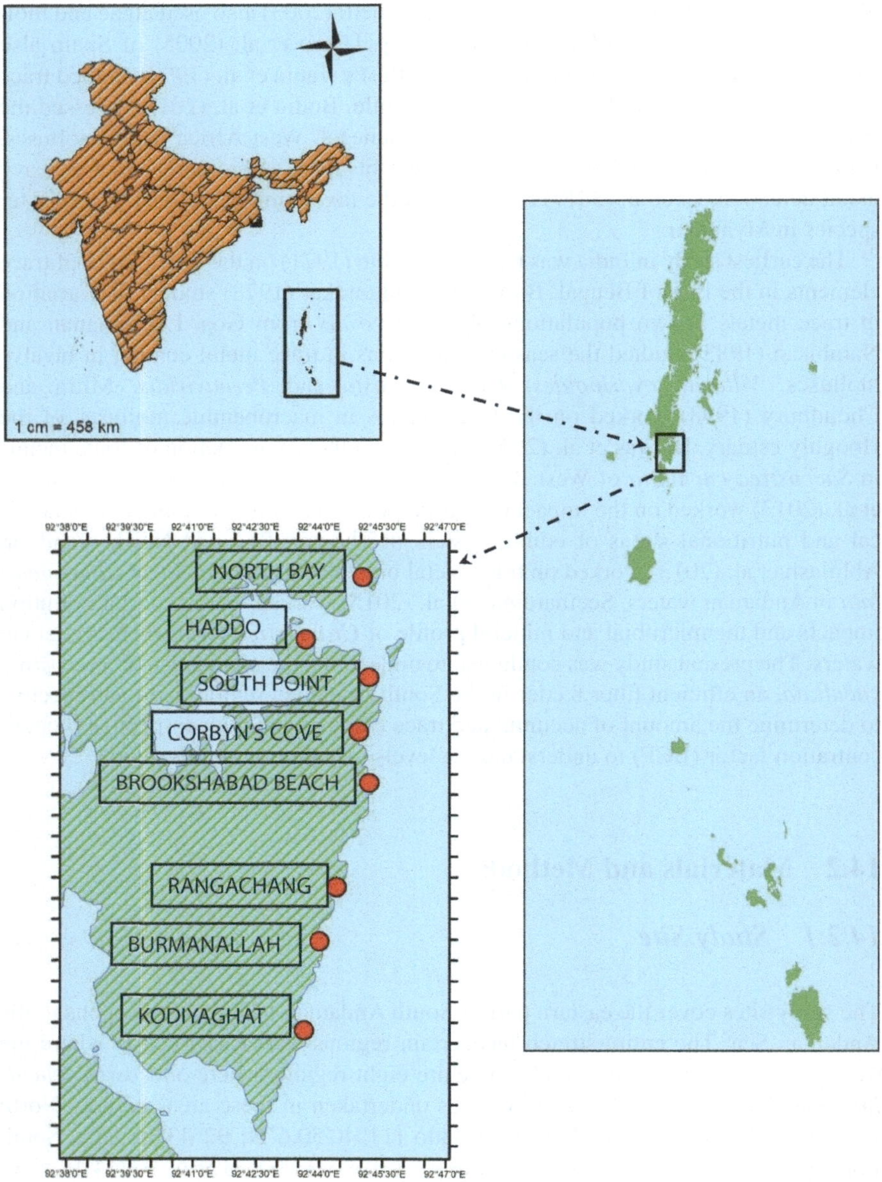

Fig. 14.1 Map showing the different study sites in South Andaman Island

14.2.2 Methodology

A sampling of *Saccostrea cucullata* was conducted during the month of January 2020 from the eight selected study sites. The surveys were carried out along three-line transects from the low tide mark to the high tide mark. Each transect was fixed at high tide, mid tide and low tide regions. In each tidal zone, three quadrats of one square meter were placed in areas where oysters were present. The total number of live oysters was counted in each quadrat.

For estimation of biomass, random samples of five oysters from each study site were collected by breaking open the shell using a hammer and knife and then transferred to a zip-lock bag. The samples were immediately transported to the laboratory and preserved at −20 °C. Further, the samples were thawed and washed with fresh water. The mantle, gills, digestive tract and soft body parts were separated. The wet weights of the samples were taken using the weighing scale. The samples were kept at 70 °C overnight to remove moisture content, after which the dry weight of each sample was noted.

The dried samples were crushed to powder using a mortar and pestle. 500 mg of each sample was taken and digested with a 10 ml mixture of Nitric acid (HNO_3) and Per-chloric acid ($HClO_4$) (ratio of 5:1) solution (Dabwan and Taufiq 2016; Usero et al. 2005). For samples that were underweight, two or more samples from the same site were mixed to make up to 500 mg. After digestion, the samples were added with more HNO_3:$HClO_4$ dropwise to maintain 10 ml and to get white fumes. The samples were transferred to a 50 ml volumetric flask and made up to mark using distilled water. The samples were filtered using Whatman no. 42 filter paper, and the readings were taken using an Atomic Absorption Spectrophotometer. The accuracy of analytical methods was monitored by repeated analysis of standard reference materials (SRM 1566b from the National Institute of Standard and Technology and BCSS-1 from the National Research Council of Canada). The values of metal content obtained were represented as µg/g of sample on a wet weight basis. Water samples were also collected from the study sites for analysing of trace metals using standard procedures. The bioconcentration factor (BCF), that is, the concentration of a particular metal in a biological tissue relative to the surrounding medium, was calculated (Abhilash et al. 2013). It can be calculated using the formula:

$$BCF = \frac{Avg.\text{concentration of the metal in tissue}\left(\mu g / g\right)}{\text{Concentration of the metal in water}\left(\mu g / g\right).}$$

The reading obtained after the experiment was then tabulated in excel, which was further used to represent the values graphically and statistically using single-factor ANOVA in excel for proper and thorough understanding.

14.3 Results

14.3.1 Abundance and Distribution

In the present study, *Saccostrea cucullata* was abundantly observed in Carbyn's Cove and was least in Kodiyaghat. The number of individuals ranged from as high as 800 per quadrat to 1 per quadrat. In their highest observed tidal range in the selected study sites. The highest number of individuals observed in a single quadrat was 842 in Corbyn's Cove (CC-T1-MT-Q1). The lowest observed of individuals in a single quadrant was 1 in Kodiyaghat (KD-T1-MT-Q2). The abundance of these organisms varied from mid-tide to high-tide. The sites with high abundance in mid-tide regions were North Bay, Haddo, Carbyn's Cove, Brookshabad beach, Burmanallah and Kodiyaghat. Whereas the regions with high abundance in high-tide regions were South Point and Rangachang. The distribution pattern across different study sites was observed to show slight variation (Fig. 14.2).

The population density of *S. cucullata* varied station-wise; starting from Carbyn's Cove followed by Haddo, Brookshabad beach, Rangachang, Burmanallah, South Point, North Bay and Kodiyaghat.

14.3.2 Analysis of Dry and Wet Weight

The size variation in these organisms ranged from 2 to 17 cm in length. The size of the individuals in North Bay ranged from 2 to 4 cm long. The individuals of Haddo ranged from 3 to 7 cm in length. The individuals of South Point ranged from 2 to 3 cm in length. The individuals of Corbyn's Cove ranged from 2 to 17 cm in length.

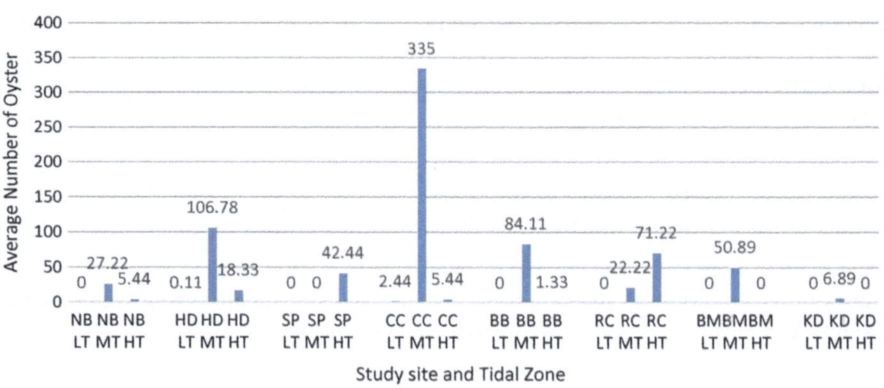

Fig. 14.2 Distribution of *Saccostrea cucullata* from all the study sites. NB North Bay, HD Haddo, SP South Point, CC Carbyn's Cove, BB Brookshabad Beach, RC Rangachang, BM Burmanallah, KD Kodiyaghat, LT Low tide, MT Mid tide, HT High tide

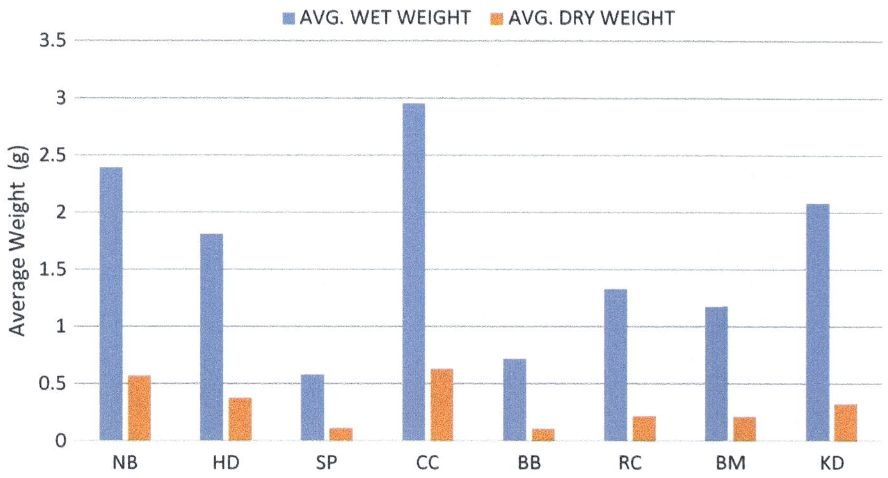

Fig. 14.3 Average values of dry and wet weight samples from all the study sites. NB North Bay, HD Haddo, SP South Point, CC Carbyn's Cove, BB Brookshabad Beach, RC Rangachang, BM Burmanallah, KD Kodiyaghat, LT Low tide, MT Mid tide, HT High tide

The individuals of Brookshabad beach ranged from 4 to 8 cm in length. The individuals of Rangachang ranged from 2 to 3 cm in length. The individuals of Burmanallah ranged from 2 to 6 cm in length, and the individuals of Kodiyaghat ranged from 5 to 12 cm in length. The heaviest sample collected was from Corbyn's Cove, which weighed 7.617 grams, whereas the sample with the least weight was collected from Brookshabad beach, which weighed 0.284 grams. The differences in the wet and dry weights were used to calculate the water content in the samples. The total water content of the individual samples collected varied from 80% to 86%. The average values of dry and wet weight were also represented graphically in Fig. 14.3.

14.3.3 Trace Metal Analysis

The trace metal concentration was analysed from the tissue and water samples from the respective study sites. Four common trace metals (Manganese, Iron, Copper and Zinc) were determined from the study. Trace metal concentration from seawater as well as the tissue samples showed considerable variation. The manganese content in the water samples varied from 0 to 0.014 μg/g, and in tissue samples varied from 0.005 to 0.618 μg/g. The manganese concentrations in tissue samples were observed to be high in northern sites, with the highest reading from Carbyn's Cove with values of 0.618 μg/g, whereas the southern parts showed lesser concentrations. The lowest reading recorded was from Haddo with 0.005 μg/g (Fig. 14.4). Iron concentration in the water samples was negligible in all the samples. In tissue samples,

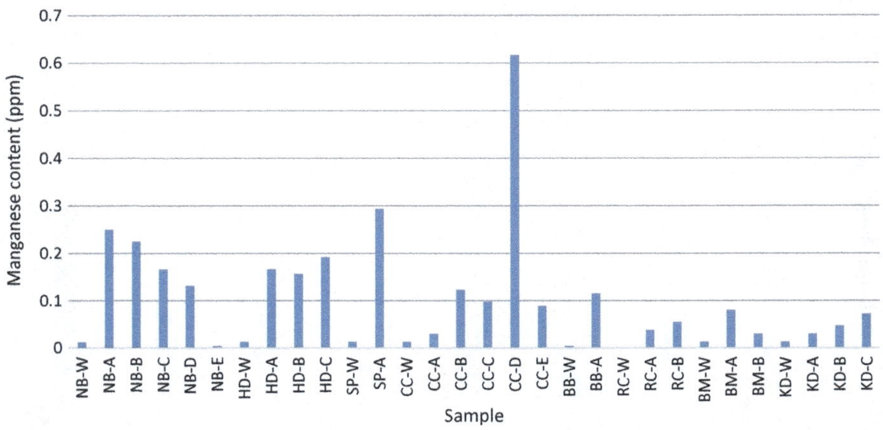

Fig. 14.4 Manganese content in all the study sites. NB North Bay, HD Hadoo, SP South Point, CC Corbyn's Cove, BB Brookshabad Beach, RC Rangahang, BM Burmanallah, KD Kodiyaghat, W Water sample, A/B/C/D/E Tissue sample

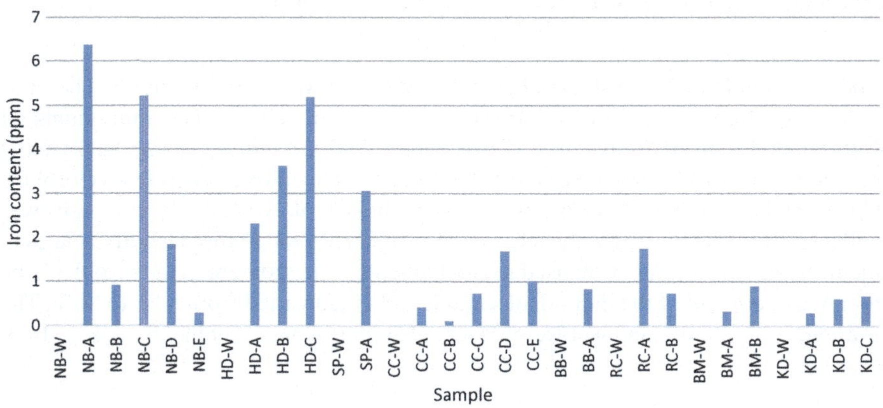

Fig. 14.5 Iron content in all the study sites. NB North Bay, HD Hadoo, SP South Point, CC Corbyn's Cove, BB Brookshabad Beach, RC Rangahang, BM Burmanallah, KD Kodiyaghat, W Water sample, A/B/C/D/E Tissue sample

values varied from 0.122 to 6.392 μg/g. Similar to manganese concentrations, the iron concentrations were high in the northern sites. The highest reading was from North Bay with 6.392 μg/g and the values decreased gradually heading south. The lowest value recorded was from Corbyn's Cove with 0.122 μg/g (Fig. 14.5). The copper concentration in the water samples was similar to iron concentration with negligible. In tissue samples, the concentrations varied from 0.364 to 1.187 μg/g. The copper concentrations in tissue samples were distributed throughout the study sites without a trend. The highest reading was from Kodiyaghat with 1.187 μg/g,

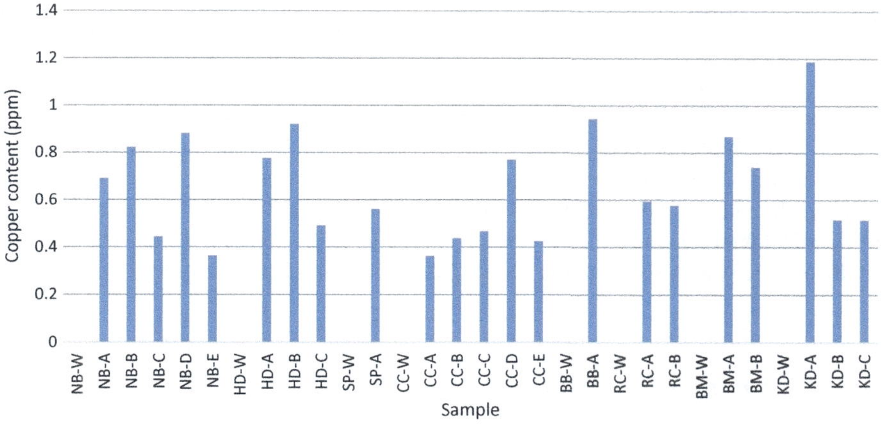

Fig. 14.6 Copper content in all the study sites. NB North Bay, HD Hadoo, SP South Point, CC Corbyn's Cove, BB Brookshabad Beach, RC Rangahang, BM Burmanallah, KD Kodiyaghat, W Water sample, A/B/C/D/E Tissue sample

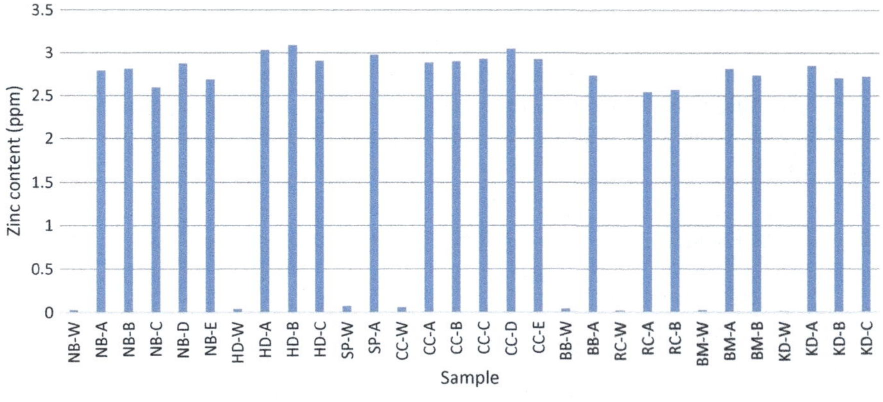

Fig. 14.7 Zinc content in all the study sites. NB North Bay, HD Hadoo, SP South Point, CC Corbyn's Cove, BB Brookshabad Beach, RC Rangahang, BM Burmanallah, KD Kodiyaghat, W Water sample, A/B/C/D/E Tissue sample

and the lowest value measured was from Corbyn's Cove with 0.364 μg/g (Fig. 14.6). The zinc concentrations in the water samples varied from 0.013 to 0.077 μg/g. The highest concentration noted was at South Point, and values decreased in both north and south directions. The lowest concentration was recorded at Kodiyaghat. In tissue samples, the concentrations varied from 2.545 to 3.096 μg/g. The zinc concentrations in tissue samples were more or less evenly distributed. The highest value recorded was from Haddo with 3.096 μg/g, and the lowest was from Rangachang with 2.545 μg/g (Fig. 14.7).

14.3.4 Bioconcentration Factor

The bioconcentration factor of *Saccostrea cucullata* in the present study from all the study sites was measured using the formula and depicted in Table 14.1. The BCF values were not obtained for iron and copper in all stations and manganese in Rangachang, because their concentration value in seawater was lower than 0.001 μg/g and was not detected by the AAS. The highest BCF for manganese was in Brookshabad beach (23.2), and the lowest was in Kodiyaghat (3.6). The highest BCF for zinc was in Kodiyaghat (212.3), and the lowest was in South Point (38.7).

14.3.5 Comparative Studies

The study conducted by Abhilash et al. (2013) and the current study have two study sites in common- North Bay and Corbyn's cove. The two studies have chosen *S. cucullata* to be the experimental organism and comparison of iron, copper and zinc BCF values in Table 14.2.

Trace metal levels obtained from the current study were compared with previous studies conducted in the Indian waters (Table 14.3) and globally (Table 14.4).

The manganese content in seawater was above the permissible value as per WHO. The values of Iron and Copper levels were well within the permissible values. The Zinc content in seawater was above the permissible value as per US EPA (Table 14.5).

Table 14.1 BCF values of *Saccostrea cucullata* from all the study sites

Study site	Manganese	Iron	Copper	Zinc
North Bay	11.2	×	×	83.5
Haddo	12.3	×	×	62.8
South Point	21	×	×	38.7
Corbyn's Cove	13.7	×	×	45.2
Brookshabad Beach	23.2	×	×	57
Rangachang	×	×	×	102.2
Burmanallah	4	×	×	92.5
Kodiyaghat	3.6	×	×	212.3

× = value unknown

Table 14.2 BCF values comparison between the 2013 study and the present study

	Abhilash et al. (2013)			Present study		
Study site	Fe	Cu	Zn	Fe	Cu	Zn
North Bay	183.97	28.65	1256.07	×	×	83.5
Corbyn's Cove	308.08	7.47	762.67	×	×	45.2
Other sites (range)	91.97–449.43	0.20–11.85	460.39–2474.72	×	×	38.7–212.3

× = unknown values

Table 14.3 Comparison of trace metal levels of molluscs from Indian water

Paper, region	Source organism	Mn (μg g^{-1})	Fe (μg g^{-1})	Cu (μg g^{-1})	Zn (μg g^{-1})
Bhosle and Matondkar (1978), Velsao, Goa	*Mytilus viridis*	17.53–71.56	539.34–2090.99	7.42–19.30	6.71–10.52
Lakshmanan and Nambisan (1983), Cochin, Kerala	*Villoreta cochinensis*	–	200.59–665.41	18.43–44.46	53.07–105.58
	Meretrix casta	–	181.22–338.82	17.35–46.85	49.82–83.35
	Perna viridis	–	220.40–940.16	11.40–30.41	67.48–100.48
Mitra and Choudhury (1993), Hooghly Estuary, West Bengal	*Nerita articulata*	14.40–15.90	238.42–303.65	19.71–21.30	63.90–81.75
	Littorina undulata	13.91–14.23	178.52–219.46	15.65–17.89	46.2–53.84
	Cymialacera	21.34–27.49	334.05–402.83	1134.20–1782.14	829.16–1004.56
	Columbella sp.	15.62–17.48	283.76–311.45	23.98–31.46	81.51–87.88
	Crassostrea cucullata	11.93–19.33	151.27–171.34	80.54–102.63	1023.36–1211.55
	Enigmonia aenigmatica	9.87	93.41	44.52	221.5
Abhilash et al. (2013), Andaman & Nicobar.	*Saccostrea cucullata*	–	11.04–114.95	0.72–154.11	467.76–3262.02
Present study	*Saccostrea cucullata*	0.005–0.618	0.122–6.392	0.364–1.187	2.545–3.096

14.4 Discussion

From the distribution chart (Fig. 14.2) it can be seen that the oysters were mostly found in the mid-tide regions. This can be due to the preferences of the oyster larvae to settle in that part due to more tidal action which in turn will help them filter-feed effectively. In sites South Point and Rangachang it was seen that the number of oysters was more in the high-tide region as compared to the mid-tide region, the presence of the sea wall could be the factor inducing this settlement pattern. The size of the oysters was comparatively smaller in the sites of South Point, Brookshabad beach, Rangachang and Burmanallah. This can be clearly seen in the graph showing the weights (Fig. 14.3). The reason for this can also be due to lack of proper settlement regions for the larvae which prevent them from filter-feed effectively which in turn affects their nutrition.

In the present study, the highest concentration measured for iron was from North Bay which was 6.3 μg/g followed by Haddo with 5.2 μg/g. This could be due to the

Table 14.4 Globally recorded values of trace metal from previous studies

Paper, region	Source organism	Mn (μg g^{-1})	Fe (μg g^{-1})	Cu (μg g^{-1})	Zn (μg g^{-1})
Conti and Cecchetti (2003), Italy	*Monodonta turbinata*	–	–	50.5–83.0	77.7–129.3
	Patella caerulea	–	–	10.2–19.2	87.4–117.1
	Mytilus galloprovincialis	–	–	5.51–11.5	123–180
Usero et al. (1997), Spain	*Ruditapes decussatus*	1.2–2.8	39–223	1.6–3.7	12–24
	Ruditapes philippieum	1.1–3.6	37–162	1.4–2.8	13–21
Bou-olayan et al. (1995), Kuwait	*Pinctada radiata*	–	–	3.67	–
Campanella et al. (1999), Mediterranean	*Monodonta turbinata*	–	–	10.9	31
	Patella caerulea	–	–	1.7	5
Szefer and Szefer (1990), Baltic	*Macoma balthica*	25–83	0.62–2.42	32–97	335–628
	Mya arenaria	245	12.63	23.8	212
	Cerastoderma glaucum	60	2.23	24.3	98
	Mytilus edulis	73.4	0.39	13.4	125
	Ariadnaria borealis	19.54	3.96	54.2	128

Table 14.5 Trace metal values obtained and permissible limits

Metal	Value in seawater (μg/g)	Value in tissue sample (μg/g)	Permissible value in seawater (μg/g)
Manganese (Mn)	0–0.014	0.005–0.618	0.01
Iron (Fe)	0.00	0.122–6.392	0.003
Copper (Cu)	0.00	0.364–1.187	0.003
Zinc (Zn)	0.013–0.077	2.545–3.096	0.01

port facilities in Haddo where ships are docked, and iron could leach in this region. In some studies, a similar condition is seen where iron is the highest recorded trace metal. In the study carried out by Bhosle and Matondkar (1978) it can be seen that the concentration of iron reaches up to 2090.99 μg/g. Zinc showed an overall high presence and almost flat graph in all the stations ranging between 2.5 and 3.1 μg/g. In most studies, zinc can be noticed as having the highest concentrations. The highest recorded is from Andaman itself, a study carried out by Abhilash et al. (2013) with a reading of 3262.02 μg/g. This leads to the conclusion that the studied sites are less polluted compared to the previous study. Zinc plays a major role in the growth and development of bivalves (Engel and Brouwer 1984). Zinc concentration is closely related to gonadal development and spawning, either through direct incorporation into gametes or through functional requirements of gonadal tissue during development (Abhilash et al. 2013). Copper concentrations are lesser compared to all the previous studies from this region. The highest copper concentration recorded was in Kodiyaghat with 1.18 μg/g. The possible reason for this could be due to higher anthropogenic activities in this region and discharge through the drainage canal. The next highest value recorded is from Brookshabad beach, with 0.944 μg/g concentration. The reason for this could be the garbage dumping around this site. Copper, along with zinc, plays a major role in the growth and development of shellfish, without which their growth will be suppressed, but in excess is toxic (Engel and Brouwer 1984). The metal with the least concentrations in this study was manganese. In this study, the highest value of manganese obtained was 0.618 μg/g. This value is less than the ones recorded in previous studies (Table 14.3).

Statistical analysis through a single-factor ANOVA test was carried out to check the significance of variation between the trace metals and water in relation to the different sampling sites. The output of the test resulted in a p-value being less than 0.05, which states that there is a significant variation among the study sites between the trace elements with respect to water (p-value < 0.05) and tissue samples (p-value < 0.05). The F-crit value for water and tissue is 2.95, which gives the option of rejecting the null hypothesis since the F statistics value of water (26.51) and tissue (25.85) is higher than the F-crit value. Thus, the F-statistic value falls in the rejection region and thus, concludes that the four factors (trace metals) are not equal and have considerable variation. The correlation factor between the trace metals from water and tissues was also calculated and observed that manganese showed a significantly less correlation ($r = 0.43$), whereas zinc showed the maximum correlation ($r = 0.73$).

Overall, the concentrations of the metals from the obtained tissue samples are much lesser than in the previous studies. This concludes that the seawater around the South Andaman coast is relatively safe for the organisms and the system is less polluted.

14.5 Conclusion

The study sites - Haddo, Carbyn's Cove, Brookshabad Beach and Rangachang, had a good abundance of *Saccostrea cucullata*. The sites – North Bay and Burmanallah had a decent abundance of *S. cucullata* and the sites – South Point and Kodiyaghat had a sparse abundance of *S. cucullata*. The iron and copper values in seawater were not detectable as their concentrations might have been less than 0.001 μg/g. The same applies to manganese seawater concentration at Rangachang. This was not observed in previous studies also. The waters around the study areas are less polluted by trace metals, and the environment is healthy. The shellfish in the region is healthy and safe for consumption, considering they do not contain other accumulated pollutants such as microplastics. Even though the species distribution is throughout the year, there were no attempts made to culture the species in open water. Thus, the study area seems to support the culture of the species. Further studies through socio-economic surveys can help promote this species as a community resource supporting gender parity. The future prospect of this study would involve long-term monitoring of spatial as well as temporal changes of trace metal content and inferring the causes of the same.

Acknowledgements We are thankful to Pondicherry University for providing the necessary facilities to carry out the present work. Sincere gratitude to Dr. K A Jayraj and Dr. T Ganesh, Pondicherry University, for supporting us through this project. Special appreciation to humbly thank Dr. Vinod Kumar, Asst. Director (Soil), and Mr. Sada Shakti, office staff, The Directorate of Agriculture for allowing us to use their laboratory. We would especially like to thank Mrs. Udaya, the Soil lab in charge, and Ms. Naaz, the Lab assistant, for helping with all the lab work.

Statement of Disclosure The authors declare that they are not aware of any affiliations or financial funding from outside sources which would influence the objective of this work. Further, the work is self-financed, the permissions were taken from requisite authorities, and the pictures were taken by the authors. The corresponding author has taken permission from the other authors before submitting the manuscript.

References

Abhilash KR, Gireesh Kumar TR, Venu S, Raveendran TV (2013) Bioconcentration of trace metals by *Saccostrea cucullata* (von Born 1778) from Andaman waters. Indian J Geo Mar Sci 42(3):326–330

Bhosle NB, Matondkar SGP (1978) Variation in trace metals in two populations of green mussel *Mytilus viridis* from Goa. Mahasagar 11:3–4

Biswas T, Bandyopadhyay PK, Chatterjee SN (2013) Accumulation of cadmium, lead, zinc and iron in edible oyster, *Saccostrea cucullata* in coastal areas of West Bengal. Afr J Biotechnol 12(24):3872–3877

Bodin N, Kâ RN, Kâ S, Thiaw OT, Morais LTD, Loc'h FL, Rozuel-Chartier E, Auger D, Chiffoleau JF (2013) Assessment of trace metal contamination in mangrove ecosystems from Senegal, West Africa. Chemosphere 90:150–157

Boening DW (1999) An evaluation of bivalves as biomonitors of heavy metals pollution in marine waters. Environ Monit Assess 55:459–470

Bordin G, McCourt J, Rodriguez A (1992) Trace metals in the marine bivalve *Macomabalthica* in the Westerschelde Estuary (The Netherlands). Sci Total Environ 127:255–280

Bou-olayan A, Al-mattar S, Al-yakoob S, Al-hazeem S (1995) Accumulation of Lead, Cadmium, Copper and Nickel by pearl oyster, Pinctada radiata, from Kuwait marine environment. Mar Pollut Bull 30(3):211–214 0025326X9400143W. https://doi.org/10.1016/0025-326X(94)00143-W

Campanella L, Conti ME, Cubadda F, Sucapane C (1999) Trace metals in seagrass, algae and molluscs from an uncontaminated area in the Mediterranean. Environ Pollut 111:117–126

Conti ME, Cecchetti G (2003) A biomonitoring study: trace metals in algae and molluscs from Tyrrhenian coastal areas. Environ Res 93:99–112

Dabwan AHA, Taufiq M (2016) Bivalves as bio-indicators for heavy metals detection in Kuala Kemaman, Terengganu, Malaysia. Indian J Sci Technol 9(9). https://doi.org/10.17485/ijst/2016/v9i9/88708

Engel DW, Brouwer M (1984) Trace metal-binding proteins in marine molluscs and crustaceans. Mar Environ Res 13:177–194

Hobden DJ (1970) Aspects of iron metabolism in a freshwater mussel. Can J Zool 48:83

Lakshmanan PT, Nambisan PNK (1983) Seasonal variations in trace metal content in bivalve molluscs, *Villorita cyprinoides* var. cochinensis (Hanley), *Meretrix casta* (Chemnitz) and *Perna viridis* (Linnaeus). Indian J Geo-Mar Sci 12(2)

Li C, Haws M, Wang H, Guo X (2017) Taxonomic classification of three oyster (Ostreidae) species from Myanmar. J Shellfish Res 36(2):365–371

Mitra A, Choudhury A (1993) Trace metals in macrobenthic molluscs of the Hooghly Estuary, India. Mar Pollut Bull 26(9):521–522

Pequegnat JE, Fowler SW, Small LF (1969) Estimates of the zinc requirements of marine organisms. J Fish Res Board Can 26(1):145

Phillips DJH, Thompson GB, Gabuji KM, Ho CT (1982) Trace metals of toxicological significance to man in Hong Kong seafood. Environ Pollut Ser B 3:27–45

Rao CV, Rao ST (1971) Distribution of trace elements (Fe, Cu, Mn & Co) in the Bay of Bengal. In: Symp Indian ocean and adjacent seas abs 56

Sarma K, Kumar AA, George G, Pandian K, Prabakaran K, Roy SD, Srivastava RC (2013) Impact of coastal pollution on biological, biochemical and nutritional status of edible oyster in Phoenix Bay Jetty and North Wandoor of Andaman. Indian J Anim Sci 83(3):321–325

Seetharaman P, Sarma K, George G, Krishnan P, Roy SD, Sankar K (2015) Impact of coastal pollution on microbial and mineral profile of edible oyster (*Crassostrea rivularis*) in the coastal waters of Andaman. Bull Environ ContamToxicol 95:599. https://doi.org/10.1007/s00128-015-1601

Szefer p, Szefer K (1990) Metals in molluscs and associated bottom sediments of the southern baltic. Helgoländer Meeresuntersuchungen 44(3-4):411–424. https://doi.org/10.1007/BF02365477

Tapia J, Vargas-Chacoff L, Bertrán C, Carrasco G, Torres F, Pinto R, Urzúa S, Szefer P, Szefer K (1990) Metals in molluscs and associated bottom sediments of the southern Baltic. Helgoländer Meeresuntersuchungen Helgoländer Meeresunters 44:411–424

Usero J, Gonzalez-Regalado E, Gracia I (1997) Trace metals in the bivalve molluscs *Ruditapes decussatus* and *Ruditapes philippinarum* from the Atlantic Coast of Southern Spain. Environ Int 23(3):291–298

Usero J, Morillo J, Gracia I (2005) Heavy metal concentrations in molluscs from the Atlantic coast of southern Spain. Chemosphere 59:1175–1181

Chapter 15
Analytical Approach of Haematology in Variation to Physical Parameters of Indian Mackerel and Yellowfin Tuna from Indian Waters

Chinmay Kar, Abhijeet Purkayastha, and S. Suresh Kumar

Abstract Study of haematology assists to understand the relationship of an organism amid the blood parameters with the habitat, as well as its susceptibility to the environment. Yellowfin tuna and Indian mackerel are two important commercial fish species in the Indian waters. Studies of haematological parameters of these two species were done from two locations, that is, the Arabian Sea and the Bay of Bengal, respectively. The purpose outlined in this study is to assess the health status of both fishes and to observe the relative similarity among the various haematological parameters. One-time sampling was done in March 2021 from Andaman and Nicobar Islands, Odisha, Kerala and Lakshadweep Islands simultaneously using requisite craft and gear. Blood samples of Indian mackerel and yellowfin tuna were taken from all these specimens and analysed for haematology. All the parameters showed a certain degree of variation, with WBC (white blood cells) in tunas dominating. Statistical analysis revealed that all the parameters were distinctly related and shared a relatively high degree of closeness. Bray-Curtis dendrogram showed maximum similarity between Andaman and Nicobar Islands and Kerala. Chlorophyll data derived through remote sensing gave a positive correlation with haematological and physio-chemical parameters. Overall, the study provides a comprehensive account of the haematology in fish which gives a positive state of the fish as well as base information for future studies from these areas. This data can be used to draw the similarities or differences between fishes from tropical water at a global scale.

C. Kar · S. Suresh Kumar (✉)
Biodiversity lab, School of Ocean Science and Technology, Kerala University of Fisheries and Ocean Studies, Kochi, Kerala, India
e-mail: suresh@kufos.ac.in

A. Purkayastha
Biological Oceanography Division (BOD), National Institute of Oceanography, Dona Paula, Goa, India

N. Jayaraju et al. (eds.), *Coasts, Estuaries and Lakes*,
https://doi.org/10.1007/978-3-031-21644-2_15

Keywords *Thunnus* · Arabian Sea · Bay of Bengal · Red blood cells · *Rastrelliger* · LINKTREE

15.1 Introduction

Coastal and oceanic waters comprise distinct ecological zones with animals with various ecological, physiological and morphological adaptations (Ingole and Koslow 2004). Fishes in the epipelagic zone represent a diverse group of about 850 species belonging to approximately 41 teleost families around the world (Miya et al. 2013). In the epipelagic region of the open waters are found two very important species of Scombrids, namely, the Indian mackerel scientifically called *Rastrelliger kanagurta* (*Cuvier 1816*) and the yellowfin tuna known as *Thunnus albacares* (*Bonnaterre 1788*). Indian mackerel and yellowfin tuna show ubiquitous distribution and contribute to major parts of fisheries at a commercial level. They are among those marine migratory fishes which generally shift across a larger area covering from sub-tropical to torrid regions in the Indian, Atlantic and Pacific Oceans (Collete and Nauen 1983). Indian mackerel is a very crucial form of food commodity among all fishes and are exclusively used in cuisines of South and South-East Asian countries, while locally it is referred to as the 'National Fish of India'. The Indian mackerels are mostly distributed in the shallow waters along the coasts of the Indian subcontinent and the Western part of the Pacific Ocean, where the surface water temperature is around 28 °C (Roxy et al. 2015). The Juveniles of Indian mackerel feed on phytoplankton (diatoms) and small zooplankton (copepods, cladocerans and ostracods). As they mature, their feed primarily includes macroplankton such as the larvae of shrimp and fish (Das et al. 2016). Yellowfin tuna are epi-thalassic fishes that mostly occupy the mixed layer on the surface of the ocean which is just over the thermocline zone. The diet of yellowfin tuna mostly includes pelagic fishes, smaller crustaceans and squids (Olson et al. 2016). Waters in the coastal regions of India are mostly affected by contaminants which are of point and non-point origin. Thus, abiotic factors such as precipitation and temperature may change which as a result interrupt the regular functioning of an aquatic ecosystem. This includes the food and feeding habits, the reproduction rate and process and the entire food chain (Weiskopf et al. 2020). Environmental contaminants can affect the aquatic life in a number of ways. Normally, the effects are gradual and eventually, the overall health of the aquatic organisms is affected. Biomarkers usage for environmental quality assessment and the robustness of fishes in an aquatic environment has vouched to be quite effective. Fishes are commonly chosen for assessing the quality of the natatory environment and are well noted as potential bio-indicators of adverse environmental stress. Fishes in their environment live in closeness, which makes them more vulnerable to any change in physical and chemical properties, which can be evident through components of their blood. A haematological analysis is of utmost importance to perceive the overall health status of aquatic organisms (Borges et al. 2004). These studies functions as valuable indicators for estimating the standing position of the fish and the environment as well (Gabriel et al. 2004). Apart from that, they are also effective in monitoring physiological and ecological changes, particularly in fish. Haematological studies also relate to fishes responding to biological and environmental aspects (Fernandes and Mazon 2003; Steinhagen et al. 1990; Haider 1973). Overall, it acts as a reliable tool in assessing the inclusive status of the fish in relation to the environment.

In the current study, a general attempt was devoted to contributing to the haematological profile of yellowfin tuna and Indian mackerel from four different locations in the Northern zone of the Indian Ocean. Also, a brief correlation between haematology and satellite-based chlorophyll data was accounted for. Since both these species are commercially important, haematological studies will provide the collective health status from four locations and assist in laying a general prospect for forthcoming studies.

15.2 Materials and Methods

15.2.1 Study Sites

Specimens of Indian mackerel and yellowfin tuna were obtained covering four spots from the northern Indian Ocean. Study sites were divided into two locations in the Bay of Bengal, that is, Andaman and Nicobar Islands (A&N) and Odisha (OD) and two locations in the Arabian Sea, that is, Kerala (KR) and Lakshadweep (LK) (Fig. 15.1). The Andaman and Nicobar Islands have ample of potential for development of fisheries due to its elongated coastal length. Odisha being an important state for maritime resources has a remarkable potential for fisheries. Kerala, which is situated on the southwestern coast covers a coastline of 589.5 km, includes 10% of the total Indian coastline. Lakshadweep group of islands are the only atoll in the country, with its unique geographical features along with rich fishery resources, particularly tuna.

15.2.2 Methodology

Sampling was done by catching the target fish species, that is, Indian mackerel and yellowfin tuna, using trawl net and longline in motorized crafts. These specimens were caught in March 2021 from four locations in the Indian waters. The specimens from different locations were sampled for the measurement of length and weight. Further, blood samples were collected from the specimen by caudal venepuncture (Zhou et al. 2009).

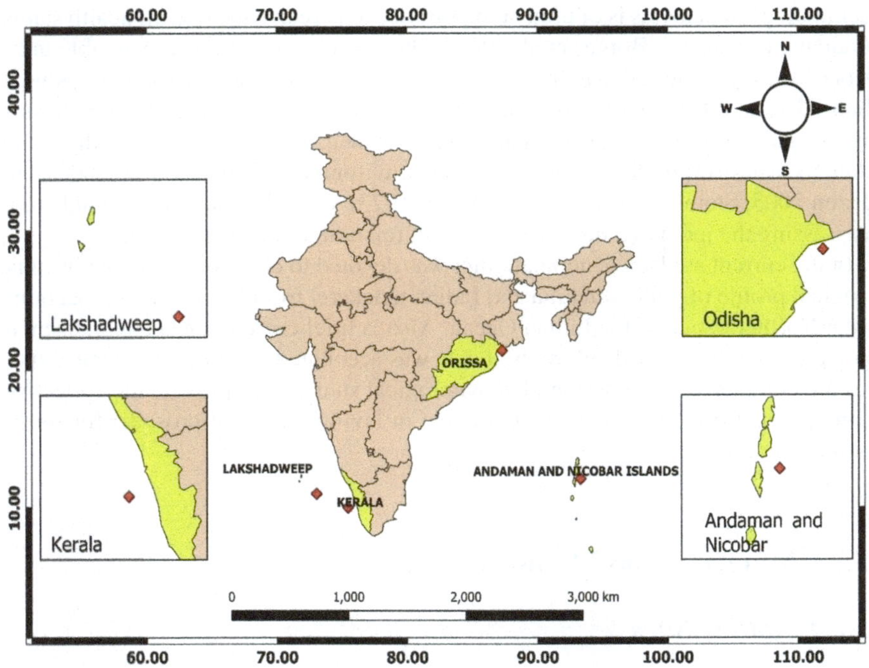

Fig. 15.1 Map showing the different study areas of the collected tuna and mackerel from the Indian Ocean

Physio-chemical parameters of all the sampling locations were recorded with appropriate instruments from the proximity of the sites. The standard thermometer was used for recording temperature, while salinity has been recorded using a handheld standard Refractometer, pH was noted through a pH meter (Laquatwin-pH-11), and dissolved oxygen (DO) was analysed vide Strickland and Parsons (1972).

The chlorophyll distribution map has been generated from NASA's Ocean Color Web portal for the month of March 2021 for the entire coast. The chlorophyll concentration of the four study regions has been computed from online sources (Giovanni data system) which is established and upheld through NASA GES DISC. The level 3 average monthly chlorophyll distribution was downloaded by using a Moderate Resolution Imaging Spectroradiometer (MODIS-Aqua sensor). The map had a resolution of 4 km.

Blood samples were taken with a syringe with heparin sodium (1%) to prevent clotting (Svobodova et al. 2008). RBC and WBC counts were evaluated from the blood samples by diluting in appropriate fluids and further counting adopting an augmented Neubauer chamber by Fisher Scientific, Loughborough, UK. Disparate leucocyte counts were recorded from smears of blood which were fixed in absolute pure methanol, followed by staining, and finally observing under a microscope connected to a video camera the count was done in the computer (Motic Images Advanced 3.2, USA) software. Haematocrit (HCT) was calculated using a microhaematocrit centrifuge (Morris and Davey 1996). Levels of haemoglobin (Hb) were achieved with cyanmethemoglobin assay using Drabkin's reagent (Drabkin and Austin 1935). Mean cell volume (MCV), mean corpuscular haemoglobin (MCH), and mean corpuscular haemoglobin concentration (MCHC) were determined using the following equation:

$$\mathrm{MCV} = \left(10 * \mathrm{HCT}\right) / \mathrm{RBC}; \mathrm{MCH} = \mathrm{Hb} / \mathrm{RBC}; \mathrm{MCHC} = \mathrm{Hb} / \mathrm{HCT}$$

Erythrocyte sedimentation rate (ESR) was attained by levels of the sediment deposited after leaving the mixed-blood sample for 60 minutes in a Westergren tube (Satheeshkumar et al. 2010).

The data obtained from the experiments were represented accordingly in tables and graphs, using excel. Statistical analyses, such as Draftsman plot, SIMPROF test Bray-Curtis similarity dendrogram and LINKTREE, were done using PRIMER package version 6.2 (Clarke and Gorley 2006) to interpret the data in a pronounced manner.

15.3 Results

In this contemporary study, samples of mackerel and tuna were gathered from four diverse sites around the Indian Ocean. A total of 78 individuals of Indian mackerel and 72 individuals of yellowfin tuna were analysed for blood samples. Andaman and Nicobar Islands (11°66′N and 92°77′E) include 21 individuals of mackerel and 20 individuals of tuna; Odisha (21°33′N and 87°07′E) 16 individuals of mackerel and 12 individuals of tuna; Kerala (09°79′N and 76°13′E) 19 individuals of mackerel and 17 individuals of tuna and from Lakshadweep Island (10°61′N and 72°61′E) are 22 individuals of mackerel and 23 individuals of yellowfin tuna.

15.3.1 Physio-chemical Parameters

Analysis of environmental parameters of the samples from varied study sites was recorded, and the data is presented in Table 15.1. Data show that a minimum temperature of 31.67 ± 0.58 °C was recorded from the sampling location of Andaman and Nicobar Islands, while a maximum temperature of 33.67 ± 0.58 °C was recorded from Odisha. pH of all the sampling stations was measured and found within the range of 6.5–8.5. All the locations were recorded to have alkaline water condition with pH variation from 7.37 ± 0.14 in Odisha to 7.96 ± 0.04 in Lakshadweep. The observed salinity revealed that all the sampling locations were saline in condition, with salinity values above 25 ppt. The highest salinity was observed from Lakshadweep (29.26 ± 0.46 ppt), whereas the lowest salinity (25.97 ± 0.95 ppt) was

Table 15.1 Physio-chemical parameters of water from the different study sites of the Indian Ocean

Parameters	A&N	OD	KR	LK
Temperature (°C)	31.67 ± 0.58	33.67 ± 0.58	32.33 ± 1.23	33.33 ± 0.58
Salinity (ppt)	25.97 ± 0.95	26.33 ± 0.58	27.73 ± 0.64	29.26 ± 0.46
pH	7.40 ± 0.02	7.37 ± 0.14	7.90 ± 0.09	7.96 ± 0.04
Dissolved Oxygen (ml/L)	5.60 ± 0.26	4.13 ± 0.25	4.43 ± 0.21	4.77 ± 0.15
Chlorophyll (mg/m^3)	0.26	0.87	0.16	1.03

recorded from Andaman and Nicobar Islands. Dissolved oxygen (DO) concentrations were recorded below 6 mg/l in all the sampling locations. However, DO was seen to be maximum in Andaman and Nicobar Islands and minimum in Odisha. The DO values recorded from the sampling locations were 5.60 ± 0.26 mg/l in Andaman and Nicobar Islands, 4.13 ± 0.25 mg/l in Odisha, 4.43 ± 0.21 mg/l in Kerala and 4.77 ± 0.15 mg/l in Lakshadweep. Chlorophyll concentration varied from 0.16 mg/m^3 in the sampling location of Kerala to 1.03 mg/m^3 in Lakshadweep. Chlorophyll concentration of 0.87 mg/m^3 in Odisha and around 0.26 mg/m^3 in Andaman and Nicobar Islands. Data obtained from the satellite show that the concentration of chlorophyll was higher in the Arabian sea when compared with that in the Bay of Bengal during the study period. In most of the coastal water of the west coast, chlorophyll concentration ranges from 2 to 10 mg/m^3, but on the eastern coast, very few places show chlorophyll concentration of more than 5 mg/m^3. In the present scenario, satellite data shows that in all the sampling sites, chlorophyll values range between 0.5 and 3 mg/m^3 (Fig. 15.2).

15.3.2 Haematological Parameters

Haematological analysis of the two species of scombrids from the four locations was also carried out. RBC (red blood cells) and WBC (white blood cells) counts were noted, followed by differential leucocyte count under a microscope. These parameters are summarized to their mean values and standard deviation and presented in Tables 15.2 and 15.3. Leucocyte cell count includes eosinophils, neutrophils, thrombocytes, lymphocytes and monocytes.

Table 15.2 provides the haematological profile of Indian mackerel, while Table 15.3 gives the haematology readings for yellowfin tuna from all the locations. The values from all the regions were very much similar with slight variation. Lakshadweep Island showed the data to be slightly on the higher side for most of the parameters. RBC values were seen to have very little variation, while WBC readings showed a reasonable degree of difference between the four sites. Haematocrit (HCT), haemoglobin (Hb) and ESR exhibit similar trends. Indices of erythrocyte including MCV, MCH and MCHC showed little variation in their numbers but MCV values wider range.

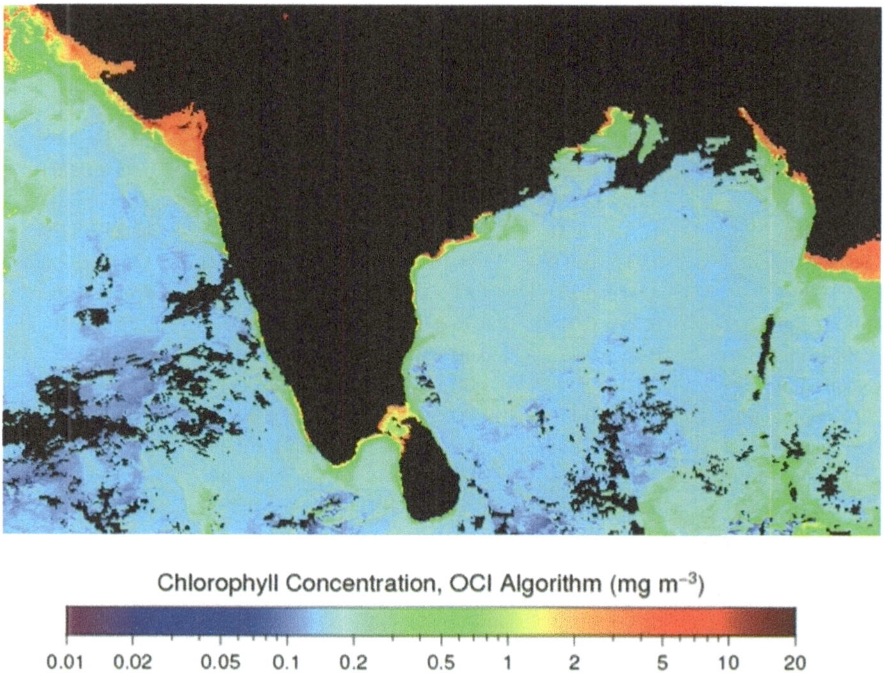

Fig. 15.2 Chlorophyll distribution map of Indian coast for March 2021 of study sites of the Indian Ocean

In this haematological study, various parameters were considered, and are represented graphically (Fig. 15.3) showing the overall variations. The graph shows that the WBC count of yellowfin tuna was dominant in all the study sites, while the WBC count of Indian mackerel was comparatively low. Mean corpuscular Volume (MCV) was seen to have large values in both yellowfin tuna as well as Indian mackerel. The ratio of RBC and WBC was seen to be the least in all the study sites for both yellowfin tuna and Indian mackerel.

Statistical analysis of haematological parameters was done to observe the systematic relationship among the same species. Draftsman plot (Fig. 15.4) reveals the correlation among the different parameters. This analysis aims to represent the vast haematological data in the form of points that highlight the relative closeness among the various parameters. From the plot, it is seen that the variables fit in the linear graphical regression thus, giving us a fair correlation among the parameters. This correlation summarises the strength of the linear relationship between the two variables. Maximum correlation is seen in the form of the points which are close to each parameter. WBC count and other parameters such as lymphocytes, monocytes, thrombocytes, eosinophils and neutrophils show a fair degree of closeness with each other.

Table 15.2 Haematological parameters and differential leucocyte count of Indian mackerel from various locations of the Indian Ocean

Parameters	A&N	OD	KR	LK
RBC (10^6 μl)	2.46 ± 0.14	2.48 ± 0.09	2.45 ± 0.18	2.48 ± 0.20
WBC (10^3 μl)	30.26 ± 3.25	29.28 ± 3.26	30.67 ± 2.30	32.52 ± 3.68
RBC/WBC (%)	0.08 ± 0.004	0.09 ± 0.007	0.08 ± 0.001	0.08 ± 0.003
HCT (%)	32.16 ± 2.69	33.25 ± 2.91	31.61 ± 2.72	33.80 ± 3.04
Hb (g/dL)	12.19 ± 1.31	12.08 ± 1.38	11.76 ± 1.34	12.23 ± 1.29
MCV (fl)	130.53 ± 3.93	134.10 ± 7.06	128.97 ± 2.21	136.03 ± 2.52
MCH (pg)	4.94 ± 0.27	4.87 ± 0.39	4.79 ± 0.22	4.91 ± 0.16
MCHC (g/dL)	0.38 ± 0.01	0.36 ± 0.01	0.37 ± 0.01	0.36 ± 0.01
ESR (mm/h)	0.43 ± 0.17	0.40 ± 0.16	0.40 ± 0.18	0.48 ± 0.21
Neutrophil (%)	3.20 ± 0.29	3.00 ± 0.29	3.06 ± 0.18	3.11 ± 0.29
Eosinophil (%)	5.00 ± 0.12	4.90 ± 0.13	4.97 ± 0.11	4.97 ± 0.20
Lymphocytes (%)	25.54 ± 2.88	24.69 ± 3.42	26.31 ± 3.09	26.15 ± 3.35
Thrombocytes (%)	25.49 ± 2.31	24.33 ± 1.83	25.26 ± 1.64	25.92 ± 2.50
Monocytes (%)	4.50 ± 0.85	4.16 ± 0.84	4.46 ± 0.79	4.47 ± 0.89

RBC red blood cells, *HCT* haematocrit, *WBC* white blood cells, *Hb* haemoglobin, *MCV* mean cell volume, *ESR* erythrocyte sedimentation rate, *MCH* mean corpuscular haemoglobin, *MCHC* mean corpuscular haemoglobin concentration

Table 15.3 Haematological parameters and differential leucocyte count of yellowfin tuna from various locations of the Indian Ocean

Parameters	A&N	OD	KR	LK
RBC (10^6 μl)	2.90 ± 0.35	2.91 ± 0.28	2.85 ± 0.29	2.97 ± 0.38
WBC (10^3 μl)	214.60 ± 17.99	208.25 ± 14.4	212.99 ± 18.18	214.24 ± 23.70
RBC/WBC (%)	0.01 ± 0.001	0.01 ± 0.0004	0.01 ± 0.0007	0.01 ± 0.0004
HCT (%)	46.33 ± 3.31	45.81 ± 2.23	45.43 ± 2.44	46.01 ± 3.58
Hb (g/dL)	19.00 ± 1.02	19.22 ± 0.85	18.79 ± 1.18	19.13 ± 0.95
MCV (fl)	160.43 ± 7.73	158.31 ± 7.97	160.30 ± 7.72	156.01 ± 8.75
MCH (pg)	6.59 ± 0.42	6.64 ± 0.36	6.63 ± 0.27	6.51 ± 0.54
MCHC (g/dL)	0.41 ± 0.01	0.42 ± 0.006	0.41 ± 0.005	0.42 ± 0.01
ESR (mm/h)	0.85 ± 0.37	0.84 ± 0.38	0.79 ± 0.34	0.73 ± 0.31
Neutrophil (%)	4.15 ± 1.11	4.02 ± 1.10	4.71 ± 1.32	4.80 ± 1.22
Eosinophil (%)	6.30 ± 3.06	6.19 ± 2.61	6.39 ± 2.57	6.67 ± 3.30
Lymphocytes (%)	55.88 ± 6.11	54.42 ± 5.89	57.72 ± 4.07	58.07 ± 7.15
Thrombocytes (%)	37.40 ± 8.09	35.49 ± 4.38	38.11 ± 6.16	37.35 ± 7.82
Monocytes (%)	4.53 ± 1.46	4.24 ± 1.10	4.46 ± 1.18	4.19 ± 1.34

RBC red blood cells, *HCT* haematocrit, *WBC* white blood cells, *Hb* haemoglobin, *MCV* mean cell volume, *ESR* erythrocyte sedimentation rate, *MCH* mean corpuscular haemoglobin, *MCHC* mean corpuscular haemoglobin concentration

Linkage tree analysis or LINKTREE (Fig. 15.5) was done to show the distinct clustering of the various parameters. The graph obtained from the analysis highlights three consecutive two-way divisions of the 14 parameters into the final four

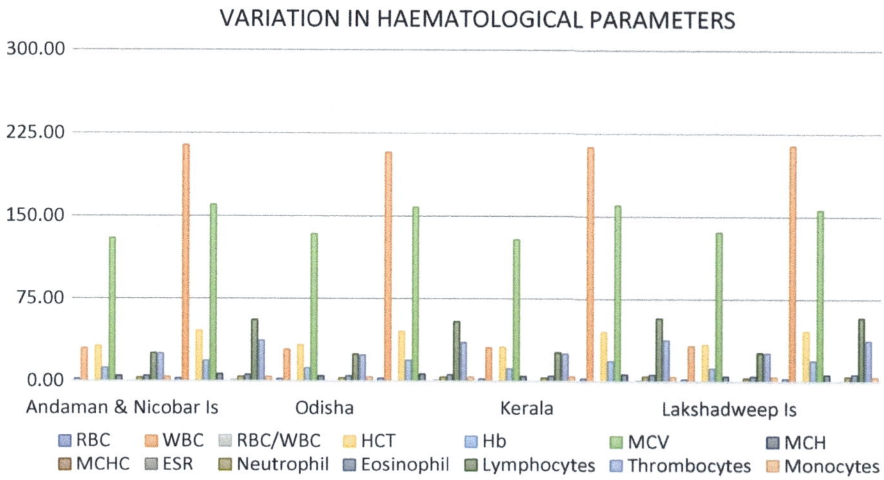

Fig. 15.3 Overall variation in haematological parameters of Indian mackerel and yellowfin tuna from the Indian Ocean

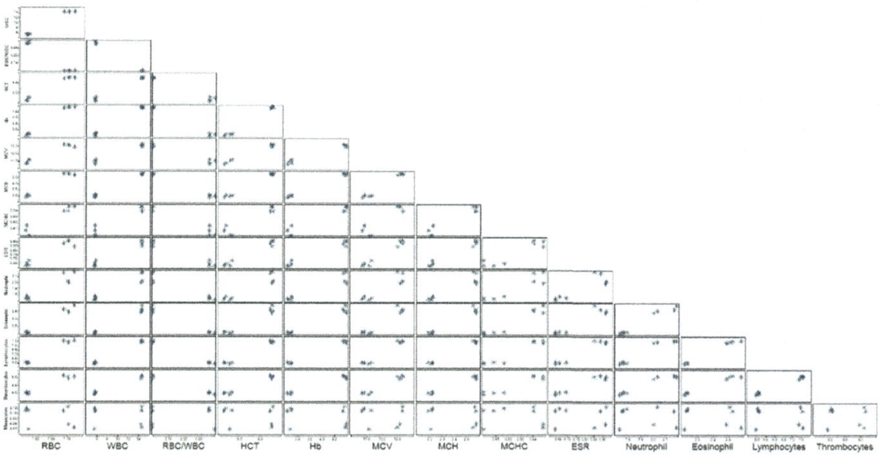

Fig. 15.4 Draftsman plot correlation between haematological parameters of all collected tuna and mackerel from the Indian Ocean

significantly different groups. The 'B%', known as the absolute group difference, serves as the index of measurement. The primary division (A) with a B% value of 100 and an *R*-value of 1 distinguishes two groups containing seven parameters each. The values of these plots are given in the graph. The next division (B), with a B% value of 50 and *R*-value of 0.78, includes a plot from Indian mackerel and (C) with a B% value of 33 and an *R*-value of 0.78 includes the data of yellowfin tuna. These divisions conclude with a clear clustering of the parameters, which indicates that these parameters are very much similar and related.

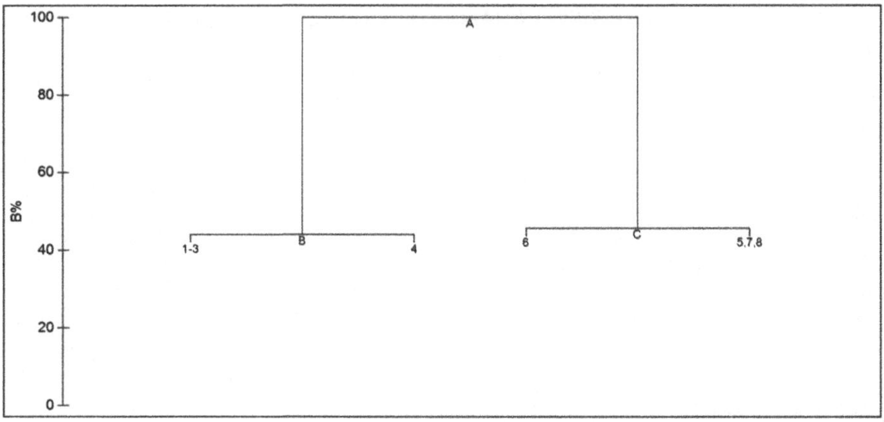

Fig. 15.5 Linkage tree analysis (LINKTREE) showing the clustered plot of haematological parameters from all the specimens collected from the Indian Ocean

A: R = 1.00; B% = 100; RBC/WBC > 0.283(<0.1) or WBC < 5.7(>14.4) or Hb < 3.5(>4.33) or MCH < 2.22(>2.55) or Lymphocytes<5.13(>7.38) or HCT < 5.81(>6.74) or RBC < 1.57(>1.69) or Eosinophil<2.24(>2.49) or Thrombocytes<5.09(>5.96) or MCV < 11.7(>12.5) or ESR < 0.693(>0.854) or MCHC<0.616(>0.64) or Neutrophil<1.79(>2)

B: R = 0.78; B% = 50; Monocytes<2.04(>2.11) or RBC/WBC > 0.3(<0.283) or Thrombocytes<4.93(>5.03) or Eosinophil<2.21(>2.23) or Neutrophil<1.73(>1.75) or Lymphocytes<4.97(>5.05) or WBC < 5.41(>5.5)

C: R = 0.78; B% = 33; Thrombocytes<5.96(>6.11) or Neutrophil<2(>2.04) or Eosinophil<2.49(>2.51) or Lymphocytes<7.38(>7.48) or WBC < 14.4(>14.6) or Hb > 4.38(<4.37) or MCH > 2.58(<2.57)

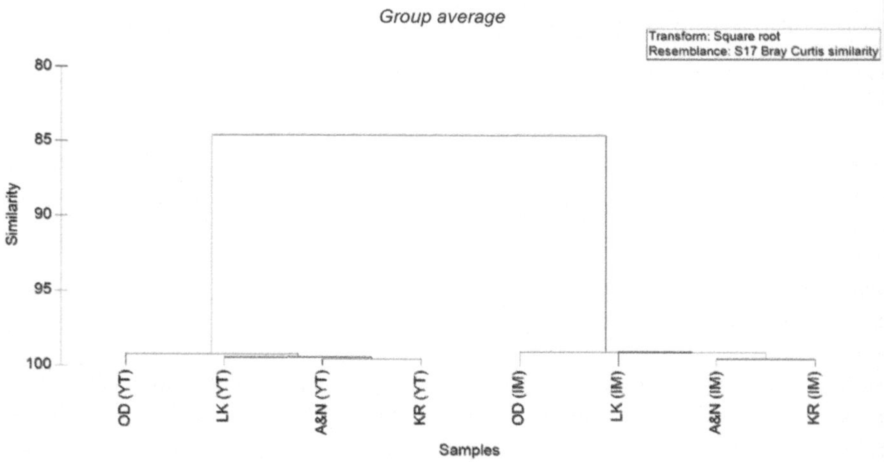

Fig. 15.6 Bray-Curtis similarity dendrogram for Tuna and Mackerel collected from the Indian Ocean. LEGENDS: OD: Odisha, LK: Lakshadweep, A&N: Andaman and Nicobar Islands, KR: Kerala, YT: yellowfin tuna, IM: Indian mackerel

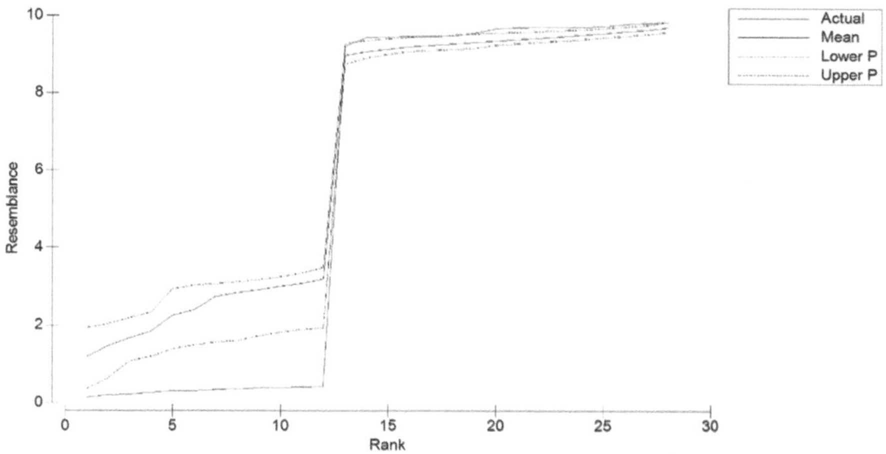

Fig. 15.7 SIMPROF test for the full set of haematological parameters of tuna and mackerel

Hierarchical clustering based on the similarities of the data is usually done using a Bray-Curtis plot. The dendrogram of Bray-Curtis similarity (Fig. 15.6) was determined to observe the similarities of the various study sites. For Indian mackerel, it was seen that Andaman and Nicobar Islands and Kerala had the maximum similarity of 99.5%, followed by 98% with Lakshadweep and Odisha, respectively. On the other hand, yellowfin tuna also showed maximum similarity between Andaman and Nicobar Islands and Kerala with 99%, followed by Lakshadweep with 98.8% and Odisha with 98.4%.

SIMPROF test or the similarity profile plot (Fig. 15.7) documents that the obtained data set has a typical pattern associated with most of the variation was both in and out of the confidence limit. This test was carried out to arbitrate the level of similarity among different parameters. The stimulated profile in the plot deciphers as 'actual', which is the similarity profile of the internal group structure of the entire set of haematological parameters, the 'mean' and the 'upper' and 'lower' limits are the factors serving as a reference. Thus, by randomly selecting samples within the data set, the similarity observed is quite high.

15.4 Discussion

The fundamental objective of the study involved the estimation of the basic haematological profile in Indian mackerel and yellowfin tuna along with the physico-chemical parameter from the waters of the Indian Ocean. Our result of physico-chemical parameters shows that there was a very slight variation observed among the study region during the pre-monsoon period. The pH fluctuations are mostly consequences of discharged organic and inorganic materials through rivers and other similar sources. Low pH concentration below 5 which is generally acidic, can lead to the death of many aquatic organisms affecting the food chain and ultimately the higher trophic organisms including fish. The optimum DO for fish growth in seawater is 6 mg/l. If DO levels are unfavourable it may cause stress due to less supply of oxygen to the brain of fish. Such conditions most likely cause the fish to die because of the unavailability of the required amount of oxygen (anoxia) in the body, as the body tissue of the fish will not be able to bind DO in the blood. The temperature of the water plays a key aspect in the distribution of fish and accordingly, 19–28° C is the ideal condition for Indian Mackerel. In our study sites, we found that the water temperature of all the study regions was higher than the designated range. The high temperature was mainly due to the pre-monsoon period of the year, where the atmospheric temperature is relatively higher, while the rate of precipitation and rainfall is quite low. In aquatic environments, phytoplankton plays a crucial part in the circulation of nutrients and its major pigment which is chlorophyll has been broadly used as an alternative to estimate biomass and primary productivity of the oceans (Westberry et al. 2008; Behrenfeld et al. 2005). The variability of chlorophyll primarily relies on physical components along with oxygen (O_2) and nutrient availability as well as light. The chlorophyll distribution map (Fig. 15.2) shows that amid pre-monsoon season, the coastal water of India was highly productive due to sufficient sunlight and low turbidity. Different authors previously reported that the productivity around the Bay of Bengal is relatively lower when brought in comparison with the Arabian sea (Radhakrishna et al. 1978; Qasim 1977). The prime reason for the low productivity levels in the Bay of Bengal is due to the domination of fresh water, which crowns the upper region of the bay, forming a halocline region below the other layers (Vinayachandran et al. 2002). There are other factors like cloud cover, suspended sediments in the water colonnade, and the continental shelf possibly being narrow, contributing to low productivity of waters in the Bay of Bengal. The high productivity around the Arabian sea may be due to the geographic sites with shallow depth of the shelf region as well as the Aeolian dust (Singh et al. 2008; Kayetha et al. 2007). From the map, it was clear that the coastal water of the Arabian sea is more dynamic and fertile compared to the Bay of Bengal during the pre-monsoon period.

The findings showed distinct variations in the levels of haematological parameters. Although the variations between the parameters were seen to be prominent, the haematological profile of fishes from different study sites was relatively similar with very little difference. Studies related to fish haematology are usually confusing and incomplete due to the lack of uniformity in morphological criteria and the unavailability of standard reference values (Alexander et al. 1980). However, few studies such as Rambhaskaran and Rao (1986) gave some reference for the present study in mackerel. In 1980, Alexander et al. and in 1998, Brill et al. gave the basic references for yellowfin tuna. Several studies on other teleosts were done by different researchers (Satheeshkumar et al. 2010; Zhou et al. 2009; Gao et al. 2007; Rough et al. 2005; Gutierrez 1967, 1970). The more recent and advanced studies in this segment of fishery science include the fish blood profile by Abdelhamid et al.

(2019) and Parrino et al. (2018) which involved comparing analogous species of teleost using the applications of flow cytometry.

Evaluation of haematological parameters in fishes is becoming an important topic to understand the physiology and normal pathological conditions within the ichthyological framework, by considering the definite fish species characterization (Hlavova 1993). Reproduction is an important process followed by all adult fishes. During the maturation of gonads, the different parameters in the blood change as the individual matures (Joshi 1982). Certain parameters such as RBC, haematocrit and haemoglobin were observed to be related to the size of the fish. An increase in these parameters generally leads the fish to be larger in size (Jawad et al. 2004). The reason for the direct relation may be due to the fact that larger fishes have a high bodily metabolic rate compared to smaller fishes (Chaudhuri et al. 1986). RBCs are also known to play a part in influencing the sex of the individuals (Luskova et al. 1995) as well as high activity, swift and fast movement (Rambhaskar and Rao 1986). Apart from RBCs, other parameters which show direct relation with the health of the fishes are WBCs and lymphocytes. They are the defense cells responsible for immunity and a high count gives a progressive boost to their health (Douglass and Jane 2010). In the present study, the RBC count of both tuna and mackerel from all the study sites was very similar with a minimum difference, whereas the WBC count showed a wide difference between tuna and mackerel. The higher WBC count observed in yellowfin tuna mainly reflects the higher immunity levels in these fishes. Thrombocytes and monocytes are said to have phagocytic activity and thus, aid the immune system (Stosik et al. 2001). The haematocrit levels are also detrimental for indicating nutritional deficiencies and other health abnormalities (Anderson 1990). Dissolved oxygen (DO) is another factor which correlates with the haematological parameters showing direct proportional relation (Lochmiller et al. 1989). On the other hand, fishes in low oxygen concentration environment tend to show an increase in erythrocyte count and haemoglobin content (Saint-Paul 1984). Haematological changes are also reflected due to the change in environment due to any form of stress. The important indicating parameter for environmental stress is haemoglobin (Cazenave et al. 2005). Toxic substances in the aquatic environment are also known to have a negative impact on the overall blood parameters in fish (van der Oost et al. 2003). Thus, haematology is an important aspect which provides the physiological status and health of living aquatic organisms as well as their exposure to stress in the environment.

Estimation of chlorophyll concentration was obtained from satellite remote sensing for the same locations where the samples were collected. Satellite data are an emerging and very much reliable source of information as they have been collected in the database through consistent monitoring using advanced techniques and designated algorithms. It provides spatially coherent data of Chl-a in the form of images, from coastal waters. However, there are certain complexities in the water bodies such as eutrophication, suspended particulate matter and other unpredictable and optical water properties, which makes retrieving chlorophyll data quite difficult. Quality control (QC) for three algorithms was carried out which ensured 83% of the dataset coverage (Lavigne et al. 2021). Further, chlorophyll assessment in waters from the coast and oceanic regions of the Bay of Bengal and the Arabian Sea using remote sensing applications has been accomplished already (Tilstone et al. 2011). A significant positive correlation was observed between chlorophyll and haematological as well as physiochemical parameters. Satellite data corresponds to the readings obtained from the present study showing high similarity for Lakshadweep Islands. Since the different parameters of Lakshadweep Island were comparatively higher, chlorophyll data obtained through remote sensing was maximum as well. A similar trend was followed by other locations which showed a significant correlation. Chlorophyll is an indicator of conditions which favours pelagic fishes in terms of their feeding (Nevárez-Martínez et al. 2003). Since most of the small pelagic fishes feed primarily on phytoplankton, higher concentrations of chlorophyll are likely to be associated with feeding. Feeding in terms is linked to the growth of the fish which includes the blood in their body. Thus, it can suggest that chlorophyll concentrations, to some extent, influence the physiology of the fish. This concludes that the haematological profile to a certain level is affected by the chlorophyll concentrations in the environment. Many elements can dispense to curb the relationship between data from the satellite with the concerned parameters. These limitations may occur due to remote sensing specifics which can range from vertical distribution, diversity and physiology of phytoplankton to aerosols and suspended sediments (Gregg and Casey 2004). Above these limitations, the chlorophyll data furnished by MODIS has proven to be very useful in establishing a definite positive correlation.

The merit of comparison can be achieved when the contrast for haematological parameters is being done between fishes of the same species, sex, physiological and environmental aspects (Abdelhamid et al. 2021). The present findings show that there is quite a range of variation in all the haematological factors. The samples were collected before the onset of monsoon, which was observed to have a steady growth in length and weight. Haematological profiles, particularly RBC involved in oxygen transport and WBC involved in the immune system, are very important. Since there are no standard values for haematological parameters, it is ambitious to conclude the exact health status of the specimen. Nonetheless, it can be understood from the statistical analysis that the environment where the samples were collected serves as a stable environment for the growth of both Indian mackerel and yellowfin tuna. Satellite data of chlorophyll suggested positive relation which may be directly linked to the food habits of the fishes.

15.5 Conclusion

The current study has revealed and provided a great deal of novel and comprehensive information on the condition and physiology of Indian mackerel and yellowfin tuna from Indian waters. Since both these species are commercially important for the Indian subcontinent, it is necessary to check the overall status and provide a thorough update on the health and welfare of these scombrids. Also, studies on the haematology of tunas and mackerels have not been conducted thoroughly, and hence the range and variation are not much known. Further, this work will deliver a reference indication for studies in future, of living fishes if exposed to any change in water quality or any form of environmental stress. The use of satellite data for the present study indicates the potentiality of remote sensing usage in biological studies. This study will help assist future research studies in establishing a reference work for conducting similar observations.

Acknowledgements The authors are highly grateful to acknowledge the cooperation and support of the owners and staff of the vessels from the fishery communities of Andaman and Nicobar Islands, Odisha, Kerala and Lakshadweep Island without whom this study would cease to exist. Further appreciation to Mr Prasanth Sai for helping with the sampling from Andaman and Nicobar Islands and Mr Sambit Singh for the technical assistance. Humble appreciation to UGC (Ref no-200510341520) as well as SERB core project (CRG/2020/004498) for the needful support during the study. We would also like to acknowledge the MODIS mission scientists and NASA personnel for the data production which was used in this research.

Disclosure Statement The authors took the very initiative for the study without any affiliation or external funding. The permission and necessary documentation were taken from the respective authorities, and the work was self-financed. The corresponding author has taken permission from the respective authors before submitting this manuscript.

References

Abdelhamid AM, Refaey MR, Salem MF, El Kattan MAM (2019) Factors affecting fish blood profile: B-Effect of environmental and genetic factors. Egypt J Aquat Biol Fish 23(2):443–459

Abdelhamid AM, Refaey MM, Mohammed El-Sayed HG, Ahmed Mo'aty G (2021) Comparison between blood haematology of Egyptian freshwater food fish and ornamental fish. J Appl Sci Eng Technol Educ 3(1):22–36

Alexander N, Laurs RM, Mcintosh A, Russel SW (1980) Haematological characteristics of albacore, *Thunnus alulunga* (Bonnaterre), and skipjack, Katsuwonus pelamis (Linnaeus). J Fish Biol 16:383–395

Anderson DP (1990) Immunological indicators: effects of environmental stress on immune protection and disease outbreaks. In: Adams SM (ed) Biological indicators of stress in fish, vol 8. American Fisheries Symposium, Bethesda, pp 38–50

Behrenfeld MJ, Boss E, Siegel DA, Shea DM (2005) Carbon based ocean productivity and phytoplankton physiology from space. Glob Biogeochem Cycles 19:1–14

Borges A, Scotti LV, Siqueira DR, Jurinitz DF, Wassermann GF (2004) Haematologic and serum biochemical values (Rhamdia quelen). Fish Biol Biochem 30:21–25

Brill RW, Cousins KL, Jones DR, Bushnell PG, Steffensen JF (1998) Blood volume, plasma volume and circulation time in a high-energy-demand teleost, the yellowfin tuna (Thunnus albacares). J Exp Biol 201:647–654

Cazenave J, Wunderlin DA, Hued AC (2005) Haematological parameters in a neotropical fish, Corydoras paleatus (Jenyns, 1842) (Pisces, Callichthyidae), captured from pristine and polluted water. Hydrobiologia 537:25–33

Chaudhuri SH, Pandit T, Benerjee S (1986) Size and sex related variations of some blood parameters of sarotheriodon massambica. Environ Ecol 4:61–63

Clarke KR, Gorley RN (2006) PRIMER v6: user manual/tutorial. PRIMER-E, Plymouth

Collete HB, Nauen CE (1983) FAO species catalogue. Scombrids of the world. An Annotated and Illustrated catalogue of tunas, mackerels, bonitos, and related species known to date, FAO Fisheries Synopsis No. 125, vol 2. FAO Press, Rome, p 137

Das I, Hazra S, Bhattacharya SB, Das S, Giri S (2016) A study on seasonal change in feeding habit, health status and reproductive biology of Indian Mackerel (*Rastrelliger kanagurta*, Cuvier) in coastal water of West Bengal. Indian J Geo-Mar Sci 45(2):254–260

Douglass JW, Jane KW (2010) Schalm's veterinary hematology. John Wiley and Sons, Blackwell Publishing Ltd, Iowa, 1232 p

Drabkin DL, Austin JH (1935) Spectrophotometric studies, II. Preparation from washed blood cells; nitric oxide haemoglobin and sulphaemoglobin. J Biol Chem 112:51–65

Fernandes MN, Mazon AF (2003) Environmental pollution and fish gill morphology. In: Val AL, Kapoor BG (eds) Fish adaptation. Science, Enfield, pp 203–231

Gabriel UU, Ezeri GNO, Opabunmi OO (2004) Influence of sex, source, health status and acclimation on the haematology of Clarias gariepinus (Burch, 1822). Afr J Biotechnol 3:463–467

Gao ZX, Wang WM, Abbas K, Zhou XY, Yang Y, Diana JS, Wang HP, Wang HL, Li Y, Sun YH (2007) Haematological characterisation of loach Misgurnus anguillicaudatus: comparison among diploid, triploid and tetraploid specimens. Comp Biochem Physiol A 147:1001–1008

Gregg WW, Casey NW (2004) Global and regional evaluation of the sea WiFS chlorophyll data set. Remote Sens Environ 93(4):463–479

Gutierrez M (1967) Estudios hematológicos en el atún, Thunnus thynnus (L.), de la costa sudatlántica de España. Investig Pesq 31:53–90

Gutierrez M (1970) Constantes eitrocíticas, velocidad de sedimentación y valores bioquímicos de la sangre del Thunnus thynnus (L.) durante las fases de maduración sexualy postfrezado. Investig Pesq 34:291–298

Haider G (1973) Comparative studies of blood morphology and haemopoiesis of some teleost. Observations on cells of the red series. J Zool 179:355–383

Hlavova V (1993) Reference values of the haematological indices in grayling (Thymallus thymallus linnaeus). Comp Biochem Physiol 105A:525–532

Ingole B, Koslow AJ (2004) Deep-sea ecosystem of the Indian Ocean. Indian J Mar Sci 34(1):27–34

Jawad LA, Al-Mukhtar MA, Ahmed HK (2004) The relationship between haematocrit and some biological parameters of Indian shad, Tenualosa ilisha (family Clupeidae). Anim Biodivers Conserv 27:47–52

Joshi BD (1982) Circannual fluctuations in some blood components of the fish Rita rita, in relation to certain eco-physiological conditions. Uttar Pradesh J Zool 2:62–66

Kayetha VK, Senthilkumar J, Prasad AK, Cervone G, Singh RP (2007) Effect of dust storm on ocean color and snow parameters. Photonirvachak – J Indian Soc Remote Sens 35(1):1–9

Lavigne H, van der Zande D, Ruddick K, Cardoso dos Santos JF, Gohin F, Brotas V, Kratzer S (2021) Quality control tests for OC4, OC5 and NIR-red satellite chlorophyll-a algorithms applied to coastal waters. Remote Sens Environ 255:112237

Lochmiller RL, Weichman JD, Zale AV (1989) Hematological assessment of temperature and oxygen stress in a reservoir population of striped bass (Morone saxatilis). Comp Biochem Physiol 93A:535–541

Luskova V, Halacka K, Lusk S (1995) Dinamics of the haemogram in the nase, Chondrostoma nasus. Folia Zool 44:69–74

Miya M, Friedman M, Satoh TP, Takeshima H, Sado T (2013) Evolutionary origin of the Scombridae (tunas and mackerels): members of a paleogene adaptive radiation with 14 other pelagic fish families. PLoS One 8(9):1–19

Morris MW, Davey FR (1996) Basic examination of blood. In: Henry JB (ed) Clinical diagnosis and management by laboratory methods, 19th edn. WB Saunders, Philadelphia, pp 549–593

Nevárez-Martínez MO, Cotero E, Martínez-Zavala MA, Félix Uraga R (2003) Recruitment of the Pacific sardine (Sardinops sagax) in Baja California, México. In: Program and Abstracts, Annual Conference 2004 CaLCOFI, 15–18 Nov 2004, p 20

Olson RJ, Young JW, Menard F, Potier M, Allain V, Goni N, Logan JM, Galvan-Magana F (2016) Bioenergetics, trophic ecology, and niche separation of tunas. In: Curry BE (ed) Adv Mar Biol. Academic Press, UK, pp 199–344

Parrino V, Cappello T, Costa G, Cannavà C, Sanfilippo M, Fazio F, Fasulo S (2018) Comparative study of haematology of two teleost fish (Mugil cephalus and Carassius auratus) from different environments and feeding habits. Eur Zool J 85(1):193–199

Qasim SZ (1977) Biological productivity of the Indian Ocean, Indian. J Mar Sci 6:122–137

Radhakrishna K, Bhattathiri PMA, Devassy VP (1978) Primary productivity of Bay of Bengal during August–September 1976, Indian. J Mar Sci 7:94–98

Rambhaskar B, Rao KS (1986) Comparative haematology often species of marine fish from Visakhapatnam coast. J Fish Biol 30:59–66

Rough KM, Nowak BF, Reuter RE (2005) Haematology and leucocyte morphology of wild caught Thunnus maccoyii. J Fish Biol 66:1649–1659

Roxy M, Ritika K, Terray P, Masson S (2015) Indian Ocean warming- the bigger picture. Bull Am Meteorol Soc 96(7):1070–1071

Saint-Paul U (1984) Physiological adaptation to hypoxia of a neotropical characoid fish Colossoma macropomum, Serrasalmidae. Environ Biol Fish 11:53–62

Satheeshkumar P, Senthilkumar D, Ananthan G, Soundarapandian P, Khan AB (2010) Measurement of hematological and biochemical studies on wild marine carnivorous fishes from Vellar estuary, southeast coast of India. Comp Clin Pathol 20:127–134

Singh RP, Prasad AK, Kayetha VK, Kafatos M (2008) Enhancement of oceanic parameters associated with dust storms using satellite data. J Geophys Res 113:C11008. https://doi.org/10.1029/2008JC004815

Steinhagen D, Kruse P, Körting W (1990) Some haematological observations on carp Cyprinus carpio L., experimentally infected with Trypanoplasma borelli Laveran & Mesnil, 1901 (Protozoa: Kitenoplastida). J Fish Dis 14:157–162

Stosik H, Deptula W, Travnicek M (2001) Studies on the number and ingesting ability of thrombocytes in sick carps (Cyprinus carpio). Vet Med 46:12–16

Strickland JDH, Parsons TR (1972) A practical hand-book of seawater analysis. Fisheries Research Board of Canada, Canada, p 310

Svobodova Z, Kroupova H, Modra H, Flajshans M, Randak T, Savina LV, Gela D (2008) Haematological profile of common carp spawners of various breeds. J Appl Ichthyol 24:55–59

Tilstone GH, Angel-Benavides IM, Pradhan Y, Shutler JD, Groom S, Sathyendranath S (2011) An assessment of Chlorophyll-a algorithms available for SeaWIFS in coastal and open areas of the Bay of Bengal and the Arabian Sea. Remote Sens Environ 115(9):2277–2291

van der Oost R, Beyer J, Vermeulen NPE (2003) Fish bioaccumulation and biomarkers in environmental risk assessment: a review. Environ Toxicol Pharmacol 13:57–149

Vinayachandran PN, Murty VSN, Ramesh Babu V (2002) Observations of barrier layer formation in the Bay of Bengal during summer monsoon. J Geophys Res 107(C12):8018. https://doi.org/10.1029/2001JC000831

Weiskopf SR, Rubenstein MA, Crozier LG, Gaichas S, Griffis R, Halofsky JE, Hyde KJW, Morelli TL, Morisette JT, Muñoz RC, Pershing AJ, Peterson DL, Poudel R, Staudinger MD, Sutton-Grier AE, Thompson L, Vose J, Weltzin JF, Whyte KP (2020) Climate change effects on biodiversity, ecosystems, ecosystem services, and natural resource management in the United States. Sci Total Environ 733:137782

Westberry T, Behrenfeld MJ, Siegel E, Boss E (2008) Carbon-based primary productivity modeling with vertically resolved photophysiology. Glob Biogeochem Cycles 22:GB2024. https://doi.org/10.1029/2007GB003078

Zhou X, Li M, Abbas K, Wang W (2009) Comparison of haematology and serum biochemistry of cultured and wild Dojo loach Misgurnus anguillicaudatus. Fish Physiol Biochem 35:435–441

Chapter 16
Geochemistry of Mollusc Shells as Proxies of Marine Pollution, East Coast of India

B. Lakshmanna, N. Jayaraju, G. Sreenivasulu, T. Lakshmi Prasad, K. Nagalakshmi, M. Pramod Kumar, M. Madakka, B. Rajender, and P. Vijayanand

Abstract The chemistry of mollusc shells was studied using SEM-EDS (scanning electron microscope energy dispersive spectroscopy) and X-ray diffraction (XRD). Their structures are composed of calcium carbonate crystallised as aragonite. Scanning electron microscope (SEM) images of two mollusc shells were taken at several magnifications. The surfaces of the shells have irregular grains. They also show flake-like features on the surface, and they vary in size at the micrometric scale. The two molluscan shells are *Cardita* (smooth) and *Gastropoda*. The analysis of energy dispersive spectroscopy (EDS) recorded the elements of *Cardita* (hard), *Cardita* (smooth), and *Gastropoda*. This is consistent with the analysis of X-ray diffraction (XRD) records. The analysis of the shells of EDS (energy dispersive spectroscopy) spectra revealed that peaks of Carbon, Calcium-Oxygen, aluminium (C, Ca-O, Al, Br, Ca, Cu) and lead (Pb) appear on the *Cardita* (hard), *Cardita* (soft), and *Gastropoda*. The energy dispersive spectroscopy (EDS) spectra make up the majority chemistry of the shells. The current study investigates pollution in brackish water as bioaccumulation to coastal surface samples. The data showed that the

B. Lakshmanna · B. Rajender · P. Vijayanand
Department of Energy and Environmental Engineering, CSIR-Indian Institute of Chemical Technology, Tarnaka, Hyderabad, Telangana, India

N. Jayaraju (✉)
Department of Geology, Yogi Vemana University, Kadapa, Andhra Pradesh, India

G. Sreenivasulu
Department of Geology, Sri Venkateswara University, Tirupati, Andhra Pradesh, India

T. Lakshmi Prasad · M. Pramod Kumar
Department of Earth Sciences, Yogi Vemana University, Kadapa, Andhra Pradesh, India

K. Nagalakshmi
Department of Geology, Government College (A), Anantapur, Andhra Pradesh, India

M. Madakka
Department of Biotechnology and Bioinformatics, Yogi Vemana University, Kadapa, Andhra Pradesh, India

N. Jayaraju et al. (eds.), *Coasts, Estuaries and Lakes*,
https://doi.org/10.1007/978-3-031-21644-2_16

fragile coastal environment had been severely harmed and important for biomaterial in orthopaedic applications.

Keywords Mollusc shells · XRD · SEM · EDS analysis · Bioaccumulation

16.1 Introduction

The mollusc shells are made up of a single layer known as periostracum and a calcified inner layer. Molluscs also have a hinge that connects the two shells at the dorsal side. Most species have a calcified surface on the dorsal side of the foot. Chitinous are distinguished from other molluscs by different features such as numerous shell plates formed spines and skinny material (Chateigner et al. 2000; Jacob et al. 2008). A thin outer ligament is attached to the two hinges. The *Gastropoda*, most of the molluscs, consists of 50,000 fossil forms and about 60,000 living species (Bandel 1990). The gastropods have an asymmetrical shell that serves as a portable shelter. The *Gastropoda* consists of several parts such as foot, hump, head, and mantle. The bifurcation of the shell by calcium carbonate ($CaCO_3$) is thought to be the prolonged liberal activity influenced primarily by several intricate and controlling environmental factors. Shell matrix is typically made up of calcium carbonate ($CaCO_3$). Microscopic canals of cell tubes are found in the shells of molluscs (Boggild 1930; Wilbur 1964). Inside the shells, they are typically pearly white, with slight tinges to dark grey shells that have been stained by something that either died with the animal or were absorbed from the sediment habitat (Canti 2017). The accumulation of chemical elements is one of the critical features of these shells (Wilbur 1972, 1973, 1976; Wilbur and Saleuddin 1983). Bioaccumulation throws light on the quality of the existing and inhabiting environment. Bioaccumulation throws light on the quality of the existing and inhabiting environment. Nowadays, the extracellular calcifying shell matrix appears as a whole integrated system, which regulates protein-mineral, and protein-protein interactions as well as feedback interactions between the biomaterials and the calcifying epithelium that synthesised them and consequently, the mollusc shell matrix may be a source of bioactive molecules that would over interesting perspectives in biomaterials and biomedical fields (Marin et al. 2007).

16.2 Study Area

The study area extends for about 110 km (Fig. 16.1). The current study is thought to be the first detailed work on the mollusc shell beneath diverse pollution outfalls and their implications for metal pollution on the east coast of India. Research is limited to mollusc shells (Hasse et al. 2000). The climate in the study area is tropical to subhumid, with an annual mean temperature of 29 °C. Temperatures in the summer

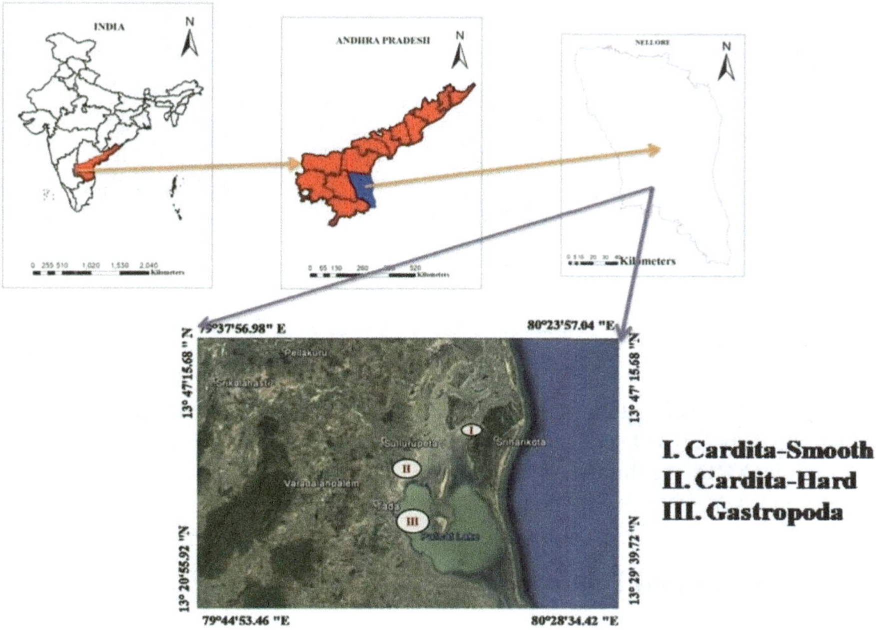

Fig. 16.1 Location of the study area at Nellore coast, Andhra Pradesh, India. (I. *Cardita* (*smooth*); II. *Cardita* (*hard*); III. *Gastropoda*)

reached 32 °C, while temperatures in the winter fell to 28 °C. The humidity ranges from 71% to 73% in the summer to 88% to 90% in the winter, resulting in a relatively fast rate of evaporation. *Gastropoda* and *Cardita* are believed to share a common biological process in the environment of the study area.

16.3 Methods

Cardita (hard), *Cardita* (smooth), and *Gastropoda* mollusc shells were collected from various locations (Fig. 16.2). The shells were thoroughly cleaned with deionised water and dried at room temperature. The mollusc shell samples were mechanically fractured after drying, and random pieces were fastened to the sample holder (stub) using carbon ribbon and gold paint to allow transversal visualisation of the fragments. The nanoparticle's surface morphologies were studied using a scanning electron microscope (SEM) after coating the powder with gold. X-ray shooting was done at the surface of the particles using energy dispersive spectroscopy (EDS) to gain information on the elemental content of the test. The varied peak types may be seen in each element captured by the energy dispersive spectroscopy (EDX) sensor and each X-ray shot on the test (shell) generates a suggestion spectrum of element

Fig. 16.2 *Cardita* shell. (**a**) *Dorsal* (*smooth*), (**b**) *Cardita* dorsal (*hard*), and (**c**) *Gastropoda* shell

content based on percentage weight (wt %) (Lakshmanna et al. 2018). For the determination of the crystal phases, the crystal powder was examined by using X-ray diffraction (XRD) machine with a CuK (1.54) wavelength.

16.4 Results and Discussion

X-ray diffraction (XRD) is used to understand the phase of the shell of the samples. The XRD patterns of samples are shown (Fig.16.3). The peaks of the given samples were primarily indicative of the orthorhombic aragonite phase ($CaCO_3$) (JCPDS card no. 86-2334), with the calcite phase. The intensity of the plane at 2.29 Å, which corresponds to the calcite phase, is greater for the *Gastropoda*. This indicates that *Gastropoda* has a higher percentage of calcite phases than the *Cardita*. The lattice parameters and cell volume of the orthorhombic aragonite phase were discovered to be, (1) A = 4.9648, (2) B = 7.9608, (3) C = 5.7497, V = 227.25 3 for *Cardita*, and (1) A = 4.9681, (2) B = 7.9690, (3) C = 5.7536, V = 227.79 3 for *Gastropoda*. Debye Scherer's formula was used for calculating crystallite sizes. The crystallite size for *Cardita* and *Gastropoda* is 60 nm and 70 nm, respectively.

The typical SEM images show a flake-like structure at various magnifications (Figs. 16.4 and 16.5) with uneven grains with micrometric shape.

Minerals are also only produced as a by-product of the cells metabolic activity or as a result of interaction with the habitat environment (Addadi and Weiner 1992). The second method, on the other hand, known as biologically controlled bioaccumulation is completely regulated and allows biota to accumulate minerals (Birchall 1989; Albeck et al. 1996; Sreenivasulu et al. 2017; Levi-Kalisman et al. 2001). Consistent with the XRD results, the EDS analysis revealed that elements such as Al, Br, Ca, and O were observed in *Gastropoda* and *Cardita* (Table 16.1). The shell EDS spectra demonstrate the elements that show on the *Cardita* shell (Figs. 16.6 and 16.7). The shell structure of *Cardita* and *Gastropoda* is composed of aluminium (55.87% and 4.29%, respectively) and oxygen (8.82% and 42.14%), respectively.

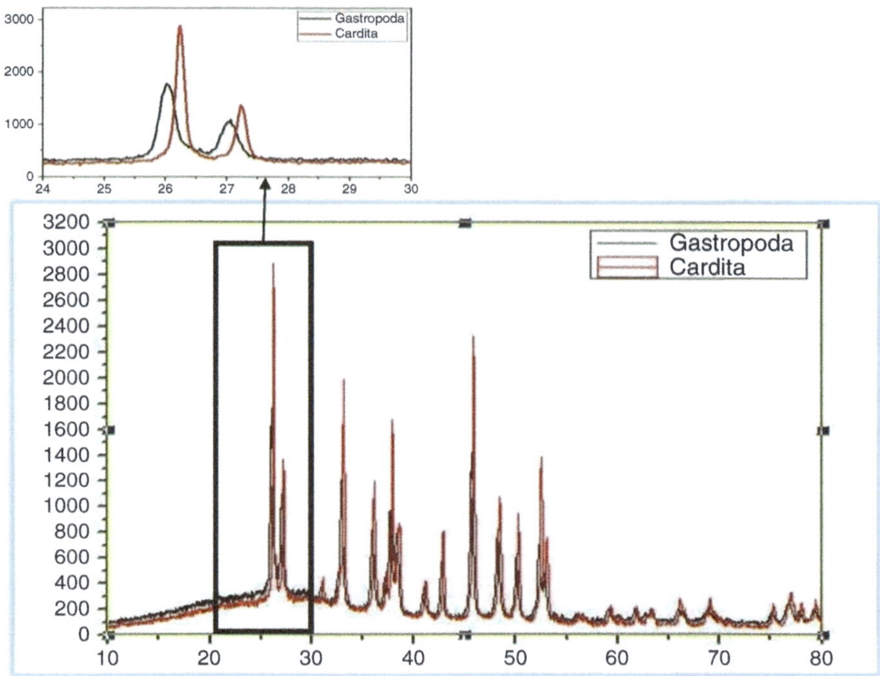

Fig. 16.3 X-ray diffraction analysis

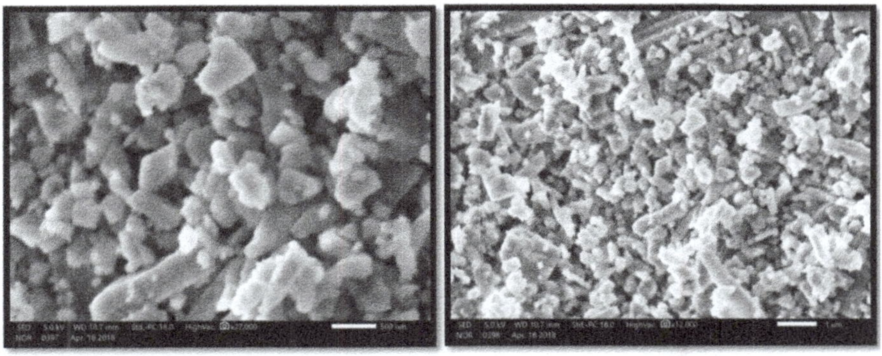

Fig. 16.4 Scanning electron microscope images of *Cardita* shell at different magnifications

These are adsorbed on the shell surface. The EDS analysis reveals that *Gastropoda* & *Cardita* are composed of atomic elements Al (59.24%) & (3.58%), O (15.78%) & (59.44%), Br (7.84%) & (3.40%), and Mg (0.92%) & (2.76%). The shell structures of *Gastropoda* and *Cardita* are chiefly made up of calcium (Silverstein and Webster 1998; Simkiss and Wilbur 1989).

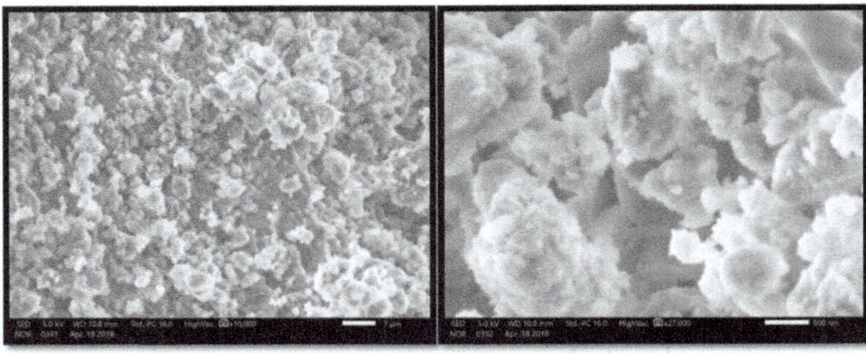

Fig. 16.5 Scanning electron microscope images of *Gastropoda* shell at different magnifications

Table 16.1 EDS data (weight and atomic percentages) of shells

Elements	*Gastropoda* (weight %)	*Gastropoda* (atomic %)	*Cardita* (weight %)	*Cardita* (atomic %)
C K	4.74	11.30	3.99	7.49
O K	8.82	15.78	42.14	59.44
Al K	55.87	59.24	4.29	3.58
Ca K	4.74	3.40	–	–
Cu L	2.99	1.35	–	–
Br K	21.89	7.84	5.26	1.49
Si K	–	–	23.74	19.07
Ca L	4.77	3.40	–	–
Mg K	0.92	1.09	2.98	2.76
Ti L	–	–	7.46	3.52
Pm M	–	–	5.86	0.91
Fe L	–	–	4.29	1.73

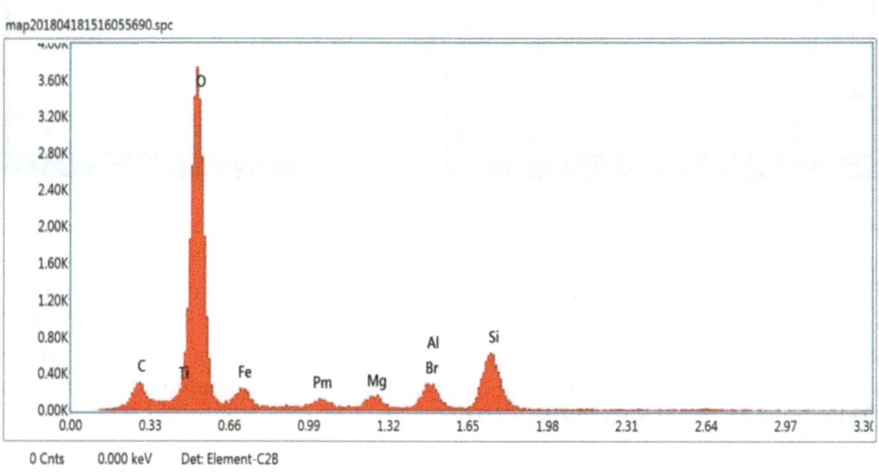

Fig. 16.6 EDS data of *Cardita* shells

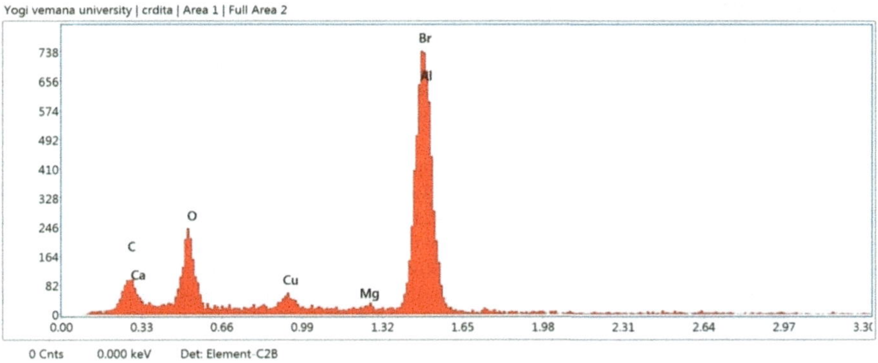

Fig. 16.7 EDS spectra of *Gastropoda*

16.5 Conclusion

The shells under study are strongly linked to the elements and have more thermal stability over bones organic matrix. This is considered to be important for biomaterial in orthopaedic applications. Elements such as Br, Ca, and Cu carbonate accumulate in the shell at relatively steady rates in comparison to others. Regardless of metabolic turnover, these processes can be thought of as organism net matter immobilisation that produces increasing amounts over time. It is interesting to comprehend the Ca and Cu pathways and regimens. This is due to the fact that both elements exist as free bivalent ions in the fragile marine ecosystem. Throughout the life of the marine biota, bioaccumulation and marine pollution are evident. Methods for assessing element concentrations using mollusc shells must be developed in order to establish links between bioaccumulation and marine pollution, despite the fact that the current study provides clear, precise indications on the key differences between abiogenic and biogenic carbonate crystal micro morphologies of shell structures.

References

Addadi L, Weiner S (1992) Control and design principles in biological mineralization. Angew Chem Int Ed, Wiley-VCH Verlag 31:153–169

Albeck S, Weiner S, Addadi L (1996) Polysaccharides of intra crystalline glycoproteins modulate calcite crystal growth in vitro. J Chem A Eur 2:278–284

Bandel K (1990) Shell structure of the *Gastropoda* excluding archaeogastropoda. In: Carter JG (ed) Skeletal biomineralisation: patterns, processes, and evolutionary trends. Van Nostrand Rheinhold, New York, pp 117–134

Birchall JD (1989) The importance of the study of biominerals to materials technology. In: Mann S, Webb J, Williams RJP (eds) Biomineralization: chemical and biochemical perspectives. VCH Verlagsgesellschaft, Weinheim, pp 491–509

Boggild OB (1930) The shell structure of the mollusks. The Royal Danish Society of Sciences' Writings Department of Science and Mathematics, Series, vol 9, 2, pp 231–326
Canti MG (2017) Burnt carbonates. In: Nicosia C, Stoops G (eds) Archaeological soil and sediment micromorphology. John Wiley & Sons, Ltd, Chichester, pp 181–188
Chateigner D, Hedegaard C, Wenk H-R (2000) Mollusc shell microstructures and crystallographic textures. J Struct Geol 22:1723–1735
Hasse B, Ehrenberg H, Marxsen JC, Becker W, Epple M (2000) Calcium carbonate modifications in the mineralized shell of the freshwater snail Biomphalaria glabrata. Chem Eur J 6:3679–3685
Jacob DE, Soldati AL, Wirth R, Huth J, Wehrmeister U, Hofmeister W (2008) Nanostructure, composition, and mechanisms of bivalve shell growth. Geochim Cosmochim Acta 72:5401–5415
Lakshmanna B, Jayaraju N, Lakshmi Prasad T, Sreenivasulu G, Nagalakshmi K, Pramod Kumar M, Madakka M (2018) Data on Molluscan Shells in parts of Nellore Coast, southeast coast of India. Data Brief 16:705–712
Levi-Kalisman Y, Falini G, Addadi L, Weiner S (2001) Structure of the nacreous organic matrix of the bivalve mollusc shell examined in the hydrated state using cryo-TEM. J Struct Biol 135:8–17
Marin F, Luquet G, Marie B, Medakovic D (2007) Molluscan shell proteins: primary structure, origin, and evolution. J Curr Top Dev Biol 80:209–276
Silverstein RM, Webster FX (1998) Spectrometric identification of organic compounds, 6th edn. John Wiley and Sons, New York
Simkiss K, Wilbur KM (1989) Molluscs-epithelial control of matrix and minerals. In: Biomineralization: cell biology of mineral deposition. Academic Press, Inc., San Diego, pp 230–260. Chap. 14
Sreenivasulu G, Jayaraju N, Sundara Raja Reddy BC, Lakshmi Prasad T, Nagalakshmi K, Lakshmanna B (2017) Organic matter from benthic foraminifera (Ammonia beccarii) shells by FT-IR spectroscopy: a study on Tupilipalem Southeast Coast of India. J Methods X 4:55–62
Wilbur KM (1964) Shell formation and regeneration. In: Wilbur KM, Yonge CM (eds) Physiology of mollusca. Academic Press, New York/London, pp 243–282
Wilbur KM (1972) Shell formation in mollusks. In: Florkin M, Scheer BT (eds) Chemical zoology VII. Academic Press, New York/London, pp 103–145
Wilbur KM (1973) Mineral regeneration in echinoderms and molluscs. Ciba Found Symp 11(newser):7–33
Wilbur KM (1976) In: Watabe N, Wilbur KM (eds) Recent studies of invertebrate mineralization in the mechanisms of mineralization in the invertebrates and plants. Univ. South Carolina Press, Columbia, pp 79–108
Wilbur KM, Saleuddin ASM (1983) Shell formation. In: Saleuddin ASM, Wilbur KM (eds) The mollusca, 4, Part 1, Physiology. Academic Press Inc., London, pp 235–287

Chapter 17
Sedimentary Structures of Tidal Flats in Recent Chandipur East Coast of Odisha, India

M. Ramachandra, B. N. Anusha, B. Pradeep Kumar, S. Jammer Ahammad, and M. Rajasekhar

Abstract Sedimentary formations of diverse coastal tropical tidal flats on India's east coast, including inner estuarine tidal point bars 30–50 km inland from the coastline, have been extensively studied under varying seasonal conditions. The current study intends to describe certain comparable physical structures that are present in both the inner estuary tidal point bars and the nearby coastal areas of Chandipur beach in Odisha. According to the findings, both habitats exhibit physical traits including flaser bedding, herringbone cross-bedding, lenticular bedding, and mud/silt couplets. In fact, during the monsoon season, the point bar facies have higher levels of flaser bedding and lenticular bedding than the coastal tidal flat habitats. Even though they are common in both environments, interference ripples exhibit a variety of architectural forms for different environmental domains. In the high-water zone, interference ripples are common on estuarine point bars, with succeeding sets that resemble wrinkle marks taking the place of the initial ripple form. When interpreting the paleoenvironments of Proterozoic rocks, take caution because the properties that have so far been identified as important structures for near-coastal tidal flats are analogous to both habitats. Despite their prevalence in both situations, interference ripples exhibit distinct architectural patterns in different environmental domains. Interference ripples with threadlike secondary set overriding the earlier ripple-form, like wrinkle marks, are distinguishing characteristics of estuarine point bars at high tide. When determining the paleoenvironment of Proterozoic rocks, one must depend only on physical sedimentary structures, because elements previously assumed to be key structures for near-coastal tidal flats are similar to both environments.

M. Ramachandra · B. N. Anusha · B. Pradeep Kumar · M. Rajasekhar (✉)
Department of Geology, Yogi Vemana University, Kadapa, Andhra Pradesh, India

S. Jammer Ahammad
KCT Group Trust, Community Development Centre (Learn, livelihood & Research), Nellore, Andhra Pradesh, India

N. Jayaraju et al. (eds.), *Coasts, Estuaries and Lakes*,
https://doi.org/10.1007/978-3-031-21644-2_17

Keywords Sedimentation · Structures · Ripples · Aeolian environment · Dunes · Bars

17.1 Introduction

One of the most dynamic regions of the earth's surface are the coastlines. As a result of several physical processes, including tidal flooding, sea level rise, erosion, and sedimentation, it is constantly experiencing both slow and abrupt changes. Coastal resources throughout the Odisha coast are the famous attraction for the tourists (FitzGerald et al. 2008; Noffke et al. 2001; Fan 2012). At the same time, sewage and solid waste pollution, deforestation, coastline erosion, and sedimentation from building operations are all significantly exacerbated by mass tourism in this area. This area is covered with various kinds of mangrove plants (Craig-Smith et al. 2006). The wave-cut dunes provide an opportunity to have a rare view of internal organization of aeolian strata and their kinds. A quit and undisturbed feeling of this 4 km-wide rippled tidal flat is almost covered with water. It gets exposed during low tides and drowned during high tides (Paton 2010; Chakrabarti et al. 2017). Climbing dunes give way to form of bars that emerge from the river mouth and separated from each other by interbar areas. Behind the bar when depositional energy drops and consequently swamp is formed (Pahari et al. 2016; Siddiqui et al. 2017; Lakhdar et al. 2021). Dagara beach, located 70 km from Chandipur and at 21°33′17″N, 87°18′35″E, will be visited to conduct a comparison study between tide- and wave-dominated beaches. In this area dominance of wave processes over the tidal regime. The wave dominant leads to form a wide beach along the coast which is entirely different from Chandipur coast (Saha and Sinha 2021).

In Chandipur, we get a few widely different environments that coexist side by side within the very small extent of about 5 sq.km. Aeolian facies, beach facies, tidal facies, bar, and lagoonal facies all lie side by side here. The tidal regime in Chandipur is semidiurnal, and twice daily, the flood tide brings in and deposits fine clastics on the tidal flat, while the ebb tide reworks them. Also, in the wave-dominated beach environment, the swash and backwash continuously deposit new sediment and rework the older ones. Here, we performed a few simple experiments on the tidal flat of Chandipur and tried to gauge the rate of coastal erosion in the area. The aim is to study the modern environment at present Chandipur coast and correlated ancient environment of Miocene age Baripada rock formation.

17.2 Study Area

The examined coastal region of Chandipur is 5 km long and has different ecosystems such as intertidal flats, narrow sandy beaches, estuaries, a very gently descending subtidal area, mud volcanoes, mud balls, and bar-interbar complexes. Here we

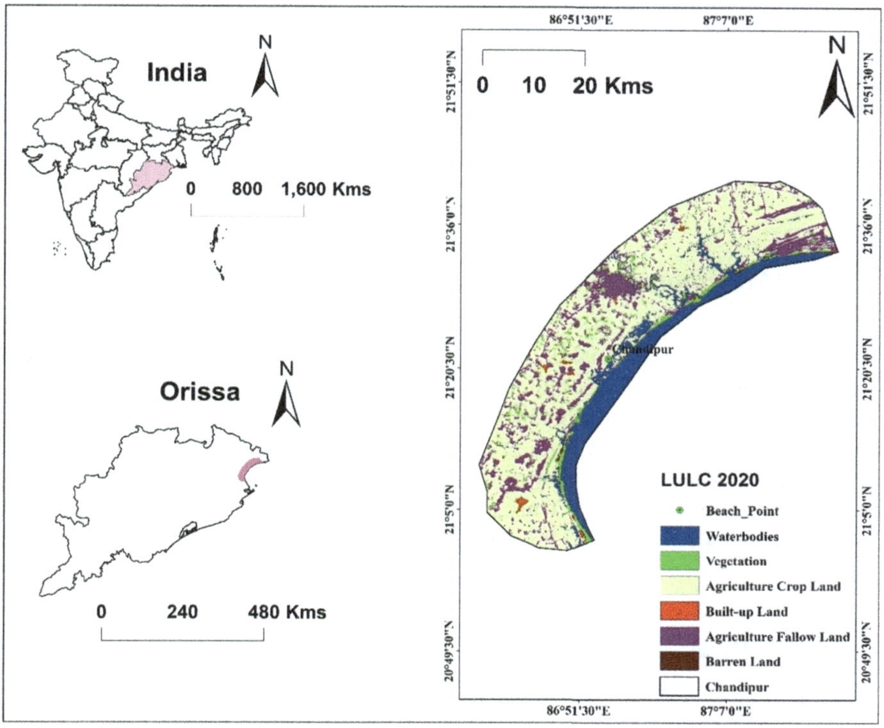

Fig. 17.1 Location Map of the study area

observe the dynamics of sedimentation and biological process. Chandipur is situated between 20° 49′ 30″ and 21° 50′ 30″ N Latitude and 86° 51′ 30″ and 87° 7′ 0″ E Longitude in the Balasore district of Odisha, on the eastern coast of India, at the confluence of the river Buribalam with the Bay of Bengal (Fig. 17.1). Chandipur is accessible from the Balasore railway station, 16 km off through a motorable road. The coast is roughly linear steep bearing wave-cut, on the west, but gently rolling down toward the confluence. The river Buribalam meets at the Bay of Bengal approaching from the west and forms an estuary. Here the river estuary and the barrier bars generated there interact with marine processes which produce a tide-dominated environment.

17.2.1 Lithology

The sediments of the intertidal flat are mainly composed of fine-grained silt and clay alternating with fine sands. Due to variation of energy distribution, distinct lithological variations were noticed in both shore parallel and shore perpendicular direction. Near the beach contact, silty sand is underlined by mud (Shang et al. 2019). The thickness of sand layers and clay decreases seaward. Near the shore coarse materials were found to accumulate in the troughs of the ripples. Accumulations of heavy minerals and charcoal matter take place along the troughs of ripples. Variation of energy distribution is mainly responsible for this type of lithology.

17.2.2 Morphology

The upper surface of the tidal flat is characterized by ripple, composed of silty sand and mud. The contact between the beach and the tidal flat is a sharp one, and there is a break in slope. Near this contact a linear trough, running parallel to the contact itself, is prominent (Sreenivasulu et al. 2017; Ganugapenta 2018). Due to the very low slope, water may flow in any direction in the tidal flat, driven by mostly winds, producing a variety of ripples. Thus, this part is so much corrugated with ripples that it may also be called a rippled flat (Habgood et al. 2003).

17.3 Methodology

The field program included beach profile surveys and visual observation of littoral environment parameters. Hydrographic survey data of Chandipur beach area collected by survey division of the Odisha state is also utilized. In this paper the focus is on sedimentary structures data collected in different sets at Chandipur beach in the year 2019. Stick methodology-red oxide was sprinkled abundantly on a stretch of the tidal flat (roughly 2 m × 2 m), situated about 200 m seaward from the foreshore-tidal flat contact near the PWD bungalow. This was done during low tide, and the spot was marked with a stick and left for the tide to come in. After an interval of 24 hours, the same spot was visited, and vertical section was retrieved using the steel corer, and the thickness of the sediments deposited above the red oxide layer was measured using a diagonal scale. It was observed there was a layer of fine clastic sediments above the red oxide layer. Its thickness was found to be 1.6 cm. One drilled and one undrilled counterpart of equal size of each two naticid species (*Natica tigrina* and *Polinices didyma*) and Turricula javana and two bivalve species (*Donax scortum* and *Mactra sp.*) were collected. The shells were washed, dried, and colored with two different colors to represent drilled and undrilled shells.

From the tidal flat, three locations were selected, each 50 m apart, and drilled shells were attached to 15-feet nylon string (of fine quality) and freely left on the surface at the time of low tide, when the tidal flat was not inundated by water. The undrilled shells were freely strewn on the surface. After 24 hours the locations were visited again, and the position of the shells was observed and measured. In order to study the shells, we made grids of 1 sq.m across the beach and registered the number of collected uniform-sized shells of *Mactra luzonica*, number of drilled and undrilled samples, and the orientation of the posterior of the studied shells with respect to the shoreline.

17.3.1 Scope of the Work

The current study proposes to examine the effect of shoreline processes and man-made structures in relation to the stability of Chandipur beach part of Balasore district. Accurate estimates of beach erosion are required for a variety of predictive, reconstructive, regulatory, and managerial purposes. For these reasons, data on short-term beach morphological changes, different sedimentary structures are made use in this paper. This study also proposes a user-friendly, economically viable solution applicable for this area to reduce the erosion or profile retreat associated with SW monsoon waves. The results would provide basic information required for coastal zone management of the study area.

17.4 Results and Discussion

17.4.1 Surface Sedimentary Structures

The tidal flat studied in Chandipur exhibits an overall monotony. It is an intertidal zone. Various types of ripples are the main sedimentary structures. The structures have been formed mainly due to mixed lateral accretion and suspension deposition. The following are the sedimentary structures encountered in the Chandipur intertidal zone:

17.4.1.1 Straight Crested and Sinuous Ripples

These ripples have straight crests; these are symmetrical in nature. These often show bifurcation of crests. These ripples are formed in lowest-energy condition (Fig. 17.2a). Sinuous ripples have crests which are gently curved (s-shaped in nature). These ripples are asymmetric in nature. Sinuous ripples are formed under slightly higher-energy condition with respect to the previously described straight crested (Fig. 17.2b).

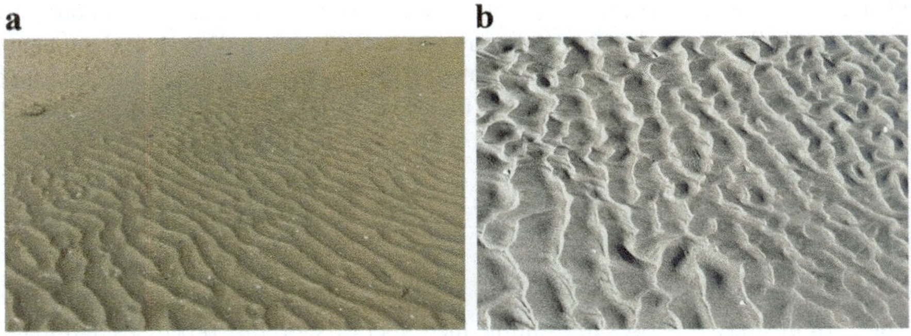

Fig. 17.2 (**a**) Straight crested and (**b**) Sinuous ripples Chandipur coast, Odisha, India

17.4.1.2 Cuspate and Lunate Ripples

This type of ripples has rounded crest and pointed troughs. These have up-current closures (Fig. 17.3a). Lunate ripples are curve-crested assemblage of asymmetric ripples with up-current closure. Some of them have a lee facing landward near the shore, and some of them have lee facing seaward near the offshore (Fig. 17.3b).

17.4.1.3 Linguoid Ripple

The abundance of linguoid ripple is less than the lunate ripples. These are also curve-crested asymmetric ripples but with down-current closure. These have amplitude and wavelength in the same range as the lunate ripples. These ripples form one after another, if flow velocity increases (Fig. 17.4).

17.4.1.4 Ladder-Back Ripple

Ladder-back ripples are formed during ebb tide when the water level gradually falls; some water gets entrapped within trough of the larger straight-crested ripples, and when flow takes place, a second set of smaller ripples are produced which migrate in a direction normal to that of the larger ripples (Fig. 17.5).

17.4.1.5 Double-Crested and Flattop Ripples

Some double-crested ripples are found here. Water forms a thin film on the crest and then reworked by air (Fig. 17.6a). Flattop ripples are of shallow origin. The crests may be curved or straight crested, but the crest is flattened due to high flowshear (Fig. 17.6b).

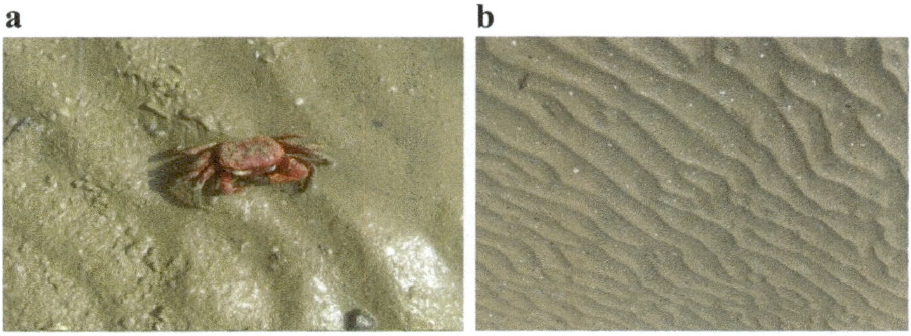

Fig. 17.3 (**a**) Cuspate and (**b**) Lunate ripples Chandipur coast, Odisha, India

Fig. 17.4 Linguoid ripple, Chandipur coast, Odisha, India

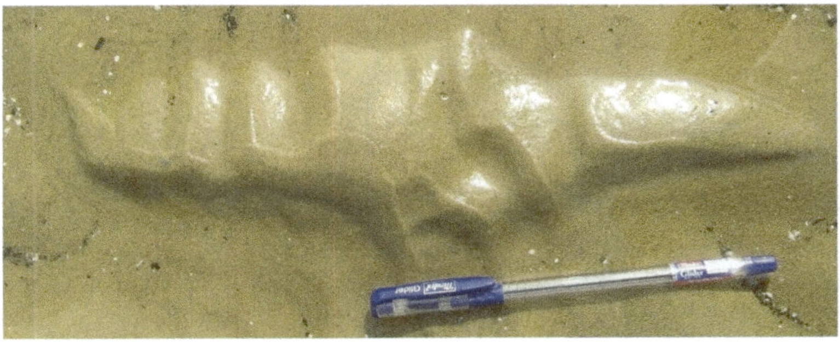

Fig. 17.5 Ladder-back ripple, Chandipur coast, Odisha, India

17.4.1.6 Superimposed Ripple and Microbial Mats

Superimposed type of ripple is formed by the action of superimposition of two waves in the same direction (Fig. 17.7a). Microbial mats are mat-like thin films of bacteria produced on the surface of sediment. They are very important tools to

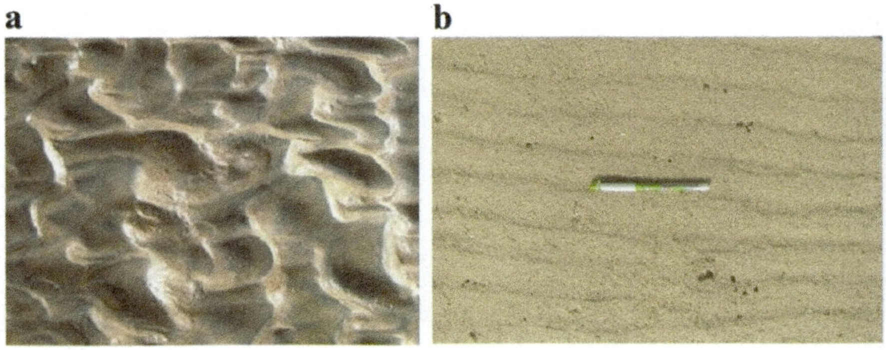

Fig. 17.6 (**a**) Double Crested and (**b**) Flat top ripples, Chandipur coast, Odisha, India

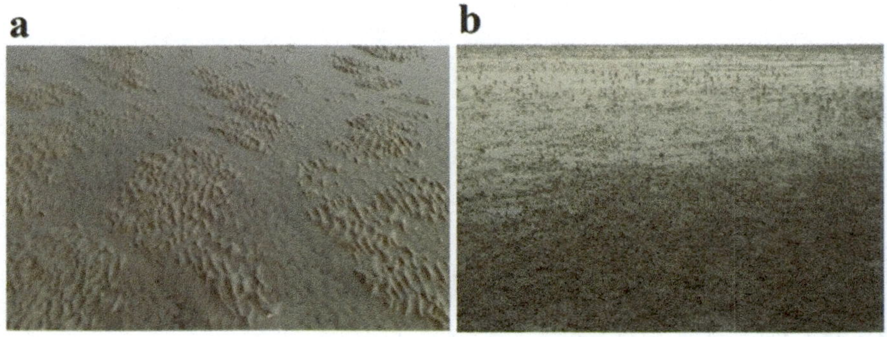

Fig. 17.7 (**a**) Superimposed Ripple and (**b**) Microbial mats, Chandipur coast, Odisha, India

interpret many ancient sedimentary features. Record of microbial mat covering was more in Proterozoic (Fig. 17.7b).

17.4.2 Beach Environment

The tides control the coast in Chandipur. Real beach has not developed yet; there is only a little piece of sand present along the shoreline. This little strip of beach was created with sand from nearby wave-cut aeolian dunes, but no crest system has grown nearby. The beach is made up of medium to fine-grained and well-sorted dune sand that is only modified by the waves.

The beach can be subdivided into two components: (a) backshore – the landward side of the beach and (b) foreshore – the seaward side of the beach. The width of the backshore region varies laterally. It is bounded by aeolian dunes and vegetation on the landward part. Water only reaches the backshore area during high-energy condition like high tides during full moon and new moon or storms (Fig. 17.8).

Fig. 17.8 Narrow stretch of beach showing its backshore and foreshore region in South east coast of Chandipur (NE- SW, direction), Odisha, India

17.4.2.1 Backshore Environment

The sands are fluffy in nature here. Heavy mineral zonations are observed. Crab burrows are also present. Those burrows are large in diameter, generally inclined in nature, but frequency of burrow occurrence is generally low compared to the foreshore part (discussed later). Burrow casting helps to identify different shapes of the burrows like I, J, U, and Y. In vertical trench section, planar to low angle cross-laminations are observed (Fig. 17.9a). The heavy mineral zonations of variable thickness, both vertically and laterally, can be easily distinguished from the sand layers in the trench sections. Occasional coarse-grained sand and shell hash bearing layers are also present. Sometimes signatures of crab burrows are preserved within the trench sections (Fig. 17.9b). As aeolian processes dominate in the backshore part of the beach, aerodynamic ripples are very common here. Coarse grain concentration is also present along those ripple crests (Fig. 17.10).

17.4.2.2 Foreshore Environment

The high-water marks divide the backshore and foreshore area of the beach. Foreshore is the seaward side of the beach. Unlike the backshore part, foreshore of the beach receives water during the entire period of neap-spring cycles, and thus a high-energy environment is formed here. Plane to low-angle cross-laminations is observed in the trench sections. The sands are finer in size compared to the backshore part. All the primary structures developed here are of characteristically high-energy environments. Swash ripples are present on the foreshore surface (average wavelength 56 cm and average amplitude is 4 mm). Their crests are sinuous in nature, and trend of the crestline is shore parallel (Fig. 17.11a, b). Along with swash

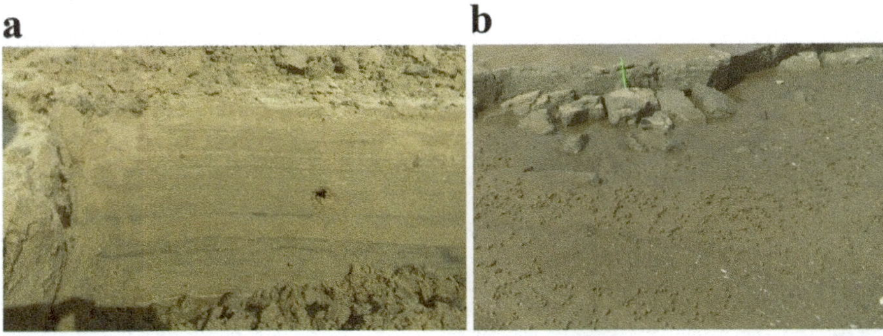

Fig. 17.9 (a) Dark colored heavy mineral zonations, (b) Crab burrows in Chandipur coast, Odisha, India

Fig. 17.10 Different shaped burrows observed in the backshore part Chandipur coast, Odisha, India

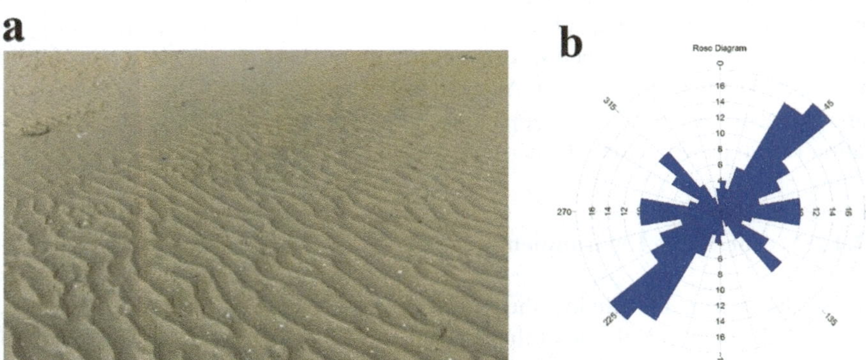

Fig. 17.11 (a) Swash ripples with sinuous crest present on the foreshore, (b) Trend of the Crestline of swash ripples in Chandipur coast, Odisha, India

ripples, current crescent structures, rill marks, rhombic ripples, and parting lineations are also observed to form on the foreshore surface. The rose diagrams are plotted in Fig. 17.12b, d, f to show the orientation of different sedimentary

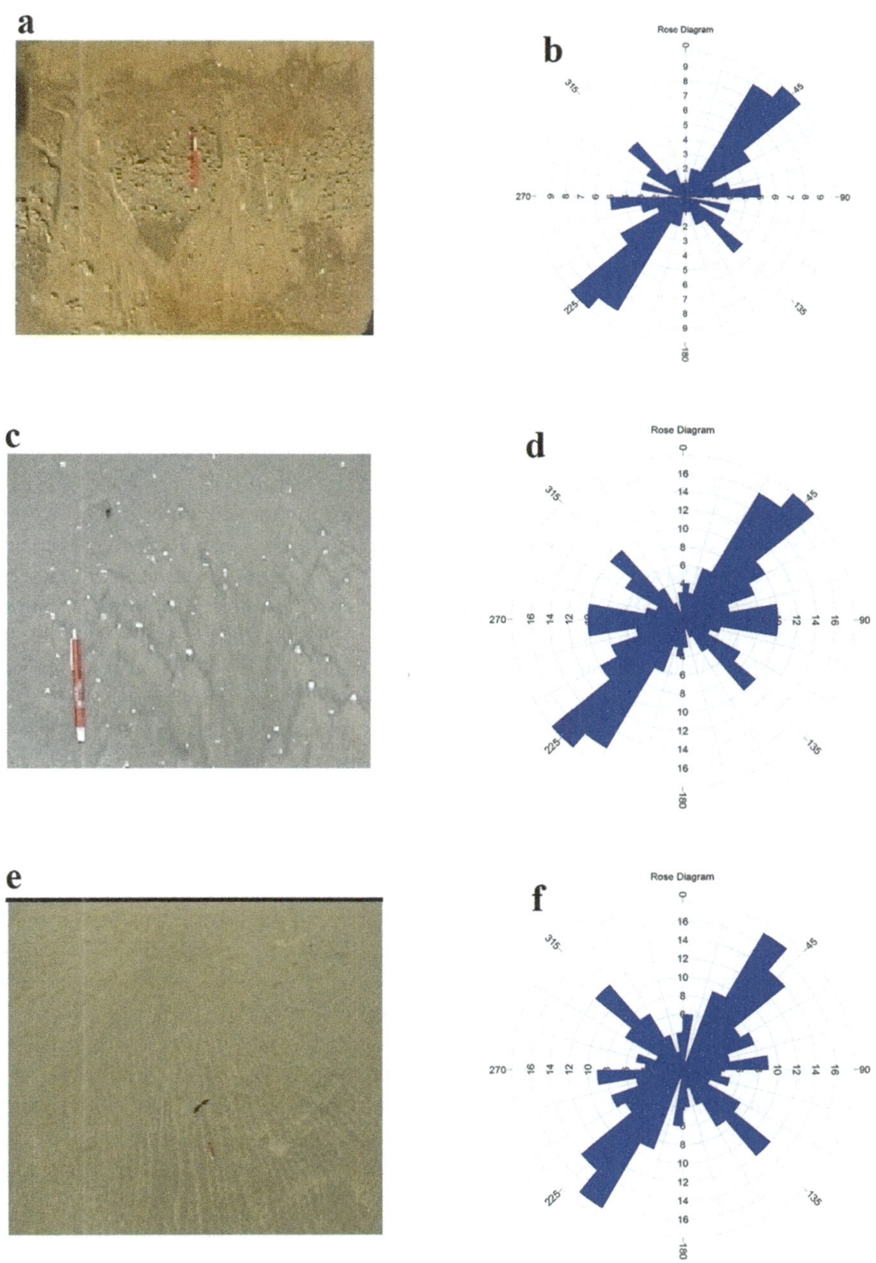

Fig. 17.12 (**a**) Rill marks, (**b**) Rill marks orientation, (**c**) Current crescent, (**d**) Crescent orientation, (**e**) Parting lineaments, (**f**) Lineaments orientation of offshore, Chandipur coast, Odisha, India

structures. The burrow frequency is very high in the foreshore region, but they are of very small in diameter and of dominantly vertical in nature. Many ophiomorpha burrows are also found to be present on the beach (Fig. 17.12a, c, e).

17.4.3 Aeolian Environments

Aeolian environment is a typical depositional environment formed by wind action, and this may be desert or coastal environment. Some unique types of facies association form in the aeolian environment due to the 800 times lesser density of air than aquatic medium. During this case study, it is observed that a vast dune field along the Chandipur coast as well as the top part of the shore is parallel elongated bar identified as aeolian deposits.

Sediment character is moderate to fine-grained well-sorted in nature. Shore parallel and shore perpendicular vertical sections are observed within trenches of different parts of the bar. Darker lighter alternating millimeter scaled lamina showing planer to low angle cross stratified with some borrowing interference observed within trenches. Darker bands are composed of black-colored minerals characterized by relatively smaller grain diameter with respect to the light-colored grains. Some darker minerals also show magnetic property. From the above observation, it can be concluded that the darker layers are formed due to heavy mineral segregation by aeolian process. Different types of bedform are observed within the aeolian subenvironment. According to their increasing diameter, major bedforms are as follows:

- Ripple (0.1 m)
- Aeolian dune (100 m)
- Draa (20–450 m)
- Erg

Aeolian ripples are shaped when wind blows across a sand bed, causing the sand particles to salt and move, creating a thin carpet of moving sand grains. The grains are very briefly suspended, and as soon as they hit the ground, they have enough energy to knock nearby grains into the open airflow, continuing the saltation process. Regions where the grains are a little more piled up will be created by irregularities in the sand's surface and the airflow's turbulence. At a same wind speed, all medium sand grains will move about the same distance every time they saltate because the grains in these heaps will be more susceptible to being swept away by the flow. The result is a series of piles of grains that are evenly spaced apart and lined up perpendicular to the wind.

Aeolian waves have wavelengths (the distance between the crests) that range from a few centimeters to several meters. The height of a ripple can range from less than a centimeter to more than 10 cm from the bottom of the trough to the apex of a crest (Zhang 2014; Bhattacharya et al. 2021; Ramesh et al. 2021). A layer of inversely graded sand may be produced by aeolian ripples as they move because coarser grains are more concentrated at the crests, while the finer grains are winnowed away by the wind. Cross-lamination can occur where well-formed grains slide into a nearby trough; however this happens less frequently in aeolian ripples than in their subaqueous counterparts. Aeolian bedforms, which can be an impact or are aerodynamic in nature, are produced in extremely low-energy environments. When determining flow characteristics from a cross section, ripple cross-lamination is frequently seen.

Ripple marks are formed by low-angle wind flow. Here, coarse lag is concentrated at the crest and wavelength/amplitude >9.1. Crest line is often bifurcated and broad sinuous (Fig. 17.13). Adhesion ripple is characterized by the entrapped loose sand particles covering the underlying wet surface. It is mainly formed due to the hitting of dry sand on the wet surface. Capillary rise of water or crystallization in many cases helps to trap further grains by consistent wetting. The thin salt moist crust adopts a hemispherical blister-like shape, so that, ripple-like surfaces generated. Windblown sand and dust adhere to hygroscopic surface of blisters. The blisters themselves may be breached and filled in by salting sand grains. Their irregular surfaces were covered by variable thicknesses of sand to form adhesion ripples with greater and very irregular amplitude. Steeper sides of the ripples tend to occur in upwind direction. They have low preservation potential as during drying up they collapsed and reworked. These are found exposed near the wet surface and had very low amplitude and wavelength.

Down-current waning of flow is commonplace in aeolian regime, and hence wind ripples often climb resulting formation of translatent strata. Very low-angle climb gives rise to planar laminae within which cross laminae are very difficult to recognize because of good sorting of grains. Translatent laminae are typically characterized by inverse grading because of concentration of coarse grains along ripple crests. It frequently develops on the windward flanks and slopes. The laminae are extremely thin, and occasionally they alternate with sand laminae and heavy laminae rich in minerals.

Fig. 17.13 Impact ripple in the aeolian system, Chandipur coast, Odisha, India

Setulf is a typical aeolian structure formed by wind action due to any obstacle present at the way of the wind. It is a structure of positive relief. Obstacle and current are the two major factors for the formation of this structure. A narrow ridge of sand is formed behind any obstacle and gradually tapering away from the obstacle (Sarkar et al. 2011). This tapering direction indicates the wind direction during the formation of setulf structures (Fig. 17.14). The larger the obstacles, the larger will be the ridge behind it. The length of ridge behind the obstacle is in proportional with the obstacle size (Sarkar et al. 2011).

In Chandipur area, we have found transverse dunes with wavelength 10–100 m, and amplitude varies from 3 to 10 m. These dunes are developed behind the backshore. They mark the boundary between the land and the beach. The seaward faces of these dunes are cut by wave, and some exposed root horizons were seen. So, these dunes are not in equilibrium with present-day marine system. In these dunes, some low-angle cross-laminations were observed. This low-angle cross lamina in low-flow regime is a characteristic of an aeolian activity. Translating strata were also observed in these dunes. Some characteristic features of these dune areas are as follows:

1. Individual lamina is reversely graded and wedging downwards.
2. Individual foreset cross-strata are inversely graded, and it is overlined by lamina which is not inversely graded.

These dunes are formed due to an oblique approach of the wind toward the dune crests, which results in the formations of small ripples which migrate upwards. After reaching the crest, they form large-scale cross-strata due to avalanching. Then wind direction changes, and the ripples start to migrate from another side of the dune forming zigzag cross-stratification. Since the wind direction changes frequently, the crest line cannot orient it perpendicular to the wind direction but does

Fig. 17.14 Setulfs directed unidirectionally in Chandipur coast, Odisha, India

so almost parallel to the resultant wind direction. When the shift in wind direction is restricted within 15°, it forms longitudinal dunes, whereas a shift of around 40°, it forms linear dunes.

In Chandipur area during our study, we have found the following internal features within the dunes. Sand laminae that are practically horizontally oriented make up the horizontal bedding of sand dunes. Usually, it is found on the windward slopes, flanks, and near the summit of a sand dune, as well as the interdra area. Laminae were very thin and were fraction of 1–2 mm. The sand was well sorted and sometimes heavy mineral poor laminae. A cross-bed is a single depositional unit or layer composed of internal lamina that is inclined toward the main surface of sedimentation. Here the inclination of the foreset lamina is very low with the principle surface of sedimentation, and then it is a low-angle cross bedding. Inclined laminae are formed due to superimposition and reworking of slip face by oblique and longitudinal winds; these laminae later produce foreset lamina and cross-bedding units which are made up of avalanche lamina of the slip face. The slip angle ranges from 8° to 15°, and the foreset laminas are slightly inclined. Foreset lamina has a tendency to flatten out and resemble horizontal lamina toward the base of the slip face. Each foreset lamina measures 5 mm thick.

The surface which separates a single depositional unit to the adjacent beds is called bounding surface. It is a shallow of nondeposition or rapid change of surrounding physical condition under which the sedimentation occurs. The lower bounding surface of this dune was horizontal to subhorizontal. In the seaward part of the dune, several additional features were also noticed. In this section we also notice the free fall of sand due to gravity. An upward increase in grain size within a single bed is known as reverse grading. Coarse grain concentrated at the top due to aeolian segregation and after reaching angle of repose its avalanches towards lee of the dune (Chakrabarti 2005; Ramanathan et al. 2010). The grain flow process of shear sorting during the flow causes coarse grains to rise to the surface, due to kinetic sieving or dispersive pressure, resulting reverse grading. This type of reverse grading is formed due to dispersive pressure. An upward decrease in grain size within a single bed is known as normal grading. This is formed due to turbidity current. The reactivation surface is formed within the set because the lee face is periodically eroded. In computer simulations, this type of bounding is formed by changes in dune migration speed much that a period of advance is punctuated by a period of erosion and dune height such that the dune grows by scour of underlying sediment. These sorts of changes are common because natural flows are rarely steady, and if the period of flow fluctuation is regular, then the surfaces are cyclic.

17.5 Conclusions

The study reveals that using features to distinguish different tidal environments in rock records arbitrarily might result in false findings. In order to distinguish transitional estuary tidal flats from nearby coastal tidal flats in the rock record, more attention should be paid to the physical sedimentary structures created in the tidal point bars under tropical conditions with high suspended load. In recent years an increase in research interest in tidal flats has increased due to the need for comparative study for fossil tidal facies interpretation. This is because coarse tidal deposits have the potential to be significant petroleum reservoirs and fine-grained tidal facies can be caprock or coal-bearing rock. Tidal flats are crucial in preventing coastal erosion and flooding brought by rising sea levels and storms that are getting stronger. The research on open coast tidal flats, tide-dominated deltas or estuaries, and estuaries is a large source of updated knowledge. The second is especially important in dispelling our previous false belief that open coast environments are unfavorable for the development of extensive tidal flats because of substantial wave impact or even dominance by waves. Recent research highlights the importance and diversity of tidal flats in modern coastal ecosystems. Based on their depositional environments, tidal flats have been divided into distinct categories: they are straight crested, sinuous, cuspate, lunate, linguoid, ladder-back, double crested, flattop, superimposed, and microbial mats of ripples which found in the surface sedimentary structures in the Chandipur coast, Odisha, India. Other features include crab burrows, swash ripples, rill markings, current crescent, and parting lineations in aeolian settings of sedimentary formations. It is important to remember that in areas where significant sand ridge/bar deposits are anticipated, some exposed tidal flats may develop on top of subtidal sand ridges, estuarine, or deltaic-distributary mouth bars.

References

Bhattacharya S, Mapder T, Fernandes S et al (2021) Sedimentation rate and organic matter dynamics shape microbiomes across a continental margin. Biogeosciences 18:5203–5222. https://doi.org/10.5194/bg-18-5203-2021

Chakrabarti A (2005) Sedimentary structures of tidal flats: a journey from coast to inner estuarine region of eastern India. J Earth Syst Sci 114:353–368. https://doi.org/10.1007/BF02702954

Chakrabarti A, Brandt SR, Chen Q, Shi F (2017) Boussinesq modeling of wave-induced hydrodynamics in coastal wetlands. J Geophys Res Oceans 122(5):3861–3883

Craig-Smith SJ, Tapper R, Font X (2006) The coastal and marine environment. In: Tourism and global environmental change. Routledge, London, pp 121–141

Fan D (2012) Open-coast tidal flats. In: Principles of tidal sedimentology. Springer, Dordrecht, pp 187–229

FitzGerald DM, Fenster MS, Argow BA, Buynevich IV (2008) Coastal impacts due to sea-level rise. Annu Rev Earth Planet Sci 36(1):601–647

Ganugapenta S, Nadimikeri J, Chinnapolla SRRB, Ballari L, Madiga R, Nirmala K, Tella LP (2018) Assessment of heavy metal pollution from the sediment of Tupilipalem Coast, southeast coast of India. Int J Sediment Res 33(3):294–302

Habgood EL, Kenyon NH, Masson DG, Akhmetzhanov A, Weaver PP, Gardner J, Mulder T (2003) Deep-water sediment wave fields, bottom current sand channels and gravity flow channel-lobe systems: Gulf of Cadiz, NE Atlantic. Sedimentology 50(3):483–510

Lakhdar R, Soussi M, Talbi R (2021) Modern and Holocene microbial mats and associated microbially induced sedimentary structures (MISS) on the southeastern coast of Tunisia (Mediterranean Sea). Quat Res 100:77–97

Noffke N, Gerdes G, Klenke T, Krumbein WE (2001) Microbially induced sedimentary structures: a new category within the classification of primary sedimentary structures. J Sediment Res 71(5):649–656

Pahari A, Mondal S, Bardhan S, Sarkar D, Saha S, Buragohain D (2016) Subaerial naticid gastropod drilling predation by Naticatigrina on the intertidal molluscan community of Chandipur, Eastern Coast of India. Palaeogeogr Palaeoclimatol Palaeoecol 451:110–123

Paton DC (2010) At the end of the river: the Coorong and lower lakes. ATF Press, Adelaide, Australia, pp 247

Ramanathan AL, Ranjan RK, Prasad MBK et al (2010) Sediment-nutrient dynamics in selected Indian mangrove ecosystems – land use and climate change implications. IAHS-AISH Publ 337:84–92

Ramesh R, Purvaja R, Rajakumari S et al (2021) Sediment cells and their dynamics along the coasts of India – a review. J Coast Conserv 25. https://doi.org/10.1007/s11852-021-00799-3

Saha K, Sinha S (2021) Grain size analysis and characterization of sedimentary process in tidal flat of Chandipur region, East Coast of India. Mar Geod 44(5):485–503

Sarkar UK, Singh AK, Jena JK (2011) Biodiversity of the freshwater fishes in the protected forest areas of Uttar Pradesh and its significance in management of riverine fish diversity. In Proceeding of national conference on forest biodiversity: earth's living treasure. Uttar Pradesh State Biodiversity Board, Lucknow (Vol. 22, pp. 36–42)

Shang X, Taizhong DUAN, Jiagen HOU, Yan LI (2019) Spatial configuration of sand and mud in the lacustrine nearshore sand bar deposits and its geological implications. Pet Explor Dev 46(5):954–968

Siddiqui NA, Rahman AHA, Sum CW, Yusoff WIW, bin Ismail, M. S. (2017) Shallow-marine sandstone reservoirs, depositional environments, stratigraphic characteristics, and facies model: a review. J Appl Sci 17(5):212–237

Sreenivasulu G, Jayaraju N, Reddy BSR, Prasad TL, Lakshmanna B, Nagalakshmi K (2017) Coastal morphodynamics of Tupilipalem Coast, Andhra Pradesh, southeast coast of India. Curr Sci 112:823–829

Zhang W (2014) Encyclopedia of marine geosciences, pp 1–6. https://doi.org/10.1007/978-94-007-6644-0

Part V
Climate Change and Anthropocene

Chapter 18
Coastal EV Index: A Case Study of Kuwaiti Coast

Subramaniam Neelamani, Dana Al-Houti, Alanoud Al-Ragum, and Abeer Hassan Al-Saleh

Abstract A method is proposed to identify the coastal areas that are expected to experience different degrees of erosion vulnerability (EV) due to future sea-level rise. The wave height, the landside slope of the beach, and the sediment size above the high-water level are the main influencing parameters for erosion, and their weightage for the contribution to erosion is assessed based on analytic hierarchy process. This concept is used to identify the EV for 162 coastal cells of the Kuwaiti coast, which is about 500 km in total length. An EV index (EVI) map is prepared. It is found that 35% of the Kuwaiti coast has very low EVI, 29% of the coast has low EVI, 12% of the coast has moderate EVI, 21% of the Kuwaiti coast has high EVI, and 3% of the Kuwaiti coast has very high EVI. Qaru and Kubbar Islands in Kuwait are having very high EVI. The coastal area north of Kuwait Bay holds very low EVI. These maps can be used by decision-makers and stakeholders to plan for suitable coastal erosion protection measures in order to protect the existing infrastructures from sea wave-induced erosion. The findings of this study would be useful for planning suitable adaptation measures and for selecting suitable coastal sites for future coastal infrastructure projects.

Keywords Coastal erosion · Vulnerability · EV index · Kuwaiti coast · Beach sediment · Wave height · Beach slope

S. Neelamani (✉) · D. Al-Houti · A. Al-Ragum
Coastal Management Program, Environment and Life Sciences Research Center, Kuwait Institute for Scientific Research, Safat, Kuwait
e-mail: nsubram@kisr.edu.kw

A. H. Al-Saleh
Crisis Decision Support Program, Environment and Life Sciences Research Center, Kuwait Institute for Scientific Research, Safat, Kuwait

N. Jayaraju et al. (eds.), *Coasts, Estuaries and Lakes*,
https://doi.org/10.1007/978-3-031-21644-2_18

18.1 Introduction

Natural and anthropogenic-induced coastal erosion is a universal problem. The rising of sea-level, time-dependent variation in sediment flow to the littoral zone, storm wave activities, wave and surge overwash on the beach, deflation of the beach sands by wind activities, sediment transport along the shore, and the sorting of beach sediments during the wave and current activities are the main natural forces for beach erosion (Shore Protection Manual 1984). The main anthropogenic causes for coastal erosions are land subsidence from the removal of subsurface resources like oil, gas, or water; interruption of transporting soil by constructing different coastal structures such as groin, offshore breakwater, marina, port, and harbor; reduction of sediment supply to the littoral zone by constructing dams in the rivers; concentration of wave energy on beaches through marine infrastructure developments; increasing the water level variations near the coast; changing the natural coastal protection by altering the coastal topography; and removal of material from the beach like black sand for processing to extract fuel for atomic power plants. Proper planning and remedial actions are possible for reducing the anthropogenic causes, but it is cumbersome to propose solutions to reduce coastal erosion caused by nature. For effective coastal erosion management, it is recommended to adopt either (a) defend or (b) adapt or (c) managed retreat (Williams et al. 2017). Coastal erosion is assessed by temporal comparison of shoreline data. This is the regular practice and is done for Kuwait also (Neelamani et al. 2011). But how to assess or identify the eroding hotspot of the coast in the future due to climate change, sea-level rise, and the associated change in environmnetal forces from sea waves? This is a million dollar question. An attempt is made to get some suitable answer and is presented in this manuscript. Kuwaiti coastal area is used as an example for this case study. It is believed that the methodology presented may be useful for other coastal countries around the world. Kuwait has about 500 km of coast. It is situated at the northwestern corner of the Arabian Gulf (Fig. 18.1).

Kuwaiti government is planning to develop many coastal areas for social and commercial activities. Hence a study to identify the vulnerable eroding area will help to choose the technically suitable area for such coastal developments such as Silk City, Boubyan Island, Failaka Island, and Sulaibikhat area. This manuscript is presented as follows. A brief literature review is provided focusing on coastal EV studies. Thousands of literatures are available on coastal vulnerability (in general physical vulnerability) and are not referred in this chapter since they are not relevant for the selected topic. A brief description is provided on sea-level rise and its possible impact on the erosion and adaptation principles. The methodology for assessment of EV index (EVI) is described. The data collection procedure for the essential data for assessment of coastal EV is then described. Then the results and discussions are revealed to identify the vulnerable locations for erosion in the future in Kuwait followed by conclusions.

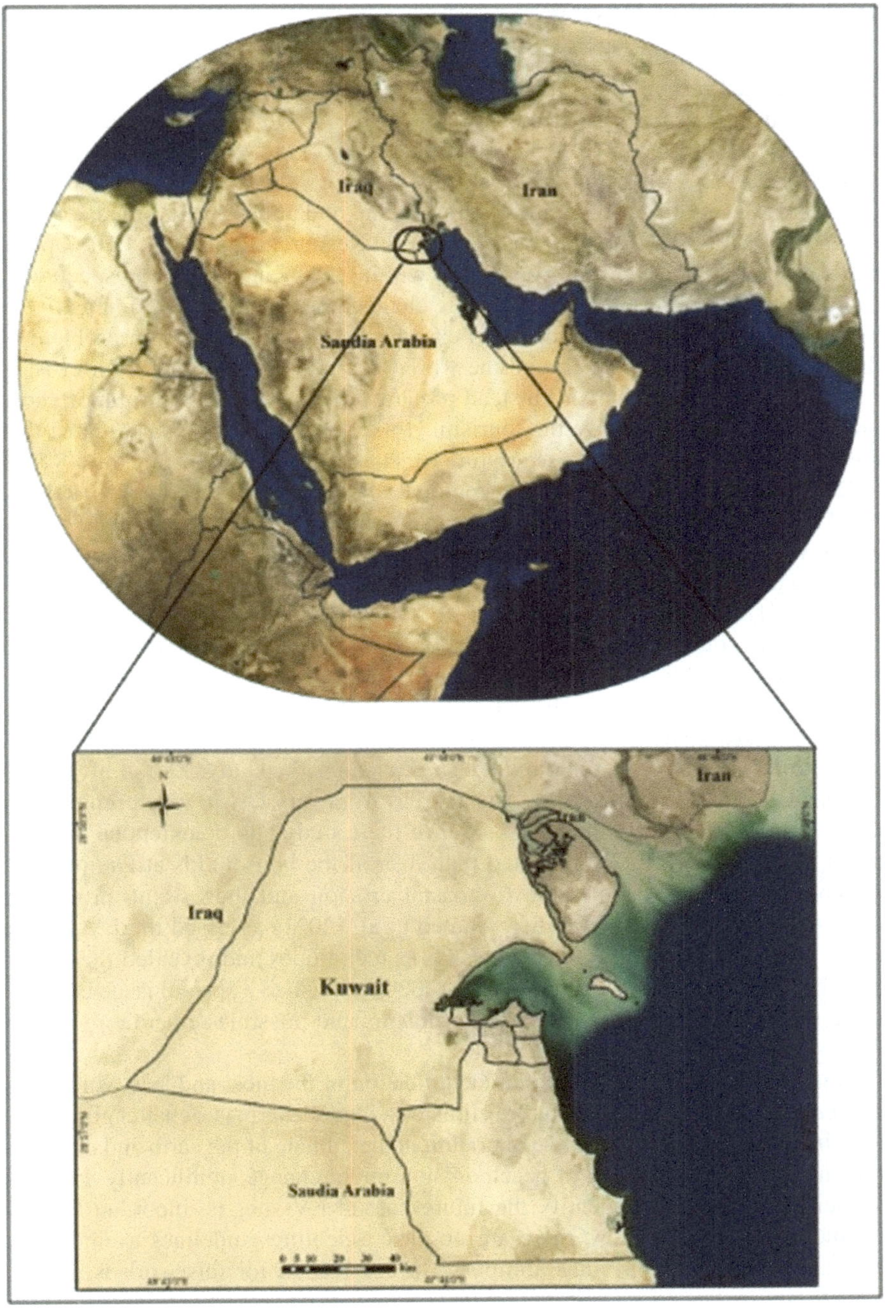

Fig. 18.1 The location map of Kuwait

18.2 Literature Review

The shoreline length of the global coast is about 1 million km. About four decades back, Bird (1985) has assessed that 20% of the world's coasts were eroding. However, it is reported in 2013 that only about 11% of the world's coasts are eroding (Hinkel et al. 2013). Barbier et al. (2008) reports that the population density in coastal area is almost two times that of inland and about 33% of the people of the earth inhabits in the area of the coast. World Resources Institute (2010) reports that 1.4 billion and 2.8 billion people live within 25 km and 100 km, respectively, from coast. It is believed that the climate change phenomena would intensify the wave actions on the beach and would exaggerate the coastal erosion problem in the future. Many countries around the world turn to use soft solution for erosion reduction (beach nourishment) instead of hard solutions (seawall, groin, offshore breakwater, etc.) to improve the tourism. Hinkel et al. (2013) has estimated that the cost of nourishment of these eroding beaches would be to the tune of US $ 65 to 220 billion. The North Atlantic regional study (ERG 2010) has carried out similar beach maintenance assessment against erosion for North Atlantic region and revealed that $3.5 billion was the public expenditure. Nordhaus (2006) assessed that the physical capital stocks in coastal erosion-vulnerable areas would be about $1.2 trillion.

There are some recent studies on different items related to coastal EV. Some of the notable publications are by Fitton et al. (2018) for Scotland, de Andrade et al. (2019) for Guarujá, in the state of São Paulo, Brazil, Zhu et al. (2019) for Xiamen Coast, southeast China, Gomez et al. (2020) for Gunjur, The Gambia, and Ahmed et al. (2021) for Bangladesh. Based on the study, Zhu et al. (2019) has identified the percentages of coastal areas from very low to very high EV and recommended that the evaluation model of their study can be extended to other areas. Gomez et al. (2020) revealed that 90% of the households at Gunjur, The Gambia were highly vulnerable to coastal erosion and there is no promising action to reduce this vulnerability. Ahmed et al. (2021) assessed the EV for the eastern coast of Bangladesh. They used GIS techniques and revealed the low to high EV zones. Recently Buitrago et al. (2020) published a special issue containing a total of 32 articles to reveal the problems and possible remedial measures for coastal hazard management.

It is clear from this literature that identification of the most and least vulnerable coastal areas of erosion for the future climate change scenario is a challenging problem. In the future, due to possible alteration in the climate of the earth and expected rise in sea level, the regime of beach erosion might change significantly. Hence a scientific assessment to identify the future coastal EV sites is important for any coastal country. As of now, there are no clear scientific guidelines available for assessing future erosion vulnerabilities. The motivation for this work is derived from these issues.

18.3 Impact on Coastal Erosion Due to Sea-Level Rise and Possible Adoption Principles

Sea-level rise (SLR) is one of the important causes for the accelerated coastal erosion. A thorough analysis is required of the future SLR and its possible effect on beach erosion. The SLR may alter the regime of coastal erosion due to permanent inundation, an increase in wave activity because of increase in local water depth, wave action on the upper part of the beach, whose existing slope may be different from the slope of the beach at present, and the soil strength or sediment particle size. One way of analyzing this problem is through an index called the coastal EV index, by suitably incorporating the parameters, which affect the intensity of coastal erosion. The beaches can be divided into many numbers of cells or segments, each with a certain known distance. The data of the influencing parameters for each cell can be collected based on fieldworks or other possible means like remote sensing. The weightage for each influencing parameter can be assessed scientifically. Then, the coastal EV for each coastal cell can be assessed. This is explained in the methodology section. A proper coastal EV assessment would help to identify the most and least erosion-vulnerable areas, which in turn would help the governments/decision-makers to select appropriate strategies for reducing the expected coastal erosion such as beach erosion protection structures, adaptation principles, and managed retreat. Normally, if a coastal structure is previously constructed closer to the high-water line (and if it cannot be moved away from the sea) and has a longer residual life, say 30–50 years, then the technically workable solution will be building impermeable seawalls and hard revetments around the eroding structure. If the property owner wishes to have the property near the high-water line and the government allows for development of coastal infrastructure, then the adaptable option is to build the structure on concrete basements, which in turn is fixed on pile groups. In this situation, the beach/seabed material below the infrastructure and around the piles may undergo erosion due to SLR and wave action, but the superstructure will be intact and serve for its design life. This is a workable adaptation option. If the SLR in the future is known (to some acceptable extent) and the beach slope is assessed, then the new infrastructure can be built after providing enough beach width as a setback zone for future erosion for the expected SLR and wave action, which is an acceptable managed retreat policy. It is important to identify the hotspots of future erosion due to sea-level rise. Identification of highly vulnerable, moderately vulnerable, and low vulnerable erosion areas will help the decision-makers to select the appropriate coastal area for future coastal development projects. A highly vulnerable erosion area needs to be avoided for the construction of beach houses, beach hotels, or other coastal industrial infrastructures, or suitable erosion protection measure should be designed and to be kept ready for construction to protect the erosion hotspots from sea wave-induced erosion.

The following section describes the methodology used for assessing the future EV areas due to SLR and provides the illustrations of this assessment for the Kuwaiti coastal area.

18.4 Methodology for Assessment of EV Index (EVI)

Fundamental analysis of sediment transport or scour around structures would help in the assessment of EVI. Seabed sediments start moving when the seawater-induced shear stress on the sediment is higher than the threshold shear stress of the sediment (Soulsby 1997). It is also proved (Soulsby 1997) that the shear stress on the bed due to steady current and wave is proportional to the square of the depth-averaged speed of the water at the seabed by current and the square of the amplitude of the bottom orbital velocity by wave, respectively. In a wave field, the wave-induced horizontal water particle velocity near the seabed is proportional to the incident wave height. Hence, the sediment movement is proportional to the square of the wave height. The longshore sediment transport is also a function of the seabed slope as well as the size of the seabed sediments (Kamphuis 2000) apart from wave period and wave direction. Kamphuis (2000) demonstrated that the sediment transport is proportional to (wave height)^2, (seabed slope)$^{0.75}$, and 1/(soil particle size)$^{0.25}$. It is clear that wave height is the most influencing parameter followed by seabed slope and soil particle size. Since the present interest is the assessment of the coastal EV of the selected coast in the future, the coastal landward slope and the landward soil particle size are appropriate to use and not the seaside slope and soil particle property of soil on the seaside from high-water level (HWL).

An appropriate weightage needs to be assigned for the wave height, coastal slope, and soil particle size. The analytic hierarchy process (AHP) can be used for this purpose (Saaty 2008). The pairwise comparison matrix between these three parameters is prepared and is provided as shown in Table 18.1.

The above table indicates that wave height is extremely important when compared to landward soil particle size and wave height is strongly important when compared to beach slope. Similarly, the beach slope is more important than soil particle size when sediment transport is the topic of discussion. The normalized matrix of the variables is prepared and is shown in Table 18.2.

From this table, the weightage for wave height is 1/3[(45/59) + (15/19) + (9/13)] = 0.7482. The weightage for beach slope is 1/3[(45/295) + (3/19) + (3/13)] = 0.1804, and the weightage for soil particle size is 1/3[(45/531) + (1/19) + (1/13)] = 0.0714.

It is required to assign the vulnerability ranking for erosion by wave height, beach slope, and soil particle size. This information is provided in Table 18.3.

Table 18.1 Pairwise comparison matrix between wave height, beach slope, and soil particle size

Variable	Wave height	Landward beach slope	Landward soil particle size
Wave height	1	5	9
Landward beach slope	1/5	1	3
Landward soil particle size	1/9	1/3	1
Sum	59/45	19/3	13

Table 18.2 The normalized comparison matrix between wave height, beach slope, and soil particle size

Variable	Wave height	Landward beach slope	Landward soil particle size
Wave height	45/59	15/19	9/13
Landward beach slope	45/295	3/19	3/13
Landward soil particle size	45/531	1/19	1/13
Sum	1	1	1

Table 18.3 Vulnerability ranking assigned for erosion by different parameters

Parameter	Very low (1)	Low (2)	Moderate (3)	High (4)	Very High (5)
Wave height, m	<0.5	0.5–0.7	0.7–0.9	0.9–1.1	>1.1
Beach slope (%)	<0.3	0.3–0.6	0.6–0.9	0.9–1.2	>1.2
Soil particle size, D_{50} mm	>1.0	0.5–1.0	0.25–0.5	0.125–0.25	<0.125

The formula for EV index (EVI) for any coastal cell "i" is given as:

$$\begin{aligned} \text{EVI}\ (\text{i}) &= (0.7482)\ \text{vulnerability ranking for wave height at location i} \\ &+ (0.1804)\ \text{Vulnerability ranking for beach slope at location i} \\ &+ (0.0714)\ \text{vulnerability ranking for soil particle size at location i} \end{aligned}$$

The highest value of EVI occurs when the coastal location "i" has a wave height greater than 1.1 m, beach slope more than 1.2%, and soil particle size less than 0.125 mm. For such situation, the value of EVI is (0.7482) 5 + (0.1804) 5 + (0.0714) 5 = 5.0. The lowest value of EVI occurs when any coastal location "i" has wave height less than 0.5 m, beach slope less than 0.3%, and soil particle size greater than 1.0 mm. For such situation, the value of EVI is (0.7482) 1 + (0.1804) 1 + (0.0714) 1 = 1.0. The length of the coast of Kuwait is about 500 km and is split into 162 coastal cells. EVI is estimated for all 162 locations based on the wave height, landside slope, and landside particle size data. The details of data collection are provided below.

18.5 Data for the Present Study

As discussed, the three important environmental data needed for the EVI study are wave height, beach slope, and sediment particle size. The details of these data collection for assessment of EVI for Kuwaiti coastal area are explained in the following sections.

18.5.1 Wave Height Data

The wave height for each cell is assessed based on the study by Al-Salem and Rakha (2005). The wave heights are derived based on hindcasted wave data. Rakha et al. (2005) have analyzed 12 years (1993–2004) of hindcasted wave heights for Kuwaiti territorial waters, and the information is provided as maximum significant wave heights for 12 years (Fig. 18.2). Hence, for the coastal vulnerability assessment for Kuwait, the maximum significant wave height is used.

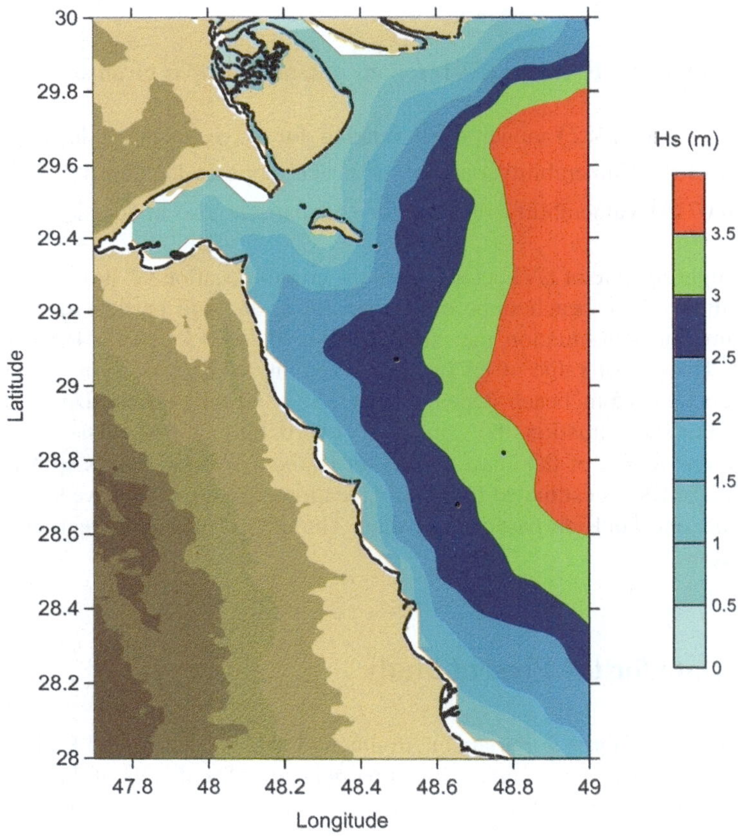

Fig. 18.2 Maximum significant wave heights over 12 Years for the Northern part of the Arabian Gulf (Rakha et al. 2005)

18.5.2 Landside Slope of the Beach

As discussed, the slope of the coast is a vital parameter for coastal EV study. If the slope of the coast is very mild (say 1:1000), then is SLR is 1.0 m, and then the permanent coastal flooding would be about 1000 m. On the other hand, if the coastal slope is 1:10, then the expected inundation will be about 10 m. However, if the coastal slope is mild, after inundation due to sea-level rise, the waves will break by spilling, and the wave erosion capacity of spilling waves is milder when compared to plunging and surging waves, which are common for steeper beach slopes. Since the assessment of coastal inundation and EV are two important aims, a detailed coastal profile assessment is carried out at all these 162 coastal locations. The beach profile survey is carried out at each location. The effort is made to get the seabed profile beyond low-water level on the seaside and at least up to 1-meter elevation on the landside, so that the area of expected future inundation can be estimated for a possible SLR of 1.0 m by the year 2100. In certain areas, the topography is so flat that we could not cover the beach profile up to 1.0-m elevations from the high-water line (HWL). This fieldwork was carried out for all 162 field locations. A typical beach profile for a location (location L1) in Ras Al-Ardh area is shown in Fig. 18.3. (Please see location L1 indicated in Fig. 18.5).

In Fig. 18.3, (0, 0) indicates the highest high-water line and is assessed during the fieldwork by marking the coastline corresponding to the highest high-water line. The seaward slope at this location is 0.129, and the landward beach slope is 0.07. The landward slope is used for EV assessment.

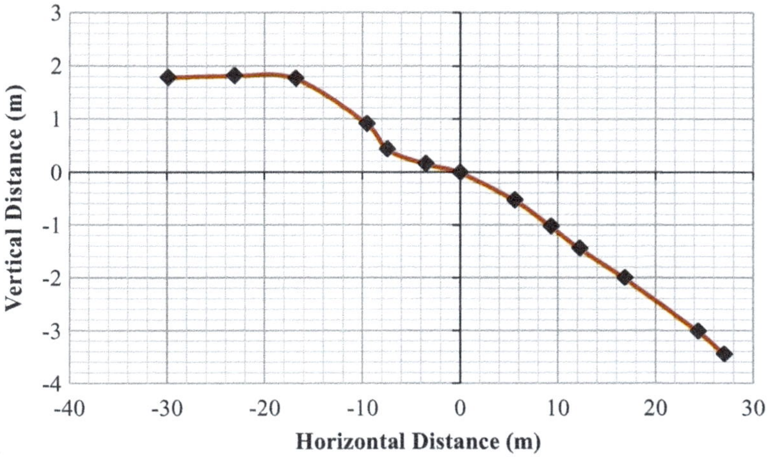

Fig. 18.3 Beach profile for location L1

18.5.3 Beach Soil Properties

The main soil property needed for coastal EV is the soil particle size, D_{50}. This corresponds to particle size, such that 50% of the soil by weight is more than D_{50} and 50% less than D_{50}. In each location, the soil is collected 25 m away from the high-water line on the land side, which represents the soil expected to experience and interact with waves during the future sea-level rise. About 2 kg of soil was collected from each location and is dried in the laboratory by keeping them exposed to atmosphere for 7 days. After 7 days, sieve analysis was carried out using the standard technique. The soil sample collected at location L1 is provided in Fig. 18.4. In this figure, the left-hand side was the soil collected on the landside, and the right-hand side was the soil collected at the seaward side of the high-water level. The D_{50} of the landside soil is 0.9 mm, and the soil looks well graded with some pebbles. The landside soil has more fine-sized materials compared to the seaside soil. This is the equilibrium condition for the existing wave regime in this location. Hence, any SLR will accelerate transportation of the smaller-sized particles of the landside soil, and this location is highly prone to erosion in the future. A similar type of analysis is carried out for all the soils collected from 162 coastal locations.

Though the soil samples are collected on both the seaside and landside in each location, as discussed earlier, the properties of the landside soil are more important for studies on coastal EV in the future due to SLR. The soil under no influence by waves and currents today is expected to experience its interaction in the coming years, when the level of the sea rises. Since D_{50} is the main parameter used for most of the engineering analyses, it is used for vulnerability ranking. Coastal land soil of

Fig. 18.4 The soil sample collected at location L1

fine size (0.12–0.25 mm) can easily be transported by waves and currents, and hence it is considered very vulnerable to erosion. On the other hand, very coarse soil of 1.0–2.0 mm can resist the wave- and current-induced shear stress better and hence has low vulnerability to erosion. This concept is used for ranking the vulnerability of the landside soil in each location. D_{50} in the range of 0.25–0.5 mm is considered as medium sand, and 0.5–1.0 mm is considered as coarse sand. The coarse sand has a better resistance capacity to move against wave- and current-induced shear stresses when compared to medium sand. With these points in mind, Table 18.4 is prepared to reflect the vulnerability of coastal landside soils.

Based on these rankings, a GIS-based map of coastal landside soil is prepared for Kuwait and is presented in Fig. 18.5. It is seen that some of the coastal locations in the southern part of Kuwait are very vulnerable to erosion. Coarse and very coarse

Table 18.4 Vulnerability ranking and color code for landside soil particles of different D_{50} sizes

D_{50} of landside soil particle (mm)	Vulnerability for erosion	Rank	Color code
1.0–2.0	Very low	1	Green
0.5–1.0	Low	2	Blue
0.25–0.5	Moderate	3	Yellow
Less than 0.25	High	4	Red

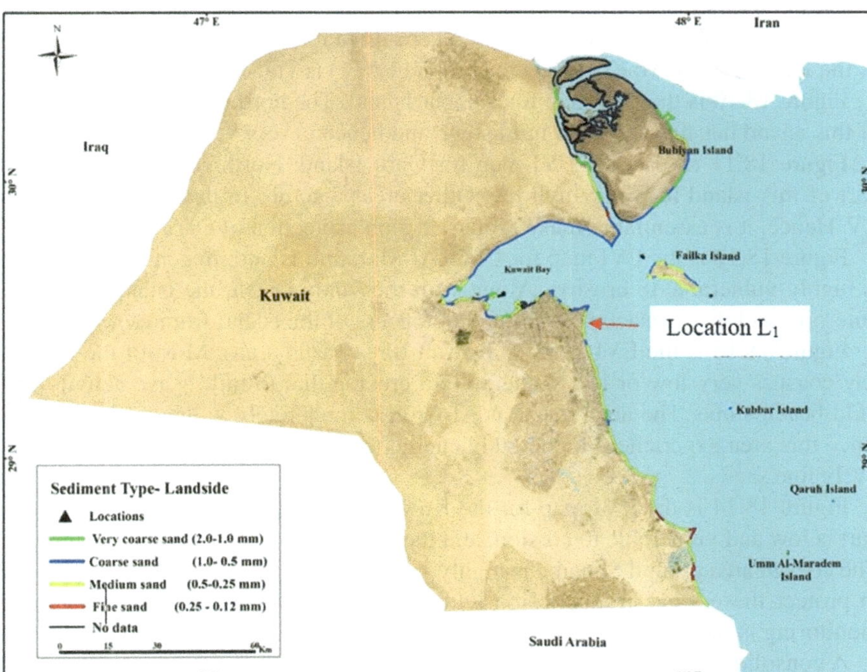

Fig. 18.5 Coastal landside soil type for Kuwaiti coast

sand is found in many locations around Kuwait Bay as shown in Fig. 18.6. In most of the southern coast, the land side of the beach has medium sand. This information is used for EV index assessment. Figure 18.6 is a flowchart to explain the methodology.

18.6 Results and Discussions

Based on the data collected for all 162 locations and as per the methods described above, the value of EVI is assessed for each location. From the EVI estimate, the coastal area of Kuwait is divided into very high, high, moderate, low, and very low erosion-vulnerable sites based on the criteria given in Table 18.5.

Figure 18.7 shows the EV index map of the complete Kuwaiti coastal area. From this plot, it is found that the future EV is very low within Kuwait Bay, around Boubyan, and most of Failaka Island, mainly due to smaller wave heights in these areas. A detailed assessment reveals that 35% of the Kuwaiti coast has very low EVI, 29% of the coast has low EVI, 12% with moderate EVI, 21% with high EVI, and 3% with very high EVI.

Figure 18.8 is the EVI map for the coasts of Boubyan Island. The EV is either very low or low mainly due to smaller wave heights acting on these coasts.

Figure 18.9 shows the EVI map for Failaka Island. Mostly, the EV is very low and low except few areas in the south, where it is moderate. The southeastern part of the Failaka Island (the coastal area around FA11) is vulnerable for erosion.

Figure 18.10 is the EVI map for Kubbar Island. The north and northeastern part of this island has fine sand and mild slope and hence is very vulnerable to erosion.

Figure 18.11 reveals the EVI map for Qaru Island. Northwestern and western part of this island has very high EV. Other areas also are in the category of high EV. Hence, it is essential to make proper erosion protection of Qaru Island.

Figure 18.12 is the EVI map for Umm Al-Maradim Island. In general, the island is highly vulnerable to erosion. Along with the sandy beach, the island has rocky outcrops, and they are acting as natural defenders of the island from erosion.

Figure 18.13 is the EVI map for Kuwait Bay coastal areas. Most of the Kuwait bay coast is very low or low vulnerable to erosion due to mild wave activity and mild beach slope. The area near Ras Al-Ardh is moderately vulnerable to erosion since this area experiences moderately energetic waves and the beaches have relatively fine sands.

Figure 18.14 is the EVI map for the Kuwait's mainland in south. The southern part is low and vulnerable to erosion, and the main reason is the mild coastal slope. The coastal area around Shuaiba is highly vulnerable to erosion, and care is needed to protect these areas from erosion. Other areas are moderately vulnerable, and monitoring is required for these areas.

A consolidated GIS map of coastal landside soil, wave height, coastal slope, and EVI for the entire Kuwaiti coast is presented in Fig. 18.15 for a clear visualization.

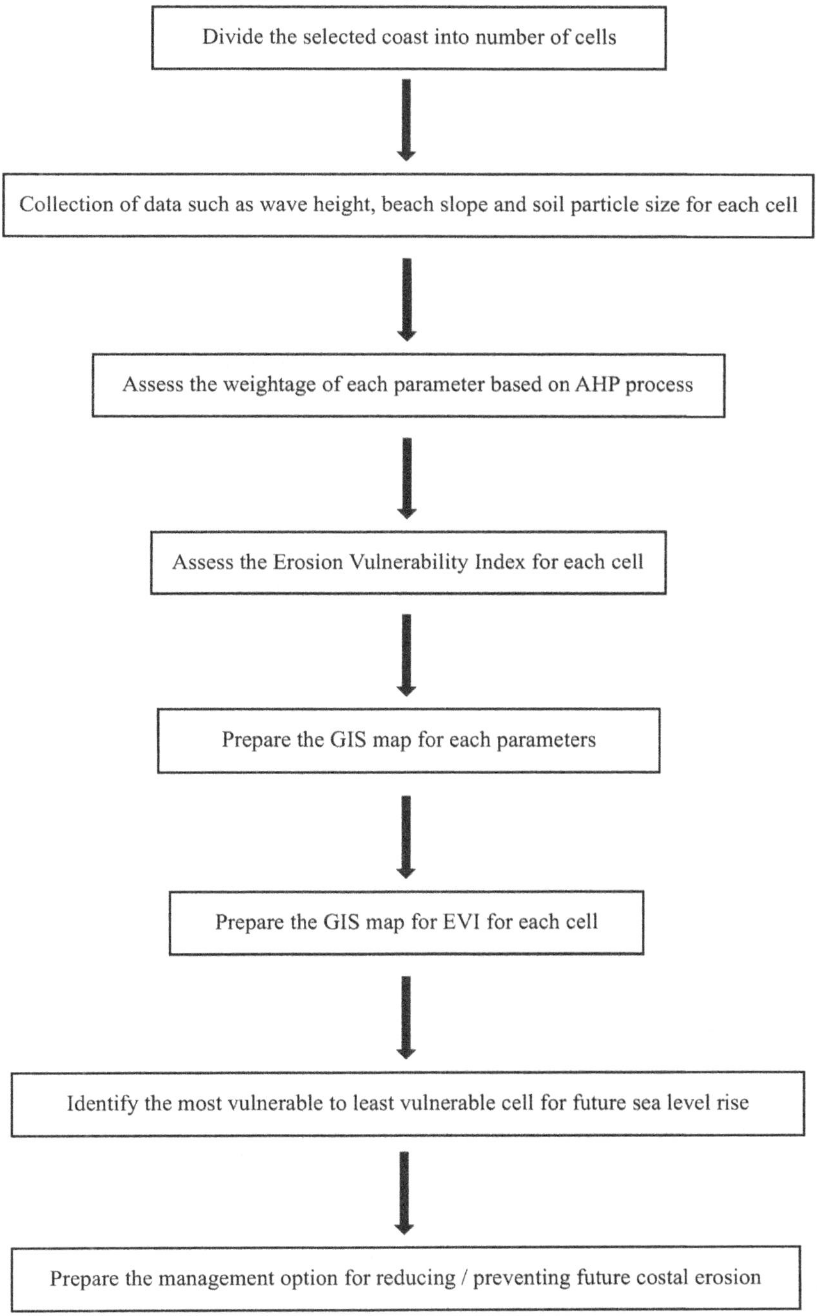

Fig. 18.6 Flow chart for assessing Coastal Erosion Vulnerability Index

Table 18.5 Ranking for erosion vulnerability index for Kuwait

EVI value	Vulnerability	Rank	Color code
Less than 2	Very low	1	Green
2–2.8	Low	2	Blue
2.8–3.6	Moderate	3	Yellow
3.6–4.2	High	4	Orange
Greater than 4.2	Very high	5	Red

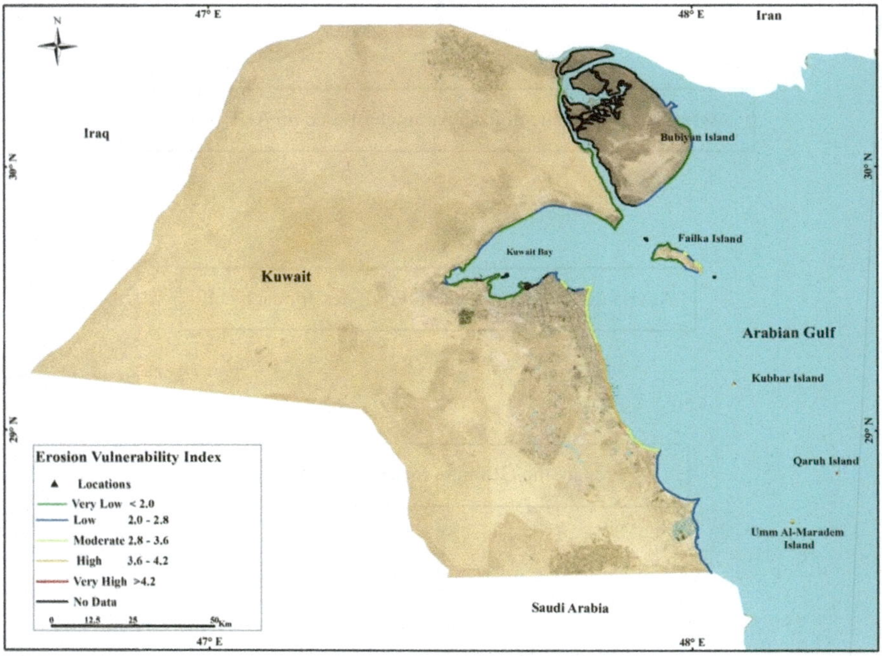

Fig. 18.7 The erosion vulnerability index map of Kuwaiti coastal area

18.7 Management Options to Deal with Coastal Erosion

There are few different management aspects to manage coastal erosion. If the structure already exists in the erosion-prone zone, either suitable hard solutions or periodic nourishment can be adopted. Not all hard solutions would work for erosion-prone locations. Out of seawall, groin field, and offshore breakwaters, the suitable one can be selected based on science and economics and the degree of acceptance by the users. Many countries adopt nourishment as the best options. If a borrow site for suitable sediments are available nearby and is economical, then periodic nourishment would be doing well. Engineers should keep in mind reduce/reuse and recycle in their works related to coastal erosion protection. Floating breakwater may be useful if the option of reduce is in mind (Neelamani and Ljubic

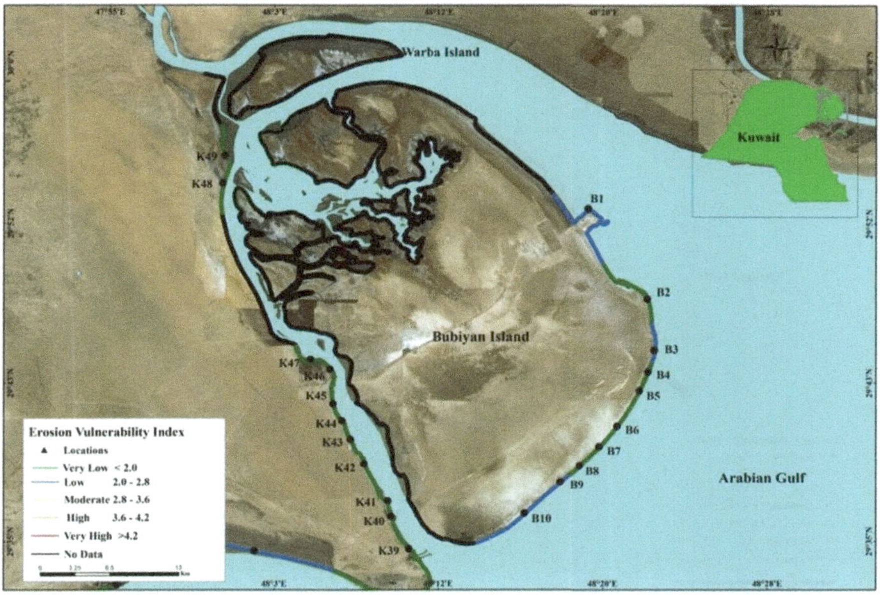

Fig. 18.8 The erosion vulnerability index map of Boubyan Island area

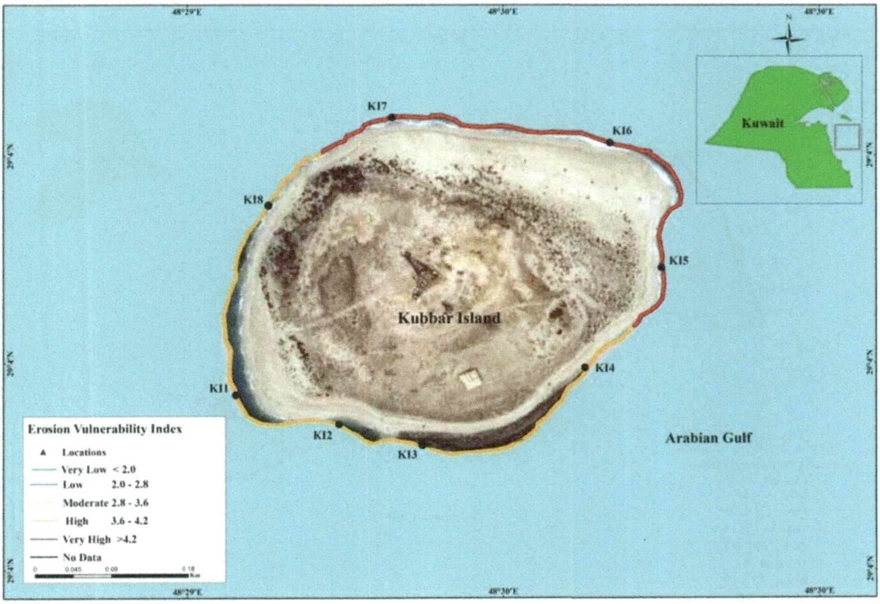

Fig. 18.9 The erosion vulnerability index map of the Failaka Island area

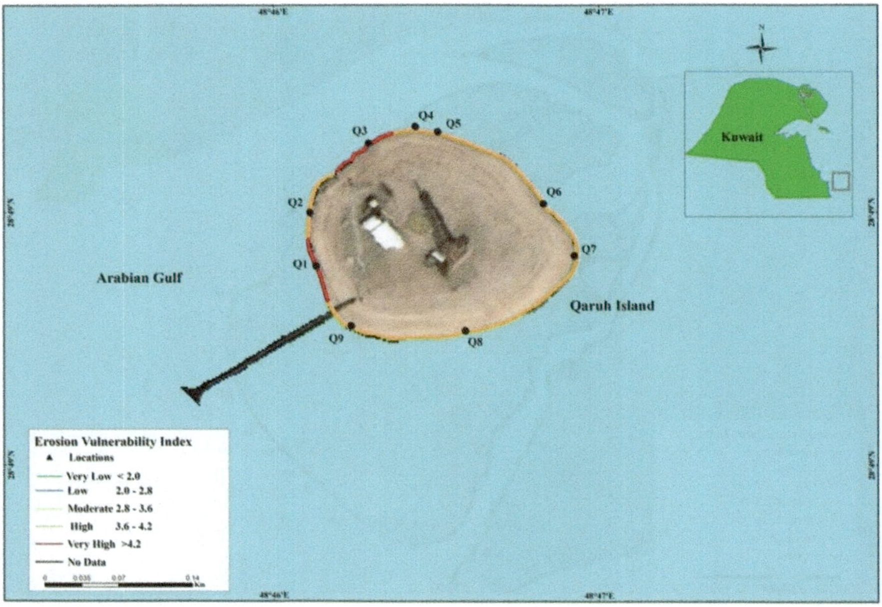

Fig. 18.10 The erosion vulnerability index map of Kubbar Island area

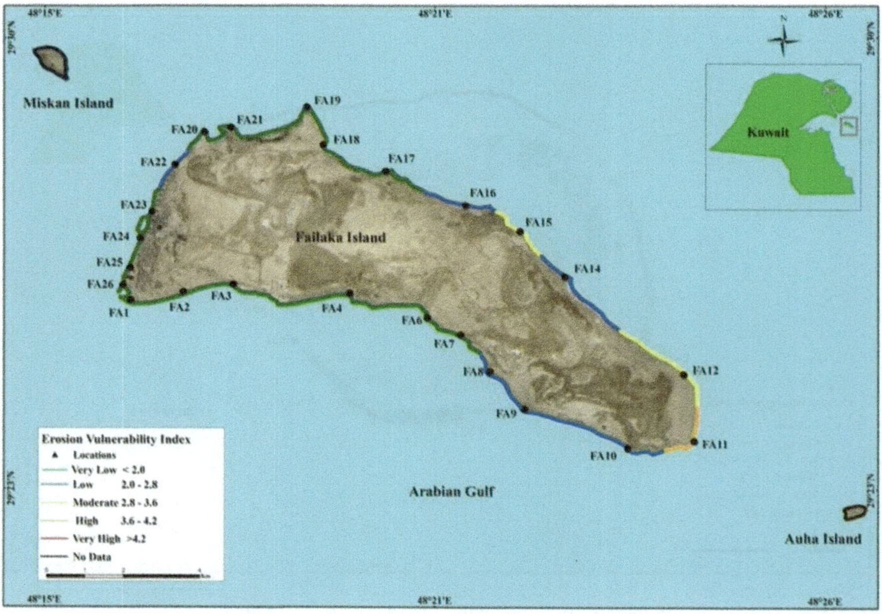

Fig. 18.11 The erosion vulnerability index map of Qaru Island area

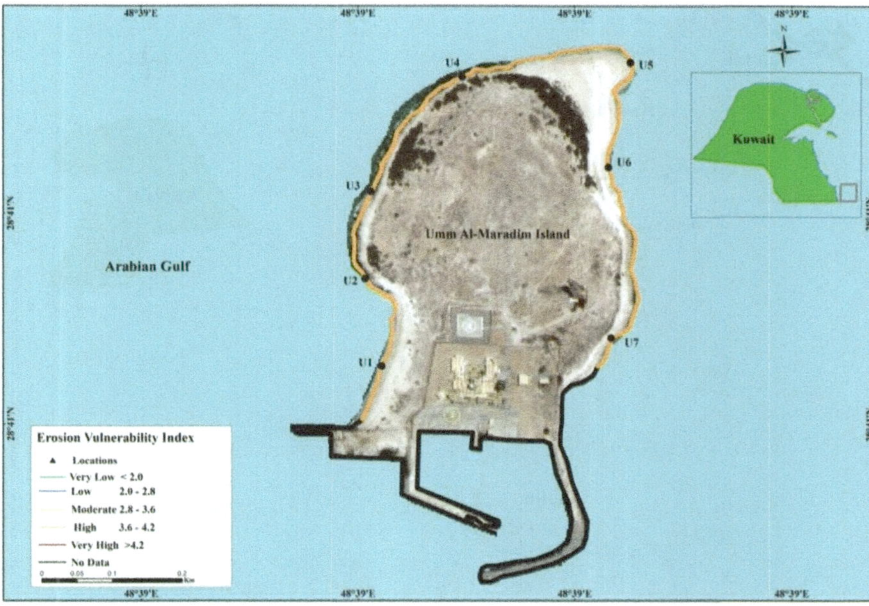

Fig. 18.12 The erosion vulnerability index map of Umm Al-Maradim Island area

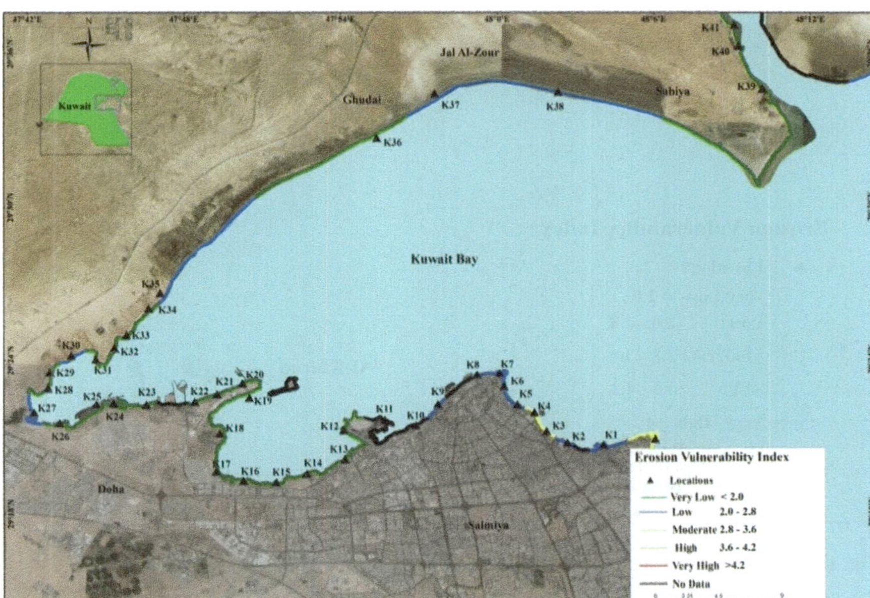

Fig. 18.13 The erosion vulnerability index map of coasts around Kuwait Bay

Fig. 18.14 The erosion vulnerability index map of Kuwait's mainland-south coasts

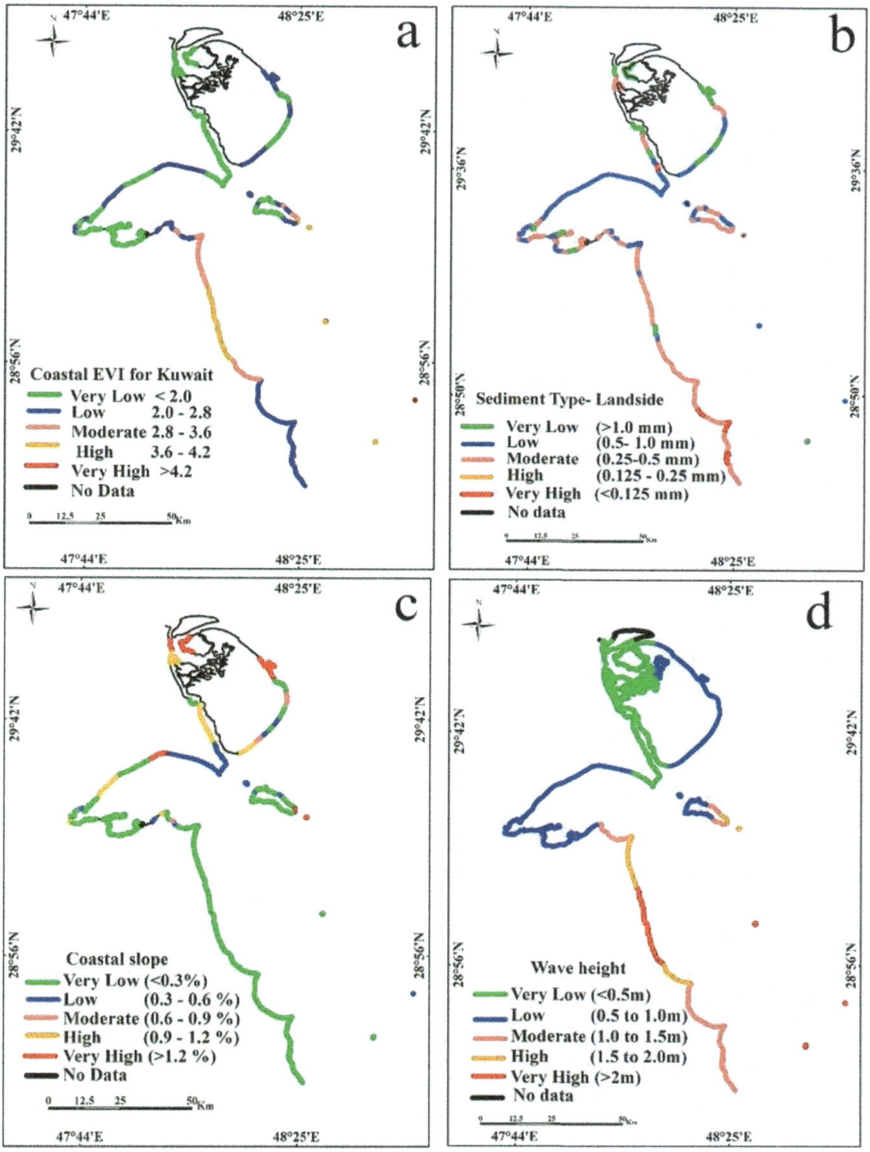

Fig. 18.15 GIS maps. (**a**) Coastal EVI for Kuwait. (**b**) Sediment type. (**c**) Coastal slope. (**d**) Wave height

2017). They are proven to be cost-effective, if the wave energy levels are moderate. Scrap tires can be used as floating breakwater (Al-Yousif et al. 2019) which falls under reuse option. An efficient arrangement can be selected for significant reduction of wave transmission. For a new type of coastal infrastructures, suitable buffer width can be provided to allow the sea to erode the beach but not scouring the planned structures.

18.8 Application of EVI Study for Preparing the Strategy for Future Adaptation Plan

Kuwait has an ambitious plan for coastal developments, which are as follows:

(a) Boubyan Island Development Plan
(b) Failaka Island Development Plan
(c) Kuwait Bay Coastal Development Plan
(d) Other Island Development Plans

The study results can be used in these development plans. For example, the areas, which are more vulnerable to erosion, can be avoided for construction of beach houses or resorts. The coastal infrastructures already developed in high erosion-vulnerable sites can be identified, and suitable erosion protection solutions can be planned.

18.9 Conclusions

An investigation is carried out to identify the hotspots of coastal erosion due to future SLR. The formula proposed by Kamphuis (2000) is used to identify the important parameters which are responsible for coastal erosion. It is found that wave height, beach slope, and sediment particle size govern the erosion. The weightage for each parameter is assessed based on the analytic hierarchy process (Saaty 2008). Kuwait is considered for the study. The 500-km coastal length of Kuwait is divided into 162 sediments. The data are collected in these cells and are used to assess the coastal EV index (EVI). From this study, it is found that 35% of the Kuwaiti coast has very low EVI, 29% has low EVI, 12% has moderate EVI, 21% has high EVI, and 3% has very high EVI. Qaru and Kubbar Islands in Kuwait are found to have very high EVI. The coastal area north of Kuwait Bay holds very low EVI. The EVI maps would be useful for decision-makers and stakeholders in Kuwait to carry out suitable coastal erosion protection measures in order to protect the existing infrastructures from sea wave-induced erosion. The study results also will be useful for planning suitable adaptation measures. Kuwait has a long-term strategy for coastal development of different areas, such as Boubyan Island, Failaka Island, Silk City, and

Sulaibikhat Bay. The results of the present study can be useful for sustainable development of these areas.

References

Ahmed N, Howlader N, Hoque M, Pradhan B (2021) Coastal EV assessment along the eastern coast of Bangladesh using geospatial techniques. Ocean Coast Manag 199:105408. https://doi.org/10.1016/j.ocecoaman.2020.105408. ISSN 0964-5691

Al-Salem K, Rakha K (2005) Verification of a WAM Model for the Arabian Gulf. Arabian Coast 2005 Conference, Dubai, 15 Oct 2005

Al-Yousif A, Neelamani S, Levinson AV (2019) Wave transmission, reflection, and dissipation of scrap tire floating breakwater in random waves. ISH J Hydraul Eng. https://doi.org/10.1080/09715010.2019.1664342

Barbier EB, Koch EW, Silliman BR, Hacker SD, Wolanski E, Primavera J, Granek E, Polasky S, Aswani S, Cramer LA, Stoms DM, Kennedy CJ, Bael D, Kappel CV, Perillo GM, Reed DJ (2008) Coastal eco system–based management with nonlinear ecological functions and values. Science 319:321–323

Bird ECF (1985) Coastline changes: a global review. John Wiley and Sons, Chichester, 219 pp

Buitrago NR, Neal WJ, Bonetti J, Anfuso G, de Jonge VN (2020) Vulnerability assessments as a tool for the coastal and marine hazards management: an overview. Ocean Coast Manag 189:105134. https://doi.org/10.1016/j.ocecoaman.2020.105134. ISSN 0964-5691

de Andrade TS, de Oliveira Sousa PHG, Siegle E (2019) Vulnerability to beach erosion based on a coastal processes approach. Appl Geogr 102:12–19. https://doi.org/10.1016/j.apgeog.2018.11.003. ISSN 0143-6228

ERG (Eastern Research Group) (2010) Economic and social effects of changing shorelines in the North Atlantic region. NOAA Coastal Services Center, Charleston

Fitton JM, Hansom JD, Rennie AF (2018) A method for modelling coastal erosion risk: the example of Scotland. Nat Hazards 91:931–961. https://doi.org/10.1007/s11069-017-3164-0

Gomez MLA, Adelegan OJ, Ntajal J, Trawally D (2020) Vulnerability to coastal erosion in The Gambia: empirical experience from Gunjur. Int J Disaster Risk Reduct 45:101439. https://doi.org/10.1016/j.ijdrr.2019.101439. ISSN 2212-4209

Hinkel J, Nicholls RJ, Tol RSJ, Wang ZB, Hamilton JM, Boot G, Vafeidis AT, McFadden L, Ganopolski A, Klein RJT (2013) A global analysis of erosion of sandy beaches and sea-level rise: an application of DIVA. Glob Planet Chang 111:150–158

Kamphuis JW (2000) Introduction to coastal engineering and management (Advanced series on ocean engineering). World Scientific Publishing Co. Pvt. Ltd., Singapore. ISBN-13: 978-9812834850

Neelamani S, Ljubic J (2017) Experimental study on the hydrodynamic performance of floating pontoon type breakwater with skirt walls. J Offshore Mech Arct Eng 140(2):021303. https://doi.org/10.1115/1.4038343, 9 pages. Paper No. OMAE-17-1029

Neelamani S, Saif ud din, Anbazagan S (2011) Kuwait coastline evolution during 1989–2007. In: Book chapter in Geoinformatics in applied geomorphology, chapter 5. CRC Press, Taylor & Francis, USA, pp 87–104

Nordhaus WD (2006) The economics of hurricanes in the United States, vol No. w12813. National Bureau of Economic Research, Cambridge

Rakha K, Al-Salem K, Neelamani S, Al-Nassar W, Al-Banaa K (2005) Interactive coastal information system for Kuwait's territorial waters, phase I: hindcasting of waves, water levels, and currents. EC026C. Final report KISR 7872

Saaty TL (2008) Decision making with the analytic hierarchy process. Int J Serv Sci 1(1):83–98

Shore Protection Manual (1984). Volume II. Department of the Army. Corps of Engineers, USA

Soulsby RL (1997) Dynamics of marine sands. A manual for practical applications. Thomas Telford, London

Williams A, Buitrago NGR, Pranzini E, Anfuso G (2017) The management of coastal erosion. Ocean Coast Manag. https://doi.org/10.1016/j.ocecoaman.2017.03.022

World Resources Institute (2010) Decision making in a changing climate. United Nations Development Programme World Bank and World Resources Institute, Washington

Zhu ZT, Cai F, Chen SL, Gu DQ, Feng AP, Cao C, Qi HS, Lei G (2019) Coastal vulnerability to erosion using a multi-criteria index: a case study of the Xiamen Coast. Sustainability 11:93. https://doi.org/10.3390/su11010093

Chapter 19
Total Suspended Matter Variability in Response to Tropical Cyclone *Titli* Along Coastal Waters of Southeast India Using Satellite Observations: Implications to Climate Change

Sravanthi Nukapothula, Ali P. Yunus, and Chuqun Chen

Abstract A tropical cyclone passing over an ocean surface can mix the water column and cause the resuspension of bottom sediments in the continental shelf regions. The impact of tropical cyclone-induced total suspended matter (TSM) concentrations in the coastal waters of India has rarely been acknowledged. Here, we used near-daily images from the ocean color monitor (OCM) onboard OceanSAT-2, to understand the variability of TSM distributions in response to tropical cyclones along the coast and in open ocean waters of southern India. Our study reports a high amount of TSM (~17 mg/l) during the cyclone overpass, which is significantly higher than the TSM during the normal period. The excess precipitation, river discharge, and resuspension of sediment by stronger winds are attributed to the increased TSM concentration. With the increasing number of tropical cyclones in the Indian Ocean due to global climatic change, it is expected to have great impacts on the variability of ocean color parameters and biogeochemical cycles.

S. Nukapothula (✉)
Marine Science and Technology, Zhejiang Ocean University, Zhoushan, China

State Key Laboratory of Tropical Oceanography, Guangdong Key Lab of Ocean Remote Sensing, South China Sea Institute of Oceanology, Chinese Academy of Sciences, Guangzhou, China

A. P. Yunus
Department of Earth and Environmental Sciences, Indian Institute of Science Education and Research, Mohali, Punjab, India

C. Chen
State Key Laboratory of Tropical Oceanography, Guangdong Key Lab of Ocean Remote Sensing, South China Sea Institute of Oceanology, Chinese Academy of Sciences, Guangzhou, China

Southern Marine Science and Engineering Guangdong Laboratory, Guangzhou, China

N. Jayaraju et al. (eds.), *Coasts, Estuaries and Lakes*,
https://doi.org/10.1007/978-3-031-21644-2_19

Keywords Total suspended matter · Tropical cyclone · River discharge · OCM · Southeast coast of India

19.1 Introduction

Tropical cyclones affect coastal regions of the world's oceans with strong winds and large amounts of rainfall on the sea and land (D'sa et al. 2011; Glenn et al. 2016). These extreme events significantly impact ocean dynamics and cause physical and biogeochemical changes in the oceans by increasing river discharge and wind forcing, causing upwelling and entrainment (Babin et al. 2004; Dhillon and Inamdar 2013). The appearance of chlorophyll blooms and sediment resuspension due to strong wind forcing are widely documented (see, e.g., Nukapothula et al. 2018). The other implications of cyclonic winds on ocean physical processes are associated with decreased sea surface temperature. For example, Lotliker et al. (2014) reported an increase in Chl-a and a decrease in SST by mixing the surface water layer to depths during Tropical Cyclone Phailin in 2013. The resuspension of total suspended matter by processes of entrainment from subsurface water results in an increasing amount of nutrient influx and dissolved organic consumption due to high vertical mixing in the euphotic zone during the cyclone (Tomasko et al. 2006; Balaguru et al. 2015). These responses, as a result of tropical cyclones, have a significant impact on marine productivity and the biochemical cycle (Gao et al. 2015).

Tropical cyclones also increase the amount of dissolved organic matter on continental shelves, gulfs, and inland waters (Fennel et al. 2013; Bianucci et al. 2018). Furthermore, such cyclones enriched by erosion along the coast and river basins and estuary mouths lead to higher amounts of particulate matter and dissolved organic carbon, resulting in significant increases in particulate matter concentrations that play an important role in biochemical oxygen (Burkholder et al. 2004). For example, the high amount of river discharge with extremely high precipitation during tropical cyclones leads to hypoxic events (Almroth-Rosell et al. 2012; Moriarty et al. 2017). Furthermore, large amounts of precipitation rapidly transport dissolved organic carbon from the terrestrial atmosphere to the coastal shelf regions. Selvaraj et al. (2015) show that a huge amount of organic carbon from terrestrial land is remineralized, which changes the net sink to a source of dissolved carbon during events. There could be more remineralization, which could lead to more organic dissolved matter consumption in coastal waters (Moriarty et al. 2018). With such changes induced during cyclonic storms and sediment fluxes that impact the nutrient cycle, monitoring SSC in coastal waters deserves attention. However, continuous monitoring of coastal waters for sediment flux by point measurements is difficult and expensive.

Alternatively, satellite remote sensing reflectance measurements are presented in this work to overcome the limitations in temporal and spatial resolution and provide

TSM distributions throughout the coastal area during and after the cyclonic storm. Since the inception of optical satellites into earth's orbit, remote sensing-based TSM retrieval has widely been adapted to monitor oceanic and inland water bodies (Nukapothula et al. 2019). Satellite sensors such as OCTS, SeaWiFs, MODIS, VIIRS, and GOCI give a near-high spatial synoptic of TSM mapping on a daily time scale (Nukapothula et al. 2018; Jie et al. 2020). However, many of these sensors are currently inactive or have sensor degradation issues due to aging. As a result, our research takes advantage of OceanSat-2 having a high spatial resolution ocean color monitor (OCM) with 360 m spatial resolution and a 2-day return period (Sravanthi et al. 2013).

Although several studies have used Chl-a satellite retrieval to study primary productivity and algae blooms, very little study has been published on the total variability of suspended matter in response to tropical storms in the Indian Ocean. In particular, ocean dynamics with the association of river runoff and rainfall remains comparatively unexplored on the southeast coast of India (Fig. 19.1). This is mainly due to the lack of cloud-free images of high-resolution satellite data during the tropical cyclone overpass. In this study work, we carried out the analysis of quantitative estimations of TSM in spatial and temporal distributions from high-resolution

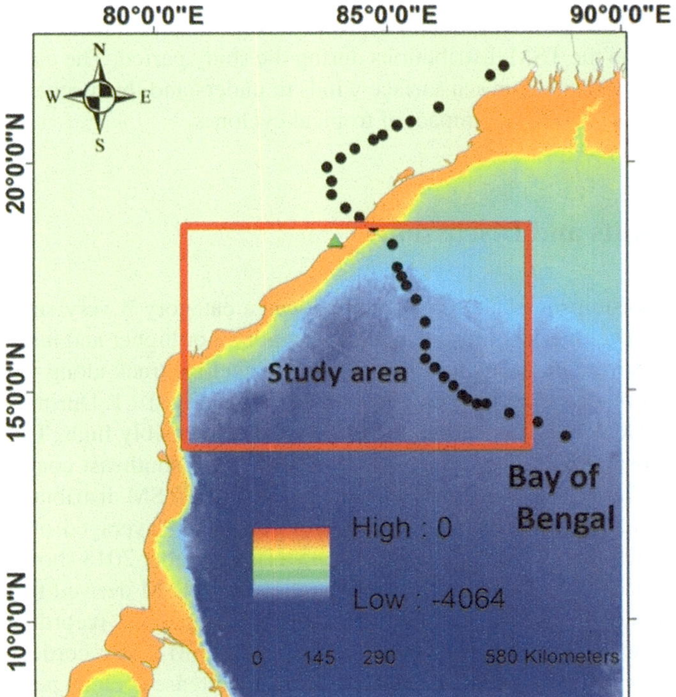

Fig. 19.1 Track of the tropical cyclone *Titli* on a bathymetry map. The green-filled triangle is presented as the location of river discharge data from field observations

OCM images in response to the cyclone *Titli* along the coastal waters of southeast India. In addition, the sediment dynamics are also examined and discussed using river runoff, rainfall, and wind forcing variables.

19.2 Data and Methods

Daily level-3 TSM images derived from ocean color monitor (OCM) data were obtained from INCOIS live las (https://las.incois.gov.in/) during the tropical cyclone period from October 5 to October 16, 2018, for the southeast coast of India. The Oceansat-2 OCM sensor provides eight spectral channels in visible and near-infrared wavelengths (412–865 nm) with 360 m spatial resolution and a 2-day review cycle (Chauhan et al. 2002). The high-resolution OCM sensor has allowed for a quantitative measurement of primary production, coastal zone management, and sediment dynamics along coastal and offshore waters at global and regional scales (Sravanthi et al. 2013). The 3-day composite images were mapped for TSM from daily OCM images to avoid cloud cover issues. The daily river discharge data products were obtained from the Central Water Commission, India, during the cyclone period. Furthermore, sea surface winds derived from an advanced scatterometer (ASCAT) with a spatial resolution of 0.25° were obtained to examine the wind forcing on the TSM distributions during the study period. The computation of wind stress is derived from sea surface winds to understand the mechanism of sediment resuspension with the impact of tropical cyclones.

19.3 Results and Discussion

On the Saffir-Simpson scale, Cyclone *Titli* was a category 3 very severe cyclone storm. Major cyclones are those that reach category 3 or higher and have the potential to cause severe damage and loss of life. The cyclone track along with the geographic location of the cyclone movement is shown in Fig. 19.1. During the cyclone period (i.e., October 8–13, 2018), the cloud cover is sensibly high. Therefore, the 3-day composite OCM-derived TSM variability of the southeast coast of India is presented to show the spatial and temporal variability of TSM distributions induced by the tropical cyclone *Titli*. The time series daily TSM averaged over the study region (80.63°E–88.11°E and 13.08°N–18.35°N) in October 2018 shows the condition of the variability of pre-cyclone and post-cyclone TSM derived from OCM in the study area (Fig. 19.2a). The high TSM value (17 mg/l) was recorded during the cyclonic storm. A very low (8 mg/l) TSM value of TSM was recorded during the pre-cyclone period (October 4–7, 2018). During the intense cyclone period, TSM in the coastal and open ocean is >15 mg/l, with relatively higher values along the coast. In addition, moderate and high levels of TSM values, which range from 10 to 17 mg/l, are recorded in the open ocean, clearly indicating the signature of sediment

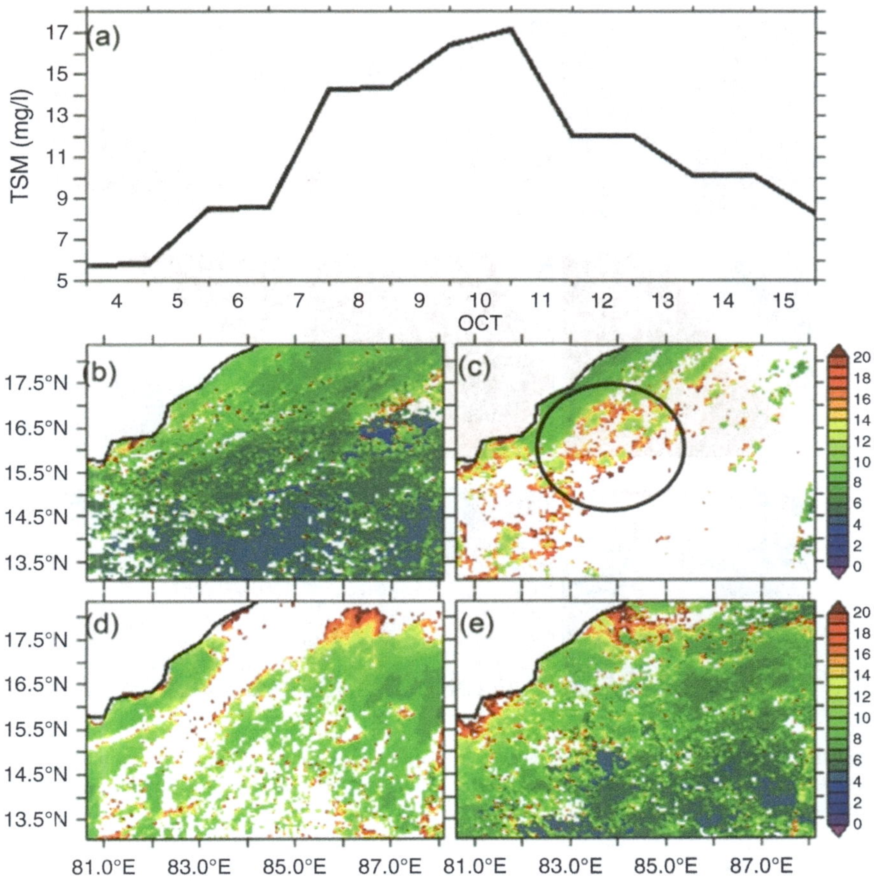

Fig. 19.2 Time series of total suspended matter (mg/l) (**a**); and spatiotemporal distribution of total suspended matter (TSM in mg/l) in coastal waters of southeast India using OCM before, during, and after the *Titli* cyclone (**b–e**)

resuspension from strong winds and extreme precipitation. After the passage of the cyclone (October 14–16, 2018), the TSM values in the study area decreased to less than 10 mg/l.

There are two main sources of surface TSM over the shelf during and after the tropical cyclone: terrestrial sediment carried into coastal waters by river discharge from major rivers and suspended sediment from bottom shear, which is generated by strong winds in response to the cyclone (Chang et al. 2001; Walker and Rabalais 2006). To understand the impact of river discharge and precipitation on sediment sources and the effects on the spatial distribution of TSM during a cyclone period, we analyzed maps of precipitation distributions before, during, and after the cyclone in the study region. The day-to-day distributions of precipitation based on GPCP in response to a cyclone are shown in Fig. 19.3.

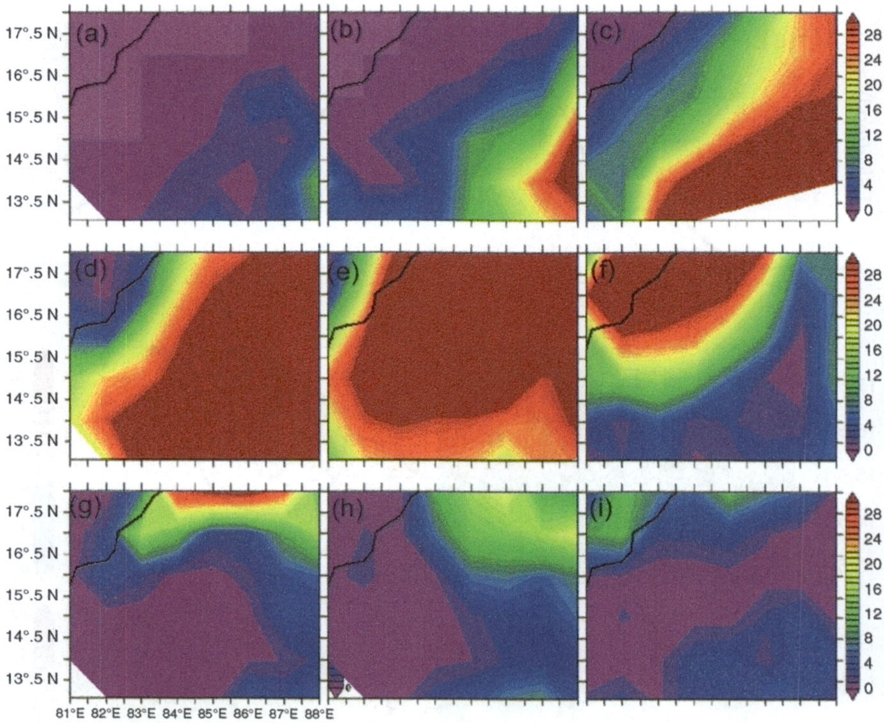

Fig. 19.3 Spatial and temporal variability of daily accumulated precipitation (mm/day) from GPCP satellite observations (**a–i**). A very high amount of precipitation was recorded during the cyclone period (**d–f**)

A high peak of precipitation (>30 mm/day) was recorded in the southeast region during a cyclone period, which is clearly seen from the GPCP data. The amount of precipitation in the open ocean was moderate on October 7, 2018 (i.e., on the day of the arrival of the tropical cyclone). There is an indication of the premature stage of the formation of a cyclone in the Bay of Bengal. On October 8, 2018, precipitation gradually increased towards the coast and high precipitation (30 mm/day) was recorded during the intense stage of the cyclone (October 8–13, 2018). From Fig. 19.3, it is clearly seen that the amount of precipitation gradually reduced in the study area after the passage of the cyclone.

To examine the dynamics and understand the mechanism of variability in TSM in response to a cyclone, we mapped the wind speeds overlaid with wind fields in the study area. Green and Coco (2014) suggested that suspended sediments are prone to resuspension by various factors along shelf waters and in the open ocean. Predominantly during extreme meteorological conditions such as tropical cyclones, strong wind forces increase their suspension of sediments from the bottom. For example, a study by Palinkas et al. (2014) reported that the wave bottom stress stretched to 7 Pa due to wave-induced bottom stresses during the intense period of

Hurricane Irene, which is typically higher than the normal value (0.2 pa). Figure 19.4 shows the day-to-day spatial distributions of wind speed before, during, and after the cyclone in the study region.

According to Fig. 19.4, a very low wind speed (i.e., <6 m/s) is recorded during the precyclonic storm, and wind speeds gradually increase from the open ocean prior to the formation of the cyclone. During the cyclone (October 8–13, 2018), stronger and higher wind speeds (>12 m/s) with a clear signature of the cyclonic circulation were recorded in the open ocean. The day-to-day movement of cyclonic structure from the open to the coastal region was clearly observed. The relatively low wind speed, with a range of 2–4 m/s in the study region, which is a signature of the effect of cyclone impact, was reduced after the cyclone dissipated.

Sediments from rivers are the main factor controlling variations in suspended sediments, and there is a nonlinear relationship between river discharge and sediments (Dai et al. 2016; Li et al. 2017). Studies have found that the effects of induced precipitation and river discharge by tropical cyclones depending on the magnitude of the cyclone (Darby et al. 2016; Yu et al. 2017). Here, we examine the time series plot of precipitation, river discharge, and wind stress along with TSM before, during, and after the cyclone. A clear signature of a high amount of discharge (750 cumecs), increased water levels (10 m), along with high wind stress, was observed

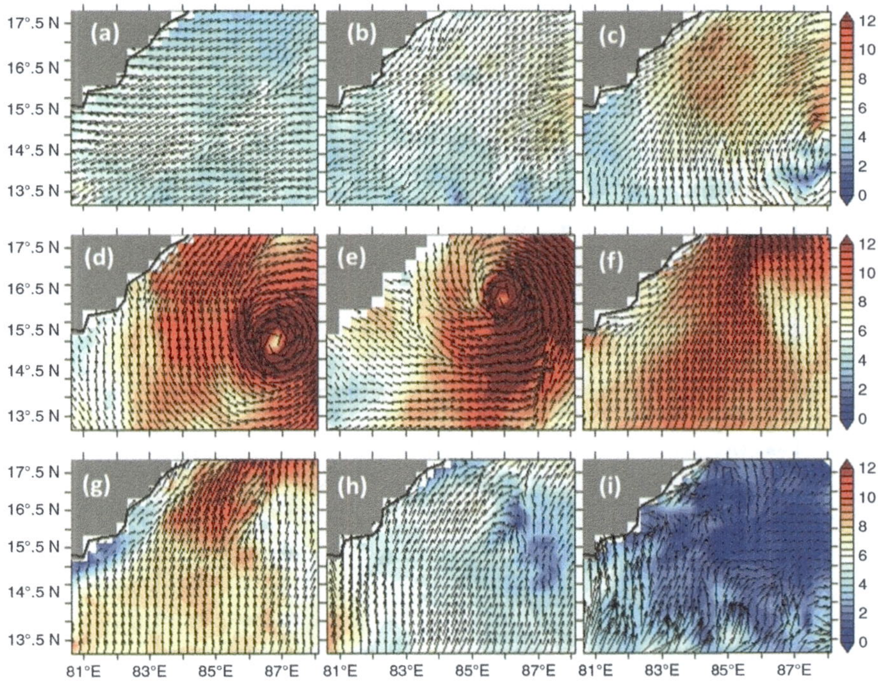

Fig. 19.4 Wind vectors plotted over wind magnitude (m/s) derived from ASCAT observations before (**a**, **b**), during (**c**–**g**), and after (**h**–**i**) the cyclone

during the cyclone period. The impact of these parameters on the TSM distributions was reduced before and after the cyclone in the study region (Fig. 19.5). Our study confirmed the results of Tang et al. (2021) that focused on the sediment dynamics induced by the tropical cyclone in the Gulf of Beibu using the available field observations. Our results indicate that sediment fluxes and dynamics in coastal and offshore waters of southeast India are driven by the tropical cyclone. Cross-shelf suspended sediment fluxes, while high precipitation and stronger wind forcing do play an important role in the discharge of river runoff, drive the terrestrial sediment export to the off-the-shelf region. The findings of Dufois et al. (2018), who highlighted the role of extreme cyclones in exporting sediments off the shelf, confirm that spatial and temporal observations indicate that there is a consistent feature of tropical cyclones on sediment dynamics and sediment resuspension. The results in the study area also suggest that a large portion of the sediments are resuspended due to the strong wind forcing in the off-shore shelf region (Fig. 19.5). During the peak of the tropical period, changes in the coastal waters of the TSM in the study area were generally found to be driven by river discharge from the main river systems and local wave-induced resuspension. However, in offshore waters, the enhancement of resuscitation dynamics is mainly induced by bottom resuspension from cyclonic storm impact.

Since the intensity and frequency of tropical cyclones have increased over the past 40 years owing to the climate warming, studies analyzing their direct impact on the ocean biogeochemical cycle are necessary and worthy of attention. A study by Schultze et al. 2020 showed that extreme events (e.g., tropical cyclones) have been

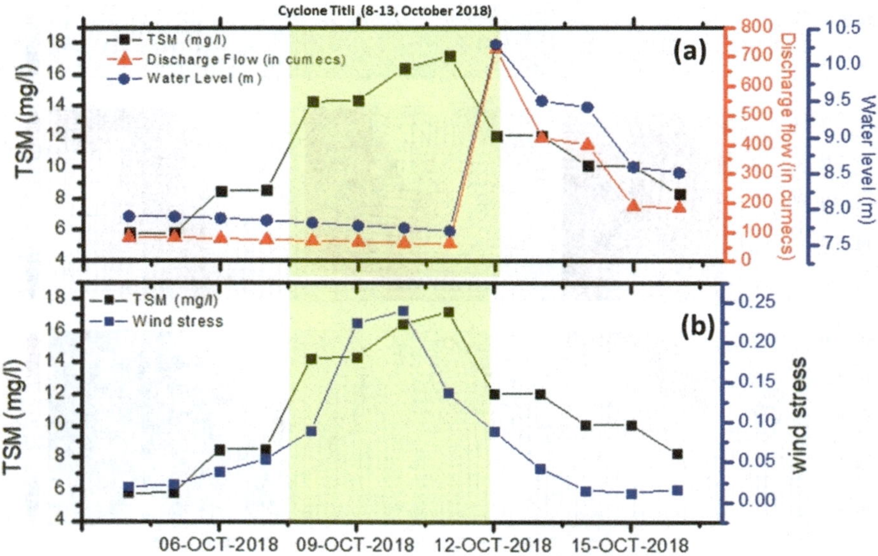

Fig. 19.5 Time series of TSM along with river discharge and water level (**a**); TSM with wind stress (**b**) over the study region

shown to modulate the ocean color variability. Our study of quantitative estimations of TSM during cyclone *Titli* based on ocean color measurements can provide vital information on primary productivity and coastal water composition in the context of climate change scenarios, as more such events will occur frequently in the near future.

19.4 Conclusions

The study analyzed the variations of total suspended matter induced by the tropical cyclone *Titli* using high-resolution satellite OCM images off the southeast coast of India. The high concentration of TSM concentrations (>17 mg/l) is higher than before and after the passage of the cyclone that was recorded during the cyclone period (October 8–13, 2018). The spatial distributions of the TSM concentrations observed by satellite indicate that an increase in TSM concentration was observed primarily along the coast and open ocean over the entire tropical cyclone-impacted region. Daily water discharge and water levels respond to the intense high amounts of precipitation in the study region. The impacts of the cyclone-induced high amounts of rainfall and river discharge of terrestrial material from rivers change the variability of TSM concentrations. Furthermore, the resuspension of TSM was induced by high wind stress and is attributed to the enhancement of TSM concentrations. Therefore, our study suggests that sediments increase by the resuspension process due to high precipitation, high amounts of sediment to the coast, and their suspension of strong winds that significantly affect the spatial distribution patterns of TSM in the area impacted in response to cyclone *Titli* on the southwest coast of India.

Acknowledgments This work was supported by the Key Special Project for the National Natural Science Foundation of China (No. U1901215) and the Introduced Talents Team of Southern Marine Science and Engineering Guangdong Laboratory (Guangzhou) (GML2019ZD0305). We thank the Live Access Server (LAS) of the Indian National Centre for Ocean Information Services (INCOIS) for the TSM data derived from Oceansat-2. River discharge is downloaded from the India Water Resources Information System. The authors also thank the Asia Data Pacific Centre for providing satellite data products for rainfall products from GPCP and wind fields from ASCAT.

References

Almroth-Rosell E, Tengberg A, Andersson S, Apler A, Hall POJ (2012) Effects of simulated natural and massive resuspension on benthic oxygen, nutrient and dissolved inorganic carbon fluxes in Loch Creran, Scotland. J Sea Res 72:38–48

Babin SM, Carton JA, Dickey TD, Wiggert JD (2004) Satellite evidence of hurricane-induced phytoplankton blooms in an oceanic desert. J Geophys Res 109:C03043

Balaguru K et al (2015) Dynamic potential intensity: an improved representation of the ocean's impact on tropical cyclones. Geophys Res Lett 42:6739–6746

Bianucci L, Balaguru K, Smith RW, Leung R, Moriarty JM (2018) Contribution of hurricane-induced sediment resuspension to coastal oxygen dynamics. 8:15740. https://doi.org/10.1038/s41598-018-33640-3

Burkholder J et al (2004) Comparative impacts of two major hurricane seasons on the Neuse River and western Pamlico Sound ecosystems. Proc Natl Acad Sci U S A 101:9291–9296

Chang GC, Dickey TD, Williams AJ (2001) Sediment resuspension over a continental shelf during Hurricanes Edouard and Hortense. J Geophys Res Oceans 106:9517–9531

Chauhan P, Mohan M, Sarangi RK, Kumari B, Nayak S, Matondkar SGP (2002) Surface chlorophyll estimation in the Arabian Sea using IRS-P4 OCM Ocean Color Monitor (OCM) satellite data. Int J Remote Sens 23(8):1663–1676

D'sa EJ, Korobkin M, Ko DS (2011) Effects of Hurricane Ike on the Louisiana–Texas coast from satellite and model data. Remote Sens Lett 2:11–19

Dai ZJ, Fagherazzi S, Mei XF, Gao JJ (2016) Decline in suspended sediment concentration delivered by the Changjiang (Yangtze) River into the East China Sea between 1956 and 2013. Geomorphology 268:123–132

Darby SE, Hackney CR, Leyland J, Kummu M, Lauri H, Parsons DR, Best JL, Nicholas AP, Aalto R (2016) Fluvial sediment supply to a mega-delta reduced by shifting tropical-cyclone activity. Nature 539:276–279

Dhillon GS, Inamdar S (2013) Extreme storms and changes in particulate and dissolved organic carbon in runoff: entering uncharted waters? Geophys Res Lett 40:1322–1327

Dufois F, Lowe RJ, Rayson MD, Branson PM (2018) A numerical study of tropical cyclone-induced sediment dynamics on the Australian North West Shelf. J Geophys Res Oceans 123:5113–5133. https://doi.org/10.1029/2018JC013939

Fennel K, Hu J, Laurent A, Marta-Almeida M, Hetland R (2013) Sensitivity of hypoxia predictions for the northern Gulf of Mexico to sediment oxygen consumption and model nesting. J Geophys Res Oceans 118:990–1002

Gao JJ, Dai ZJ, Mei XF, Ge ZP, Wei W, Xie HL, Li SS (2015) Interference of natural and anthropogenic forcings on variations in continental freshwater discharge from the Red River (Vietnam) to sea. Quat Int 380-381:133–142

Glenn SM et al (2016) Stratifed coastal ocean interactions with tropical cyclones. Nat Commun 7:10887

Green MO, Coco G (2014) Review of wave-driven sediment resuspension and transport in estuaries. Rev Geophys 52:77–117. https://doi.org/10.1002/2013RG000437

Jie W, Chuqun C, Sravanthi N (2020) Atmospheric correction of GOCI using quasi-synchronous VIIRS data in highly turbid coastal waters. Remote Sens 12(1):89. https://doi.org/10.3390/rs12010089

Li SS, Dai ZJ, Mei XF, Huang H, Wei W, Gao JJ (2017) Dramatic variations in water discharge and sediment load from Nanliu River (China) to the Beibu Gulf during 1960s–2013. Quat Int 440:12–23

Lotliker AA, Srinivasa Kumar T, Reddem VS, Nayak S (2014) Cyclone Phailin enhanced the productivity following its passage: evidence from satellite data. Curr Sci 106(3):360–361

Moriarty JM et al (2017) The roles of resuspension, diffusion and biogeochemical processes on oxygen dynamics of shore of the Rhône River, France: a numerical modeling study. Biogeosciences 14:1919–1946

Moriarty JM, Harris CK, Friedrichs MA, Fennel K, Xu K (2018) Impact of seabed resuspension on oxygen and nitrogen dynamics in the northern Gulf of Mexico: a numerical modeling study. J Geophys Res Oceans. https://doi.org/10.1029/2018JC013950

Nukapothula S, Chen C, Yunus Ali P, Wu J (2018) Satellite-based observations of intense chlorophyll-a bloom in response of cold core eddy formation: a study in the Arabian Sea, Southwest Coast of India. Reg Stud Mar Sci 24:303–310

Nukapothula S, Chen C, Wu J (2019) Long-term distribution patterns of remotely sensed water quality parameters in Pearl River Delta, China. Estuar Coast Shelf Sci 221:90–103. https://doi.org/10.1016/j.ecss.2019.02.03

Palinkas CM, Halka JP, Li M, Sanford LP, Cheng P (2014) Sediment deposition from tropical storms in the upper Chesapeake bay: field observations and model simulations. Cont Shelf Res 86:6–16

Schultze LKP, Merckelbach LM, Carpenter JR (2020) Storm-induced turbulence alters shelf sea vertical fluxes. Limnol Oceanogr Lett. https://doi.org/10.1002/lol2.10139

Selvaraj K et al (2015) Stable isotopic and biomarker evidence of terrigenous organic matter export to the deep sea during tropical storms. Mar Geol 364:32–42

Sravanthi N, Ramana IV, Yunus Ali P, Ashraf M, Ali MM, Narayana AC (2013) An algorithm for estimating suspended sediment concentrations in coastal waters of India using remotely sensed reflectance and its application to coastal environments. Int J Environ Res 7(4):841–850

Tang R, Dai Z, Zhou X, Li S (2021) Tropical cyclone-induced water and suspended sediment discharge delivered by mountainous rivers into the Beibu Gulf, South China. Geomorphology 389:107844. https://doi.org/10.1016/j.geomorph.2021.10784

Tomasko DA, Anastasiou C, Kovach C (2006) Dissolved oxygen dynamics in Charlotte Harbor and its contributing watershed, in response to hurricanes Charley, Frances, and Jeanne—impacts and recovery. Estuar Coasts 29:932–938

Walker N, Rabalais N (2006) Relationships among satellite chlorophyll a, river inputs, and hypoxia on the Louisiana continental shelf, Gulf of Mexico. Estuar Coasts 29:1081–1093

Yu ZF, Wang YQ, Xu HM, Davidson N, Chen YD, Chen YM, Yu H (2017) On the relationship between intensity and rainfall distribution in tropical cyclones making landfall over China. J Appl Meteorol Climatol 56(10):2883–2901

Chapter 20
Climate Change and Its Impact on Depletion of Oxygen Levels on Coastal Waters and Shallow Seas

Mohammad Afsar Alam

Abstract If there is insufficient oxygen on our globe, aquatic habitats, people's health, and the biosphere as a whole may be jeopardized. All water bodies are facing the problem of oxygen depletion because of the rising temperatures. Ocean's biological and biogeochemical activities need oxygen. According to actual measurements taken at various locations worldwide, oxygen-minimum zones have spread many million square kilometers and disrupted the nutrient cycle at plenty of coastal sites where the amount of oxygen is critical to support animal population dispersal. This study is primarily based on secondary sources of data gathered from government agencies, academic journals, books, and other magazines, as well as browsing via the Internet. Human-caused global warming is the leading source of marine oxygen depletion, according to a study. Toxic wastes from humans are also a factor in coastal zones since they are dumped into the sea. Both the Arctic and equatorial oceans noticed an abrupt fall in the amount of oxygen. This research is concerned with the global spatial distribution of oxygen in coastal waters and shallow seas. It also explains why oxygen loss is unevenly distributed over the earth. Oceanic oxygen levels, we believe, will be the next great fatality of climate change. To reverse this trend, we must expedite ocean-based climate solutions that increase oxygen levels, such as those outlined at COP26.

Keywords Climate change · Coastal water · Oxygen level · Shallow seas

20.1 Introduction

There must be enough supply of oxygen for the existence of life on the planet otherwise Ocean ecosystems, people's health, and the ecological balance as a whole could be at risk. For the three billion people who rely on coastal fishing as a source

M. A. Alam (✉)
Department of Social Sciences, Fiji National University, Suva, Fiji
e-mail: mohammad.alam@fnu.ac.fj

N. Jayaraju et al. (eds.), *Coasts, Estuaries and Lakes*,
https://doi.org/10.1007/978-3-031-21644-2_20

of income, the decline in ocean oxygen levels is devastating. We are entering a period where climate change is causing a chain reaction in our environment, not merely an increase in temperature. This has implications for the oceans as a food source and environment for the ocean's ecosystem. People that depend on the sea for their livelihood would be adversely affected by the rising incidence of oxygen-depleted dead zones. There has been a great deal of discussion about global warming, precipitation, sea-level rise, and ocean acidification as a result of human-caused climate change, but little has been said about the influence of these factors on air and water oxygen levels. CO_2 traps radiation in the atmosphere, warming the air, but it also warms water by doing the same. In the Anthropocene, the oceans have absorbed 90% of the additional heat created by global warming, which is a complicated and intertwined relationship. Scientists studying the ocean and atmosphere climate expect ocean oxygen levels to plummet as a result of rising temperatures. We need to build on the enthusiasm of the recent COP26 meeting and widen our focus on the perilous status of oceanic oxygen levels, the life-support system of our planet.

As mentioned above, oceans have absorbed around 90% of the surplus heat generated by climate warming throughout the Anthropocene. Several variables contribute to the environmental challenges. Of these is the heating of the seas, rivers, and oceans, which results in a decrease in oxygen content. A decrease in water's oxygen content causes the death of hundreds and thousands of marine organisms globally. Say for instance, in July 2021 in Madeira Beach of Florida plenty of dead fish are found floating in the Bay of Boca Ciega. Deoxygenation is the main cause of such deaths.

More than 100 miles of Florida's coastal seas became an oxygen-depleted dead zone during the summer of 2020, littered with fish that could be seen in Tampa Bay. On the other side of the country, Dungeness crabs were washing up on the Oregon coast, unable to flee water that had become seasonally depleted of oxygen in dramatic occurrences during the preceding two decades. Larger marine organisms need the availability of oxygen even though just 0.6% of atmospheric oxygen comes from marine sources. In addition, organisms' respiration requires oxygen constantly throughout the water. Since 1960, an average of 2% oxygen loss has been found in global seas, as per the recent investigation of oxygen fluctuations. There is a wide variation in the amount of deoxygenation among oceans, according to the research. There has been a dramatic decrease in oxygen levels throughout the tropics and Arctic. (Pullen and Goodkin 2021).

Health and environmental quality are just as vital as meeting the most basic needs of people. Several variables contribute to environmental challenges. Water bodies such as lakes and oceans are becoming warmer and less oxygen-rich as a result (Alam and Alam 2020). In recent studies, the reduction of marine oxygen is attributed mostly to human-caused global warming. It's hard to tell which of the various systems at play is responsible for human activity in coastal locations, where humans provide nutrients to the seas.

Global warming affects the ocean and its dissolved oxygen levels in several ways. Other properties of this chemical include its ability to be dissolved in water. As the temperature rises, so does the quantity of gas that may be dissolved in the water. Ocean surface layers that have just recently come into contact with air have been the principal focus of this process up to this point in time. As much as 20% of marine oxygen loss has been attributed to this phenomenon, with 50% of that loss happening in the ocean's top 1000 m. Climate change is causing surface waters to be mixing with those that lack oxygen, resulting in an imbalance. As a result, it affects the quantity of marine oxygen needed by organisms, amongst many other things.

Temperature rises have a direct impact on the export of biomass that may be used for respiration over the ocean's surface, which in turn affects nutrient availability and productivity (Oschlies et al. 2018).

20.2 Objectives of the Study

20.2.1 The Present Study Has the Following Objectives

1. To get a deeper understanding of how climate change influences the oxygen levels in coastal waters and shallow seas.
2. To investigate the loss of oxygen in coastal waters and shallow seas and their effect on marine life.
3. To estimate the emerging outcomes in oxygen levels, it is necessary to determine the required mitigating methods.

20.3 Research Methodology

This research is descriptive and relies on secondary data sources. The main data was gathered from unpublished and published books and also by browsing the internet. Besides, some research papers in various national and international journals and articles by climate scientists published in the newspapers were also used for this purpose.

20.4 Background Literature

The oxygen solubility in the ocean decreases as it warms. Solubility decline is predicted to contribute to >50% of oxygen loss in the ocean's top thousand meters and 15% of global oxygen loss (Schmidtko et al. 2017; Helm et al. 2011). In marine ecosystems, ocean deoxygenation is one of the most significant trends (Rabalais et al. 2010; Levin and Breitburg 2015; Zhang et al. 2010). Biogeochemical cycles, productivity, and biodiversity are all influenced by the amount of oxygen in the ocean. Warm temperatures and oxygen-deficient waters have been linked to significant extinction events in Earth's history (Norris et al. 2013), and on existing trajectories, during the next thousand years, the oceans might face an acute shortage of oxygen because of irrational human activities (Watson 2016). Over the last 50 years, 77 billion metric tons, that is, about 2% of oxygen lost, have been observed in the open oceans (Schmidtko et al. 2017) (Fig. 20.1).

The volume of oxygen-depleted water (anoxic) has more than doubled, and open-ocean oxygen minimum zones (OMZs) have grown by approximately equal to the whole of Europe, that is, 4.5 million square kilometers (Stramma et al. 2010; Chan et al. 2008). Oxygen-depleted water upwelling has become more frequent and severe along specific beaches, with potentially disastrous biological implications (Riemann et al. 2016). Health and environmental quality are just as vital as meeting people's most basic needs. Several variables contribute to environmental challenges. Water bodies such as lakes and oceans are becoming warmer and less oxygen-rich (Alam and Alam 2020). It is not only nutrient loads that have been reduced but also chlorophyll levels that are susceptible to nutrient enrichment. Oxygen diminishes in particular coastal systems (Rabalais et al. 2010).

Low or non-existent levels of oxygen in the atmosphere are caused by inadequate long-term oxygen supply and photosynthesis. These systems include OMZs, deep fjords, coastal upwelling zones, deep basins, eddies, and shallow, productive seas with the limited flow (Karstensen et al. 2015). Oxygen extraction, transport, and storage have developed in Eukaryote organisms living in low-oxygen environments to sustain aerobic metabolism and minimize energy consumption (Seibel 2011). Adaptation to hypoxic and anoxic settings is facilitated by metabolic depression (Gallo and Levin 2016) and high H2S tolerance.

Metabolic rates rise as a result of increased temperature, which also increases oxygen use. Remineralization moves to shallower depths due to the faster decomposition of sinking particles. However, the amount of oxygen loss does not alter due to the geographic redistribution. Human activity has created some of the most

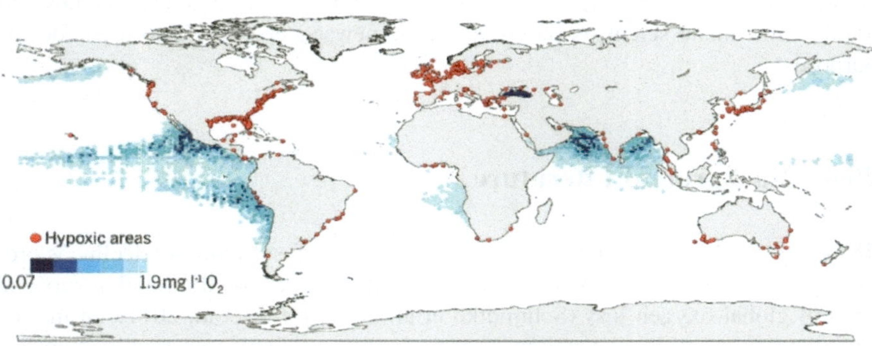

Fig. 20.1 Coastal locations with O_2 below 2 mg l^{-1} (63 mmol l^{-1}) due to anthropogenic fertilizers (red dots) and ocean oxygen minimum zones (blue shaded patches) (blue shaded areas). (*Source*: Breitburg et al. (2018). Also available on World Ocean Atlas 2009)

dangerous anthropogenic threats to marine ecosystems across the globe, including coastal dead zones caused by low dissolved oxygen levels. Every decade since the mid-1900s, the number of such incidents has doubled (Rabalais et al. 2010; Gooday et al. 2009; Vaquer-Sunyer and Duarte 2008; Diaz and Rosenberg 2008; Diaz 2001). When you consider how rapidly dead zones are growing in quantity as well as how large and destructive they are, it's clear that limiting eutrophication is a good idea (Rabalais et al. 2010; Gooday et al. 2009; Diaz and Rosenberg 2008). There have been several grazer population extinctions as a result of reduced phytoplankton consumption at the same time fertilizers are driving up primary output (Lotze et al. 2006; Jackson et al. 2001).

Water column stratification and runoff, primary production, microbiological activity, as well as respiration by organisms all affect hypoxia severity. There are several indicators suggesting climate change is to blame for the observed increase of dead zones, including increasing temperatures, acidification of the oceans, rising sea level, rainfall, winds, and frequency of storms (Fig. 20.2). In the initial days of climate research, several studies looked at how changes in precipitation may alter the freshwater flow and the dynamics of related nutrients. Even while these studies show a strong connection between runoff and eutrophication, which might have serious consequences for hypoxia, there are numerous other ways that climate change could affect nutrients or the creation of dead zones (Justić et al. 2005; Struyf

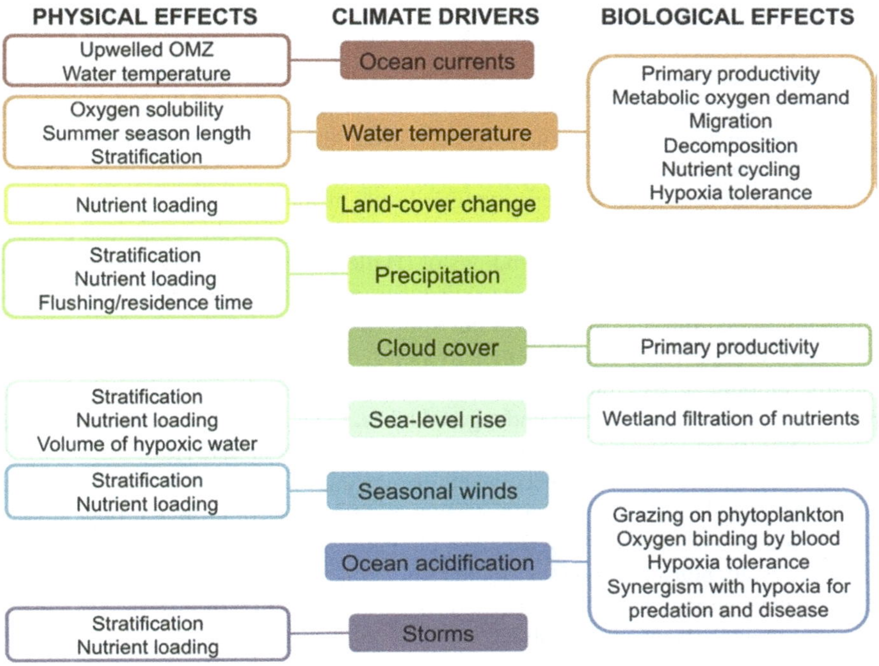

Fig. 20.2 A schematic diagram depicting how climatic forces interact with physical and biological processes to affect hypoxia severity. (*Source*: Andrew and Gedan (2015))

et al. 2004; Howarth et al. 2000). Eutrophication rates can be maintained under control, but evidence is growing from certain ecosystems that climate influences are severe enough to aggravate the severity of dead zones (Carstensen et al. 2014; Villate et al. 2013; Meier et al. 2011a).

20.5 The Effect of Climate Warming on Lakes and Seas' Oxygen Levels

It is among the most significant challenges mankind has encountered in terms of the environment in recent memory. Something more complex is going on here than hot air and ice melting. In addition to affecting where people are living and where their food is grown, it has far-reaching repercussions that go well beyond that. There is less apparent uncertainty in people's minds concerning the dangers of climate change when they have a greater understanding of the subject matter (Alam et al. 2021). Over the twentieth century, temperatures in several lakes across the world rose significantly (Austin and Colman 2008; Schneider and Hook 2010). The less often the water in lakes changes the deeper it gets (Pilla 2021). Worldwide, lakes and reservoirs face environmental threats from dissolved oxygen. Deep-water oxygen levels are dropping due to growing nutrient loads as a result of climatic changes in the mixing and stratification of lakes (Saros et al. 2012; Savin 1977). Warming of the upper ocean and greater stratification are anticipated to result in ocean deoxygenation (ocean depletion), which has a negative impact on ocean productivity, nitrogen cycle, and carbon cycling as well as on marine habitats. Over the next century, ocean models forecast a 1–7% decrease in the world ocean's O2 inventory, which is expected to continue for a thousand years or longer (Keeling et al. 2009).

Some lakes have seen a rise in oxygen levels at the surface. This may lead to algae blooms, which in turn can create toxic compounds, as the temperature rises. In a study by Karlson et al., The findings of (Hallegraeff 2003; Karlson et al. 2021) Climate change is suffocating the world's lakes, according to a new study. Species are being wiped out and drinking water sources are at risk because of increasing temperatures, according to researchers. It's been 40 years since a study discovered that the amount of oxygen in lakes is dropping 3–9 times faster than they were before. One of Germany's most endangered lakes, Lake Ammersee, is located in the state of Lower Saxony (Kraemer et al. 2021; Woolway et al. 2021).

There is a direct correlation between decreasing oxygen levels and decreased habitat for a wide variety of species (Jane et al. 2021). Average animal populations have fallen by 84% since 1970. Besides, climate warming and greenhouse gasses, too much water consumption in agriculture is a contributing factor. Freshwater ecosystems have already been badly impacted by the destruction of fish, insects, birds, and mammals in their natural habitats.

In the investigation of tropical and temperate lakes, a falling level of oxygen was discovered. According to the study, oxygen loss was noticed regardless of location

(Climate Research Committee and National Research Council 1996). Oxygen levels in upper water rose in about one-fifth of the lakes investigated, and pollution was found in virtually every one of them. There are signs of massive algal blooms, Rose claims. "We can't be sure without taxonomy data, but there is nothing else we can think of that might explain this trend."

Maintaining current levels of global warming would entail the cleansing of freshwater bodies, even if global average temperatures remain stable. New York's Oneida Lake was cleaned up, allowing oxygen-producing algae to produce more photosynthesis, according to Rose (Dale et al. 2006).

If the global temperature rises by 1.5 °C in the next 80 years, just 10% of the species studied would be at risk of extinction. 60% of the fish species studied will be unable to flourish in their current locations by 2100, if global temperatures increase by 4–5 °C over pre-industrial levels according to research published in Science Advances (7.2 °F). According to new research, a wide range of fish species from both freshwater and saltwater are in danger of being killed or badly harmed when water temperatures rise (Dahlke et al. 2018).

It is possible that even if this scenario is taken to its logical conclusion, the precious Atlantic Cod, Swordfish, Pacific Salmon, and Alaska Pollock needed to make frozen fish sticks might still be depleted. To accurately predict the repercussions of a 10% loss in species, it is necessary to take into account the whole ecosystem "stem." Think of the North Sea, where we expect the temperature to rise so high that cod will be unable to reproduce by the end of this century. That is a distinct possibility! Removal of such fish species, according to researchers from Alfred-Wegener Institute, Helmholtz Centre for Polar and Marine Research, and Bremen University, would have a substantial impact on the ecosystem as a whole, as well as all activities and interactions among species (Dahlke et al. 2018).

Many fishes are now dwelling in areas where they have achieved their maximum temperature endurance of regions with temperatures of 40°. His next words were: "As a consequence of our actions, the planet's temperature is rapidly approaching uninhabitable levels, and we are rapidly losing suitable habitat. I think it is worth it to help meet the 1.5 C goal" (Pörtner 2002).

It was only through repeated hydrographic observations over the past half-century that scientists discovered widespread oxygen loss in the open ocean. Hydrographic measurements indicated oxygen decreases in places spanning from the northeast Pacific to the Gulf of Mexico, allowing this finding to be made (Whitney et al. 2007) and the northern Atlantic to tropical oceans (Stendardo and Gruber 2012; Stramma et al. 2008). The continuing deoxygenation in many sections of the open ocean is most certainly caused by global warming caused by man-made greenhouse gas emissions (Bopp et al. 2013). Since the 1980s, there has been a strong correlation between ocean heat content and deoxygenation, spanning the period 1958–to 2015 in the upper oceans (Ito et al. 2017). The pace at which oxygen is used increases as a result of an increase in metabolic rate as a result of warming. To reduce oxygen loss, the rate at which sinking particles decompose and remineralize is increased, but the amount of oxygen lost remains the same (Brewer and Peltzer 2017).

Temperature stratification is directly influenced by warming, while salinity-driven stratification is facilitated by ice melt and precipitation. The deep overturning circulation is slowed by increased stratification, which changes the mostly wind-driven top ocean circulation (Schmidtko et al. 2017). Deutsch et al. (2014) observed that low- to mid-latitude oceanic depths (100–300 m) show considerable indicators of decreased subsurface ventilation, whereas high-latitude oceans show less obvious symptoms down to a few thousand meters. Decade-to-multidecade variations in meteorological forcing patterns may have an impact on this. A drop in oxygen levels has occurred in some systems, such as the California Current, as the relative effect of different currents has altered and recrystallization has formed (Nam et al. 2015; Gilbert et al. 2005).

According to Diaz and Rosenberg (2008), approximately 400 dead zones have been documented globally and their number has grown dramatically. Temperature anomalies from year to year in these habitats were used to estimate the severity of climate change they would experience throughout this century. Because 94% of reported oxygen-depleted dead zones are located in areas expected to warm by more than 2 °C by the end of the century, it is critical to better understand climate change to better predict and manage coastal dead zones.

20.6 Climate Warming and Warming of Coastal Seas and Estuaries

When it comes to water temperature, estuaries and coastal waterways tend to have a stronger correlation with air temperature than open ocean seas. There is a 4-degree Celsius difference in summer surface water temperature between the neighboring ocean and the Rio de La Plata Estuary in Argentina. This is visible when comparing estuaries to the adjacent ocean (Simionato et al. 2010). Temperature changes in coastal waters and estuaries may be seen over long periods (Fig. 20.3). It is 6–13 times faster than the current global air temperature increase that the Baltic Sea has warmed by an average of 0.03 °C every year since 1985 (MacKenzie and Schiedek 2007). Warming of the Chesapeake Bay with an average of 0.03 °C annually since 1960s (Najjar et al. 2010). Several temperature-dependent impacts on hypoxia are projected to accompany rising coastal water temperatures as a result of atmospheric climate change. Climate warming has been extending oxygen-depleted zones and expanding their area in numerous ways in addition to causing biological responses and hypoxic conditions to become more severe.

Sea surface temperatures (SSTs) in the Baltic and North Seas have risen to historic highs in the last decade, and they're warming at a rate that's outpacing that of the open ocean. According to global climate models, summer SSTs are growing three times faster than air temperatures in the summer months (MacKenzie and Schiedek 2007). Twenty years of sweeping curves are being smoothed out by bold lines.

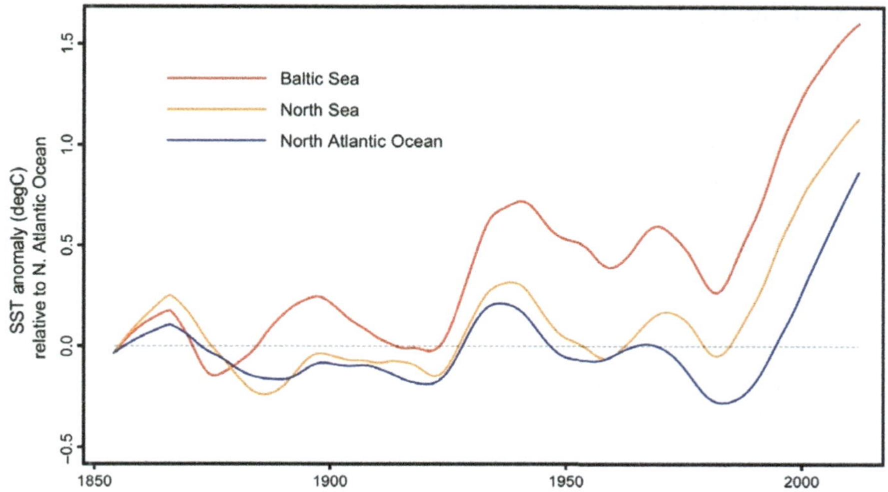

Fig. 20.3 Changes in SST in three hypoxic seas (North Atlantic at 62°N, 2°E; North Sea 56°N, 4°E & Baltic Sea 56°N, 18°E) from 1859 to 2000. (*Source*: Data are from NOAA Extended Reconstructed Sea Surface Temperatures (ERSST 3b, http://nomads.ncdc.noaa.gov/las/getUI.do))

20.7 Climate Warming and Continental Waters

Increased stratification, lower oxygen solubility, and faster respiration are all projected to aggravate oxygen depletion in many nutrient-rich coastal systems as a result of warming (Carstensen et al. 2014). Increased nutrient concentrations and warmer shallow water are likely to blame for coastal oxygen depletion being quicker than in the open ocean (Gilbert et al. 2010). In eutrophic environments, increased air temperatures may affect the periodic stratification and the rate of oxygen depletion, which could contribute to hypoxia developing sooner and persisting longer. According to a Baltic Sea modeling study, precipitation increase, reduced atmospheric oxygen transfer, and enhanced internal nutrient cycling all contribute to a decline in the sea's oxygen levels. When it comes to climate change, nutrient reduction would be greatly minimized even in the most extreme circumstances (Meier et al. 2011a, b). For each unique coastal water body, the rate of change in oxygen concentration is likely to vary because of differences in precipitation and wind patterns in different regions (Rabalais et al. 2014; Altieri and Gedan 2015). There is a strong correlation between increased precipitation and increased stratification and nutrient releases, with the opposite being true in areas with decreased precipitation levels. Seasonal variations in precipitation and evaporation rates may also have a significant impact. However, the rate at which wetland flooding occurs and the capacity of wetlands to relocate landward will vary with the rise in sea levels, making it difficult for wetlands to filter nutrients before they reach the open oceans.

20.8 Climate Warming and Rise in Sea Level

Glacier melt and the expansion of water due to heat have resulted in an increase in bays and estuary volume. As a result, the total volume of shallower coastal water susceptible to hypoxia increases, as well as the water column stratifies, resulting in low oxygen in the bottom waters of bays, especially those with sills near their mouths (Davies and Xing 2007; Van der Zwaan and Jorissen 1991). Furthermore, when sea levels rise, the diminished ability of wetlands to remove nitrogen might lead to greater eutrophication (Kemp et al. 2005).

20.9 Warming of the Climate and Oxygen Solubility in Water

Just as gas solubility reduces when temperatures rise, the amount of oxygen accessible to aquatic organisms will be reduced as a result of warming (Weiss 1970). Temperature and oxygen solubility has a nonlinear connection, with lower temperatures having a higher effect on the latter. High-latitude temperate and arctic estuaries will be affected more by rising temperatures since the water temperature is lower and warming rates are expected to be greater. If an estuary's surface waters become warmer and less oxygen-rich due to surface currents, this will reduce the availability of oxygen to deeper waters as well (Meier et al. 2011b).

20.10 Climate Change and Stratification of Water

By reducing the ability of oxygenated upper waters to mix with the bottom waters where hypoxia is most widespread, climate change will increase stratification and hence promote hypoxia (Cloern 2001). Stratification, which occurs in places like the Baltic Sea and the Chesapeake Bay, has been linked to an increase in dead zones, according to studies (Murphy et al. 2011; Conley et al. 2007). The North Sea is expected to see a rise in hypoxia as a result of stratification changes caused by climate change (Meire et al. 2013).

20.11 Proper Monitoring of Oceans and Seas

In many sections of the world's oceans, particularly the Pacific, oceanic oxygen depletion is becoming recognized as a severe hazard to ocean life and changing habitat conditions. A higher temperature is associated with a greater likelihood of deoxygenation feedback on climate, particularly the production of powerful greenhouse gases such as N2O and methane in low-oxygen conditions. As a result, bridging the gap between observations and models, which is required for effective future forecasting, is crucial. To address these problems, we propose that more wide-ranging and globally synchronized ocean observations be done. Furthermore, multidisciplinary process studies are required to better understand the delicate balance between oxygenation and oxygen absorption in the seas, which is always changing. Another benefit of developing marine oxygen budget models is that oxygen is a useful indicator for evaluating models that calculate the ocean's CO2 uptake. We would have a better grasp of the carbon cycle as a result of this experiment (Oschlies et al. 2018).

20.12 Prediction of Decline in Oxygen

Reliable projections and knowledge of the underlying uncertainties are vital to successful marine ecosystem management. Numerical modeling may help evaluate the influence of global warming and algae blooms on the open coastal and marine ecosystems' oxygen supply, which can be forecasted. By the end of the century, the most recent global models all indicate that oxygen depletion will be barely a few percent, which might have substantial biogeochemical and ecological ramifications (Bopp et al. 2013). Across models, there is no agreement on the present and future distribution of low-oxygen zones (Cabré et al. 2015) or the spatial patterns of O2 fluctuations over a long timeframe (Stramma et al. 2012). Currently, our capacity to anticipate the consequences of global warming on open ocean OMZs and oxygen-sensitive biogeochemical processes like nitrogen budgets is severely constrained. Other than CO2-driven global warming, such as atmospheric nutrient deposition and long-term climate variability, more realistic and detailed characteristics may be possible to boost model consistency (particularly changes in wind patterns).

Marine oxygen levels may be estimated by taking into account a wide range of factors including interactions with the atmosphere, the land, sediment, and sea waters at a variety of spatial and temporal dimensions. Existing three-dimensional coupled hydrodynamic–water quality models that are particular to estuaries and regions may help with this (Rose et al. 2017); these and other cutting-edge modeling techniques deserve to be used more widely. Forced data such as river flows and atmospheric conditions, on the other hand, might have an impact on model performance if they lack appropriate spatial and temporal resolution. As a consequence of different climate change and nutrient management scenarios, future deoxygenation estimates need correct data on changes to key factors and interactions, and they benefit from methodologies that explicitly simulate linkages throughout the estuary close to sea or ocean continuum.

20.13 Suggestions for Work Direction in Future

Many concerns must be addressed immediately to accurately predict the future of dead zones, even though there is abundant evidence that climate change, particularly warming, will have a significant impact on dead zone dynamics.

There may be a relationship between climate change and the disappearance of marine ecosystems. How much of an impact do phytoplankton blooms have on marine carbon sinks? The fact that dead zones produce nitrous oxide has ramifications for our understanding of climate management and how it relates to dead zones (Naqvi et al. 2010).

How does temperature impact the macro fauna that controls microbial activity linked with hypoxia in a community-level manner? Microorganisms have to adjust to the changed conditions due to irrigation and sediment turnover induced by diverse crab and amphipod species in the coastal environment, lowering their resource availability (Laverock et al. 2011; Lohrer et al. 2004). Deposits of bivalves and shells have a substantial impact on the nitrogen cycle and primary production because they contain organisms (Welsh and Castadelli 2004; Newell et al. 2002). They are heat-sensitive, which means that any changes in their activity might have major effects on oxygen and nutrient fluxes, which could lead to algal blooms or dead zones.

Does changing ranges allow organisms to avoid habitats where the combined impacts of climate change and hypoxia have exceeded their tolerance limits at both the population and species levels?" Due to the complexity of ecosystem dynamics and organismal reactions, it is difficult to anticipate how coastal habitats will change in the future. Due to a combination of factors, including the geographical mosaic of ecological change predicted in coastal habitats, this is the case (Harley et al. 2006). This might have major effects on the humans that depend on foundation species, such as hypoxia and climate change (Altieri and Van De Koppel 2013). A resident species' physiological resistance to temperature change and dead zones alone is not enough to forecast these indirect consequences.

How do temperature and OA impact hypoxia adaption mechanisms and restrictions at the organismal level? To keep pace with climate change, would organisms' rates and ranges of response to numerous stressors be sufficient? When it comes to hypoxia and climate change and the various stresses they bring with them, how do they interact in different places and overtime? Only by investigating metabolic pathways and the functional genomes that regulate them can these issues be resolved. Since we've witnessed temperature tolerance limits approaching in certain species and groups, should we think that their capacity to adapt to climate change would be limited?

Climate change components' probable synergisms are a major source of uncertainty, particularly when a single cause might have several effects. Bays and estuaries, as well as hypoxic water volume, depth, and wetland persistence, will all be affected by rising sea levels, which will change the quantity of land-based inputs. These components may interact differently based on the specifics of the system.

Given the number of stresses that the marine environment faces (Breitburg and Riedel 2005; Crain et al. 2008), it's important to think about the long-term consequences of any one of these stresses.

20.14 Conclusion

The worsening of oxygen-depleted dead zones might be one of the most devastating consequences of climate change on ocean ecosystems. The response of ecosystems and species to hypoxic conditions and dead zones is strongly influenced by temperature. As with other elements of climate change, including changes in weather patterns, sea-level rise, and acidification, dead zone dynamics are highly responsive to temperature changes (OA). There is a strong correlation between the increase of dead zones and climate change, according to this research. When a multifactor view is taken into account, it is probable hypoxia thresholds established decades ago have been decreased, which means that nutrient reductions will have to be more severe when a climate change perspective is taken into account. Only by considering all of the climate factors that influence dead zones as a whole can reliable predictions of climate warming interactions with eutrophication and other human consequences be made and measures are taken.

It has been half a century since human actions like global warming and fertilizer runoff into coastal waterways have caused the ocean and coastal seas to lose oxygen. As a result of these modifications, oxygen consumption by microorganisms has increased, water solubility has decreased, and the rate at which oxygen is replenished from the atmosphere to the ocean's core has decreased, all of which have a variety of ecological and biological repercussions. There is a need for further research on the long-term effects of regional and global-scale oxygen changes on marine and estuarine fisheries and ecosystems.

20.15 Recommendation

An alternate system is necessary to preserve the rising and falling condition of hydrospheric oxygen levels to avert the deteriorating repercussions of climate change. Many inventors are creating technologies or business strategies that create less carbon than conventional ways, such as mushroom leather instead of cow leather and electric automobiles instead of fossil fuels. Furthermore, researchers are investigating novel technologies that capture the already existing carbon in the atmosphere to develop new strategies to battle climate warming from various perspectives. These approaches offer a lot of potential advantages, but they also have a lot of drawbacks at the moment.

References

Alam MA, Alam M (2020) Climate rising and falling state of hydrospheric oxygen level: a grim global concern. Int J Environ Health Res:8. https://doi.org/10.47062/1190.0204.02

Alam MA, Alam M, Kundra S (2021) Knowing climate change: knowledge, perceptions and awareness (KPA) among higher education students in Eritrea. Disaster Adv 14(1):30–39

Altieri AH, Gedan KB (2015). Climate change and dead zones. Glob Chang Biol 21(4):1395–1406

Altieri AH, Van De Koppel J (2013) Foundation species in marine ecosystems. In: Bertness MD, Bruno JF, Silliman BR, Stachowicz JJ (eds) Marine community ecology and conservation. Sinauer Associates, Sunderland, pp 37–56

Andrew H, Gedan KB (2015) Climate change and dead zones. Glob Chang Biol 21(4):1395–1406

Austin J, Colman SA (2008) A century of temperature variability in Lake Superior. Limnol Oceanogr 53(6):2724–2730

Bopp L, Resplandy L, Orr JC, Doney SC, Dunne JP, Gehlen M, Halloran P, Heinze C, Ilyina T, Séférian R, Tjiputra J, Vichi M (2013) Multiple stressors of ocean ecosystems in the 21st century: projections with CMIP5 models. Biogeosciences 10(10):6225–6245

Breitburg DL, Riedel GF (2005) Multiple stressors in marine systems. In: Marine conservation biology: the science of maintaining the sea's biodiversity. Island Press, Washington

Breitburg D et al (2018) Declining oxygen in the global ocean and coastal waters. Science 359(6371):eaam7240

Brewer PG, Peltzer ET (2017) Depth perception: the need to report ocean biogeochemical rates as functions of temperature, not depth. Philos Trans R Soc A Math Phys Eng Sci 375(2102):20160319

Cabré A, Marinov I, Bernardello R, Bianchi D (2015) Oxygen minimum zones in the tropical Pacific across CMIP5 models: mean state differences and climate change trends. Biogeosciences 12(18):5429–5454

Carstensen J, Andersen JH, Gustafsson BG, Conley DJ (2014) Deoxygenation of the Baltic Sea during the last century. Proc Natl Acad Sci U S A 111:5628–5633. Casualty of Climate Change, Opinion Article in Scientific American

Chan F, Barth JA, Lubchenco J, Kirincich A, Weeks H, Peterson WT, Menge BA (2008) The emergence of anoxia in California's current large marine ecosystem. Science 319(5865):920–920

Cloern JE (2001) Our evolving conceptual model of the coastal eutrophication problem. Mar Ecol Prog Ser 210:223–253

Conley DJ, Carstensen J, Ærtebjerg G, Christensen PB, Dalsgaard T, Hansen JL, Josefson AB (2007) Long-term changes and impacts of hypoxia in Danish coastal waters. Ecol Appl 17(sp5):S165–S184

Crain CM, Kroeker K, Halpern BS (2008) Interactive and cumulative effects of multiple human stressors in marine systems. Ecol Lett 11(12):1304–1315

Dahlke FT, Butzin M, Nahrgang J, Puvanendran V, Mortensen A, Pörtner HO, Storch D (2018) Northern cod species face spawning habitat losses if global warming exceeds 1.5 C. Sci Adv 4(11):eaas8821

Dale B, Edwards M, Reid PC (2006) Climate change and harmful algal blooms. In: Ecology of harmful algae. Springer, Berlin, Heidelberg, pp 367–378

Davies AM, Xing J (2007) On the influence of stratification and tidal forcing upon mixing in sill regions. Ocean Dyn 57(4):431–451

Deutsch C, Berelson W, Thunell R, Weber T, Tems C, McManus J, Crusius J, Ito T, Baumgartner T, Ferreira V, Mey J (2014) Centennial changes in North Pacific anoxia linked to tropical trade winds. Science 345(6197):665–668

Diaz RJ (2001) Overview of hypoxia around the world. J Environ Qual 30(2):275–281

Diaz RJ, Rosenberg R (2008) Spreading dead zones and consequences for marine ecosystems. Science 321(5891):926–929

Gallo ND, Levin LA (2016) Fish ecology and evolution in the world's oxygen minimum zones and implications of ocean deoxygenation. Adv Mar Biol 74:117–198

Gilbert D, Pettigrew B, Sundby B, Gobeil C, Mucci A, Tremblay GH (2005) A seventy-two-year record of diminishing deep-water oxygen in the St. Lawrence estuary: the northwest Atlantic connection. Limnol Oceanogr 50:1654–1666

Gilbert D, Rabalais NN, Diaz RJ, Zhang J (2010) Evidence for greater oxygen decline rates in the coastal ocean than in the open ocean. Biogeosciences 7(7):2283–2296

Gooday AJ, Jorissen F, Levin LA, Middelburg JJ, Naqvi SWA, Rabalais NN, Scranton M, Zhang J (2009) Historical records of coastal eutrophication-induced hypoxia. Biogeosciences 6(8):1707–1745

Hallegraeff GM (2003) Harmful algal blooms: a global overview. In: Hallegraeff M, Anderson DM, Cembella AD (eds) Manual on harmful marine microalgae. Monographs on Oceanographic Methodology, 2nd Edition, IOC-UNE-SCO, Paris, 25–49.

Harley CD, Randall Hughes A, Hultgren KM, Miner BG, Sorte CJ, Thornber CS, Rodriguez LF, Tomanek L, Williams SL (2006) The impacts of climate change in coastal marine systems. Ecol Lett 9(2):228–241

Helm KP, Bindoff NL, Church JA (2011) Observed decreases in oxygen content of the global ocean. Geophys Res Lett 38(23):23602

Howarth RW, Swaney DP, Butler TJ, Marino R (2000) Rapid communication: climatic control on eutrophication of the Hudson River estuary. Ecosystems 3(2):210–215. https://worldresearchersassociations.com/Archives/DA/Vol(14)2021/March2021.aspx

Ito T, Minobe S, Long MC, Deutsch C (2017) Upper-ocean O2 trends: 1958–2015. Geophys Res Lett 44(9):4214–4223

Jackson JB, Kirby MX, Berger WH, Bjorndal KA, Botsford LW, Bourque BJ, Bradbury RH, Cooke R, Erlandson J, Estes JA, Hughes TP (2001) Historical overfishing and the recent collapse of coastal ecosystems. Science 293(5530):629–637

Jane SF, Hansen GJ, Kraemer BM, Leavitt PR, Mincer JL, North RL, Pilla RM, Stetler JT, Williamson CE, Woolway RI, Arvola L (2021) Widespread deoxygenation of temperate lakes. Nature 594(7861):66–70

Justić D, Rabalais NN, Turner RE (2005) Coupling between climate variability and coastal eutrophication: evidence and outlook for the northern Gulf of Mexico. J Sea Res 54(1):25–35

Karlson B, Andersen P, Arneborg L, Cembella A, Eikrem W, John U, West JJ, Klemm K, Kobos J, Lehtinen S, Lundholm N (2021) Harmful algal blooms and their effects in coastal seas of Northern Europe. Harmful Algae 102:101989

Karstensen J, Fiedler B, Schütte F, Brandt P, Körtzinger A, Fischer G, Zantopp R, Hahn J, Visbeck M, Wallace D (2015) Open ocean dead zones in the tropical North Atlantic Ocean. Biogeosciences 12(8):2597–2605

Keeling RF, Kortzinger A, Gruber N (2009) Ocean deoxygenation in a warming world, review in advance, pp 463–493. ANRV399-MA02-17.tex (noaa.gov)

Kemp WM, Boynton WR, Adolf JE, Boesch DF, Boicourt WC, Brush G, Cornwell JC, Fisher TR, Glibert PM, Hagy JD, Harding LW (2005) Eutrophication of Chesapeake Bay: historical trends and ecological interactions. Mar Ecol Prog Ser 303:1–29

Kraemer BM, Pilla RM, Woolway RI, Anneville O, Ban S, Colom-Montero W, Devlin SP, Dokulil MT, Gaiser EE, Hambright KD, Hessen DO (2021) Climate change drives widespread shifts in lake thermal habitat. Nat Clim Chang 11(6):521–529

Laverock B, Gilbert JA, Tait K, Osborn AM, Widdicombe S (2011) Bioturbation: impact on the marine nitrogen cycle. Biochem Soc Trans 39(1):315–320

Levin LA, Breitburg DL (2015) Linking coasts and seas to address ocean deoxygenation. Nat Clim Chang 5(5):401–403

Lohrer AM, Thrush SF, Gibbs MM (2004) Bioturbators enhance ecosystem function through complex biogeochemical interactions. Nature 431(7012):1092–1095

Lotze HK, Lenihan HS, Bourque BJ, Bradbury RH, Cooke RG, Kay MC, Kidwell SM, Kirby MX, Peterson CH, Jackson JB (2006) Depletion, degradation, and recovery potential of estuaries and coastal seas. Science 312(5781):1806–1809

MacKenzie BR, Schiedek D (2007) Daily ocean monitoring since the 1860s shows record warming of northern European seas. Glob Chang Biol 13(7):1335–1347
Meier HM, Andersson HC, Eilola K, Gustafsson BG, Kuznetsov I, Müller-Karulis B, Neumann T, Savchuk OP (2011a) Hypoxia in future climates: a model ensemble study for the Baltic Sea. Geophys Res Lett 38(24):6
Meier HM, Eilola K, Almroth E (2011b) Climate-related changes in marine ecosystems simulated with a 3-dimensional coupled physical-biogeochemical model of the Baltic Sea. Clim Res 48(1):31–55
Meire LKER, Soetaert KER, Meysman FJR (2013) Impact of global change on coastal oxygen dynamics and risk of hypoxia. Biogeosciences 10(4):2633–2653
Murphy RR, Kemp WM, Ball WP (2011) Long-term trends in the Chesapeake Bay seasonal hypoxia, stratification, and nutrient loading. Estuar Coasts 34(6):1293–1309
Najjar RG, Pyke CR, Adams MB, Breitburg D, Hershner C, Kemp M, Howarth R, Mulholland MR, Paolisso M, Secor D, Sellner K (2010) Potential climate-change impacts on the Chesapeake Bay. Estuar Coast Shelf Sci 86(1):1–20
Nam S, Takeshita Y, Frieder CA, Martz T, Ballard J (2015) Seasonal advection of Pacific Equatorial Water alters oxygen and pH in the Southern California Bight. J Geophys Res Oceans 120(8):5387–5399
Naqvi SWA, Bange HW, Farías L, Monteiro PMS, Scranton MI, Zhang J (2010) Marine hypoxia/anoxia as a source of CH4 and N2O. Biogeosciences 7(7):2159–2190
National Research Council, Climate Research Committee (1996) Natural climate variability on decade-to-century time scales. National Academies Press, Washington, DC
Newell RI, Cornwell JC, Owens MS (2002) Influence of simulated bivalve deposition and microphytobenthos on sediment nitrogen dynamics: a laboratory study. Limnol Oceanogr 47(5):1367–1379
Norris RD, Turner SK, Hull PM, Ridgwell A (2013) Marine ecosystem responses to Cenozoic global change. Science 341(6145):492–498
Oschlies A, Brandt P, Stramma L, Schmidtko S (2018) Drivers and mechanisms of ocean deoxygenation. Nat Geosci 11:467–473
Pilla RM (2021) Lake vertical ecosystem responses to climate and environmental changes: integrating comparative time series, modeling, and high-frequency approaches. Doctoral dissertation, Miami University
Pörtner HO (2002) Climate variations and the physiological basis of temperature-dependent biogeography: systemic to molecular hierarchy of thermal tolerance in animals. Comp Biochem Physiol A Mol Integr Physiol 132(4):739–761
Pullen J, Goodkin N (2021) Marine oxygen levels are the next great casualty of climate change. Opinion article in Scientific American
Rabalais NN, Diaz RJ, Levin LA, Turner RE, Gilbert D, Zhang J (2010) Dynamics and distribution of natural and human-caused hypoxia. Biogeosciences 7(2):585–619
Rabalais NN, et al (2014) The economics of dead zones: causes, impacts, policy challenges, and a model of the Gulf of Mexico hypoxic zone. Rev Environ Econ Policy 8(1):58–79
Riemann B, Carstensen J, Dahl K, Fossing H, Hansen JW, Jakobsen HH, Josefson AB, Krause-Jensen D, Markager S, Stæhr PA, Timmermann K (2016) Recovery of Danish coastal ecosystems after reductions in nutrient loading: a holistic ecosystem approach. Estuar Coasts 39(1):82–97
Rose KA, Justice D, Fennel K, Hetland RD (2017) Numerical modeling of hypoxia and its effects: synthesis and going forward. In: Modeling coastal hypoxia. Springer, Cham, pp 401–421
Saros JE, Stone JR, Pederson GT, Slemmons KE, Spanbauer T, Schliep A, Cahl D, Williamson CE, Engstrom DR (2012) Climate-induced changes in lake ecosystem structure inferred from coupled neo-and paleoecological approaches. Ecology 93(10):2155–2164
Savin SM (1977) The history of the Earth's surface temperature during the past 100 million years. Annu Rev Earth Planet Sci 5(1):319–355

Schmidtko S, Stramma L, Visbeck M (2017) Decline in global oceanic oxygen content during the past five decades. Nature 542(7641):335–339

Schneider P, Hook SJ (2010) Space observations of inland water bodies show rapid surface warming since 1985. Geophys Res Lett 37(22):1–5

Seibel BA (2011) Critical oxygen levels and metabolic suppression in oceanic oxygen minimum zones. J Exp Biol 214:326

Simionato CG, Tejedor MLC, Campetella C, Guerrero R, Moreira D (2010) Patterns of sea surface temperature variability on seasonal to sub-annual scales at and offshore the Río de la Plata estuary. Cont Shelf Res 30(19):1983–1997

Stendardo I, Gruber N (2012) Oxygen trends over five decades in the North Atlantic. J Geophys Res Oceans 117(C11): 1–18

Stramma L, Johnson GC, Sprintall J, Mohrholz V (2008) Expanding oxygen-minimum zones in the tropical oceans. Science 320(5876):655–658

Stramma L, Schmidtko S, Levin LA, Johnson GC (2010) Ocean oxygen minima expansions and their biological impacts. Deep Sea Res Part 1 Oceanogr Res Pap 57(4):587–595

Stramma L, Oschlies A, Schmidtko S (2012) A mismatch between observed and modeled trends in dissolved upper-ocean oxygen over the last 50 yr. Biogeosciences 9:4045–4057. https://doi.org/10.5194/bg-9-4045

Struyf E, Van Damme S, Meire P (2004) Possible effects of climate change on estuarine nutrient fluxes: a case study in the highly nutrified Schelde estuary (Belgium, The Netherlands). Estuar Coast Shelf Sci 60(4):649–661

Van der Zwaan GJ, Jorissen FJ (1991) Bio facial patterns in river-induced shelf anoxia. Geol Soc Lond, Spec Publ 58(1):65–82

Vaquer-Sunyer R, Duarte CM (2008) Thresholds of hypoxia for marine biodiversity. Proc Natl Acad Sci 105(40):15452–15457

Villate F, Iriarte A, Uriarte I, Intxausti L, de la Sota A (2013) Dissolved oxygen in the rehabilitation phase of an estuary: influence of sewage pollution abatement and hydro-climatic factors. Mar Pollut Bull 70(1–2):234–246. Volume 14, No. 3, pp 30–39

Watson AJ (2016) Oceans on the edge of anoxia. Science 354:1529–1530. https://doi.org/10.1126/science.aaj2321. pmid: 28008026

Weiss RF (1970) The solubility of nitrogen, oxygen, and argon in water and seawater. In: Deep-sea research and oceanographic abstracts, vol 17(4). Elsevier, Amsterdam, pp 721–735

Welsh DT, Castadelli G (2004) Bacterial nitrification activity is directly associated with isolated benthic marine animals. Mar Biol 144(5):1029–1037

Whitney FA, Freeland HJ, Robert M (2007) Persistently declining oxygen levels in the interior waters of the eastern subarctic Pacific. Prog Oceanogr 75(2):179–199

Woolway RI, Sharma S, Weyhenmeyer GA, Debolskiy A, Golub M, Mercado-Bettín D, Perroud M, Stepanenko V, Tan Z, Grant L, Ladwig R (2021) Phenological shifts in lake stratification under climate change. Nat Commun 12(1):1–11

Zhang J et al (2010) Natural and human-induced hypoxia and consequences for coastal areas: synthesis and future development. Biogeosciences 7:1443–1467. https://doi.org/10.5194/bg-7-1443-2010

Chapter 21
Nanoparticle-Based Bioremediation for Crude Oil Removal from Marine Environment

Sonal Bhandari, Meesa Saraswathi, Ballari Lakshmanna, and M. Madakka

Abstract The crude oil impurities have a constant detrimental effect on the environment and human health in the marine ecosystem. Thus, over several decades many safe and effective techniques have been used as a remediation method. The remediation method in combination with nanotechnology has been explored by researchers for the eradication of crude oil from the ocean matrices. Studies are going on to understand the interaction between the impurity, microbes, and nanoparticles to see the positive and negative impacts such as some nanoparticles act as toxic for the microbes while others act as stimulants. The marine microorganism used spilled oil as a primary source of energy in a technique known as bioremediation. Other methods such as in situ burning, skimmers, and sorbents can be problematic to use in extreme weather conditions. Nowadays nano-size particles have become an emerging method for the eradication of crude oil and other contaminants from the marine environment. Nanoparticles act as magnetic sorbents, and also show the qualities of emulsifiers by enhancing the bioaccessibility of the oil and giving the surface to the microorganism to which they can attach and ease the

S. Bhandari
National Institute for Pharmaceutical Education and Research, Balanagar, Hyderabad, Telangana, India

M. Saraswathi
Applied Biology, CSIR-Indian Institute of Chemical Technology, Tarnaka, Hyderabad, Telangana, India

Academy of Scientific and Innovative Research (AcSIR), Ghaziabad, Uttar Pradesh, India

B. Lakshmanna
Department of Energy and Environmental Engineering, CSIR-Indian Institute of Chemical Technology, Tarnaka, Hyderabad, Telangana, India

M. Madakka (✉)
Department of Biotechnology and Bioinformatics, Yogi Vemana University, Kadapa, Andhra Pradesh, India

N. Jayaraju et al. (eds.), *Coasts, Estuaries and Lakes*,
https://doi.org/10.1007/978-3-031-21644-2_21

spread. The nano-enhanced bioremediation is the most prominent technique used for the eradication of crude oil from the marine environment and has less impact by weather change. The nanoparticle remediation technique gives a sustainable approach rather than traditional methods which are toxic and ineffective. This chapter deals to determine the principles of bioremediation aided by nano-size particles, and their interaction with the marine matrices. The researchers are looking for a balance between essential bioremediation methods in order to make marine life free from organic matter contamination and economic development of nano-sized particles remedy. Hence, this chapter deals with the effective method for the eradication of petroleum oil from marine life and its benefits to human well-being.

Keywords Nanoparticle · Marine ecosystem · Bioaugmentation · Bioremediation · Crude oil

21.1 Introduction

Petroleum oil consists of heavy metals and natural gas that are distributed in huge amount all over the ocean like other unsafe chemical agents, thus seemed to be a serious marine ecosystem alarm. Land vehicle, pipelines, accidental spills, and inappropriate water disposal pattern is the major source of contamination of marine water. More crucially, these accidental spillages cause major chemical, biological and physical loss to the marine environment (Raimi et al. 2018). The International Tanker Owner Pollution Federation Ltd. (ITOPF), anticipated the total persevering and non-preserving hydrocarbon leakages to the marine environment to be around 1000 tons by 2019, and according to recent studies on two medium and one large spill location estimated to reach from 7 to 700 tons. Currently, many countries have established a bioremediation technique for the contaminated ocean and sea water (Cirer-Costa 2015). For centuries, the marine environment is becoming polluted, but the clean-up process remains unchanged over the course. The techniques useful for the clean-up are dependent upon the spill location, such as a spill occurring at sea and water site needs different action than those on land. The life of boats, fish, and birds are affected by the leakage of oil into the sea. Various remediation techniques have been employed in order to clean the marine environment such as physical remediation technique include sorbents, skimmers, and booms. Booms comprise an oil spill, through dispersal which is then removed by the use of skimmers or sorbent materials. The dispersants and solidifiers are the chemical remediation technique to clean the oil spill (Dave et al. 2011a, b). The process by which a microorganism speeds up the destruction of an oil spill is known as the biological remediation technique.

There are two main reasons by which oil spill in the ocean and sea happens, that is, natural cause and pollution due to human activity. Consequently, the leakage of oil from the beds of the ocean and the release of oil from a submissive rock at the sea is the natural cause to pollute marine water. Another cause is due to the drastic accidents of tankers which removes millions of tons of petroleum oil into the ocean water (Jernelöv 2010). For instance, there is a major damage to the environment due to the oil spill incidence 2011, in the Penglai and Gulf of Mexico in which each year almost 1.3 million of petroleum hydrocarbon and heavy metals enter into the marine environment (Ventikos and Sotiropoulos 2014). A deep-water prospect oil spill is a famous example of oil leakage due to accidents by human activity. Oil spillage during the storage and handling of crude oil and its by-products also comes under accidental spillage (Zhang et al. 2019). The petroleum components and the waste discharge from the factories are intentionally thrown into the ocean and sea which makes the marine matrices contaminated (Vlaev et al. 2011; Lucas and Macgregor 2006). The wide destruction of the ecosystem happened because of the announcement of ten million tons of oil spills into the Arabian Gulf (Murad et al. 2007).

21.1.1 Advancement in the Properties of Spilled Oil

The consequent flows of petroleum oil on the ocean surface form a thin oil layer known as an oil sleek. A number of chemical, biological, and physical changes happen when organic matter is released into the marine environment (Amro 2004). The partition, dispersal, absorption, photo-oxidation, sedimentation, emulsification with water, and microbes degradation are the changes that happen during the weathering process (Annunciado et al. 2005; Dave and Ghaly 2011a, b; Jordan and Payne 1980). The weathering procedure is complicated and happens concurrently (Wypych 2018). The weather at the accident site and the characteristics of the spilled oil decide the exact weathering process (Kapoor and Rawat 1994). The thickness of the oil, the speed of the wave, and the location of the oil flow decide the dispersal of oil (Elliott 1986). Due to the low molecular weight and volatile matters, the majority of the oil evaporates whereas the more molecular weight oil is not much affected by evaporation as they are made up of non-volatile matters. Thus, petroleum oil sleek has a high viscosity and density and has the capability to emulsify with water (Hussein 2009). The dispersion process speedup up the biodegradation process by enhancing the rate of spreading with decreasing oil thickness and with the increase in environmental commotion (Delvigne and Sweeney 1988; Rogowska and Namieśnik 2010). The dispersal process leads to the increment of the polycyclic aromatic hydrocarbons (PAH) which is harmful to the organism living in the ocean (Schein 2009). Although the photo-oxidized matters such as pyrene, phenanthrene, and anthracene are more harmful compared to the organic compounds and other oxygenated matters present in the ocean which leads to cancer (Rogowska and Namieśnik 2010). The thickness of the sleek is increased by the formation of emulsion with water. Dissolution is the process by which the matter of oil is dissolved in the water (Parker and Pitt 2012), likely the cleaning procedure becomes unwieldy due to the oxidation process. Sedimentation is the process that is caused by the union of dense matter and the evaporation of the light gears. It leads the oil to become

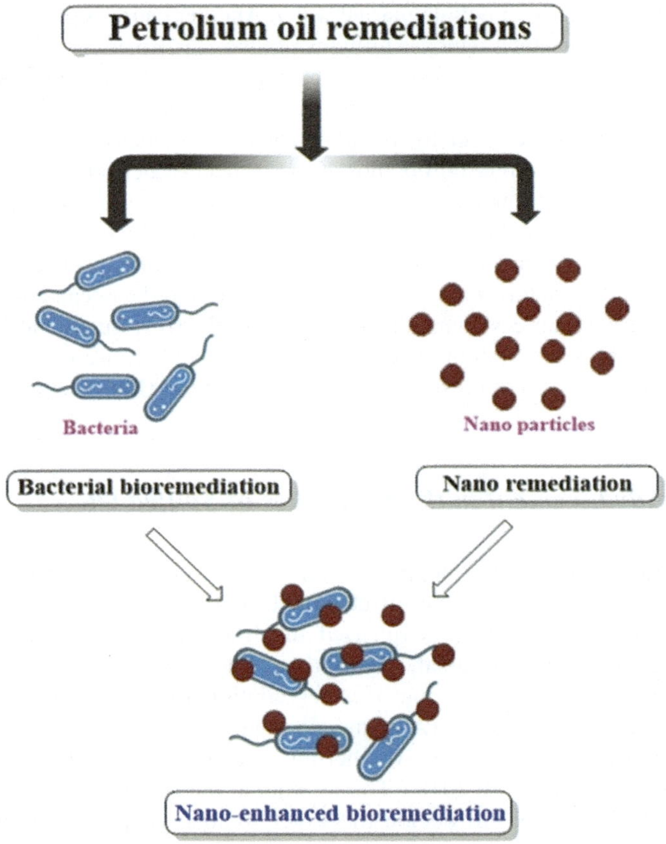

Fig. 21.1 Process of petroleum oil remediation

deeper than water and consecutively descends. The destiny of sediment matters is similar to dispersed and dissolved particles (Gill and Robotham 1989).

The procedure by which microorganism changes the particles of oil into carbon dioxide and water is known as the biodegradation process (Fig. 21.1). Due to the enhanced phosphorus and nitrogen content in the ocean leads to the increase in the salinity, and the oxygen present leads to the development of biodegradation process. Though, the aromatics and asphaltenes are either resilient or degrade at a very slow rate. Because of the weathering process, sleek oil density and thickness became the most concerning topic. Thus, by seeing all these factors, a strong measure for the oil spill remedy should be taken (Lessard and DeMarco 2000).

21.2 Nanoparticles Enhancing Bioremediation Process

Nanotechnology has become one of the emerging technologies for the bioremediation of the biological, chemical, and physical issues associated with the removal of crude oil from the marine system. The nanoparticles have been classified into three types: (1) complexing nanoparticle which attracts the particle by electrostatic charges, (2) Conjugating nanoparticles that link to particles by a covalent bond. (3) Encapsulating nanoparticles are the nanosponges which are the nanoparticles consist of holes that carry molecules and nanocapsules are the particles that carry molecules in their aqueous core (Ealia and Saravana Kumar 2017).

Nanotechnology is the science in which the size of the particle can be engineered and designed in a nanoscale <100 nm, which are called engineered nanomaterials (ENMs) or nanoparticles (NPs) (Valavanidis and Vlachogianni 2016). Nowadays, researchers are more focused on the nanoscale level which has novel applications for the bioremediation of oil from the ocean environment, due to the nanostructure such as sponges and diatoms constructed of coral reefs and silica (Hoek et al. 1995). Conventional approaches such as mechanical and chemical methods used for the improvement of marine life are not well ahead to reduce the unavoidable next oil spill. In recent years scientists are working on eco-friendly and cost-effective methods for the identification and elimination of harmful compounds from the marine environment (Mishra et al. 2022). Nanoscience is the twenty-first era technology that has a wide range of pharmaceutical uses such as immunology, endocrinology, oncology, and cardiology. Minimizing the particles to nano-size results in the formation of nanomaterial comprising a wide range of applications such as textile fictionalization, drug delivery, enzyme immobilization, biochemical engineering, and microbial degradation (Yezdani et al. 2018). The nanoparticles are present in various forms such as solid-lipid nanoparticles, carbon nanotubes, nanosponges, nanoemulsion, polymeric nanoparticles, micellar, and dendrimers systems (Maravajhala et al. 2012). In the last 10 years, nanotechnology is considered as one of the best techniques to combat chemical pollution from marine ecosystems. Advantages of nano-size particles include the following: (i) increase solubility, (ii) increase surface area, (iii) increase surface area, (iv) increase the oral bioavailability and rate of dissolution, and (v) requires less amount of dose in the pharmacy.

21.2.1 Types of Nano-size Particles

According to Siegel the types of nanoparticles are zero-dimensional, one-dimensional, two-dimensional, and three-dimensional. Polymeric biomaterials used in the formation of nanoparticles are classified as nano-spheres and nano-capsules. Based on the arrangement, nanoparticles are classified as (Norhasri et al. 2017):

- Inorganic nanoparticles such as copper, silver, gold, copper oxide, zinc oxide ceria, and titania.
- Organic–inorganic hybrid such as polymethyl methacrylate as the organic component and aluminum oxide as the inorganic component.
- Carbon nanoparticles.
- Combined nanoparticles such as graphene–polymeric-based nanostructure.

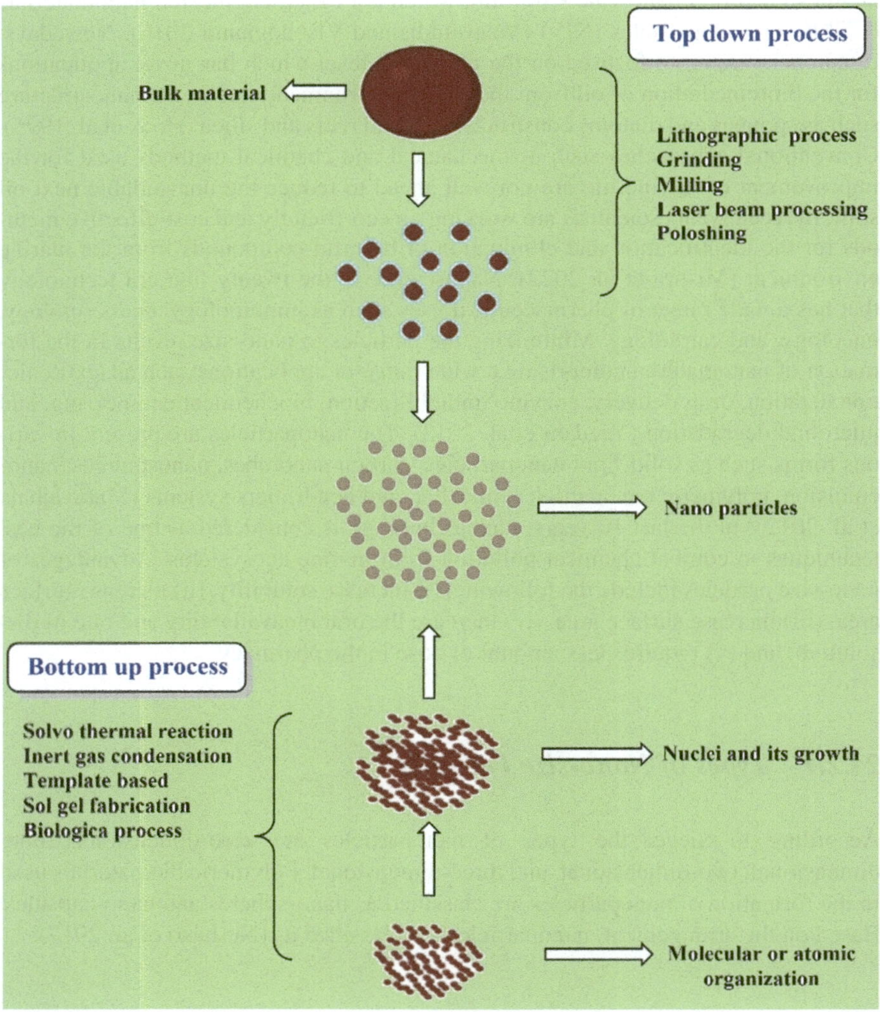

Fig. 21.2 Synthesis of nanoparticles

21.2.2 Approaches for the Formation of Nanoparticles

The approaches for the formation of nanoparticles are physical and chemical methods. A physicochemical method are the mostly used techniques such as chemical reduction, solvothermal, and sol-gel procedures (Alqahtani et al. 2018). The two different methods for the synthesis of nanostructure particle depends upon the size of the raw material such as top-down approach which makes the nanoparticle by using varied lithographic tools and the bottom-up approach by which nanosized particles can be synthesized by combining molecules and atom (Aryal et al. 2019). The synthesis of the nanoparticle is shown in Fig. 21.2. These conventional approaches are very common and extensively used for the synthesis of nano-size particles but have limitations such as being costly, use of toxic matters, and needs high energy. The researchers are more focused on developing an eco-friendly method for the generation of desired shape and size nanoparticles (Yaqoob et al. 2020).

21.2.3 Physiochemical and Biological Approach for the Formation of Nanostructured Particles

Several methods used for the formation of nanoparticles are physical, chemical, and biological. Physical approaches include lithography, high-energy radiation, and laser excision whereas chemical methods include phytochemical, electrochemical, and chemical reduction processes. The chemical process is less time-consuming and uses toxic chemicals which results in the formation of non-eco-friendly by-products (Din et al. 2017). The use of a chemical method for the synthesis of nanoparticles causes adverse effects on human health and the ecosystem. Thus, to evade risky things biological methods have been used for the synthesis of nanoparticles in an eco-friendly manner (Rafique et al. 2017).

The biological things such as plants and microorganisms for the formation of a nanoparticle are the bottom-up process. The nanoparticle of flexible shape and size can be synthesized by using different plants, but the garnering of rare plant species can be a risk. Hence researchers are more focused on marine microbes isolated from varied surroundings for the synthesis of nanoparticles and gained huge success with a variety of marine species (Nadaroglu et al. 2017). The biologically active compounds present in marine water have a notable impact in the area of biotechnology and pharmaceutical goods.

21.2.4 Plant-Based Marine Nanoparticle

The physical and chemical approaches for the synthesis of nanoparticles are highly expensive thus in order to reduce the certain costs for the synthesis of nanostructured materials, a scientist is more focused on the biological organism for the generation of nanoparticles. Nature has invented many inorganic materials which are useful for the biosynthesis of nanoparticles. Plants are easily accessible and safe sources for the generation of nanoparticles with varied metabolites. The phytochemicals in the plant such as ketones, aldehydes, amides, and terpenoids have been examined for the synthesis of nanoparticles. The bioactive agents present in marine plants are totally different from terrestrial plants. The bioactive agents produced by marine plants are alkaloids, polyphenols, tannins, and flavonoids due to the different environmental conditions of marine ecosystem (Gnanadesigan et al. 2011; Ravikumar et al. 2010). Mangroves are a well-known source for the synthesis of nanoparticles. One-pot synthesis of silver nanoparticles by using mangrove leaf shoots has been documented (Umashankari et al. 2012). To date, 26 coastal plants are known for the synthesis of silver nanoparticles (Asmathunisha 2010). The formation of nanoparticles by using plant extracts is eco-friendly and cost effective (Casida and Quistad 2005).

21.2.5 Marine-Microbes-Based Nanoparticles

Marine microbes are known to have unique identification such as they can tolerate extreme weather conditions and can tolerate high salt concentrations thus have vast potential for the synthesis of nanoparticles (Baker et al. 2013). A marine organism is in the beginning stage of the production of nanoparticles. Microbes are considered to be the most efficient method for the purification of water and the impurity from other organic sources. With the help of the bioaugmentation process (i.e., the process of adding cultured microorganisms to remove desired contaminants), biostimulation (i.e., it stimulates the microorganism which is involved in the bioremediation process), and combined the innate microbial concentration can be maintained (Becker and Seagren 2010). With the use of a silver mine in 1984 Pseudomonas stutzeri AG259 was discovered for the synthesis of nano-size particles. In 1989, Mullen et al. conducted an experiment on the biosorption of metals bacteria (Arun 2016). The biological route with unique shape and size are more facile compared to the other biological units such as biomolecules (amino acids, proteins, carbohydrates, etc.), plant resources (leaves, roots, barks, roots, and fruits), and different microorganism (fungi, algae, and bacteria) for the synthesize nanoparticles (Huang et al. 2009). The interaction between microbes and metals has various biological applications such as biomineralization, bioremediation, biocorrosion, and bioleaching. The nano remediation technique is more effective than the conventional method as the nanometric particles show high reactivity,

Table 21.1 Biological remediation of oil spill by using nanoparticle process

Microbes	Type of nanomaterials	Size
Bacillus subtilis	Au	10 nm
Lactobacillus sp.	Au, Ag, Au-Ag	20–50 nm
Aspergillus flavus	Ag	8–9 nm
Aspergillus fumigatus	Ag	5–25 nm
Thermonospora sp.	Au	8 nm
Fusarium solani	Ag	16–20 nm
Schizosaccharomyces pombe	CdS	1–1.5 nm
Sargassum algae	SiO2	50–100 nm
Fusarium acuminatum	Ag	5–40 nm
Fusarium oxysporum	Ag	16–20 nm

ability to capture target specific area, and has high surface area and volume ratio. The synthesis of nanoparticles in a cost-effective, non-toxic and eco-friendly way is one of the major challenges for recent nanotechnology (Ealia and Saravana Kumar 2017). Biotechnology has become one of the emerging routes for the synthesis of the nanoparticle by using biological resources such as plants and fungi. Natural phenomenon such as forest-fire smoke, ocean spray, volcanic activity, food cooking, and from vehicle exhausts all these natural phenomena consist of nanoparticles (Sonwani et al. 2021).

The microbes used for the formation of nanoparticles have been measured as a very eco-friendly and non-toxic method for the preparation of nano-size materials. The micro-organism has capability to grow at high metal ion concentrations due to its resistance power. The nanoparticles synthesized from the microbes have the ability to control the size of the particle by operating on the parameters such as temperature, pH, time limits to substrate, and substrate concentration. Over millions of years, marine organism stays at the bottom of the sea, thus decreasing the wide amount of inorganic substance from the sea. Moreover, the peptides such as phytochelatins stabilize the nanoparticles which are synthesized from the microbes by preventing accretion. Nanotechnology has emerged as one of the well-known areas of research over the past few decades due to its increasing demand in the petroleum oil, and bioremediation process of the marine ecosystem. Microorganisms derived from mangroves such as *Aspergillus niger*, *Thraustochytrids*, *Escherichia coli*, and *Penicillium fellutanum* have the ability to reduce the silver ions with various antimicrobial activities. The mangrove-derived yeast such as *Rhodosporidium diobovatum* and *Pichia capsulata* also have the ability to synthesize nanoparticles. *Idiomarina* sp. PR58-8 are the nanoparticles synthesized by marine bacteria that possess a high silver tolerant capacity. Table 21.1 shows biological remediation of oil spills by using the nanoparticle process.

Table 21.2 Types of nanoparticles used in bioremediation process

Sr no.	Nanotechnology used	Modification	Microbes used	Mechanism	Characteristics
1.	Cobalt and cobalt oxide nanomaterials	Microwave and reductive chemical heating	…	Radiation and large surface area.	Cost-effective, and easy to make the photocatalytic degradation efficiency
2.	NiO and MgO nanoparticles	Silica-fixed	..	Fast, endothermic, and physical adsorption of Cu^{2+} and Cr^{3+} and exothermic and chemical of Zn^{2+}	Renewal and reusable proved sustainability
3.	Zirconia nanoparticles	Synthesis from microbial cell-free culture supernatant	*Pseudomonas aeruginosa*	Chemisorption and strong electrostatic interaction among Zwitter ions	Green synthesis of nanomaterial and sustainable biological remediation
4.	Silica nanoparticles	Produced from actinomycetes	*Actinomycetes*	Photocatalytic decomposition	Cost-effective and sustainable
5.	Graphene oxide and carbon nanotubes	Nano-sized nickel metal-organic structure	…	Hydrophobic/pi–pi interaction, more surface area, and the wide morphological features of mixed nanomaterials	High interaction of nanomaterials

21.3 Types of Nanoparticles Involved in Petroleum Degradation

Different physical chemical and biological properties of the nanoparticles such as surface charge, specific surface area, morphology, and small particle size distribution make them appropriate for the degradation of crude oil from the marine ecosystem. Based on their synthesis technique metals and their oxide nanoparticle, that is, Fe_2O_3, Fe_3O_4, MgO, ZnO, $BaSO_4$, and TiO_2), metal nanoparticles that is Rh, Pd, Ir, Ag, Au, ZnS, and carbon-based nanoparticles such as nanotubes and nanocomposites are the efficient method to for the removal of petroleum oil from the ocean environment. A nanoparticle used in the bioremediation process is shown in Table 21.2. The bioremediation method involves photocatalytic degradation, adsorption, filtration through nanoparticles, and removal of contaminated waste from marine water. The nanoparticles synthesized from the microorganism were able to remove 90% of the heavy metals such as Cu(II), Zn(II), Ni(II), and Pb(II).

21.4 Bioaugmentation and Its Application to Bioremediation Process

The bioaugmentation process is one of the most significant techniques for the bioremediation of crude oil from the marine ecosystem (El Fantroussi and Agathos 2005). The environmental conditions at the contaminated site decide the existence and degrading ability of a microorganism. The strain isolated from site A may not be applicable for site B, and not necessarily that the organism synthesized under laboratory conditions can be effective in situ for the removal of oil (Sayler and Ripp 2000). The bioaugmentation process works in areas where the inherent microbial groups are incapable of deteriorating all organic matters, or where due to oil exposure all the native species promote a lag phase (Tyagi et al. 2011., Bento et al. 2005; Leahy and Colwell 1990). The biodegradation of crude oil present in the marine ecosystem can be improved by genetically modified organisms. The microorganism present range from 0.003% to 100% of the microbial community enhance the bioremediation process (Omarova et al. 2012; Zhang et al. 2018).

21.4.1 Cell Immobilization

Cell immobilization is one of the most efficient processes in biotechnology which limits the mobility of the particular cells that are confined to a particular region, without affecting their biological activity (Adebajo et al. 2003; Li et al. 2017; Xue et al. 2017; Oh et al. 2000). The bioencapsulation or microencapsulation terms are used for the encapsulated cell in an immobilized cell system. Nowadays, the nanoencapsulation of biological molecules is more in trend (John et al. 2011). Many techniques for cell immobilization have been developed in the last 3 years. The microbes have the ability to attach to the surfaces of the solid or liquid which forms biofilms to protect from the environmental stress which is tense to the formation of microbial communities. Cell immobilization has many advantages such as (Martins et al. 2013):

- A high flow rate allows high volumetric productivity.
- Providing high biomass.
- Defense against sheer damage.
- High resistance to toxic materials, pH, temperature, heavy metals, and solvents.
- Improving genetic stability.
- Providing suitable micro-environmental conditions.

Scientists are focusing on the use of biofilms for external protection by cell immobilization and to increase bioremediation. The microbial cells that are immobilized are organism whose movement is completely limited by attaching to a surface by entrapment, adsorption, or encapsulation. The immobilized microbial cells benefit in bioremediation oil spills and wastewater management. The cell immobilization bioremediation process is cost-effective and efficient due to the use of biocatalyst—minimizes the risk of genetic mutations, enhances the biocatalyst resistance towards adverse environment conditions, provides a stable microenvironment for enzymes and cells, assures resistance to shear forces present in the bioreactors, increase the tolerance to high pollutant concentration.

21.4.2 Carrier Attached to the Biofilms

The carriers which are used for cell immobilization should be non-biodegradable, non-polluting and non-toxic, highly stable, cost-effective, long shelf life, and facilitate the easy removal of cells from the media. The materials studied include inorganic clays, organic polymer, and inorganic polyelectrolyte coatings. Cell immobilization provides bioremediation of crude oil in an eco-friendly and cost-effective manner. The studies on cell immobilization on the mesocosm scale are going on to further enhance the scope. The techniques for cell immobilization are adsorption, encapsulation, entrapment, and covalent binding. Among all the methods entrapment technique is used the most (Dzionek et al. 2016).

21.4.3 Uses of Cell Immobilization

The Rhodococcus corynebacterioides QBTo immobilized on chitosan flakes and chitin was discovered by Gentili et al. (2006) which enhances the biodegradation of crude oil. Radwan et al. (2005) use natural immobilization which immobilizes oil-utilizing bacteria in biofilms coating macroalgae and also provides oxygen, nitrogen, and probably phosphorous compounds for oil-utilizing bacteria. Díaz et al. (2002) demonstrated the bioremediation of organic compounds in water with salinities vary from 0 to 108 g L^{-1} with the use of bacterial consortium MPD-M on polypropylene fibers. They found that the immobilized cells enhance the deterioration of crude oil as compared to free-living cells. Li et al. (2018) found the organic matter deteriorating bacteria immobilized on peanut body powder as biocarrier that removes the oil spill from the marine ecosystem. The degradation of the oil spill by using biocarrier offers strong adsorption capability and a large surface area which improves oxygen supply and increases the dehydrogenase activity in the marine ecosystem to prevent an oil spill.

21.4.4 Surfactants That Enhance the Bioremediation Process

Surfactants are the particles that are produced by microorganism for the absorption and degradation of organic compounds. Biosurfactants are a group of microorganisms produced by living organisms. They gained attention as the organic material dissolving agent and likely replace synthetic surfactants such as sulfonates, carboxylates, and sulfate acid esters which are present in the pharmaceutical, food, and oil factories. Surfactants play a key role in cleaning the contaminated oil surface of the marine ecosystem, which is greatly affected by organic material pollutants (Singh and Cameotra 2004). In this approach cleansing the organic matter from the marine system, the surfactant enhances the surface area and reduces the size. It reduces the oil-water interfacial tension and thus decreases the bacterial adhesion to the droplets. The micro-organism releases the surfactants which are adsorbed on the surface of the cells and thus enhance the hydrophobicity. By increasing the amount of surfactant in a water/oil or oil/water area decrease the surface tension of the bacteria up to a critical point known as critical micelle concentration (CMC) and forms bilayers, vesicles, or micelles. The polyelectrolyte coatings on the microbial cells form a thin film around the cell. This is one of the developing methods to increase the bioremediation process by enhancing the cell adherence to the oil-water interface thus rise the degradation strengths. Surfactants are amphiphilic molecules having hydrophilic and hydrophobic portions. The hydrophilic part contains positively or negatively charged normally a carbohydrate, an amino acid, or a phosphate and the hydrophobic part usually contains a long fatty acid chain. The bacteria and yeast have been studied for the production of biosurfactants (Mohanty et al. 2013).

21.5 Biostimulation

Biostimulation is the process in which growth-controlling nutrients are added for the native microbe's group and also involves the use of oleophilic fertilizers which enhance the inorganic nutrients such as phosphorus, potassium, and nitrogen. This process led to an increase in the growth of the oil-decomposition organism. The biostimulation process in situ depends upon the environmental conditions such as pH, temperature, and nutrient accessibility (such as phosphorus and nitrogen). Based on the changes in weather, native microbes, and oil composition, the biostimulation process results in a change of microbial group in a site-specific manner (Villela et al. 2019). In Exxon Valdez oil spill, the coastlines were treated with olefins fertilizers Inipol EAP 22 (0.7% phosphorus and 7.4% nitrogen) and slow release of customblen fertilizer (3.5% phosphorus and 28% nitrogen). It was observed that the location was treated with 63% decomposition of total visible organic compound and 44% decomposition of polynuclear aromatic organic compound over 109 days. In comparison, the sites which are not treated with fertilizers showed no remarkable decomposition even after 70 days. When the fertilizers were added after 70 days, decomposition same as that happen initially happened, which proves the efficacy of biostimulation by the addition of fertilizer (Hassanshahian et al. 2014).

Sometimes the biostimulation process seems to be ineffective due to the dilution of the additional nutrients, thus by the addition of the water-insoluble uric acid, the effectiveness can be increased. Due to the less water solubility of uric acid in the water, it is used as a source of nitrogen near the oil–water interface. The oil droplets are used as a source for the release of nutrients which proves helpful in the oil decomposition microbe's process. Usually, all the microbes need phosphorus and nitrogen, thus duplication of the unwanted microbes, which leads to competition around for organic-derogatory group (Wang et al. 2012). Thus, the biostimulation process is considered as one of the most effective methods for the removal of an oil spill from the ocean environment. More research for the betterment of this process needs to be done. There are wide range of developments, such as surface change, immobilization, and advanced method of biostimulation which makes a better understanding of the microbe's behavior at oil–water surface. The nanoscience utilization along with the marine microbes increases the efficiency of the bioremediation process (Koren et al. 2003).

21.6 Future Prospects and Conclusion

This chapter deals with the progress made in the nano-size particles and its remediation method for the eradication of oil from the marine ecosystem (Archakunakorn et al. 2015). The researchers are working on the field of nano-enhanced methods as it provides novel chance and functionalities such as:

- Increase microbial spread on oil particles compared with toxic materials.
- Environment and sustainable friendly methods have been used for the crude oil spill response.

Many limitations and tasks such as:

- The biological and environmental inferences of the nanoparticles to the marine environment.
- Nanomaterial separation from the marine environment.
- Due to the wide range of native populations and the difference in pH, salinity, temperature, and other environmental situations, there is difficulty in selecting the appropriate approach for increasing the biodegradation process.

Thus, the researchers are more focused on the eco-friendly approach as a remediation process such as the use of natural polymers and a clay as an emulsifier. The emulsifier used to increase the microbial spread includes zein, chitosan, and lignin. Nanotechnology combined with microbes gives a greener approach to the eradication of organic matter from the marine environment. Nanoparticles combined with microorganism provides an effective and cost-effective remediation. Nanoparticle provides long-lasting, stability in harsh conditions, can be used multiple times, provides more surface area, easily do changes in materials, and are more efficient method than the traditional methods for the removal of oil from the marine environment. The nanomaterials designed with the use of natural precursors are cost-effective, biologically compatible, and eco-friendly which leads to the generation of a new class of organic matter. Moreover, the risk associated with the chemically synthesized particles can be reduced by the use of microbes. Remediation combined with biology is one of the most effective methods for the removal of crude oil from marine matrices but the nanotechnology approach leads to the development of the natural process. Nano-enhanced bioremediation method plays a significant role in the removal of crude oil from the marine environment. These include enhancing the oil recovery, decomposition of crude oil, reduces the economic and ecological impact on the oil spill place. Researchers are working on an effective bioremediation method combined with nanoparticles to increase the crude oil cleanup from marine systems. The greatest achievement for oil spill management is the use of the Pickering emulsifier which enhances the microbes spread onto the oil droplets and increases the decomposition rate. Further the effects of nanoparticle synthesis on large scale have been studied, it is performed by using mesocosm study for bioremediation, which shows results in a condition similar to the original oil spill surroundings such as temperature, pH, agitation, and salinity. This chapter gives a broad outline of the nanotechnology approach and its effective bioremediation and brings awareness to the present need for more research in the field of bioremediation combining nanotechnology for the oil spill cleanup process. A major focus is on the cost-effective alternative for the bioremediation process, after conducting cost analysis studies it is concluded that the nanoparticle remediation process is a less-cost alternative compared to traditional methods (Feng and Lee 2016).

References

Adebajo MO, Frost RL, Kloprogge JT, Carmody O, Kokot S (2003) Porous materials for oil spill cleanup: a review of synthesis and absorbing properties. J Porous Mater 10(3):159–170

Alqahtani MM, Ali AM, Harraz FA, Faisal M, Ismail AA, Sayed MA, Al-Assiri MS (2018) Highly sensitive ethanol chemical sensor based on novel Ag-doped mesoporous α–Fe 2 O 3 prepared by modified sol-gel process. Nanoscale Res Lett 13(1):1–13

Amro MM (2004) Factors affecting chemical remediation of oil contaminated water-wetted soil. Chem Eng Technol: Ind Chem Plant Equip Process Eng Biotechnol 27(8):890–894

Annunciado TR, Sydenstricker THD, Amico SC (2005) Experimental investigation of various vegetable fibers as sorbent materials for oil spills. Mar Pollut Bull 50(11):1340–1346

Archakunakorn S, Charoenrat N, Khamsakhon S, Pongtharangkul T, Wongkongkatep P, Suphantharika M, Wongkongkatep J (2015) Emulsification efficiency of adsorbed chitosan for bacterial cells accumulation at the oil–water interface. Bioprocess Biosyst Eng 38(4):701–709

Arun N (2016) Isolation and identification of microfungi from Ariyankuppam mangrove soil and their application as nanomaterials. In: Doctoral dissertation, Department of Botany, KMCPGS, Pondicherry University

Aryal S, Park H, Leary JF, Key J (2019) Top-down fabrication-based nano/microparticles for molecular imaging and drug delivery. Int J Nanomedicine 14:6631

Asmathunisha N (2010) Studies on silver nanoparticles synthesized by mangroves, salt marshes and associated plants. In: Doctoral dissertation, M. Phil Thesis, CAS in marine biology, Annamalai University, India
Baker S, Harini BP, Rakshith D, Satish S (2013) Marine microbes: invisible nanofactories. J Pharm Res 6(3):383–388
Becker JG, Seagren EA (2010) Bioremediation of hazardous organics. Environ Microbiol 177:212–230
Bento FM, Camargo FAO, Okeke BC, Franken Berger WT (2005) Comparative bioremediation of soils contaminated with diesel oil by natural attenuation, biostimulation and bioaugmentation. Bioresour Technol 96(9):1049–1055
Casida JE, Quistad GB (2005) Serine hydrolase targets of organophosphorus toxicants. Chem Biol Interact 157:277–283
Cirer-Costa JC (2015) Tourism and its hypersensitivity to oil spills. Mar Pollut Bull 91(1):65–72
Dave DAEG, Ghaly AE (2011a) Remediation technologies for marine oil spills: a critical review and comparative analysis. Am J Environ Sci 7(5):423
Dave D, Ghaly AE (2011b) Meat spoilage mechanisms and preservation techniques: a critical review. Am J Agric Biol Sci 6(4):486–510
Delvigne GAL, Sweeney C (1988) Natural dispersion of oil. Oil Chem Pollut 4(4):281–310
Díaz MP, Boyd KG, Grigson SJ, Burgess JG (2002) Biodegradation of crude oil across a wide range of salinities by an extremely halotolerant bacterial consortium MPD-M, immobilized onto polypropylene fibers. Biotechnol Bioeng 79(2):145–153
Din MI, Arshad F, Hussain Z, Mukhtar M (2017) Green adeptness in the synthesis and stabilization of copper nanoparticles: catalytic, antibacterial, cytotoxicity, and antioxidant activities. Nanoscale Res Lett 12(1):1–15
Dzionek A, Wojcieszyńska D, Guzik U (2016) Natural carriers in bioremediation: a review. Electron J Biotechnol 23:28–36
Ealia SAM, Saravana Kumar MP (2017) A review on the classification, characterisation, synthesis of nanoparticles and their application. IOP Conf Ser: Mater Sci Eng 263(3):032019. IOP Publishing
El Fantroussi S, Agathos SN (2005) Is bioaugmentation a feasible strategy for pollutant removal and site remediation. Curr Opin Microbiol 8(3):268–275
Elliott ET (1986) Aggregate structure and carbon, nitrogen, and phosphorus in native and cultivated soils. Soil Sci Soc Am J 50(3):627–633
Feng Y, Lee Y (2016) Surface modification of zein colloidal particles with sodium caseinate to stabilize oil-in-water pickering emulsion. Food Hydrocoll 56:292–302
Gentili AR, Cubitto MA, Ferrero M, Rodriguéz MS (2006) Bioremediation of crude oil polluted seawater by a hydrocarbon-degrading bacterial strain immobilized on chitin and chitosan flakes. Int Biodeterior Biodegrad 57(4):222–228
Gill RA, Robotham PWJ (1989) Composition, sources and source identification of petroleum hydrocarbons and their residues. In: The fate and effects of oil in freshwater. Springer, Dordrecht, pp 11–40
Gnanadesigan M, Anand M, Ravikumar S, Maruthupandy M, Vijayakumar V, Selvam S, Dhineshkumar M, Kumaraguru AK (2011) Biosynthesis of silver nanoparticles by using mangrove plant extract and their potential mosquito larvicidal property. Asian Pac J Trop Med 4(10):799–803
Hassanshahian M, Emtiazi G, Caruso G, Cappello S (2014) Bioremediation (bioaugmentation/biostimulation) trials of oil polluted seawater: a mesocosm simulation study. Mar Environ Res 95:28–38
Hoek C, Mann D, Jahns HM, Jahns M (1995) Algae: an introduction to phycology. Cambridge University Press, Cambridge
Huang DW, Sherman BT, Zheng X, Yang J, Imamichi T, Stephens R, Lempicki RA (2009) Extracting biological meaning from large gene lists with DAVID. Curr Protoc Bioinformatics 27(1):13–11

Hussein A (2009) The use of triangulation in social sciences research. J Comp Soc Work 4(1):106–117
Jernelöv A (2010) The threats from oil spills: now, then, and in the future. Ambio 39(5):353–366
John RP, Tyagi RD, Brar SK, Surampalli RY, Prévost D (2011) Bio-encapsulation of microbial cells for targeted agricultural delivery. Crit Rev Biotechnol 31(3):211–226
Jordan RE, Payne JR (1980) Fate and weathering of petroleum spills in the marine environment: a literature review and synopsis. Ann Arbor Science, Ann Arbor
Kapoor S, Rawat HS (1994, January) Indian west coast oil spills: a remedial preparedness. In: SPE health, safety and environment in oil and gas exploration and production conference. OnePetro
Koren O, Knezevic V, Ron EZ, Rosenberg E (2003) Petroleum pollution bioremediation using water-insoluble uric acid as the nitrogen source. Appl Environ Microbiol 69(10):6337–6339
Leahy JG, Colwell RR (1990) Microbial degradation of hydrocarbons in the environment. Microbiol Rev 54(3):305–315
Lessard RR, DeMarco G (2000) The significance of oil spill dispersants. Spill Sci Technol Bull 6(1):59–68
Li Y, Gong H, Cheng H, Wang L, Bao M (2017) Individually immobilized and surface-modified hydrocarbon-degrading bacteria for oil emulsification and biodegradation. Mar Pollut Bull 125(1–2):433–439
Li Z, Wu H, Yang M, Jiang J, Xu D, Feng H, Lu Y, Kang W, Bai B, Hou J (2018) Spontaneous emulsification via once bottom-up cycle for the crude oil in low-permeability reservoirs. Energy & Fuels 32(3):3119–3126
Lucas Z, MacGregor C (2006) Characterization and source of oil contamination on the beaches and seabird corpses, Sable Island, Nova Scotia, 1996–2005. Mar Pollut Bull 52(7):778–789
Maravajhala V, Papishetty S, Bandlapalli S (2012) Nanotechnology in development of drug delivery system. Int J Pharm Sci Res 3(1):84
Martins SCS, Martins CM, Fiúza LMCG, Santaella ST (2013) Immobilization of microbial cells: a promising tool for treatment of toxic pollutants in industrial wastewater. Afr J Biotechnol 12(28):4412–4418
Mishra S, Chauhan G, Verma S, Singh U (2022) The emergence of nanotechnology in mitigating petroleum oil spills. Mar Pollut Bull 178:113609
Mohanty S, Jasmine J, Mukherji S (2013) Practical considerations and challenges involved in surfactant enhanced bioremediation of oil. Biomed Res Int 2013:328608
Murad AA, Al Nuaimi H, Al Hammadi M (2007) Comprehensive assessment of water resources in the United Arab Emirates (UAE). Water Resour Manag 21(9):1449–1463
Nadaroglu H, Güngör AA, Selvi İNCE (2017) Synthesis of nanoparticles by green synthesis method. Int J Innov Res Rev 1(1):6–9
Norhasri MM, Hamidah MS, Fadzil AM (2017) Applications of using nano material in concrete: a review. Constr Build Mater 133:91–97
Oh YS, Maeng J, Kim SJ (2000) Use of microorganism-immobilized polyurethane foams to absorb and degrade oil on water surface. Appl Microbiol Biotechnol 54(3):418–423
Omarova EO, Lobakova ES, Dolnikova GA, Nekrasova VV, Idiatulov RK, Kashcheeva PB, Perevertailo NG, Dedov AG (2012) Immobilization of bacteria on polymer matrices for degradation of crude oil and oil products. Mosc Univ Biol Sci Bull 67(1):24–30
Parker H, Pitt GD (2012) Pollution control instrumentation for oil and effluents. Springer Science & Business Media, Graham & Trotman Inc., Norwell, MA
Radwan SS, Dashti N, El-Nemr IM (2005) Enhancing the growth of Vicia faba plants by microbial inoculation to improve their phytoremediation potential for oily desert areas. Int J Phytoremediation 7(1):19–32
Rafique M, Sadaf I, Rafique MS, Tahir MB (2017) A review on green synthesis of silver nanoparticles and their applications. Artif Cells Nanomed Biotechnol 45(7):1272–1291
Raimi MO, Nimisngha D, Odipe OE, Olalekan AS (2018) Health risk assessment on heavy metals ingestion through groundwater drinking pathway for residents in an oil and gas producing area of Rivers State, Nigeria. Open J Yangtze Oil Gas 3:191–206

Ravikumar B, Sarkar S, Davies JE, Futter M, Garcia-Arencibia M, Green-Thompson ZW, Jimenez-Sanchez M, Korolchuk VI, Lichtenberg M, Luo S, Massey DC (2010) Regulation of mammalian autophagy in physiology and pathophysiology. Physiol Rev 90(4):1383–1435

Rogowska J, Namieśnik J (2010) Environmental implications of oil spills from shipping accidents. Rev Environ Contam Toxicol 206:95–114

Sayler GS, Ripp S (2000) Field applications of genetically engineered microorganisms for bioremediation processes. Curr Opin Biotechnol 11(3):286–289

Schein EH (2009) The corporate culture survival guide, vol 158. John Wiley & Sons, Hoboken

Singh P, Cameotra SS (2004) Enhancement of metal bioremediation by use of microbial surfactants. Biochem Biophys Res Commun 319(2):291–297

Sonwani S, Madaan S, Arora J, Suryanarayan S, Rangra D, Mongia N, Vats T, Saxena P (2021) Inhalation exposure to atmospheric nanoparticles and its associated impacts on human health: a review. Front Sustain Cities 3:690444

Tyagi M, da Fonseca MMR, de Carvalho CC (2011) Bioaugmentation and biostimulation strategies to improve the effectiveness of bioremediation processes. Biodegradation 22(2):231–241

Umashankari J, Inbakandan D, Ajithkumar TT, Balasubramanian T (2012) Mangrove plant, Rhizophora mucronata (Lamk, 1804) mediated one pot green synthesis of silver nanoparticles and its antibacterial activity against aquatic pathogens. Aquat Biosyst 8(1):1–7

Valavanidis A, Vlachogianni T (2016) Engineered nanomaterials for pharmaceutical and biomedical products new trends, benefits and opportunities. Pharm Bioprocess 4(1):13–24

Ventikos NP, Sotiropoulos FS (2014) Disutility analysis of oil spills: graphs and trends. Mar Pollut Bull 81(1):116–123

Villela HD, Peixoto RS, Soriano AU, Carmo FL (2019) Microbial bioremediation of oil contaminated seawater: a survey of patent deposits and the characterization of the top genera applied. Sci Total Environ 666:743–758

Vlaev L, Petkov P, Dimitrov A, Genieva S (2011) Cleanup of water polluted with crude oil or diesel fuel using rice husks ash. J Taiwan Inst Chem Eng 42(6):957–964

Wang X, Wang Q, Wang S, Li F, Guo G (2012) Effect of biostimulation on community level physiological profiles of microorganisms in field-scale biopiles composed of aged oil sludge. Bioresour Technol 111:308–315

Wypych G (2018) Handbook of material weathering. Elsevier, San Diego

Xue JL, Wu YN, Liu ZX, Li ML, Sun XY, Wang HJ, Liu B (2017) Characteristic assessment of diesel-degrading bacteria immobilized on natural organic carriers in marine environment: the degradation activity and nutrient. Sci Rep 7:8635

Yaqoob AA, Parveen T, Umar K, Mohamad Ibrahim MN (2020) Role of nanomaterials in the treatment of wastewater: a review. Water 12(2):495

Yezdani U, Khan MG, Kushwah N, Verma A, Khan F (2018) Application of nanotechnology in diagnosis and treatment of various diseases and its future advances in medicine. World J Pharm Pharm Sci 7(11):1611–1633

Zhang Y, Gao W, Lin F, Han B, He C, Li Q, Gao X, Cui Z, Sun C, Zheng L (2018) Study on immobilization of marine oil-degrading bacteria by carrier of algae materials. World J Microbiol Biotechnol 34(6):1–8

Zhang B, Matchinski EJ, Chen B, Ye X, Jing L, Lee K (2019) Marine oil spills—oil pollution, sources and effects. In: World seas: an environmental evaluation. Elsevier Ltd. Academic Press, pp 391–406

Part VI
Socio-economic Scenarios Related to Sustainable Development

Chapter 22
Impact of COVID-19 Pandemic on Coastal Tourism of Andaman Isles, India: Sustainable Development Scenario

N. Jayaraju, G. Sreenivasulu, M. Madakka, B. Lakshmanna, K. Nagalakshmi, M. Pramod Kumar, T. Lakshmi Prasad, and M. Swarna Pragathi

Abstract The idyllic tourist atmosphere in the beautiful Andaman and Nicobar Islands (ANIs) nature's floating paradise, which relatively had fewer cases of the COVID-19 virus, as compared to mainland India, was suddenly shattered. The locals depend on the tourists for their livelihood and with the sudden surge of COVID-19 cases in the country and other nations, the government imposed a lockdown and the ensuing travel ban put everything on hold. The psycho-economic fallout had just begun as people were grappling with what was happening all around and what lay ahead. Coastal tourism, which is considered to be the world's key economic area, is a huge economy booster for the nation's GDP, and the Island tourism suddenly shut down along with the deficit revenue to the government. The worst affected were those whose livelihoods were involved with tourists on a daily basis.

N. Jayaraju (✉)
Department of Geology, Yogi Vemana University, Kadapa, Andhra Pradesh, India

G. Sreenivasulu
Department of Geology, Sri Venkateswara University, Tirupati, Andhra Pradesh, India

M. Madakka
Department of Biotechnology and Bioinformatics, Yogi Vemana University, Kadapa, Andhra Pradesh, India

B. Lakshmanna
Department of Energy and Environmental Engineering, CSIR-Indian Institute of Chemical Technology, Hyderabad, Telangana, India

K. Nagalakshmi
Department of Geology, Government College (Autonomous), Anantapur, Andhra Pradesh, India

M. Pramod Kumar · T. Lakshmi Prasad
Department of Earth Science, Yogi Vemana University, Kadapa, Andhra Pradesh, India

M. Swarna Pragathi
Department of Economics, Yogi Vemana University, Kadapa, Andhra Pradesh, India

N. Jayaraju et al. (eds.), *Coasts, Estuaries and Lakes*,
https://doi.org/10.1007/978-3-031-21644-2_22

Thus, the initiatives of special packages by the government and NGOs are released in order to boost the tourism activities. This work mainly addresses the effect of this deadly pandemic on the lives of the people, more so its governmental implications on the psychological and socioeconomic conditions of the Islanders. The government subsidies and necessary measures for sustainability can help them tide over the crisis. Now the government has administrated the vaccination program and initiated the opening of many recreation places for the public. Gradually, the government keenly executed the necessary steps to bring life to normal.

Keywords COVID-19 pandemic · Coastal tourism · Governance · Andaman Islands · India

22.1 Introduction

The rocks, landforms, and sands of Andaman and Nicobar Islands (ANIs) are linked closely with most other aspects of the unique heritage found here. Also known as the Emerald Islands of India for its spectacular natural beauty in the form of pristine beaches, exotic flora and fauna, and rare marine life, the ANI makes for an attractive tourist destination for mainland Indian and international tourists. Being an eco-friendly tourist hotspot makes it all the more appealing to the travelers. Physical processes drive many habitat changes, as in river and coastal environments. They also moderate other reactions and energy flows that impinge on ecological processes, as in the retention or release of pollutants in soils. Rocks, landforms, and soils constitute economic resources that are utilized by the minerals industry, for food and timber production, and for recreation, tourism, and education. At a time when interest is growing in sustainable use of natural resources, landscape interpretation, coastal tourism, and integrated management based on knowledge of physical processes, it is time to address more explicitly the broader links of the earth sciences to ecosystems and landscape. Even as the Island tourism was bustling with activity, the global COVID-19 pandemic struck and life and livelihoods came to a standstill. It caused tremendous havoc in the lives of people of all economic strata across the nations. People lost their lives and jobs while healthcare and livelihoods were badly affected due to the outbreak. With no exemption, nearly every industry has been badly affected. And for those in the hospitality industry, especially Island tourism, which is largely dependent on tourists, it has been a double setback.

The Andaman and Nicobar Islands are a breathtaking Archipelago of over 500 islands located in the Bay of Bengal. Exotic, sandy beaches, many of them remote and unpopulated, surrounded by rugged forests, the Andamans is a nature lover's delight and an attractive tourist destination. The coral reefs, mangroves, tropical forests, stunning marine life, and rare flora and fauna make the Islands a hugely popular, affordable holiday getaway for tourists and nature lovers (Sharma et al. 2019). The thriving tourism sector across the world took a huge hit owing to the massive outbreak of this global deadly COVID-19 pandemic (Mulder 2020). More so, in small islands like ANIs where tourism contributes significantly to the economy, it took a toll on the lives of people, leaving families stranded and desperate in terms of earnings and also impacting mental health (Sangeetha 2006).

The global pandemic caused a massive dent in the economic activities and supply chains (Lenzen et al. 2020). This has a direct impact on agricultural products, labor force, and food supply chains (Marchant-Forde and Boyle 2020). The gross revenue of the tourism sector is key for the sustainable development of society. It generates revenue from foreign (12.8%) and domestic (87.2%) tourists (Ministry of Tourism, Govt. of India, 2020–21). This sector also provides massive employment for people to derive their livelihood from it. The tourism industry in India has also been adversely affected by COVID-19 (UNCTAD 2020; Chandel et al. 2021). The sector is expected to crash a revenue loss of Rs 1.25 trillion in 2020. This is because of the closure of hotels and suspension of air traffic operations during the outbreak of the coronavirus (COVID-19) pandemic across the globe (Mohit 2020). About a 40% decline in revenue has been recorded in 2019. From April to June 2020, the tourism sector in India is projected to book a loss of Rs 69,400 crore (The Economic Times 2020).

22.2 Location

The Andaman and Nicobar Islands, a union territory of India, lie in the Indian Ocean, over 600 km east of the coast of Myanmar. It is located between the Bay of Bengal and the Andaman Sea (Fig. 22.1). Most of the 572 Islands are uninhabited and covered by thick forest areas. They are a rare blend of geodiversity and biodiversity, supported by amazing marine life, coral reefs, rare species of birds and animals, and sparkling beaches. It is a tourist haven, for pleasure and fun as well as the nature explorer. There is something unique about the Islands for every tourist soaking in the peaceful environs of its verdant natural beauty. Some of the remote Islands inhabited by the Indigenous tribes are closed to visitors.

22.3 Materials and Methods

The statistics of foreign/domestic tourists of different countries/states to ANI, from March 2020 to June 2021 (15 months) is generated. The data sets are obtained from the CMIE (Centre for Monitoring Indian Economy) and the annual reports of the Ministry of Tourism, Govt. of India. In addition, to analyze the impact of tourism on the economy, monthly foreign exchanges are obtained for the above period.

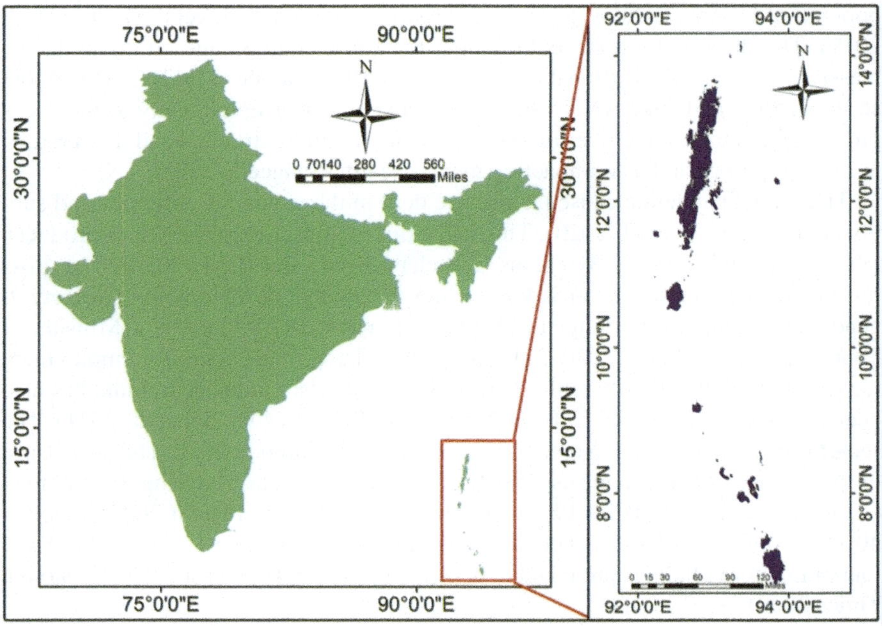

Fig. 22.1 Location map of the study area

22.4 Tourist Attractions in the ANI

The beautiful locales in the Islands are a travelers' delight. We have highlighted some of the popular tourist destinations – nature in all its splendor – whether it is the beaches, parks, or captivating marine life (Fig. 22.2). The archipelago is mostly surrounded by forest cover and is breathtaking on all sides.

22.5 Havelock Islands

Havelock Islands is one of the most popular tourist attractions, for its untouched natural beauty. The sandy beaches, sparkling blue waters, jungle animals, and birds around the jungles, make it a delightful tourist spot. It is famous for Radhanagar Beach, which is the most visited tourist site in the ANI for its stunning locale. Also, one can do any of the fun activities like Scuba diving, Snorkeling, and Kayaking

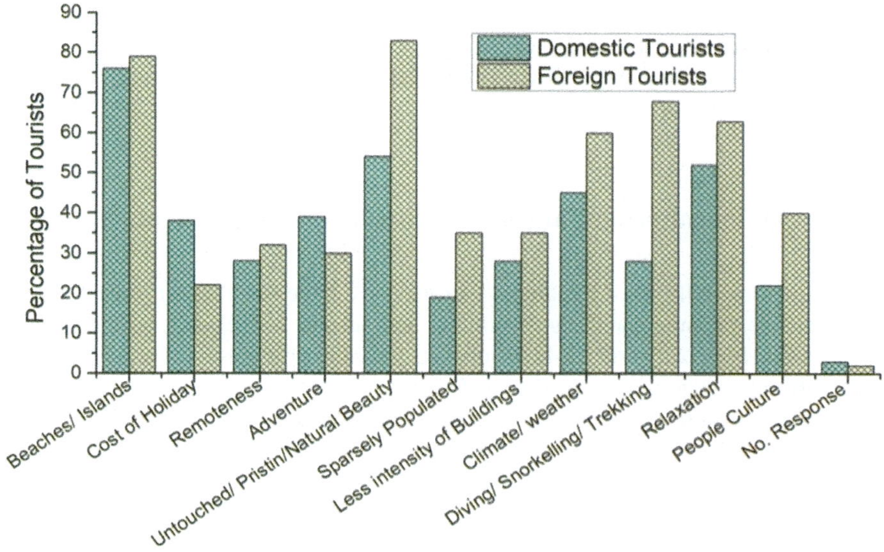

Fig. 22.2 Tourist attractions and reasons to visit ANI

22.6 Baratang Island

This Island is famous for its indigenous Jarawa Tribes and its Limestone Caves and Mud Volcanoes.

22.6.1 Neil Island

This Island is famous for coral reefs and people who enjoy snorkeling. It is also known as the coral capital of Andamans. One can also visit Bharatpur Beach and Natural Bridge on the Island.

22.6.2 Port Blair

Port Blair has a charm of its own, with its famed architecture and historical background. The Cellular Jail, which is a national memorial now, was built by the British and is notoriously famous for imprisoning the Indian freedom fighters for their dissidence against British rule. It was also known as "Kala Pani" or black waters from where the prisoners could not escape.

The stories of the freedom fighters are proudly displayed soldiers on the walls of the Jail, showcasing their struggles.

The light and sound show thrice a week, depicting the freedom struggle is one of the key attractions of the Cellular Jail.

22.7 Socioeconomic Fallout Versus Government Involvement

The immediate fall-out of the global Coronavirus is the travel and tourism sectors that bring in revenue for the government, and sustain livelihoods and businesses (Table 22.1; Figs. 22.3 and 22.4). It supports jobs and provides livelihoods and creates a strong psychological mindset and socioeconomic structure in society. Due to the sudden travel restrictions as a result of the government-imposed lockdown and total ban on flying within and between countries, Island tourism came to a complete standstill. This in turn grossly affected the nation's GDP, as it has for other small island economies that are mostly dependent on tourism for revenue (Table 22.2; Figs. 22.5 and 22.6). Apart from this, the livelihood and businesses of people that centered around the tourists came to a grinding halt. The total foreign tourists in 2019 are 10.9 million and the foreign exchange earnings were INR 210,971 crore. The states like Maharashtra, Tamil Nadu, Uttar Pradesh, and Delhi accounted for 60% of foreign tourist arrivals (The Economic Times 2020).

The fall of GDP rate from 8.2% in January–March 2018 to 3.1% in January–March 2020 is a sure indicator of economic fallout. The numbers were on the

Table 22.1 Month-wise COVID-19 confirmed, recovered, and death cases in ANI from March 2020 to June 2021

Month and year	Confirmed	Recovered	Deaths
Mar 20	36	0	0
Apr 20	485	211	0
May 20	1023	969	0
Jun 20	1352	1075	0
Jul 20	6372	3809	8
Aug 20	63,866	38,101	794
Sep 20	105,810	96,860	1526
Oct 20	126,128	118,562	1726
Nov 20	136,011	129,659	1817
Dec 20	149,755	145,485	1901
Jan 21	154,187	151,473	1922
Feb 21	140,209	138,309	1736
Mar 21	156,054	153,888	1922
Apr 21	160,355	155,404	1905
May 21	218,176	211,365	3743
Jun 21	203,089	193,479	2728

Source: https://nidm.gov.in/covid19/

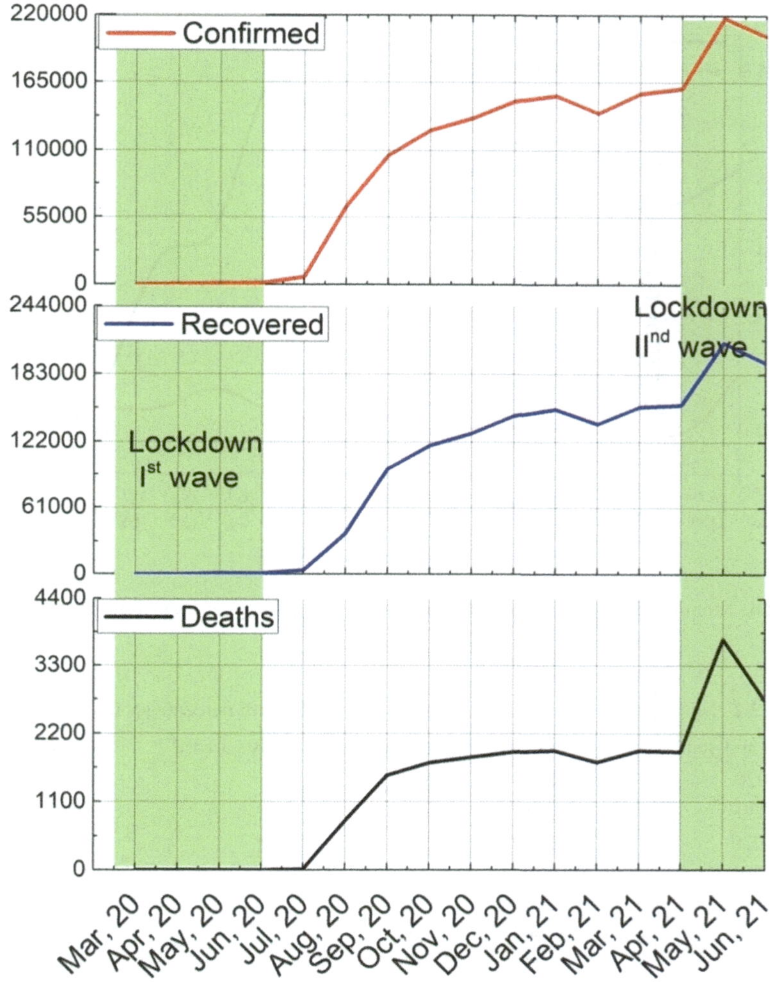

Fig. 22.3 Month-wise COVID-19 confirmed recovered and death cases in ANI from March 2020 to June 2021. (Source: https://nidm.gov.in/covid19/)

decline early in the years 2020–2021. The GDP all-time low growth rate (23.9%) was recorded for April–June 2020. Crucial sectors like construction, manufacturing, trade, tourism, and the hotel industry saw a crunch and declined further down. The tourism and hotel industry growth stood at −47% (The Economic Times 2021).

India is one of the popular holiday destinations for many international tourists and this in turn has created multiple employment opportunities and generates enormous taxes. The tourism industry in India can be classified into three categories, namely, (a) domestic tourism, (b) international tourism, and (c) outbound tourism. Thus, it generated a huge job market tool (87.5 million) with 12.75% of

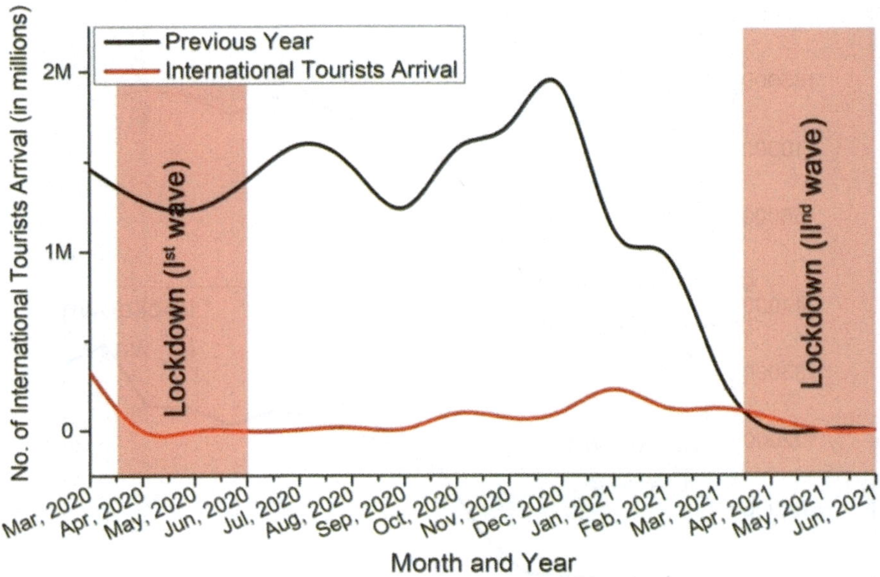

Fig. 22.4 Month-wise international tourist arrivals in India. (Source: India Tourism Statistics)

Table 22.2 Month-wise international tourist arrivals and Growth percentages in India

Month and year	Previous year	International tourists arrival	Growth (%)
Mar 2020	1,462,838	328,462	−77.54
Apr 2020	1,282,355	470	−99.96
May 2020	1,233,736	0	−100
Jun 2020	1,395,170	0	−100
Jul 2020	1,593,705	5974	−99.62
Aug 2020	1,474,406	19,253	−98.69
Sep 2020	1,241,027	11,532	−99.07
Oct 2020	1,569,480	98,439	−93.72
Nov 2020	1,699,316	72,791	−95.71
Dec 2020	1,916,815	106,956	−94.42
Jan 2021	1,118,150	230,871	−79.35
Feb 2021	978,236	127,984	−86.91
Mar 2021	328,462	123,179	−62.49
Apr 2021	470	69,442	14674.9
May 2021	0	0	0
Jun 2021	0	0	0

Source: India Tourism Statistics

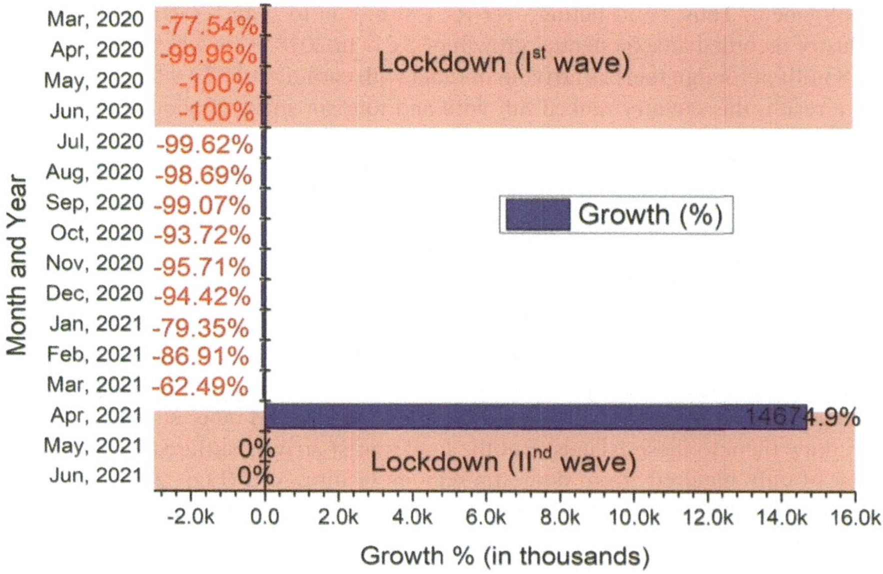

Fig. 22.5 Month-wise growth of international tourist arrivals in India. (Source: India Tourism Statistics)

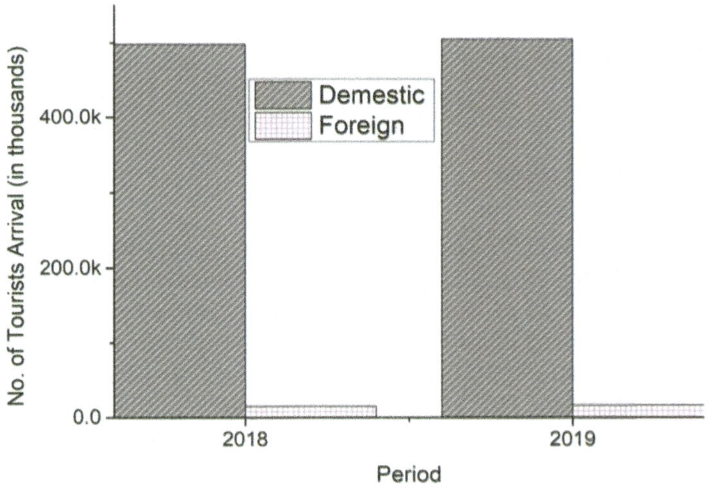

Fig. 22.6 Domestic and foreign tourist visits to Andaman and Nicobar Islands during 2018 and 2019

employment. Thus, contributing over Rs 194 billion to India's GDP. Further, this industry recorded an exponential growth of 3.2% in 2018. This was generated with 10.8 million foreign tourists arriving in India with earnings of $29.9 billion in 2019. As a result, this country ranked 8th with and tourism share of about $108 billion (FICCI 2020).

Compared to the early year, there is a decline in overseas tourists' (66.4%) arrivals in March 2020 compared to the previous year (TAN 2020). The estimations show that there are more than 50 million job losses in India, with an annual loss of about $17 billion (FICCI 2020; Scroll 2020). Every year the income generated by this tourism industry is estimated to be $1000 million.

However, owing to the COVID-19 pandemic, it has suffered a massive loss. It is estimated that every year a total of half a million tourists visit ANI for retreat and relaxation. The untouched, pristine natural beauty with beaches is the top attraction for tourists. Diving, snorkeling, scuba diving, trekking, and other adventure activities draw them to these Islands. But the total tourist arrivals suffered a terrible setback of only hundreds from domestic regions visiting ANI (Figs. 22.3 and 22.4). India recorded over 32 million COVID cases in the study period with deaths of half a million. However, ANI reported 7482 cases and a death count of 128 people, considered to be relatively low. But owing to lockdown restrictions tourism suffered substantially (Table 22.3). Now the government has eased the passage of tourists by administering vaccination to all the people for free and lifting up the sanctions on the free flow of the people. Mobile vaccination drives, constant monitoring of social distancing, proving sanitizers, and adopting internally (WHO) approved protocols to safeguard the public from the COVID-19 pandemic. The untiring role of the government has made tourism to think of visiting Andaman and Nicobar Islands now-a-day with ease and limited risk.

22.7.1 Cluster Analysis

The available data from 15 months are subjected to Cluster Analysis. This yielded three different Clusters (Figs. 22.7 and 22.8). From March to June 2020 constitutes Cluster I. This period was a nationwide lockdown clamped by the government in order to check the spread of the COVID-19 first wave. Cluster II includes a period from July 2020 to March 2021, considered to be the safe period in ANI. The second wave period (April–June 2021) is classified in Cluster III. This analysis is on par with the original data sets. Some of the preparations and precautions are now being carried out in order to revive tourism.

Table 22.3 Opinions of several researchers on the impact of COVID-19 pandemic on the tourism industry

S. No.	Studies	Country/ region	Author's opinion
1.	Abbas et al. (2021)	China	The tourist and leisure industry has been severely impacted by the COVID-19 tourism consequences and is one of the most devastating worldwide industries.
2.	Aburumman (2020)	Abu Dhabi (UAE)	The limitations on travel and the closure of borders. The UAE MICE business is experiencing a significant drop in demand. Significant substantial losses have been suffered by Emirates Airlines, hotels, and other tourism-related enterprises.
3.	Al-Hasni (2020)	Oman	COVID-19 is responsible for Oman's tourist industry's poor performance. Unemployment has risen, wages have decreased, and consumer purchasing power has decreased.
4.	Arshad et al. (2021)	India	The tourism and hotel business has been hit the hardest by the nationwide shutdown. The Indian tourism business is one of the most important sectors of the Indian economy, contributing around 10% of the GDP.
5.	Bouarar et al. (2020)	Algeria	The COVID-19 epidemic continues to have an influence on all aspects of human existence and all sectors of the world economy. Perhaps one of the most noticeable consequences was on the tourist industry, as flight bans and restrictions on mobility and travel were among the first measures taken.
6.	Chandel et al. (2021)	India	The administration must now focus on the third phase of COVID-19. Experts and academics say it will be difficult for the children since the virus is rapidly evolving and impacting a huge population. So, the only option is to vaccinate everyone. As of now, the government has begun vaccinations for major age groups, but not for youngsters. When COVID-19 instances cease spreading, people will feel safe, the economy will improve, commercial activities will increase, the hotels and hospitality industries will recover, and the tourism industry will be revived.
7.	Chang and Wu (2021)	Taiwan	Tourism has grown to be one of the world's most significant businesses. Due to visitor movement, tourism has also resulted in disasters and crises such as viruses and infectious illnesses. It has an impact on economic advantages, employment prospects, and international trade. These consequences are becoming increasingly relevant. Many conversations have taken place concerning the COVID-19 pandemic, as well as substantial disputes in academic studies. The modeling constructs of stakeholders in the tourist sector, on the other hand, have received little attention.

(continued)

Table 22.3 (continued)

S. No.	Studies	Country/ region	Author's opinion
8.	Dogra (2020)	India	The pandemic impacts every industry. Tourism is severely affected since it is driven by people who are highly concerned about security. Tourists will be reluctant to go during the pandemic since their safety is essential. In order to travel again after the pandemic, travel and tourism firms will need to regain public trust.
9.	Guha and Gandhi (2020)	India	Globally, the tourism and hotel sector is under trouble. Millions of visitors visit India every year, contributing significantly to the country's GDP. To overcome the present eco-tourism slump, certain early actions must be implemented.
10.	Hafsa (2020)	Bangladesh	Bangladesh's economy has completely collapsed, and the tourist sector, like other industries, is in the worst shape it has ever been in. People are unable to travel from one location to another due to the lockout.
11.	Khan et al. (2020)	China	COVID-19's fast proliferation has had far-reaching effects on people's daily lives in nearly every country. Millions of people have lost their employment, particularly in the tourist industry, and hundreds of airlines are on the verge of bankruptcy.
12.	Lagos et al. (2021)	Greece	The COVID-19 pandemic has had direct economic and societal effects. The duration and intensity of the resulting tourist malfunction are unknown, thus it is early to estimate the yearly financial losses.
13.	Sah et al. (2020)	Nepal	The collapse of the tourism industry, job losses in the informal sector, and rising costs of vital goods are all contributing to Nepalis slipping into extreme poverty.
14.	Shivakoti (2021)	Nepal	Tourism has experienced an unparalleled setback as a result of COVID-19, and its effects can be observed in the world economy. It has resulted in the unemployment of millions of people, as well as the freezing of large investments that are unable to create business and drive the economy.

22.8 Conclusion

The picturesque palm-fringed serene beaches, dense forests, and spectacular marine life and its ecosystem, showcasing the Islands' raw natural beauty, drew thousands of tourists throughout the year to the ANI. The sudden outbreak of the Pandemic has dented the global economy, brought the tourism and hospitality industries to a screeching halt, and left thousands of vulnerable households and livelihoods in shambles, more so in small islands like ANI. Since this sector brings in the bulk of government revenue through its constant inflow of tourists from all over the world,

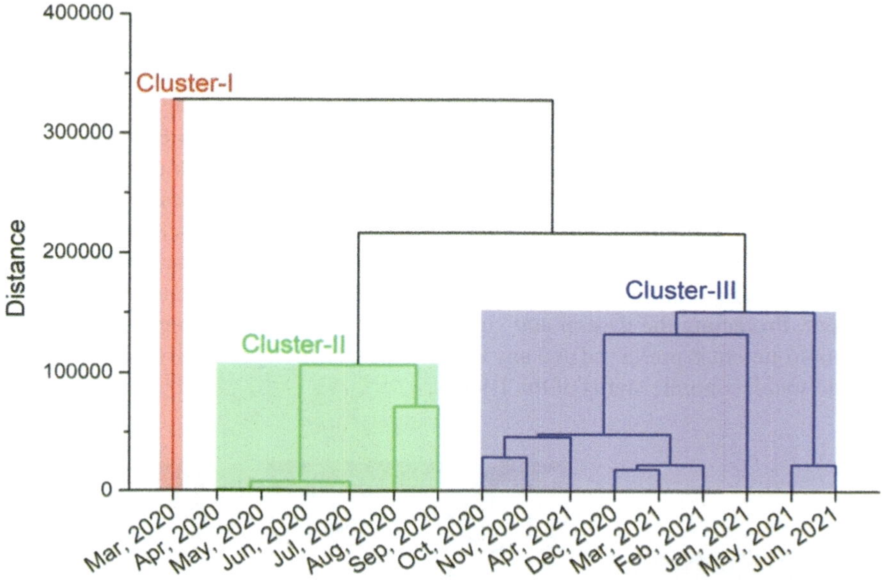

Fig. 22.7 Cluster plot showing the relation among the months of the studied period

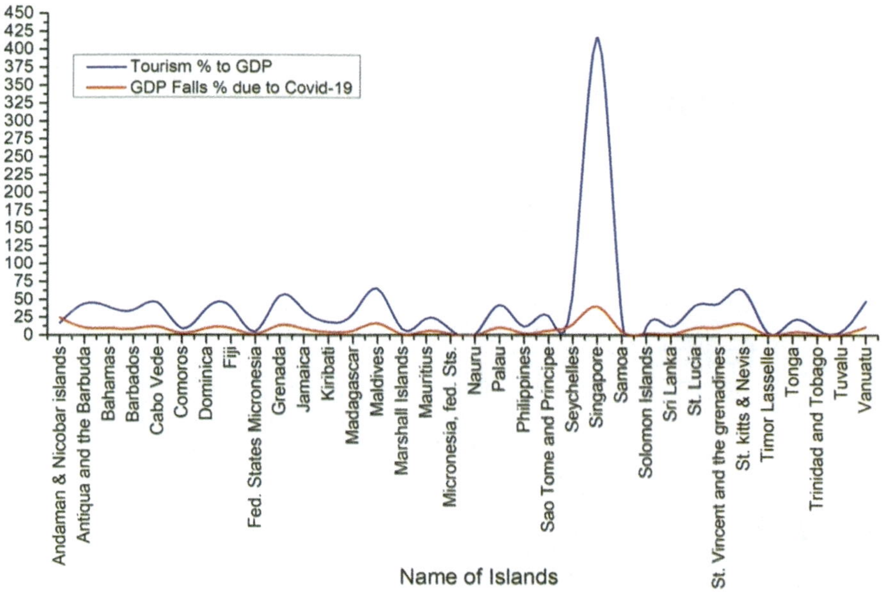

Fig. 22.8 GDP fall out % due to COVID-19 during the pandemic

it is only imperative that it should provide effective relief, psychological mindset, and economic measures to sustain the Islanders' livelihood and gradually improve their psychological, and socioeconomic conditions. Not short-term but long-term policies that are sustainable and aid in development, as the pandemic effect could be felt in future endeavors too. The tourism industry in India has to be revived with effective planning that will benefit the sector and society as a lot of people are wholly dependent on it. Challenging times call for a pragmatic yet sensitive approach by the policymakers that will boost Island Tourism albeit slowly, along with promoting eco-tourism, that will be sustainable in the long run irrespective of any threat of natural disaster to this beautiful and fragile biodiversity cum holiday hotspot. Protecting the natural and cultural heritage is of prime importance so that the environment is preserved and any tourism-based activity will have a huge impact on the socioeconomic status of the Islands.

Fig. 22.9 Pictorial silent evidence of COVID-19 impact on ANI

Reviving economies and livelihoods after any natural disaster or health outbreak is no mean task. With the Tourism sector being reopened in the Andaman Islands now, it will be a much-needed boost for the government and locals (Fig. 22.9). Rebuilding Tourism and sustaining it to stabilize the economy calls for structured, reliable measures. Accordingly, tourists and visitors have to follow the government guidelines for their safe and happy travel. It may be a long time before things could begin to look up and people can travel as they once did due to the present vulnerable situation but effective planning can help in the recovery of the economy through sustainable tourism. The government's initiatives like vaccination, lending financial loans, and providing infrastructure to the industry have suddenly started to increase the free flow of tourists. Travel sanctions have been slowly lifted up and local government has taken all necessary steps to surge the industry once again by the end of this year. The vital role of the government is never ending and this is warranted in all future endeavors.

References

Abbas J, Mubeen R, Iorember PT, Raza S, Mamirkulova G (2021) Exploring the impact of COVID-19 on tourism: transformational potential and implications for a sustainable recovery of the travel and leisure industry. Curr Res Behav Sci 2:100033. https://doi.org/10.1016/j.crbeha.2021.100033

Aburumman AA (2020) COVID-19 impact and survival strategy in business tourism market: the example of the UAE MICE industry. Humanit Soc Sci Commun 7:141. https://doi.org/10.1057/s41599-020-00630-8

Al-Hasni ZS (2020) The economic impact of COVID-19 on the Omani tourism sector. Psychol Educ 58(2):824–830

Arshad MO, Khan S, Haleem A, Mansoor H, Arshad MO, Arshad ME (2021) Understanding the impact of Covid-19 on Indian tourism sector through time series modeling. J Tour Futures. https://doi.org/10.1108/JTF-06-2020-0100

Bouarar AC, Mouloudj K, Mouloudj S (2020) The impact of coronavirus on tourism sector – an analytical study. J Econ Manag 20(1):323–335

Chandel RS, Shruti K, Singh SK (2021) Impact of COVID-19 on tourism sector: a case study of Rajasthan, India. AIMS Geosci 7(2):224–243. https://doi.org/10.3934/geosci.2021014

Chang DS, Wu WD (2021) Impact of the COVID-19 pandemic on the tourism industry: applying TRIZ and DEMATEL to construct a decision-making model. Sustainability 13:7610. https://doi.org/10.3390/su13147610

Dogra T (2020) Impact of Covid-19 on the tourism industry in India. Int J Adv Res 8(11):273–278

FICCI (2020, June) Travel and tourism – survive, revive and thrive in times of COVID-19

Guha C, Gandhi MK (2020) Covid-19 effect on tourism industry in India. Eur J Mol Clin Med 7(11):47–56

Hafsa S (2020) Impacts of COVID-19 pandemic on Tourism & Hospitality Industry in Bangladesh (July 23, 2020). Available at SSRN: https://ssrn.com/abstract=3659196 or https://doi.org/10.2139/ssrn.3659196

Khan SAR, Janjua LR, Yu Z (2020) Opportunities and obstacles in the global tourism industry: a story of post-Covid-19 [online first]. IntechOpen. https://doi.org/10.5772/intechopen.93683. Available from: https://www.intechopen.com/online-first/73254

Lagos DG, Poulaki P, Lambrou P (2021) COVID-19 and its impact on tourism industry. Adv Exp Med Biol 1318:815–824. https://doi.org/10.1007/978-3-030-63761-3_45

Lenzen M, Li M, Malik A, Pomponi F, Sun YY, Wiedmann T, Faturay F, Fry J, Gallego B, Geschke A, Gomez-Paredes J, Kanemoto K, Kenway S, Nansai K, Prokopenko M, Wakiyama T, Wang Y, Yousefzadeh M (2020) Global socio-economic losses and environmental gains from the Coronavirus pandemic. PLoS One 15(7):e0235654. https://doi.org/10.1371/journal.pone.0235654

Marchant-Forde JN, Boyle LA (2020) COVID-19 effects on livestock production: a one welfare issue. Front Vet Sci 7:585787. https://doi.org/10.3389/fvets.2020.585787

Ministry of Tourism, Govt. of India, Annual report- 2020–21

Mohit A (2020) Changing dynamics of hospitality industry in India amid Covid'19. In: Proceedings of online national level conference on innovative business strategies to cope with the present era of globalisation, JournalNX- A Multidisciplinary Peer Reviewed Journal. ISSN: 2581-4230

Mulder N (2020) The impact of the COVID-19 pandemic on the tourism sector in Latin America and the Caribbean, and options for a sustainable and resilient recovery, International Trade series, No. 157 (LC/TS.2020/147). Economic Commission for Latin America and the Caribbean (ECLAC), Santiago

Sah R, Shailendra S, Akihiko O, Yasuhiro K, Divya B, Priyanka R, Ali AR, Rachana M, Mahesh A, Namrata R, Kuldeep D, Tetsuya T, Alfonso JR, Rachana D (2020) Impact of COVID-19 on tourism in Nepal. J Travel Med 27(6):105. https://doi.org/10.1093/jtm/taaa105

Sangeetha M (2006) The great Indian Ocean tsunami of 2004. An overview of a national disaster. National Institute of Advanced Studies, Bangalore. ISBN: 81-87663-62-6

Scroll (2020) India's Covid-19 lockdown may cause 38 million job losses in the travel and tourism industry, https://scroll.in/article/959045/indias-covid-19-lockdown-may-cause-38-million-job-losses-in-the-travel-and-tourism-industry

Sharma D, Bijoor S, Ramesh M (2019) Tourism today in the Andaman Islands: an assessment of challenges through two case studies. Dakshin Foundation, Bengaluru. 45 pages

Shivakoti A (2021) Impact of COVID-19 on tourism in Nepal. Gaze J Tour Hosp 12(1):1–22

TAN (2020) Foreign tourist arrivals to India tumble over 66% in March owing to coronavirus pandemic, https://travelandynews.com/foreign-tourist-arrivals-to-india-tumble-over-66-in-march-owing-to-coronavirus-pandemic

The Economic Times impact of coronavirus on Indian tourism could run into thousands of crores of rupees, retrieved from: https://economictimes.indiatimes.com/industry/services/travel/impact-of-coronavirus-on-indian-tourism-could-run-intothousands-of-crores-of-rupees/articleshow/74592482.cms?from=mdr. Access on 5 Apr 2020

The Economics Times (2021) Coronavirus tourism affect News and Updates from The Economic Times – Page 1. https://economictimes.indiatimes.com/topic/coronavirus-tourism-affect/news

United Nations Conference on Trade and Development (UNCTAD) (2020) COVID-19 and Tourism: assessing the economic consequences, UNCTAD, 1211 Geneva 10, Switzerland. https://unctad.org/system/files/official-document/ditcinf2020d3_en.pdf

Chapter 23
Spatial Planning for Sustainable Resource Use with a Special Reference to Aquaculture Development

M. Jayanthi

Abstract Commercial aquaculture has grownup manifold in the past four decades and is considered an important food-producing sector worldwide. However, the sector has received equal criticism for unplanned development and its impact on important coastal resources. Coastal aquaculture in India is regulated through the Coastal Aquaculture Authority Act 2005 with restrictions on land classes and mandatory buffer zone provisions. Hence aquaculture development can happen only in a permitted land class adhering to the regulations of the Government of India. However, unregulated and unlicensed aquaculture operations continue to exist. The advancement of geospatial techniques coupled with satellite data products can offer a wide range of choices in aquaculture development regarding site selection, impact assessment, macro-level planning, and aquaculture zoning. The important criteria in aquaculture planning were land use, distance from the water source, source water characteristics, soil quality, soil texture, adjoining land type, and accessibility. These factors can be integrated through the geographical information system (GIS) to assess the zones for aquaculture expansion. Likewise, multi-temporal data analysis capability helps to quantify the impacts of aquaculture on other resources and to direct the sector toward sustainable development without multiuser conflicts. This study highlights the issues in coastal aquaculture development, criteria selection in spatial planning, and spatial decision support systems to manage aquaculture development in a sustainable and responsible manner.

Keywords Spatial planning · Aquaculture · Coastal resources · Sustainability

---

M. Jayanthi (✉)
ICAR-Central Institute of Brackishwater Aquaculture (Ministry of Agriculture and Farmers Welfare), Chennai, Tamil Nadu, India
e-mail: m.jayanthi@icar.gov.in

N. Jayaraju et al. (eds.), *Coasts, Estuaries and Lakes*,
https://doi.org/10.1007/978-3-031-21644-2_23

23.1 Introduction

The increasing global population demands increased fish production to meet the protein requirement and nutritional stability. Fisheries and aquaculture are the most important food-producing sectors, which contribute immensely to the economic and social well-being of developing countries. The demand for aquatic foods is expected to increase many folds in the decades to come, and there are several forecasts on demand for aquatic foods, and it is expected that it will be 183 million tonnes by 2030 (FAO 2020). Stationary or even declining supply of fish from captured fisheries requires aquaculture to assume a more important role in satisfying the increasing future fish demands induced by rapid population and income growth (FAO 2020). The economic contribution of aquaculture has been growing and is likely to contribute more in the coming years. But, the rapid growth has confronted the sector with many environmental issues and multiuser conflicts, apart from the viral disease outbreaks. Major shrimp farming countries of the world have faced environmental problems due to unplanned development, which has resulted in several complex social and environmental problems that have led to litigation at national and international levels. Sustainable use of natural resources is one of the important needs of today's world. The unregulated expansion of aquaculture farms for exploiting temporary revenues without considering sustainability has generated several apprehensions worldwide, predominantly in the major shrimp-growing countries, including India.

Aquaculture farming in the country is almost synonymous with Penaeid shrimp, namely, *Penaeus vannamei* since 2009, and *Penaeus monodon* initially. The early 1990s witnessed a phenomenal growth of the sector entirely dependent on the tiger shrimp, *P. monodon*. During this period, the shrimp culture was a low-risk, high-profit venture. In the late 1990s, severe problems of viral diseases, particularly white spot syndrome virus, and environmental safety issues, which arose mainly because of the lack of planning and regulation, made *P. monodon* culture a high-risk, low-profit venture. As an alternative species, SPF P. vannamei was introduced in 2009 by the Government of India after the risk assessment study carried out by CIBA. However, of 1.2 million ha of potential area available for aquaculture, 14% is only utilized, and the remaining vast land is still available. In spite of the limited utilization of resources, it has confronted many problems such as environmental issues, quality input, lack of facilities, lack of schemes, and sector competition. Aquaculture planning without affecting other coastal resource users and coping with changing climate is very much needed in the present context. Figure 23.1 shows the trend of aquaculture development in India. The area has stagnated at around 2 lakh ha, but the productivity has increased due to the introduction of SPF *P. vannamei* (MPEDA 2021). The stagnated total aquaculture area despite many new regions indicates the presence of vast disused or abandoned farms.

The abandonment was unregulated encroachment due to improper planning, disease outbreak influenced by exceeding carrying capacity of source water bodies, and overcrowding of farms adjacent to particular source water bodies. Comprehensive

Fig. 23.1 (**a**) Development of shrimp aquaculture in India. (**b**) Shrimp aquaculture farm in India

development of aquaculture is still rare, and the evidence that numerous farms are abandoned or are struggling to manage viral disease problems proves the necessity for spatial planning.

23.2 Environmental Issues Due to Uncontrolled Aquaculture Development

Recent events in the shrimp culture sector of various countries have demonstrated the conversion of different productive ecosystems to aquaculture, deterioration of shrimp farms and culture sites, increasing incidence of diseases, abandonment of farms, and especially conflicts among common users of land and water resources. Major environmental issues are

Improper and Unrestrained Progress Shrimp aquaculture has been established haphazardly without any planning by converting the available spaces. But, confronted with unexpected and abrupt collapse due to viral diseases worldwide. Disused shrimp farms in disease-hit areas were a common scenario (Jayanthi et al. 2019).

Use of Mangrove Forests Mangrove ecosystems are considered important to maintain biodiversity and protect coastal regions from natural disasters. Besides, they are the nursery regions for fish breeding. But the unregulated expansion of shrimp aquaculture has made tremendous changes in mangrove forests globally. Of the manmade activities, aquaculture was the primary factor in the loss of mangroves at a global level (Giri et al. 2014).

Use of Agricultural Fields Shrimp aquaculture first progressed in lands which are not fit for agriculture. However, the good profit in a short period and growing requirement in foreign markets tempted the other productive resource users to convert their lands (Jayanthi et al. 2020) which created concern across the globe.

Salinization of Lands and Water Resources Penaeid shrimp *P. monodon* grows well in medium saline water. So, the brackish water was used to grow shrimp in many converted agricultural fields, which increased the soil salinity and nearby drinking water resources due to seepage from shrimp farms and prolonged stagnation of saltwater for the entire crop period.

Multi-Resource Use Conflicts As a majority of the coastal regions are thickly populated, the demand for the limited coastal resources is high. The fisher's accessibility to the sea has become an issue due to corporate farms and hatcheries occupying the entire coast.

Increasing Adversity of Climate Change Impacts The lakes and other reservoirs used to hold water during heavy rains have been changed to shrimp farms in certain regions. This led to frequent flooding in densely populated coastal regions.

These environmental issues necessitated the comprehensive planning and monitoring of the aquaculture development using advanced spatial tools and techniques.

23.3 Spatial Planning of Aquaculture

ERDAS Imagine 2018 and Arc GIS 10.6 were used for satellite image processing and spatial analysis, respectively. Image processing techniques such as layer stacking, subsetting, radiometric correction, and projection to the UTM zone were carried out in ERDAS. Visual interpretation, digitization, and spatial analysis were made by ArcGIS. The satellite data was downloaded from the USGS website, and topographical maps were obtained from the Survey of India. High tide lines were drawn from the satellite images and also checked with the lines given in the topographical maps. The land classes such as water bodies, aquaculture farms, abandoned aquaculture, mudflats, scrubland, and abandoned salt pans were delineated using the image interpretation keys. Coastal Aquaculture Authority (CAA) Act 2005 demands a buffer zone of 50–100 m between various land types (mangroves, agricultural lands, settlements, and common water bodies) and aquaculture (CAA 2014). A buffer of 200 m from the high tide line was provided to eliminate the classes. A buffer of 100 m was given between the suitable areas and adjacent productive land types such as agricultural and mangroves. The areas liable to be flooded were removed. The flood-free wastelands with appropriate water quality and support facilities will be the correct choice for commercial aquaculture (Hossain and Das 2010). The regions with 500 m from the source water bodies have been marked for potential aquaculture development. The model (Fig. 23.2a) was evaluated for Nagapattinam District of Tamil Nadu, which lies on the east coast of India between 10.10° and 11.20° north longitude and 79° 15° and 79.50° east longitude with an area of 2585 km^2 and coastline of 187 km.

The spatial analysis indicated that the existing aquaculture area was 4172 ha, and the suitable areas available for expansion were 3000 ha from mudflats, abandoned salt pans, and scrublands (Fig. 23.2b). Zoning regions for aquaculture need to be the first step to protect the ecosystems and also to develop aquaculture sustainably. However, this alone is not sufficient as we need to make sure that licensing permission has been obtained and renewed properly. In spite of spatial restrictions and licensing guidelines in most countries, aquaculture farms without a license are operated as no regular monitoring system is in place. The aquaculture expansion without site selection may modify or disturb ecological conditions and aquatic biodiversity (Gimpel et al. 2018). The study has eliminated flood-prone regions to prevent flooding in nearby regions. If not, the rainwater flows into the ocean will be blocked due to aquaculture farm construction, which will lead to frequent flooding in human settlements and other nearby resources. As per IPCC 2014, the low-lying regions and people living in those regions will bear the impact of flooding to minuscule sea level rise. Besides, the use of unused lands for aquaculture development will benefit the policymakers to draw livelihood options without altering nearby ecosystem functions.

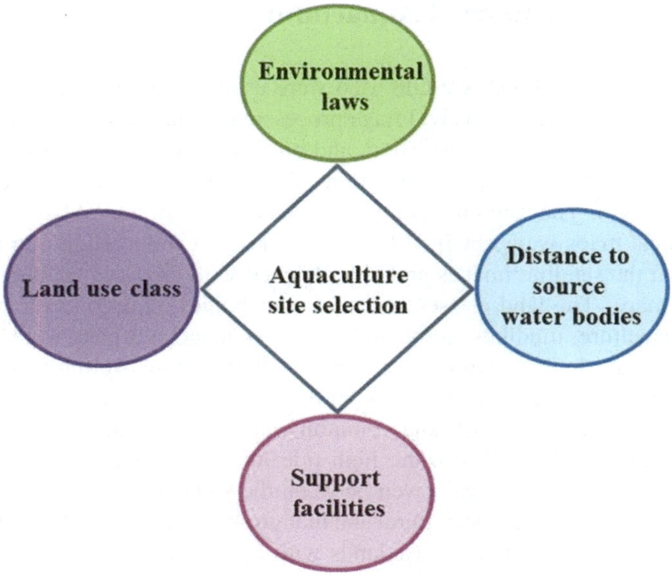

Fig. 23.2 (**a**) Criteria considered in site selection. (**b**) Suitable regions for aquaculture in Nagapattinam District, Tamil Nadu, India

23.4 Monitoring Aquaculture Development Using Spatial Tools

To demonstrate the capability of spatial tools in monitoring aquaculture development, the part of Coringa mangrove regions, Andhra Pradesh, agricultural lands in West Bengal, and Kolleru lake, Andhra Pradesh were taken for the study. Landsat satellite images of 1988 were compared with Sentinel images of 2022 to assess the changes in aquaculture development. The supervised classification of images and overlay analysis quantified the extent of conversion for aquaculture.

The changes in Coringa mangroves to aquaculture has shown in Fig. 23.3a, indicating 660 ha mangrove loss to aquaculture. Studies indicated that the loss of mangroves to natural classes is more in mangrove regions of India than anthropogenic activities (Jayanthi et al. 2019). The Kolleru lake has lost its ecological characteristics as aquaculture expanded in the Kolleru lake area of 180 km^2 (Fig. 23.3b) and restructured the region's topography (Kolli et al. 2022). These fish ponds, once restricted to the shoreline and shallows, are now moved into 75% of the lake area. The conversion of agricultural lands to aquaculture was a common scenario before the Coastal Aquaculture Authority Act 2005. Though it has been reduced, it could not stop completely due to a lack of monitoring.

The sustainable use of natural resources for developmental activities is one of the most crucial issues in today's world. The unplanned proliferation of aquaculture

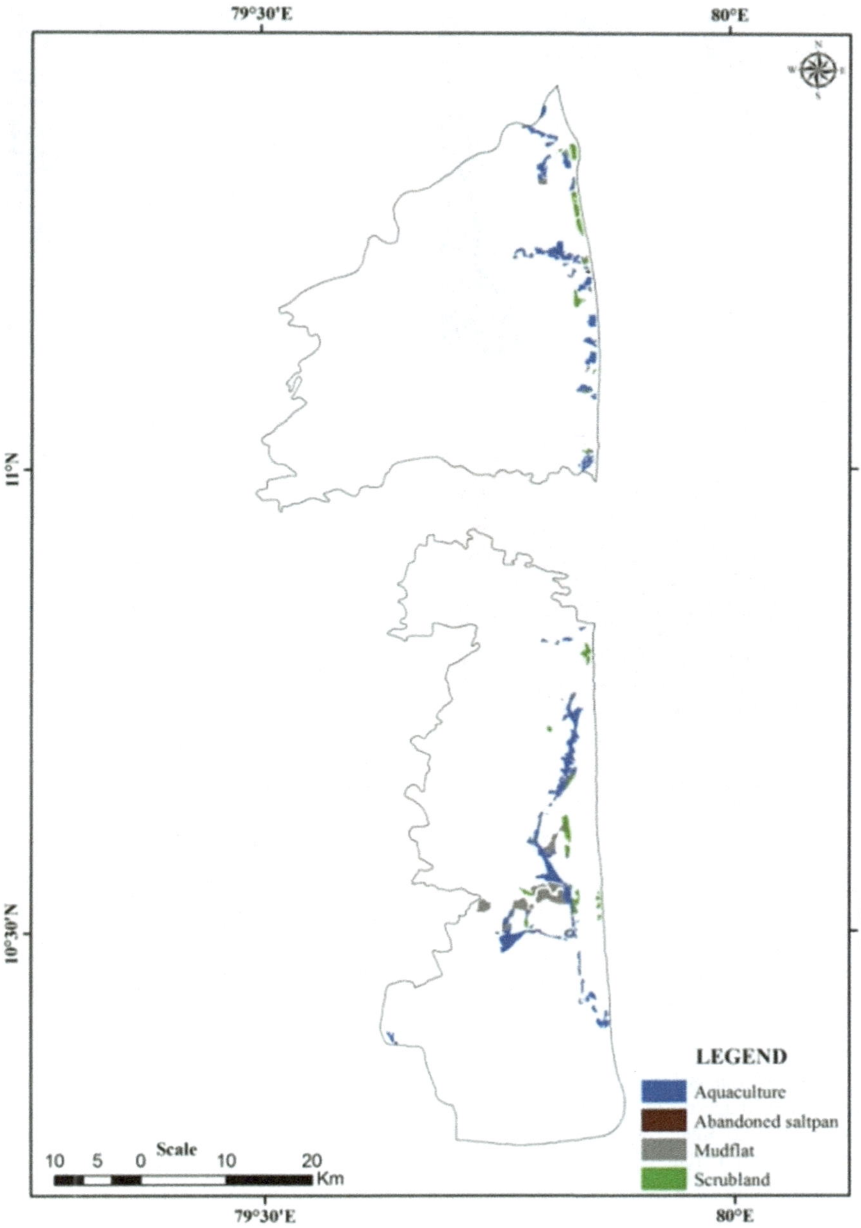

Fig. 23.2 (continued)

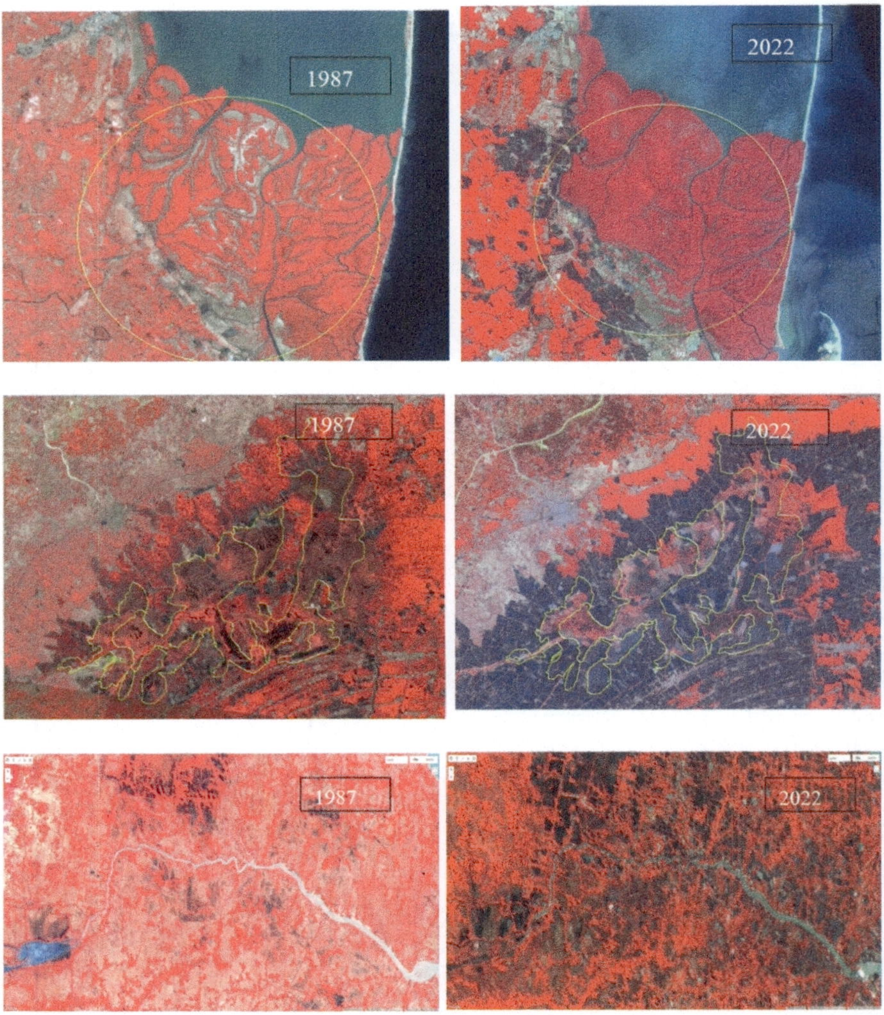

Fig. 23.3 (**a**) Changes in the mangroves of Coringa are indicated by the Landsat image of 1987 and Sentinel image of 2022. (**b**) Conversion of Kolleru lake, Andhra Pradesh to fish ponds as shown by Landsat image of 1987 and Sentinel image of 2022. (**c**) Conversion of agricultural lands to aquaculture in Purba, Medinipur, West Bengal

farms for maximizing short-term returns without addressing sustainability has created many concerns worldwide, such as the conversion of mangroves, agricultural lands, salinization of agricultural lands, multi-resource user conflicts, etc., particularly in Southeast Asian countries and the need to be addressed logically with advanced spatial tools. Comprehensive planning for aquaculture is still uncommon, and the fact that many farms are abandoned or are suffering significant disease problems demonstrates the need for such improvement in planning.

The available coastal resources form the basis for the development of aquaculture in the country, but most countries, including India, do not have enough databases on their resources. The coastal area faces challenges and many problems due to population pressure, environmental degradation, urbanization, and industrial development. Policymakers and planners seldom incorporate current and accurate information based on scientific tools in developing countries. Aquaculture demands production space that is extremely appropriate and in very restricted supply. Remote sensing techniques and Geographic Information systems are indispensable tools in resource assessment due to their far-ranging functions and analytical capabilities of handling large spatial data, analytical and mapping capabilities. The best mechanism for coastal resource use is to assess the resources available in the country and then derive the plan considering the other resource users and the sustainability of the ecosystem.

23.5 Conclusion

Spatial aquaculture planning is an important management measure that can mitigate the negative impacts of aquaculture development and protect nearby important ecosystems and resources. The site selection, monitoring, and impact assessment of aquaculture need integration of land and water resources conditions, regulations, and support facilities that can be carried out using spatial techniques. Spatial techniques can help integrate and interpolate the data to assess the climate change vulnerability and identify the vulnerable coastal area, which demands special resilient management measures for successful aquaculture. The district-level spatial database on existing and potential aquaculture zones will also help the easy implementation of government schemes to make use of unused lands to support livelihood opportunities.

References

Coastal Aquaculture Authority (CAA) (2014) Compendium of acts, rules, regulations and other notifications, p 236. ISBN 81-901701-0-5. http://www.caa.gov.in/uploaded/doc/COMPUPD/2014.pdf. Accessed 3 Jan 2022

FAO (2020) The state of world fisheries and aquaculture 2020. Sustainability in action. Rome. https://doi.org/10.4060/ca9229en. Accessed 10 Apr 2021

Gimpel A, Stelzenmüller V, Töpsch S, Galparsoro I, Gubbins M, Miller D, Murillas A, Murray AG, Pınarbaşı K, Roca G, Watret R (2018) A GIS-based tool for an integrated assessment of spatial planning trade-offs with aquaculture. Sci Total Environ 627:1644–1655. https://doi.org/10.1016/j.scitotenv.2018.01.133

Giri C, Long J, Abbas S, Murali RM, Qamer FM, Pengra B, Thau D (2014) Distribution and dynamics of mangrove forests of South Asia. J Environ Manag 148:101–111. https://doi.org/10.1016/j.jenvman.2014.01.020

Hossain MS, Das NG (2010) GIS-based multi-criteria evaluation to land suitability modelling for giant prawn (*Macrobrachium rosenbergii*) farming in Companigonj Upazila of Noakhali, Bangladesh. Comput Electron Agric 70(1):172–186. https://doi.org/10.1016/j.compag.2009.10.003

Jayanthi M, Ravisankar T, Nagaraj G, Thirumurthy S, Muralidhar M, Saraswathy R (2019) Is aquaculture abandonment a threat to sustainable coastal resource use? – a case study of Andhra Pradesh, India, with options for reuse. Land Use Policy 86:54–66. https://doi.org/10.1016/j.landusepol.2019.04.034

Jayanthi M, Thirumurthy S, Samynathan M, Manimaran K, Duraisamy M, Muralidhar M (2020) Assessment of land and water ecosystems capability to support aquaculture expansion in climate-vulnerable regions using analytical hierarchy process based geospatial analysis. J Environ Manag 270:110952. https://doi.org/10.1016/j.jenvman.2020.110952

Kolli MK, Opp C, Karthe D, Pradhan B (2022) Automatic extraction of large-scale aquaculture encroachment areas using Canny Edge Otsu algorithm in Google earth engine – the case study of Kolleru Lake, South India. Geocarto Int:1–21. https://doi.org/10.1080/10106049.2022.2046872

MPEDA (2021) The Marine Product Development Authority (MPEDA) State-wise details of shrimp and scampi production. http://mpeda.gov.in. Accessed 13 Feb 2022

Chapter 24
Sustainable Aquaculture and Economic Development in Coastal Areas: The Case of Andhra Pradesh, India

M. Swarna Pragathi, M. Anitha, G. Sreenivasulu, and N. Jayaraju

Abstract Aquaculture farming systems are the predominant source in coastal regions of the country. India, which accounts for 7.56% of global production, is the second-largest producer of marine foods. This industry accounts for around 1.24% of India's Gross Value Added (GVA). Seafood (fish, prawns, etc.) continues to provide a vital source of healthy and nutritious food, revenue, and financial assistance to millions of people throughout the world, particularly in India. The sector employs around 28 million people at the primary level and over 50 million individuals along the value chain. Over the previous 3 years, the yearly average growth rate in this industry has been 7%. Seafood is one of the best alternatives for reducing hunger and nutrient shortage since it is a low-cost, high-protein source. The main difficulties in culture fisheries are low average production because of poor technological adoption, illness prevalence, and a lack of high-quality seed and affordable feed for targeted species. Better governance, aquaculture farming, seafood, policy support, and capacity building are all critical activities for the country's blue economy ambitions. This diagnostic chapter seeks to comprehend the current level of production and product development, notably in the aquaculture industry in the Indian state of Andhra Pradesh. Shrinkage of the fertile land-based agriculture industry gave up a new avenue for growth in Andhra Pradesh, India. Obviously, coastal zones are the state's bountiful natural resources. The state of Andhra Pradesh has identified the marine food sector as one of the primary growth sectors of the newly established Andhra Pradesh's socio-economic development.

Keywords Economic development · Aquaculture sector · Trends production · Andhra Pradesh

---

M. Swarna Pragathi (✉) · M. Anitha
Department of Economics, Yogi Vemana University, Kadapa, Andhra Pradesh, India

G. Sreenivasulu
Department of Geology, Sri Venkateswara University, Tirupati, Andhra Pradesh, India

N. Jayaraju
Department of Geology, Yogi Vemana University, Kadapa, Andhra Pradesh, India

N. Jayaraju et al. (eds.), *Coasts, Estuaries and Lakes*,
https://doi.org/10.1007/978-3-031-21644-2_24

24.1 Introduction

In India, the fishing industry is expanding quickly. It already employs over 30 million people and provides income and food security for a vast community. About 10% of the total variety in fish and shellfish species, which now number 3137 and include 462 exotic species, is supported by the country's abundant aquatic genetic resources, which range from deep oceans to rivers and lakes in the Himalayas (Lal and Jena 2019). In the past 40 years, the aquaculture industry has evolved from a traditional source of subsistence in the 1950s and 1960s to a business driven by science and technology (Ayyappan 2012). India is the second rank in world fish and aquacultural production, accounting for more than 7%, producing 13.4 million t in 2018–2019, including 3 million t derived from the seafood sector and about 10 million t from the mainland thrust areas, respectively. The Blue economy sector is identified as a potential engine for generating both employment and economy for the rural poor. Since Seafood is labeled as nutritious, healthy, and soft accessible for both high and low-class societies of the world over. This sector's export revenues were about 5 million in 2019–2020 (Year Ender Review 2021). The fisheries sector contributes significantly to the country's socioeconomic growth, with export profits of Rs. 46,662.85 crores in 2019–2020. Seafood production is increasing at an annual pace of 8%. Half of the Seafood is consumed by humans. This sector has been classified as a revenue-generating wing because it stimulates the growth of several sectors such as industries and is a source of revenue, food. However, for the great majority of the economically poor, it is a source of income. This industry is crucial to socioeconomic prosperity. The production of seafood is expanding at an alarming rate of 8% per annum (Edwards et al. 2019).

24.2 Methodology

The present study is based on secondary data collected for 5 years (2015–2016 to 2020–2021) from the Statistical Abstract of Andhra Pradesh. The State Primary Census Abstract 2011, Socio-economic survey 2020–2021, Agricultural statistics at a glance 2020, and Hand Book on Fisheries Statistics 2018 to 2020. Factor analysis (FA), one of the key standard multivariate statistical methods, was employed to retard datasets. Thus useful to derive a little number of hidden factors for understanding relationships among the district-wise marine and inland fish productions during 5 years (2015–2016 to 2019–1020). Growth rate analysis was performed to study the trends in the development of production and productivity of the coastal zone.

24.3 Geographical Profile of State Andhra Pradesh

The state of Andhra Pradesh is geographically located on the East Coast of India thus considered as the natural gateway to East and South East Asia. The State is classified as Coastal Andhra and Rayalaseema. It has 13 districts of which 9 are in Coastal Andhra and 4 in the region of Rayalaseema (Fig. 24.1). The study area occupied a geographical area of 1,62,970 km^2 with a population of about 5 million. The state is ornamented with fertile and productive river basins, large canal systems, and conductive agro-tropical weather conditions best suited for aquaculture. Situated in a tropical region, this state is bestowed with a huge coastline (975 km), thus the largest producer of marine products of economic importance. The state's long coastline is also dotted with several ports and has a long history of sea trading. The state has 13 revenue districts of which 9 are in coastal (Srikakulam, Vijayanagaram, Visakhapatnam, East Godavari, West Godavari, Krishna, Guntur, Prakasam, and Nellore). It is primarily agriculture and allied sector state that contributes significantly to national food grain output and grows all essential crops. It is one of the primary states that supplies the country with seafood. It is also known as the river-state since it has two major inter-state rivers, such as Godavari and Krishna. These rivers, together with their numerous tributaries, run through the core of the state. These huge water bodies offer significant opportunities for the sustainable development of inland aquaculture and fisheries (Madhu and Venkata Rao 2021).

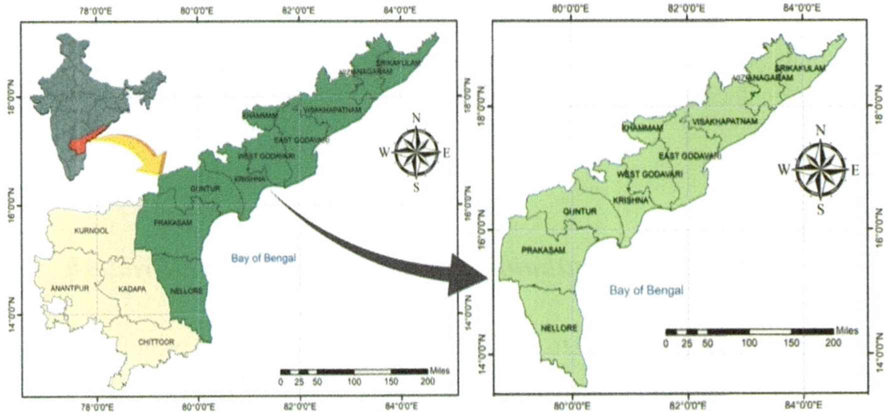

Fig. 24.1 Location map of the study area

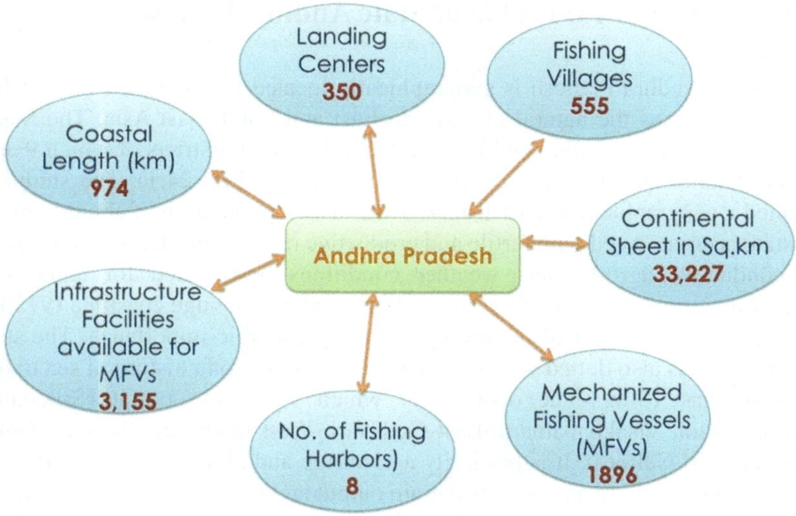

Fig. 24.2 Fisheries and Infrastructure Statistics of Andhra Pradesh (2015–2020)

24.4 Results and Discussion

24.4.1 Development of Aquaculture in Andhra Pradesh

Andhra Pradesh has 555 Marine Fishing Villages spread throughout 65 Mandals, as well as 350 Marine Fish Landing Centers. There are 3,01,956 marine fishermen in the world. According to a study (2019–2020), 2,00,000 lakhs are actively involved in fishing and aquaculture along Andhra Coast. Visakhapatnam, Kakinada, and Nizampatnam have established landing facilities for motorized boats. The graphical representation of fisheries and Infrastructure statistics of Andhra Pradesh is given (Fig. 24.2).

24.4.2 Fishermen Population in the State of Andhra Pradesh

In Andhra Pradesh in the year 2019–2020, the total population who depends on fisheries is 1.5 million out of 50 million people. By district wise total population is 0.1 million. During 2019–2020, both marine and inland fishermen's population rate is high in male population that is 0.5 million and 0.4 million female populations is 0.4 and 0.2 million. The details of the fishermen population of Andhra Pradesh during 2019–2020 are furnished (Table 24.1).

Table 24.1 Fishermen population of Andhra Pradesh 2019–2020

S. no	Description		Population details (in millions)
1.	Fisher population		1.5
2.	Whole population		50
3.	Fishermen population per district		0.2
1.4.	Marine fishermen population	Male	0.46
		Female	0.42
5.	Inland fishermen population	Male	0.34
		Female	0.29

Source: Hand Book on Fisheries Statistics (2020)

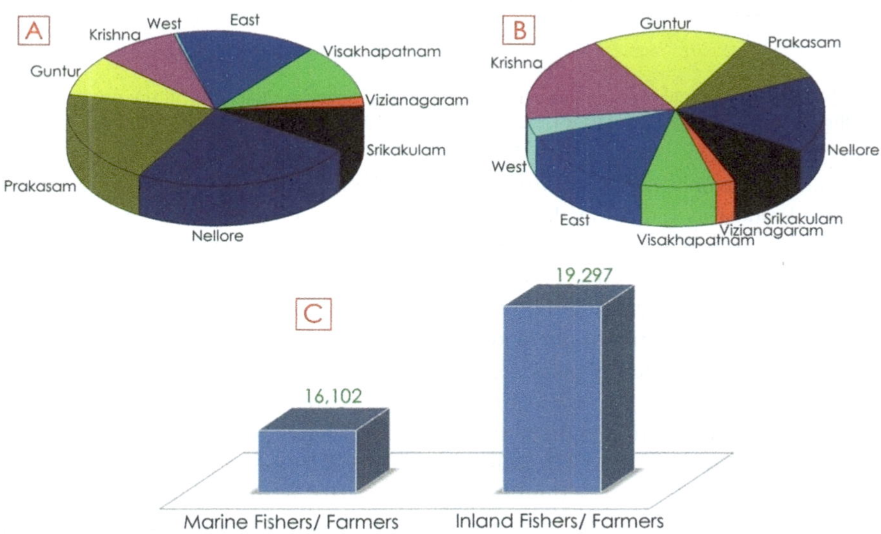

Fig. 24.3 District-wise total number of marine (**a**), inland (**b**), and total number (**c**) of fishers/farmers of the study area. (*Source*: Hand Book on Fisheries Statistics 2020)

24.4.3 Marine and Inland Fishery Resources

Andhra Pradesh has abundant marine and inland fisheries resources. Compared to agriculture, aquaculture has grown dramatically in recent years. This is because it generates greater profit than agriculture in a shorter period of time while requiring less work per unit area (Sushmakrishna et al. 2019). The state region contributes 25% of the total fish output in India. In recent years, fisheries have witnessed major changes in terms of output, value addition, job creation, and input costs in the marine industry. The present status of the total number of fishers/farmers engaged in marine fish production is high in the Nellore district (3931) and inland fishers/farmers are high in the Krishna district (3485). District-wise current status of the total number of fishers/farmers of Andhra Pradesh is mentioned (Fig. 24.3).

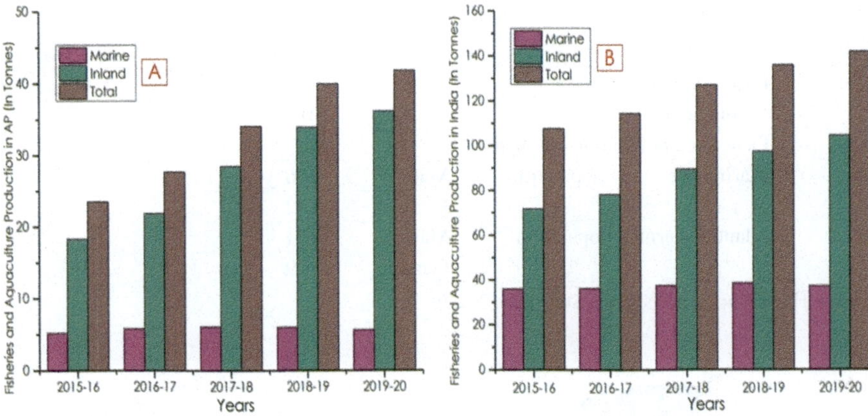

Fig. 24.4 The graphical representation of Marine, Inland fisheries, and aquaculture production in Andhra Pradesh (**a**) and India (**b**) for the study period (2015–2020)

Aquaculture and fisheries constitute the key components of the Sea Food sector. Andhra Pradesh is India's leader in aquaculture (shrimp, prawn, etc.) farming and production, as well as second in inland fish output. Aquaculture is practiced in both fresh and brackish waters. Recently, marine farming is widely popular. Hence, some of fin and shellfish species including seaweeds are now being cultivated. The graphical representation of marine, inland fisheries, and aquaculture production for the study period is presented (Fig. 24.4).

During 2015–2016, the value of marine products reached 2.4 million in Andhra Pradesh and 11.62 million in India. Fish and aquaculture production increased from 2.8 million in 2016–2017 to 3.4 million by 2018–2019 in Andhra Pradesh and 1.14 million in 2016–2017 to 12.70 million by 2017–2018 in the country. 4.0 and 13.5 million in 2018–2019 in the state and India. The fisheries production of Andhra Pradesh and India are 4.2 and 14.16 million tonnes, respectively, during 2019–2020 (Hand Book of Fisheries Statistics 2018 and 2020). The district-wise marine and inland fish production in Andhra Pradesh during 2015–2020 are presented (Tables 24.2 and 24.3).

In 2015–2016, marine fish production is 0.5 million and inland fish production is 1.5 million. It increased up to 0.6 tonnes and inland fish production by 2.9 million in 2019–2020. Fish production by district wise during 2015–2016, the value of marine products reached the highest (0.1 million) in the East Godavari district. The value of inland products reached a high (0.6 million) in the West Godavari district. The marine fish production during 2016–2017 was high in Visakhapatnam and East Godavari (0.12 million) and inland production is high in West Godavari and Krishna districts (0.78 and 0.71 million, respectively). The value of marine and inland fish products increased during 2017–2018 and reached high in Visakhapatnam and East Godavari (0.12 million) and West Godavari district (1.0 million). Marine fish production increased in the year 2018–2019 in Visakhapatnam and East Godavari (0.13 and 0.11 million, respectively). The inland production is high in West Godavari and

Table 24.2 District-wise marine fish production in Andhra Pradesh (in million tonnes)

S. no	District	2015–2016	2016–2017	2017–2018	2018–2019	2019–2020
1.	Srikakulam	0.059	0.059	0.060	0.047	0.053
2.	Vizianagaram	0.017	0.019	0.020	0.014	0.010
3.	Visakhapatnam	0.103	0.125	0.127	0.132	0.121
4.	East Godavari	0.111	0.124	0.126	0.117	0.106
5.	West Godavari	0.012	0.012	0.014	0.001	0.001
6.	Krishna	0.041	0.048	0.055	0.050	0.047
7.	Guntur	0.040	0.047	0.053	0.055	0.060
8.	Prakasam	0.041	0.046	0.039	0.066	0.055
9.	Nellore	0.091	0.095	0.105	0.114	0.107
Andhra Pradesh		0.520	0.580	0.604	0.600	0.564

Source: Statistical Abstract of Andhra Pradesh (2015–2020)

Table 24.3 District-wise inland fish production in Andhra Pradesh (in million tonnes) GoAP (2015)

S. no	District	2015–2016	2016–2017	2017–2018	2018–2019	2019–2020
1.	Srikakulam	0.023	0.027	0.034	0.070	0.078
2.	Vizianagaram	0.014	0.017	0.020	0.035	0.020
3.	Visakhapatnam	0.017	0.024	0.023	0.050	0.037
4.	East Godavari	0.055	0.080	0.114	0.208	0.217
5.	West Godavari	0.668	0.783	1.004	1.126	1.257
6.	Krishna	0.611	0.712	0.914	0.950	1.021
7.	Guntur	0.030	0.031	0.046	0.079	0.088
8.	Prakasam	0.023	0.019	0.037	0.032	0.037
9.	Nellore	0.112	0.128	0.182	0.241	0.219
Andhra Pradesh		1.556	1.826	2.378	2.796	2.979

Source: Statistical Abstract of Andhra Pradesh (2015–2020)

Krishna districts (1.12 and 0.95 million, respectively). Marine fish production increased (0.1 million) in the Visakhapatnam district and inland fish production in the West Godavari district (1.2 million) in 2019–2020. During the study period (2015–2020), both marine and inland fish production rate is higher in the West Godavari district. The marine and inland fisheries production rates have also significantly increased during 2015–2020 in other districts of AP.

24.4.4 *Factor Analysis (FA)*

The data on marine and inland fish production during 2015–2020 in the coastal districts of Andhra Pradesh were subjected to statistical treatment. The analysis generated three factors for the marine fish production from the study area, namely, Factor 1 (97.96%), Factor 2 (1.59%), and Factor 3 (0.21%) (Fig. 24.5a, b). Factor 1 was represented by Visakhapatnam (3.304) and West Godavari (−3.118) and was

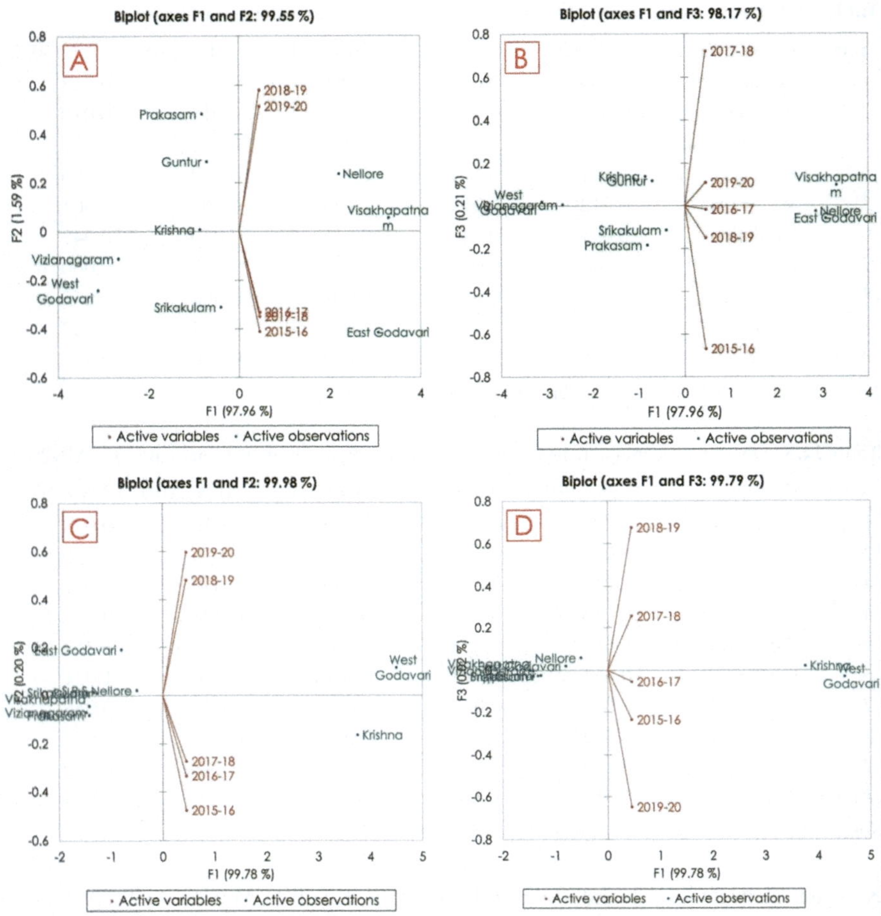

Fig. 24.5 Factor loadings of fisheries production in coastal districts of Andhra Pradesh. (**a**) Marine fisheries production (factors 1 and 2). (**b**) Marine fisheries production (factors 1 and 3). (**c**) Inland fisheries production (factors 1 and 2). (**d**) Inland fisheries production (factors 1 and 3). (*Source*: Statistical Abstract of Andhra Pradesh (2015–2020))

assigned as Visakhapatnam–West Godavari assemblage. The Visakhapatnam district showed the highest marine fish production 0.125 and 0.127 million tons during 2016–2017 and 2017–2018, respectively. However, West Godavari showed the lowest productions of 0.012 and 0.014 million tons during 2016–2017 and 2017–2018, respectively. Factor 2 was enveloped with East Godavari (−0.414) and Prakasam (0.483) districts and termed as East Godavari–Prakasam assemblage. During the years 2018–2019 and 2019–2020, both East Godavari and Prakasam districts showed the highest marine fish production. Factor 3 was composed of Krishna (0.139) and Guntur (0.118) districts and is called as Krishna–Guntur assemblage.

During the year 2015–2016, Krishna and Guntur districts showed the lowest marine fish production of 0.041 and 0.040 million tons, respectively, whereas in another year, these districts showed moderate production.

For Inland fish production, factor loadings indicate three dominant assemblages: Factor 1 (99.78%), Factor 2 (0.20%), and Factor 3 (0.02%) (Fig. 24.5c, d). Factor 1 is represented by two districts, West Godavari (4.503) and Krishna (3.750) assigned as West Godavari–Krishna assemblage. West Godavari showed the highest inland fish production 0.783 and 1.004 million tons during 2016–2017 and 2017–2018, respectively. And also the Krishna district showed the highest production 0.712 and 0.914 million tons during 2016–2017 and 2017–2018, respectively. Factor 2 is composed of two districts, East Godavari (0.191) and Nellore (0.023). Therefore, it is called as East Godavari- Nellore assemblage. The East Godavari and Nellore districts showed the lowest inland fish productions of 0.055 and 0.112 million tons respectively during 2015–2016. The highest productions were during 2019 and 20 (0.217 and 0.219 million tonnes, respectively). However, Factor 3 includes, Vizianagaram (0.002) and Guntur (−0.022) and it is called as Vizianagaram–Guntur assemblage. These two districts showed the highest inland fish production during 2018–2019.

However, cross plots of factors 1 and 2 and factors 1 and 3 for both marine and inland fish productions gave distinct information about districts wise fish production during different periods.

24.4.5 *Analysis of Growth Rate in Fish Production*

The present study utilized conventional tools and techniques like percentage of growth rate analyzing the production of marine and inland fisheries by district wise of Andhra Pradesh. The growth rate between the years 2015–2016 and 2019–2020, both in marine and inland growth records a higher rate both in quantity and value. Performance in dimension has been expressed as applying the normalization formula:

$$\text{Percentage of growth rate} = \left(\text{Y1} - \text{Y2} / \text{Y1}\right) \times 100$$

$$\text{Percentage of growth rate} = \frac{\text{Previous year}\left(\text{Y1}\right) - \text{Present year}\left(\text{Y2}\right)}{\text{Previous year}\left(\text{Y1}\right)} \times 100$$

The current trend of increasing production can be maintained either through intensification or expansion of area under aquaculture production. In recent years, fisheries production recorded a significant hike. Thus thrust area is growing at an alarming speed covering both marine and inland areas respectively during 2015–2020 (Fig. 24.6).

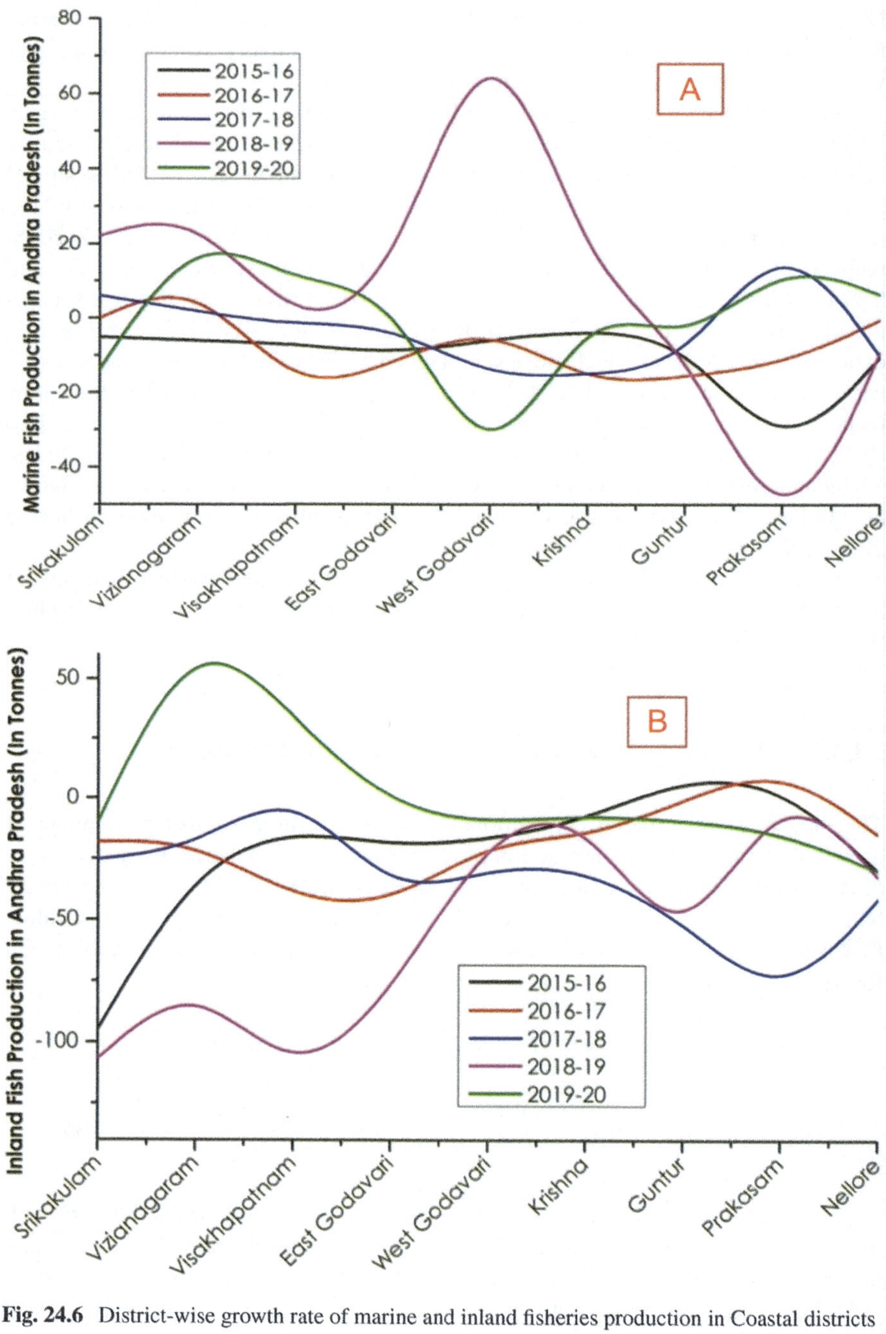

Fig. 24.6 District-wise growth rate of marine and inland fisheries production in Coastal districts

24.5 Summary

The government has taken many steps to encourage the growth of fisheries and the food security of fishermen. They include the issuance of registration certificates to aquaculturists (17,000 fishermen and dependents). Furthermore, a sales tax exemption on engine fuels, mechanization, the purchase of solar pumps, night lighting, and other necessary machinery. Obviously, the establishment of MatsyMitra teams and societies for the fishing community, infrastructure development through the use of numerous available grants, and the construction of massive, expansive facilities are all involved (CESS Report 2016). The policy should cover several fisheries activities such as production processing and technology in prone. Resource depletion and oversights are considered to be very serious threat to sustainability. A policy shift to promote sustainable fishing, in the long run, is the need of the hour. Furthermore, the government should consider providing incentives for the use of cost-effective and environmentally friendly machinery designed at the Central Institute of Fisheries Technology (CIFT) to increase the abundance of fish resources, mostly within the Exclusive Economic Zone (EEZ) off the study area (CESS Report 2016). Another formidable concern is the oppression of fishermen, which leads to precarious livelihoods. This must be handled in a framework that includes growth. As a result, a policy shift in marine fisheries is required. This is critical to focusing on procedural adaptation by community development direction rather than aquaculture production and expansion. The sustainable increase of aquaculture output is intended to be attained by the use of an appropriate fisheries policy framework, enhanced management, and strong governance, as well as possible cooperation among beneficiaries and public. The issues related to the blue economy, marine environment, sustainability, and fishermen societal business must be addressed quantitatively (Lakra and Gopalakrishnan 2021).

24.6 Conclusion

It is obvious that the aquaculture industry in this state is under significant stress from sea fishery and other associated activities. It has recently undergone significant alteration in terms of increasing output, resource estimate, job creation, and input difficulties. Between 1985 and 2020, output more than doubled. Fishermen's socio-economic resiliency is likewise quite gloomy and questionable. Population, insufficient shelter, a scarcity of drinking water, cleanliness, and health care make them more susceptible. They are vulnerable to natural calamities such as floods and cyclones. Fishermen Societies have limited access to inland natural resources for subsistence. Out-migration appears to be a critical concern in fishermen's coping mechanisms. Poor infrastructure and insufficient basic facilities make the fishing community more exposed to natural disasters at sea. Currently, the fishing community is dealing with a number of issues in their fishing operations. They are as follows:

1. Depletion of fish resources owing to near-shore overfishing.
2. Harsh competition with mechanized boats.
3. Coastal pollution caused by anthropogenic sources.
4. Unfair prices for their catch as a result of unethical exploitation of fishermen by brokers, middlemen, and business tycoons.
5. Infrastructural constructions like ports and water theme parks result in the general public and biosphere in the relocation of a fragile ecosystem, disrupting the food cycle.

Finally, other issues such as climate change, dubious management, unhygienic preservation, and processing are to be addressed in the future. In order to conserve fossil fuel and reduce maintenance costs, it is also necessary to gradually forsake the traditional methods and procedures for fishing and using machines. Additional research in this area demonstrates the strong connections and intensities between fishing communities, the natural environment, and the delicate, evolving marine marginal ecology.

References

Ayyappan S (2012) Indian fisheries: issues and the way forward. Natl Acad Sci Lett 35:1–6

Edwards P, Zang W, Belton B, Little DC (2019) Misunderstandings, myths and mantra in aquaculture: its contribution to world food supplies has been systematically over reported. Mar Policy 106:103547. https://doi.org/10.1016/j.marpol.2019.103547

GoAP (2015) Fisheries present status, future plans and issues. Presentation made by Department of Fisheries at the Centre for Economic and Social Studies Hyderababd on 31st-08-2015, Hyderababd.

Hand Book on Fisheries Statistics 2018 & 2020, Department of Fisheries, Ministry of Fisheries, Animal Husbandry & Dairying, Government of India, New Delhi.

Lakra WS, Gopalakrishnan A (2021) Blue revolution in India: status and future perspectives, 2021. Indian J Fish 68(1):137–150

Lal KK, Jena JK (2019) Fish genetic resources – India. In: Tyagi RK, Munasinghe DHN, Ashoka Deepananda KHM, Niranjan F, Khetarpal RK (eds) Regional workshop on Underutilized fish and marine genetic resources and their amelioration – Proceedings and recommendations. Asia-Pacific Association of Agricultural Research Institutions (APAARI), Bangkok, 55 pp

Madhu A, Venkata Rao P (2021) Fisheries sector in India and Andhra Pradesh: an overview. IJCRT 9(4):423–433. ISSN: 2320-2882

Report of the Commission on Inclusive and Sustainable Agricultural Development of Andhra Pradesh, Government of Andhra Pradesh, Centre for Economic and Social Studies (CESS), August 2016.

Statistical Abstract of Andhra Pradesh, Government of Andhra Pradesh from 2015–2020.

Sushmakrishna KS, Cherishya DR, Leela MS (2019) Aquaculture, a replacement of agriculture in Andhra Pradesh, India. Acta Scientific Agriculture 3(6):122–124

Year Ender review 2021 on highlight key Initiatives and achievements pertains to Department of Fisheries, Ministry of Fisheries, Animal Husbandry and Dairying for the year 2021.

Chapter 25
Marine and Coastal Ecosystem Services for Sustainable Development

Meesa Saraswathi, Sonal Bhandari, M. Madakka, R. S. Prakasam, and Sunil Misra

Abstract A marine and coastal ecosystem offers various services to human civilisation, such as provisioning, supporting, cultural, and regulating services. These services have a positive impact on human well-being both directly or indirectly through their commodities. However, rising extensive constraints like poaching, water contamination, destruction of coastal habitat, and overall deterioration of biodiversity and biogeochemical processes lead to marine species and their ecosystems being at risk. These threats mainly obstruct the flow of ecological services on a broad scale, whose long-term value lowers the current economic benefits to society. Hence, Sustainable Development Goals (SDGs) are implemented to uplift coastal and marine life for the betterment of society. This chapter deals with marine biota, ecological services, climatic effects on marine species, and SDGs for improving the marine environment.

M. Saraswathi · S. Misra (✉)
Applied Biology, CSIR-Indian Institute of Chemical Technology, Hyderabad, Telangana, India

Academy of Scientific and Innovative Research (AcSIR), Ghaziabad, Uttar Pradesh, India
e-mail: smisra@iict.res.in

S. Bhandari
National Institute for Pharmaceutical Education and Research, Hyderabad, Telangana, India

M. Madakka
Department of Biotechnology and Bioinformatics, Yogi Vemana University, Kadapa, Andhra Pradesh, India

R. S. Prakasam
Organic Synthesis and Process Chemistry Division, CSIR-Indian Institute of Chemical Technology, Hyderabad, Telangana, India

N. Jayaraju et al. (eds.), *Coasts, Estuaries and Lakes*,
https://doi.org/10.1007/978-3-031-21644-2_25

Keywords Marine biota · Marine and Coastal ecosystem services · Climate impacts · Sustainable Development Goals (SDGs).

25.1 Introduction

25.1.1 Ecosystem

The ecosystem is a natural/biological community comprising all living (biotic) and non-living (abiotic) factors. All biotic species (plants, microbes, and animals) living in a particular region may interact with non-living (abiotic) environmental components such as air, soil, water, fire, and sunlight of an ecosystem (Balasubramanian 2011) which can maintain continuous nutrient cyclings and energy flows. The term 'ecosystem' was first used by Arthur Tansley, a British ecologist, in 1935. Later, the term was developed by Arthur Roy Clapham. Tansley proposed an idea to emphasise the significance of material exchanges between organisms and their surroundings (Tansely 1935). He eventually modified the word and included 'the complete system, comprising not just the biological complex but also the entire complex of physical variables defined as the environment' (Odum and Odum 1959; Odum and Barrett 1971). As shown in (Fig. 25.1) the ecosystems are classified into two groups, one is an aquatic/ water-based ecosystem, and another one is a terrestrial/ land-based ecosystem. Aquatic ecosystems are further categorised into two types, that is, fresh water and marine water

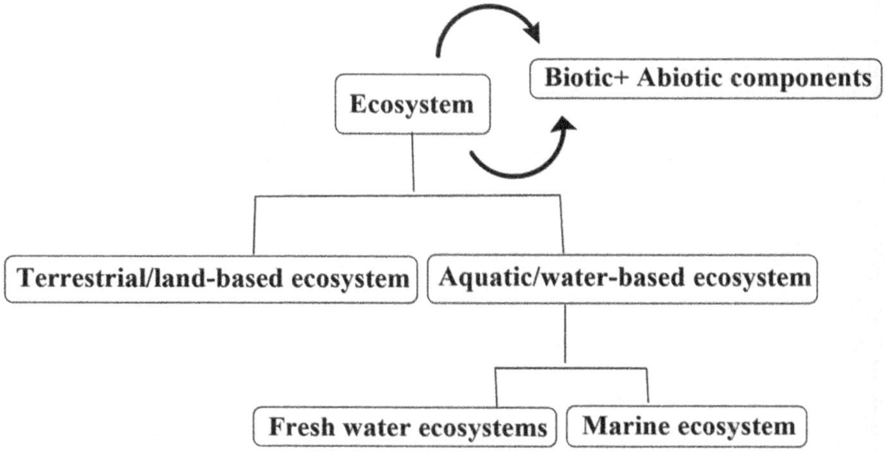

Fig. 25.1 Classification of the ecosystem and its types emphasising the interaction between living (biotic) and non-living (abiotic) components

ecosystems, which are deviated based on the quality of water involved in a particular habitat. Freshwater ecosystems are a subgroup of aquatic ecosystems on Earth. For instance, Lakes, rivers, wetlands, ponds, marshes, streams, and springs are among them. They differ from marine habitats, which exclusively possess a higher salt content in their ecosystem (Balasubramanian 2011).

25.2 Marine Ecosystem

Marine ecosystems situated near the ocean, containing a significant number of dissolved salts, are termed marine ecosystems and consist of unique biotic (living) and abiotic (non-living) components (Belkin and Cornillon 2007). Sunlight is among the most predominant abiotic component in marine ecosystems. Marine ecosystems can be classified into three zones based on the quantity of light that marine species receive. The uppermost layer of an aquatic ecosystem is the Euphotic zone, protruding towards 200 meters (656 feet) just under the surface, where sufficient sunlight reaches for the photosynthesis activity at this depth, where the majority of marine life may be found. The Dysphotic zone is situated just under the euphotic zone and may stretch between 200 and 1000 m (656 to 3280 ft) below the surface. At these depths, the Dysphotic zone represents still sunlight is available but does not facilitate photosynthesis, whereas the Aphotic zone was present just below the Dysphotic zone where there is no sunlight, it was extending from 1500 to 10,000 m just below the Dysphotic zone (Fig. 25.2) (Lalli and Parsons 1997; Civera et al. 2013).

Marine biota dominates the most significant ecological system of the planet, which includes seas, oceans, and lagoons. The ocean is a complex chemical system controlled by physical, chemical, and biological processes (Islam and Tanaka 2004). Approximately 70% of the earth's surface's covered by water, most of which is marine. Among the earth's aquatic ecosystem marine species are expressing rich biological wealth. Marine organisms, including brown algae, red algae, green algae, bryozoans, molluscs, tunicates, dinoflagellates, cephalopods, sharks, echinoderms, and corals reefs, have the ability to detect sound in the water over far distances through the specialised environment (Tyack et al. 2011; Barnes and Hughes 1999) and also produces many structurally unique bioactive products which are helpful for human sustainability. Marine plants release oxygen into the atmosphere, and CO_2 is absorbed by the ocean. The process, such as volcanic activity, erosion of rock and sediment, the metabolic activity of the organism, and the exchange of gas is affected by the chemical composition of the ocean and sea (Valavanidis and Vlachogianni 2010). The macronutrients other than carbon, phosphorus, and nitrogen are essential for the marine species, similarly, other elements such as silicon (used in the skeleton of diatoms) and calcium (used in the skeleton of fish and coral marine organisms), which are normally present in low amounts because they are rapidly taken up by the marine organism. (Harrison and Hurd 2001).

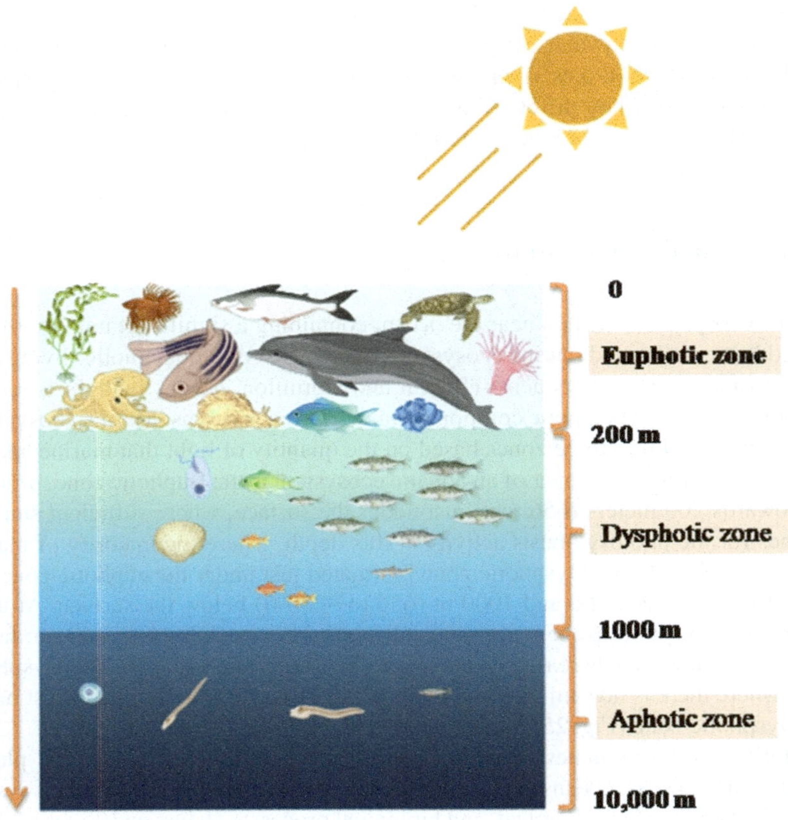

Fig. 25.2 Major zones of marine ecosystem

25.2.1 Classification of Marine Species

The marine environment has very high biodiversity due to the presence of 31 animal phyla. Marine species are more diversified in terms of size, shape, form, and weight. According to Costello et al., there are a total number of 1.5 million extant species are known (Castello and Stewart 2010). The marine species are classified into three groups: plankton, nekton, and benthos.

1. *Plankton:* Plankton consists of animals with extended body parts that are small and microscopic in size. They do not have self-propulsion or locomotion. These are further subdivided into two groups. (i) Phytoplankton: Have features of microscopic diatoms, dinoflagellates, and blue-green algae. (ii) Zooplankton: Many protozoans, small molecules, crustaceans, a few worms, and a host of larvae.

2. *Nekton:* Nekton refers to both tiny and large organisms, such as dolphins, porpoises, and whales, which comes under this category. These species of fish actively partake in the swim.
3. *Benthos:* The term 'benthos' refers to the colony of organisms that live at the bottom of the ocean. The creatures that dwell on the seafloor are known as benthic epifauna (Buskey et al. 2016).

25.2.2 Components of Marine Diversity

The photosynthetic species in the sea appear to be low compared to the land; nearly 308,000 living species of the vascular plant have appeared. Appeltans et al. described 0.23 million eukaryotic species so far, but the total number is around 0.3–1.0 million. Fisher et al estimated that coral reefs support around 0.8 million multicellular species. According to Tett and Barton, there are an estimated 5000 species of marine phytoplankton and around 9000 species of seaweeds are found. There are two major categories of marine organisms, namely pelagic organisms (that occupy the water column) and benthic organisms (that inhabit the sea floor). It is assumed that more than 95% of marine animal species are benthic rather than pelagic (Probert 2017).

The increased human population, which reached 7 billion in 2011 and is expected to reach 8 billion in 2024 and 9.3 billion in 2050, is exerting pressure on the marine ecosystem (Goulletquer et al. 2014). Around 60% of the population lives near the coastal areas which depend on the marine habitat for food, shelter, waste disposal, etc. The non-coastal population covers the rivers and other waterways thus indirectly affecting marine biodiversity.

An array of organisms derived from the marine ecosystem and their products are beneficial for human life. Fishes are the major source of many therapeutic and pharmaceutical important substances. Asian and African people are dependent on fish for dietary protein (FAO 1990) which prized between $50 billion to $100 billion. In 1990 approximately 14 million tons of freshwater fish were harvested, its value of about $8.2 billion even exceeded $16 billion. The sport fishing services, gain around $46 billion from fishing activities (Fedler and Nickum 1992). Due to habitat destruction, the population of fish has decreased.

25.3 Marine and Coastal Ecosystem Services

Diverse marine habitats provide an extensive array of vital ecological services. (a) The group of species present in the various marine habitats helps to stabilise sediment and shallow coastlines and provide nurseries and forage sites for fish. (b) Mangrove trees offer timber, shelter for young fish, recycle terrestrial runoff, protect coasts from wave and disaster damage, and supply food for commercial shrimp production. There are vast numbers of plants, animals, and various microorganisms that offer these complex functions. (c) Coral reefs and kelp forests are the most prolific ecological unit on the earth. These habitats provide recreational, economical, and sustenance fishing, as well as protection for the coastline and tourist money (Palumbi et al. 2009).

The ocean performs a vital role in climate control. According to the studies, the oceans absorb over a third of the carbon dioxide generated each year. Furthermore, marine and coastal environments are habitats for diverse species of plants and animals that provide a wide variety of services to people. For example, mangroves assist to maintain friable, or crumbly, soil on the coastline, which helps to avoid coastal erosion. They also act as a natural barrier to water currents, providing a favourable environment for the birth and growth of many fish species. As a result, mangroves aid in the preservation of existing fishery resources. On the other hand, the faecal matter of whales contains an abundant amount of iron, which is a necessary component for photosynthesis. The amount of iron in the water has a direct influence on the growth of phytoplankton, which is an essential component of carbon storage.

25.3.1 Types of Marine and Coastal Ecosystem Services

The Millennium Ecosystem Assessment (MEA) classified marine and coastal ecosystem services into four major types (Fig. 25.3). They are (1) provisioning, (2) regulating, (3) cultural, and (4) supporting services.

1. *Provisioning Services*

   The aids derived from the marine environment and used by mankind are provisioning services. Significant benefits provided by nature to humans are food, fibre, fisheries, natural medicines, fuel, biomass, and building materials that are directly obtained from the ecosystem.

2. *Regulating Services*

   Nature has provided vital ecosystems to the living planet, which makes the active and healthy lifestyle of humans on the earth. Things such as clean water and air are provided by the plant, erosion can be prevented by the tree roots, and bacteria decompose waste, carbon sequestration, storage, and moderation of extreme events are interconnected to make the earth sustainable to live.

3. *Cultural Services*

   The role of the ecosystem in national, global, and local cultures leads to the development and advancement of the people. The creativity which comes from the interaction with nature (architecture, music, and art) is cultural services. For instance, tourism, recreational, aesthetic, and spiritual benefits come under cultural services.

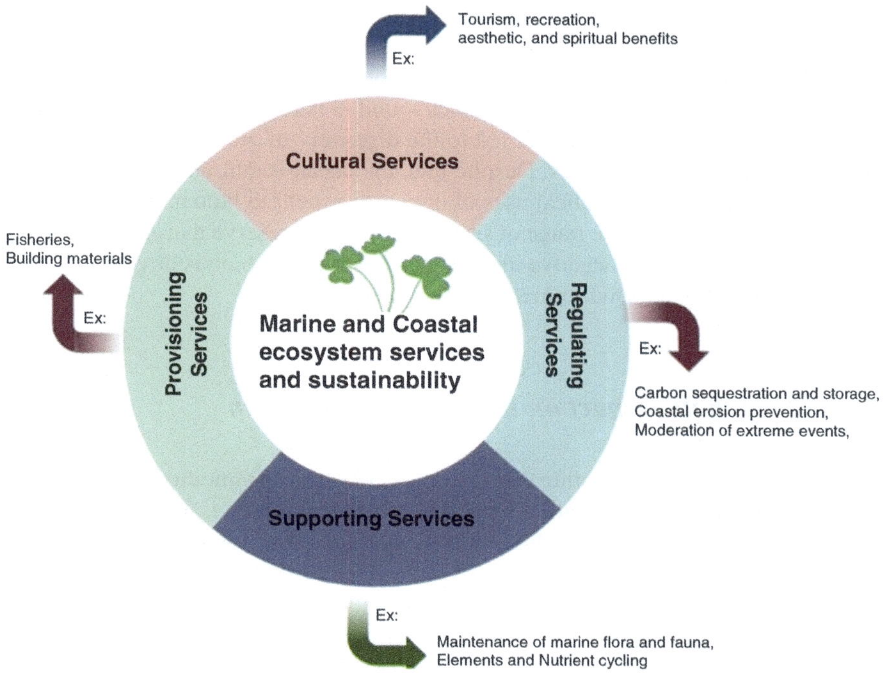

Fig. 25.3 Types of marine and coastal ecosystem services

4. *Supporting Services*

The ecosystem could not be maintained without the integrity of flora and fauna, element and nutrient cycling, water cycling, etc. The basic life of human beings on the earth is sustained through supporting services without this ecosystem could not exist on earth.

The overall value of marine and coastal ecosystem services is estimated to be US$29.5 trillion per year, which is more than the US GDP in 2015. However, the resilience and amount of protection of ecosystems determine the quality of these services. A damaged ecosystem provides fewer services. For example, in the Mediterranean Sea, Posidonia meadows blooming plants were present underwater. These plants are in danger due to human activities, growing coastal development, as well as enhancing the number of boats whose anchors rip Posidonia plants out of the ground; these are some of the consequences which pose natural ecosystems at risk. Although these Posidonia meadows perform a vital role against coastal erosion. Additionally, since they shelter immature fish from predators, these plants provide a safe refuge for fish to breed and develop nurseries for their progeny. The removal of Posidonia plants diminishes the total fish supply, which has a detrimental effect on recreational divers and fishermen. Moreover, ecosystems can provide tremendous unrecognised benefits to society. Though cultural services are difficult to measure, they are often disregarded when calculating the entire worth of the ocean. In many nations, the fishing industry is a significant commercial sector, yet we often overlook the importance of the sea as a huge cultural forum. The ocean is also a source of economic development, notably in biotechnologies, which involves creating products based on biological principles found in nature. The study of the toxins found in cone snails, for example, led to the discovery of ziconotide, an analgesic commonly used in therapeutics. Despite this, the ocean is still understudied. There may still be many more practical uses of immense benefit to humanity to be uncovered. Ocean supplies a wide range of services; we must preserve marine biodiversity by restricting the human negative impacts on ecosystems which is directly propositional to protecting the environment and mankind.

25.3.2 Climate Impact on the Marine Ecosystem

Climate change poses many threats to marine ecosystem commodities and services. These include ocean acidification (specific to marine systems), fluctuations in climate and precipitation, ultraviolet radiation, the occurrence and spread of storms, and changes in pH at the ocean level (Halpern et al. 2008; Harley et al. 2006).

Coastal regions are gradually becoming a foundation of the worldwide human civilisation (McGranahan et al. 2007; Neumann et al. 2015). Some of the world's most minor island governments are among the least developed, relying primarily on marine ecosystem services (Neumann et al. 2015; Guillaumont 2010). Changes in the climate may impede the achievement of SDGs established in the 2030 Sustainable Development Agenda. Climate change impacted negatively ecosystem services via expert elicitation and literature analysis, even though most research has focused on warming and the consequences for various species in the ecosystem. This leads to a detrimental influence on ecosystem services by enhancing local/regional stressors like freshwater outflow and pollutant load, as well as global stressors like acidification of ocean and greenhouse effect. Experts predicted that climatic effects on ecosystem services will have a significant negative impact on all SDGs, with hunger alleviation being one of the most immediately impacted. In spite of these obstacles, the SDGs aim to change consumption and production patterns and develop sustainable energy systems which have been proven to be the least impacted by climate change.

25.3.3 Impacts of Human Activities on Global Marine Environment

Every person on the planet is connected to the ocean, and very remote areas of the ocean have been affected by human pressure. Ocean sustainability is crucial for the balance of the earth's ecosystem and supports the wealth of marine wildlife. Human pressure has a wide range of complex effects on the ocean, both directly (for instance, oil spill destroys sessile benthic biota and seabirds) and indirectly. (The stratification of seawater is impacted by climate change, which has negative effects on the nitrogen cycle and the production of plankton that fish rely on for survival.) Changes in marine ecosystems and their socioeconomic aspects as a result of human activities (Crain et al. 2008; Matin et al. 2008) are shown in Table 25.1. Therefore,

Table 25.1 Impact of human activities on the marine environment and its impact on the socio-economy

S. No	Due to human activities	Impacts on environmental aspects of the marine environment	Impacts on socioeconomic aspects of the marine environment
1	Climate change	The reduction of sea-ice cover in the Antarctic will impair species dependent on the habitat. The ecosystem productivity and community structure variations are due to changes in species distribution. If the sea surface temperature arises stratification also increases which affects the nutrient cycle and productivity.	Fisheries and aquaculture potential changes due to adverse changes in weather.
2	Acidification of the ocean	Reduction in reproductive growth and survival of some species especially those with exoskeletons (shells). Reduced resilience of coral reefs thus causes the loss of other species.	Loss of production of some commercial fisheries which affects the livelihood of small fisheries. Effect of CO_2 emission.
3	Noise pollution	Affects the natural behaviour of cetaceans and many other marine species. Causes mass beaching events.	Limiting economic activities by reducing noise emissions.
4	Plastic pollution	A vast range of animals consumes plastic on a regular basis. Millions of plastic entered over the past 100 years.	Costs for the clean-up of plastics, lost fishing gear, etc. are very high. Potential effect on fish shellfish stocks through changes in the food web.
5	Changes in the salinity of seawater	The surface salinity of the ocean is the key factor in climate change. The density of ocean water changes as its temperature and salinity changes.	The changes in density are the main source of the power that drives ocean circulation.
6	Increased direct mortality of the marine animal population	An increase in temperature in the ocean can directly impact the survival rates of marine mammals. Risks to aquatic creatures are most often from human interventions, including the accidental catch of fishing gear, habitat degradation, exploitation, contamination, intimidation, and ship strikes.	Affect the efficiency and productivity of commercial fisheries and aquaculture, as well as presenting a direct danger to fish populations.

(continued)

Table 25.1 (continued)

S. No	Due to human activities	Impacts on environmental aspects of the marine environment	Impacts on socioeconomic aspects of the marine environment
7	Increased UV radiation	Because ozone is an effective absorber of UV light, depletion of the ozone layer results in an increase in ground-level ultraviolet radiation due to human activities. The Sun transmits radiation with a broad range of energies, with high-energy ultraviolet (UV) radiation.	During extreme weather events, there is an increase in death and morbidity.
8	Input of hydrocarbons	Carbon dioxide (CO_2) and other greenhouse gases are released during the combustion of hydrocarbon fuels, contributing to air pollution and climate change. CO_2 is an inevitable consequence of hydrocarbon burning, unlike impurities in fossil fuels that result in by-product emissions. Hydrocarbon products, particularly oils, harm the community of organisms near the water's surface, including fish eggs and larvae, such as the plaice (Rhombus maeoticus), phytoplankton, zooplankton, nectonic species, including adult fishes (via direct damage or causing them to emigrate), and a variety of benthic organisms.	Hydrocarbon pollution is a new, negative ecological component that has the potential to permanently alter the biological structure of the seas and coastal waters, reducing their productivity.
9	Physical alteration of sea bed habitats	Light accessibility, water flow, oxygen supply, salinity, density, and pH are all factors to consider. These circumstances might differ greatly from one habitat to the next, and they can either promote or hinder the life processes of the marine species that live there.	Climate-related losses in ocean health are expected to cost the world economy $428 billion per year by 2050, and $1.98 trillion per year by 2100, according to the Intergovernmental Panel on Climate Change.

the current scientific research on ocean sustainability is oriented toward mitigating the negative effects of human actions.

25.3.4 Climates Partial Stabilisation

Greenhouse gasses (GHGs) absorb heat in the atmosphere similarly that heat trapped in a horticulture glasshouse. Balanced, properly working ecosystems and natural sources such as soils, plants, and oceans can minimise or stabilise this rate of change through long-term carbon capture and storage, this process is called carbon sequestration (Doney et al. 2011). On a smaller level, vegetation can reduce climate unpredictability and extremes by generating local microclimates via shade and increased humidity, making the area more hospitable to people and animals. Our natural resources may help to stabilise rapidly changing climatic circumstances by storing carbon and adjusting local climate conditions (Pickard et al. 2015).

25.3.5 Protection of Wave Erosion on the Coast

Coastal erosion and flooding are becoming major global concerns (Paterson et al. 2015). Coastal erosion is mainly caused by human practices and changes in the natural environment, which provokes the coastal dynamic activity (waves, tides, currents, and wind) to lose balance, resulting in the long-term loss of sediments in the coastal zone, which leads to the destruction of coastline recession and beach erosion. Extensive utilisation of coastal regions for human activities (living, leisure, or commercial) poses significant pressure on coastal ecosystems, which are already among the world's most endangered ecosystems (Barbier et al. 2011). Coastal protection defends against floods and erosions. A failure to avoid erosion and floods may lead to loss/damage of property/life. The coastline may become more vulnerable as a result of rising sea levels and more intense weather caused by climate change.

25.3.6 Preservation of Biodiversity

Biodiversity preservation refers to the protection, restoration, and management of ecosystems for the benefit of the current and upcoming generations. There are three primary goals for the conservation of biodiversity as follows:

1. To maintain and preserve species diversity.
2. To ensure species and ecosystem management.
3. To prevent and restore ecological processes and life support systems.

Climate change may be mitigated by the conservation of biodiversity. Conservation of ecosystems may aid in reducing the amount of carbon dioxide (CO_2) released into the environment. The protection of mangroves and other coastal ecosystems may aid in reducing the effects of climate change, such as flooding and coastal flooding. Projects that reduce species and ecosystems' vulnerability to the effects of climate change have the potential to save vital ecosystem services such as food production, pollination, air and water purification, and carbon sequestration.

25.4 Sustainable Development Goals (SDGs)

The SDGs were introduced by the UN in 2017. The SDGs go beyond halting environmental deterioration and reducing poverty by establishing objectives for 'the future we want' (Assembly 2015). The SDGs address a wide range of issues, such as eradicating poverty: SDG 1, preventing hunger: SDG 2, promoting health: SDG 3, ensuring minimum standards of education: SDG 4, improving gender equality for women: SDG 5 and minorities: SDG 10, increasing access to fresh water: SDG 6 and energy: SDG 7, economic growth and create job opportunities: SDG 8, built infrastructure: SDG 9 and cities: SDG 11, reorganise supply and demand systems: SDG 12, creating restrictions for climate change mitigation and adaptation: SDG 13, protect marine: SDG 14 and terrestrial: SDG 15 systems sustainably, establish institutional frameworks: SDG 16 and improve policy coordination and partnerships: SDG 17. The detailed SDGs are displayed in (Table 25.2).

In 2017, the United Nations approved a resolution on the (SDGs), which included reduced marine pollution as a quantitative target under Goal 14. The international community has decided that eliminating international community has determined that eradicating pollutants in the seas is a priority, which is monitored by SD Goal 14, which dynamically pursues to eliminate these human-caused effects on the oceans (MacFeely 2020). A target of 14.1 indicates that, by 2025, all marine pollution, especially land-based, and marine debris and pollutants, will be significantly reduced or avoided (United Nations 2017).

25.4.1 Effects of Ecosystem Services on SDGs

The amendments in the marine ecosystem due to climate change leading to the progress of sustainable development have been described in the 16 SDGs (Partnerships for the Goals), which relate to international policy cooperation (Singh et al. 2019).

25.4.2 Ocean Sustainability and Sustainable Development

Sustainable Developmental Goal number 14 is one among the 17 goals of SDGs (Fig. 25.4) which was established by the UN IN 2015, defined as 'Conserve and sustainably use the oceans, seas and marine resources for sustainable development' (MacFeely 2020) An SDG for Oceans and Coasts a list of specific targets and the implementation of key performance indicators must be included in a sustainable ocean management strategy.

Table 25.2 Depicts sustainable developmental goals (SDGs)

SDGs goal No.	Sustainable development emblem	Sustainable development goals (SDGs) deal with
SDG 1		No poverty
SDG 2		Zero hunger
SDG 3	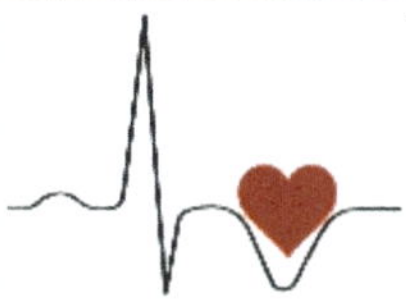	Good health and well-being
SDG 4		Quality education
SDG 5		Gender equality
SDG 6		Clean water and sanitation
SDG 7		Affordable and clean energy
SDG 8	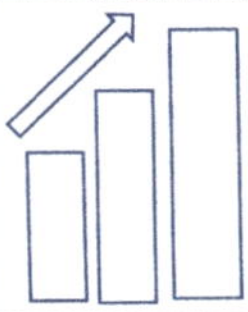	Decent work and economic growth

(continued)

Table 25.2 (continued)

SDGs goal No.	Sustainable development emblem	Sustainable development goals (SDGs) deal with
SDG 9		Industry, innovation, and infrastructure
SDG 10	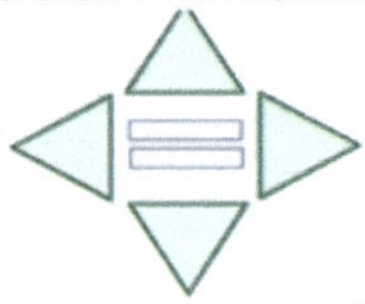	Reduce inequalities
SDG 11		Sustainable cities and communities
SDG 12	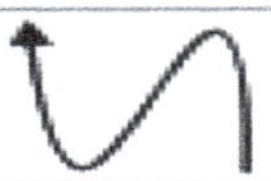	Responsible for consumption, and production
SDG 13		Climate action
SDG 14		*Life below water*
SDG 15		Life on land
SDG 16		Peace, justice, and strong institutions
SDG 17		Partnerships

Fig. 25.4 The Sustainable Developmental Goal number 14 (SDGs on Oceans and Coasts), which is *Life below the water*

Coasts, exclusive economic zones (EEZs), and oceans should be included in the sustainable development objective and indicator set. The notion of polluter pays also be included in an SDG for oceans and coasts, along with an ecological approach. The ecosystem approach aims to manage whole natural processes by combining land, water and vital resources. It was adopted as the main framework for action by the Convention on Biological Diversity (CBD 1992). Ocean sustainability can support all areas of sustainability outlined in the Sustainable Development Goals (SDGs) (Singh et al. 2019). People benefit from healthy oceans in a variety of ways, including by providing raw materials and food (Pauly et al. 1998), controlling local temperatures (Charlson et al. 1987), and providing development and employment opportunities (Oviedo and Jeanrenaud 2007). Although projected climate change affects marine ecosystems in various ways, including changing marine food webs and increasing ocean levels, which might have an impact on the advantages humans gain from the ocean (Harley et al. 2006). Sustainable marine ecosystems perform various services like food security, feed for animals, construction materials from coral rock and sand, raw materials for medicines, and natural defences against threats such as coastal erosion and flooding. The species found in the marine environment may also be used in fine chemicals, nutritional supplements, bioactive substances for treating illness, and industrial developments.

25.4.3 Sustainable Development Goals (SDGs) for Marine and Coastal Ecosystem

The following fundamental objectives are included in the proposed SDG ocean and coasts:

1. Maintain the core life-supporting and regulating activities of the seas (key practices in the hydrological cycle, production of oxygen, and climate change).
    - Reduce CO_2 emissions to prevent further acidification, deoxygenation of the oceans, and Ocean warming. Reduce the activities that affect these processes.
2. To preserve and ensure the robustness of the marine ecosystem for sustainable services of oceans and coasts including provisioning and non-provisioning services.
    - Utilisation of all living resources responsibly within the biological limitations, according to the precautionary principles of an ecosystem.
    - Utilisation of non-living resources using the approaches of the ecosystem and its precautionary principle
    - Restriction on marine environment deterioration and exploitation.
    - Promote and improve technical knowledge for long-term ocean resource conservation.
    - Increase worldwide capability for ocean assessment and management by providing access to marine knowledge and data.
    - Provide regular updates on the condition of the seas and coastlines based on a set of ocean and coastal indicators.
3. Introduce adaptive coastal communities by implementing resilience and mitigation strategies, innovation and environmentally sustainable development through the distribution of benefits and tasks.
4. Participate in integrated, multilevel ocean governance.
    - Establish an MSP system for EEZs and territories that are outside of national authority.
    - Improve and unify ocean and coast legal frameworks to account for existing and future usage.
    - Improve and unify ocean and coastal regime governance.

To achieve these goals, specific targets must be developed and agreed upon at the local, national and international levels. Therefore, the proposed SDGs for oceans and coasts can be implemented using three measurements (ocean services, ocean health, and coastal resilience). Progress towards these goals should be assessed using a set of related indicators to evaluate the sustainability of marine and coastal development.

25.4.4 Ocean and Coastal Sustainability Indicators

The establishment of ocean and coastal sustainability required some appropriate reference tools. Nevertheless, an overall and comprehensive evaluation of the marine ecosystem's condition and future growth remains in its early stages. The World Ocean Assessment, which is a common practice for reporting and evaluating the marine ecosystem as well as its socioeconomic components globally, is one of the projects that are currently in progress. Another is the United Nations general assembly resolution (UNGA) 64/71), which aims to improve ocean life. Both the Global Ocean Partnership (GOP), an alliance of more than 100 governments, academic institutions, societal groups, and international organisations, and the Ocean Compact, which was introduced by the secretary-general to improve UN-wide coordination in the dispensation of ocean-related ordinances and encourage collaboration within the UN system, were created to address issues with the health, productivity, and resilience of the world's oceans. Halpern et al. (2015) developed a new method for assessing the relationship between people and the ocean. The creation of a comprehensive set of sustainability indicators for the ocean-human system prompted a proper examination of how the term 'sustainability' could be functionalised. Sustainable development may be quantified economically by assessing whether the economic growth and overall productivity are preserved or growing, such as future generations' wealth is not diminished (Smith et al. 2001; Arrow et al. 2012). The Earth develops an incredibly complex system with its seas, tides, continents, ice formations, and living and non-living species, whose disparate systems and interconnections are less explored (Van der Sluijs 2012). To examine the effects of non-market products and capital stocks, Ekins et al. (2003) proposed that it could begin with a notion of relatively strong sustainability and, if necessary, move gradually toward a concept of weaker sustainability. Furthermore, it is essential to explore how the ocean should be seen not just as a tool for humanity but also as a valuable resource. As a result, the indicators set have to signify the degree to which marine biological practices in the oceans were protected, biodiversity is retained, and the use of ocean service is preserved.

Finally, data availability limits the choice of relevant indicators, which is always a formal decision with significant consequences for the outcomes (Krellenberg et al. 2010). These constraints must be weighed against the necessity that the human ocean system's condition and development be assessed to determine suitable management actions. As a result, in addition to the specialists, many stakeholders must be included in choosing, weighing, and aggregating good indicators. Hence, the UN Ocean Initiative must be retained to improve Ocean life.

25.5 Conclusion

One of the most popular study areas is now the benefits provided by the marine environment to human civilisation. Due to a lack of human understanding and incorrect social behaviour, the marine ecosphere has been significantly affected, resulting in ecological damage. The scientific community has offered a clear explanation of the ecosystem's functions. Sustainable Development Goals (SDGs) have been adopted as a proactive step to mitigate climate harm and minimise disaster risk cost-effectively. Scientists have concentrated on the criteria for bettering human behaviour toward the water and ocean to preserve the land. Hence, this chapter discusses the different factors which affect ecological services and the appropriate measures taken to get rid of the problems related to the ecosystem and to attain sustainable development for the betterment of society.

Acknowledgements The authors acknowledge the Director of CSIR-IICT for providing the research facilities. Meesa, the first author, thanks for the financial support granted by DST New Delhi, India, in the form of the Inspire Fellowship (**IF160172**). All the authors express deep gratitude toward Dr. N. Jayaraj and Dr. Madakka for their constant support and encouragement. All the authors are thankful to all the reviewers for their insightful comments for IICT Manuscript Communication **No. IICT/Pubs./2022/155.**

References

Arrow KJ, Dasgupta P, Goulder LH, Mumford KJ, Oleson K (2012) Sustainability and the measurement of wealth. Environ Dev Econ 17(3):317–353

Assembly UG (2015) Transforming our world: the 2030 agenda for sustainable development, 21 October 2015 (Vol. 16301). A/RES/70/1

Balasubramanian A (2011) Aquatic ecosystems-freshwater types. University of Mysore

Barbier EB, Hacker SD, Kennedy C, Koch EW, Stier AC, Silliman BR (2011) The value of estuarine and coastal ecosystem services. Ecol Monogr 81(2):169–193

Barnes RSK, Hughes RN (1999) An introduction to marine ecology. Wiley

Belkin IM, Cornillon PC (2007) Fronts in the world ocean's large marine ecosystems. ICES CM 500(130):21

Buskey EJ, White HK, Esbaugh AJ (2016) Impact of oil spills on marine life in the Gulf of Mexico: effects on plankton, nekton, and deep-sea benthos. Oceanography 29(3):174–181

Castello L, Stewart DJ (2010) Assessing CITES non-detriment findings procedures for Arapaima in Brazil. J Appl Ichthyol 26(1):49–56

CBD (Convention on Biological Diversity). 1992. CBD convention text. http://www.cbd.int/convention/text/[accessed on 20 November 2012]

Charlson RJ, Lovelock JE, Andreae MO, Warren SG (1987) Oceanic phytoplankton, atmospheric sulphur, cloud albedo and climate. Nature 326(6114):655–661

Civera JI, Miró NL, Breijo EG, Peris RM (2013) Secchi depth and water quality control: measurement of sunlight extinction. In: Mediterranean Sea: ecosystems, economic importance and environmental threats, pp 91–114

Crain CM, Kroeker K, Halpern BS (2008) Interactive and cumulative effects of multiple human stressors in marine systems. Ecol Lett 11(12):1304–1315

Doney SC, Howarth R, Chan F, Conley DJ, Garnier J, Marino R, Billen G (2011) Coupled biogeochemical cycles: eutrophication and hypoxia in temperate estuaries and coastal marine ecosystems. Front Ecol Environ 9(1):18–26

Ekins P, Simon S, Deutsch L, Folke C, De Groot R (2003) A framework for the practical application of the concepts of critical natural capital and strong sustainability. Ecol Econ 44(2–3):165–185
FAO (United Nations Food and Agriculture Organization) (1990) Soilless Culture for Horticultural Crop Production. Rome: FAO. Grifo, F. and J. Rosenthal, editors. 1997. Biodiversity and Human Health. Island Press, Washington, DC
Fedler AJ, Nickum DM (1992) The 1991 economic impact of sport fishing in the United States. Sport Fishing Institute
Goulletquer P, Gros P, Boeuf G, Weber J (2014) The importance of marine biodiversity. In: Biodiversity in the marine environment. Springer, Dordrecht, pp 1–13
Guillaumont P (2010) Assessing the economic vulnerability of small island developing states and the least developed countries. J Dev Stud 46(5):828–854
Halpern BS, Frazier M, Potapenko J, Casey KS, Koenig K, Longo C, Lowndes JS, Rockwood RC, Selig ER, Selkoe KA, Walbridge S (2015) Spatial and temporal changes in cumulative human impacts on the world's ocean. Nat commun 6(1):1–7
Halpern BS, Walbridge S, Selkoe KA, Kappel CV, Micheli F, D'Agrosa C, Bruno JF, Casey KS, Ebert C, Fox HE, Fujita R (2008) A global map of human impact on marine ecosystems. Science 319(5865):948–952
Harley CD, Randall Hughes A, Hultgren KM, Miner BG, Sorte CJ, Thornber CS, Rodriguez LF, Tomanek L, Williams SL (2006) The impacts of climate change in coastal marine systems. Ecol Lett 9(2):228–241
Harrison PJ, Hurd CL (2001) Nutrient physiology of seaweeds: application of concepts to aquaculture. Cah Biol Mar 42(1–2):71–82
Islam MS, Tanaka M (2004) Impacts of pollution on coastal and marine ecosystems including coastal and marine fisheries and approach for management: a review and synthesis. Mar Pollut Bull 48(7–8):624–649
Krellenberg K, Kopfmüller J, Barton J (2010) How sustainable is Santiago de Chile? Current performance, future trends, potential measures. Synthesis report of the Risk Habitat Megacity research initiative (2007–2011). UFZ-Report, Vol. 4 (2010), UFZ, Leipzig
Lalli C, Parsons T (1997) Biological oceanography: an introduction. Elsevier
MacFeely S (2020) Measuring the sustainable development goal indicators: an unprecedented statistical challenge. J Off Stat 36(2):361–378
Matin A, Azimul SK, Matiu, AKM, Shamianaz S, Shabnam JH, Islam T (2008) Maternal socioeconomic and nutritional determinants of low birth weight in urban area of Bangladesh. J Dhaka Med College, 17(2):83–87
McGranahan G, Balk D, Anderson B (2007) The rising tide: assessing the risks of climate change and human settlements in low elevation coastal zones. Environ Urban 19(1):17–37
Neumann B, Vafeidis AT, Zimmermann J, Nicholls RJ (2015) Future coastal population growth and exposure to sea-level rise and coastal flooding-a global assessment. PLoS One 10(3):e0118571
Odum EP, Barrett GW (1971) Fundamentals of ecology, vol 3. Saunders, Philadelphia, p 5
Odum EP, Odum HT (1959) Fundamentals of ecology. (No. 574.5), Saunders
Oviedo G, Jeanrenaud S (2007) Protecting sacred natural sites of indigenous and traditional peoples. In: Protected areas and spirituality, pp 77–99
Palumbi SR, Sandifer PA, Allan JD, Beck MW, Fautin DG, Fogarty MJ, Halpern BS, Incze LS, Leong JA, Norse E, Stachowicz JJ (2009) Managing for ocean biodiversity to sustain marine ecosystem services. Front Ecol Environ 7(4):204–211
Paterson RRM, Kumar L, Taylor S, Lima N (2015) Future climate effects on suitability for growth of oil palms in Malaysia and Indonesia. Sci Rep 5(1):1–11
Pauly D, Christensen V, Dalsgaard J, Froese R, Torres F Jr (1998) Fishing down marine food webs. Science 279(5352):860–863
Pickard BR, Daniel J, Mehaffey M, Jackson LE, Neale A (2015) EnviroAtlas: a new geospatial tool to foster ecosystem services science and resource management. Ecosyst Serv 14:45–55
Probert PK (2017) Marine conservation. Cambridge University Press
Singh GG, Hilmi N, Bernhardt JR, Cisneros Montemayor AM, Cashion M, Ota Y, Acar S, Brown JM, Cottrell R, Djoundourian S, González-Espinosa PC (2019) Climate impacts on the ocean are making the Sustainable Development Goals a moving target travelling away from us. People Nat 1(3):317–330

Smith R, Simard C, Sharpe A (2001) A proposed approach to environment and sustainable development indicators based on capital. In: Prepared for The National Round Table on the environment and the economy's environment and sustainable development indicators initiative, Canada

Tansely AG (1935) The use and abuse of vegetational concepts and terms. Ecology 16(3):284–307

Tyack PL, Zimmer WM, Moretti D, Southall BL, Claridge DE, Durban JW, Clark CW, D'Amico A, DiMarzio N, Jarvis S, McCarthy E (2011) Beaked whales respond to simulated and actual navy sonar. PLoS One 6(3):e17009

UN (2017) Resolution adopted by the General Assembly on 6 July 2017, A/RES/71/313, Work of the Statistical Commission pertaining to the 2030 Agenda for Sustainable Development

United Nations (2017) Resolution adopted by the General Assembly on 6 July 2017, Work of the Statistical Commission Pertaining to the 2030 Agenda for Sustainable Development (A/RES/71/313 Archived 28 November 2020 at the Wayback Machine)

Valavanidis A, Vlachogianni T (2010) Metal pollution in ecosystems. Ecotoxicology studies and risk assessment in the marine environment. Department of Chemistry, University of Athens University Campus Zografou, p 15784

Van der Sluijs J (2012) Uncertainty and dissent in climate risk assessment: a post-normal perspective. Nat Cult 7(2):174–195

Part VII
Application of Geospatial Tools

Chapter 26
Advanced Remote Sensing Methods for High-Resolution, Cost-Effective Monitoring of the Coastal Morphology Using Video Beach Monitoring System (VBMS), CoastSnap, and CoastSat Techniques

M. Ramesh, L. Sheela Nair, V. Amrutha Raj, S. G. Sarankumar, S. Akhildev, and R. P. Arya

Abstract The dynamic behaviour of the coastal environment demands for continuous monitoring to understand the impacts of various natural and anthropogenic activities on the coastal geomorphology. This will be achieved through the application of three recent remote sensing techniques together for coastal monitoring, namely, Video Beach Monitoring System (VBMS), CoastSnap, and CoastSat, are presented in this chapter with a pilot study along the Thiruvananthapuram coast in the southwest (SW) coast of India. The study aims to illustrate the application and capabilities of these techniques for understanding the coastal processes and the related changes in the morphology along the coast. Data from two VBMS stations (Valiyathura and Varkala), one CoastSnap station (Adimalathura), and six CoastSat locations (Adimalathura, Kovalam, Valiyathura, Shangumukham, Muthalapozhi, and Varkala) along Thiruvananthapuram coast, have been used. Details of installations of VBMS and CoastSnap stations and a discussion on the preliminary results with their interpretation are presented. Coastal morphological parameters like beach width, and surf zone width (at daily and hourly intervals) have been derived from the video imagery and CoastSnap photographs and validated with field measure-

M. Ramesh (✉)
National Centre for Earth Science Studies (NCESS), Ministry of Earth Sciences, Thiruvananthapuram, Kerala, India

Cochin University of Science and Technology, Cochin, Kerala, India
e-mail: ramesh.madipally@ncess.gov.in

L. Sheela Nair · V. Amrutha Raj · S. G. Sarankumar · S. Akhildev · R. P. Arya
National Centre for Earth Science Studies (NCESS), Ministry of Earth Sciences, Thiruvananthapuram, Kerala, India

N. Jayaraju et al. (eds.), *Coasts, Estuaries and Lakes*,
https://doi.org/10.1007/978-3-031-21644-2_26

ments (correlation >85%). The results are then used to study the short-term shoreline changes as it is ideal to understand the abrupt morphological changes due to the impact of cyclones and also to observe the time taken for recovery, particularly in critical/vulnerable areas which are not accessible for taking in situ measurements during such ephemeral events. In addition, CoastSat method has been used to study the historical shoreline changes in recent years (since 2015) from satellite imagery. To summarise, the results obtained from this pilot study are quite encouraging and show that the combination of the three techniques can be employed for successful monitoring of any critical/vulnerable location along the Indian coast. The long-term site-specific primary data once collected from these methods can be used for the development of a high-resolution database for the Indian coast, which can provide valuable information on the coastal processes and related changes, thereby, providing reliable and timely information to the coastal scientists/engineers for taking appropriate decisions for disaster mitigation as well as management and also planning of coastal development and protection related activities.

Keywords Video monitoring · CoastSnap · CoastSat · Beach-surfzone dynamics · Shoreline changes

26.1 Introduction

For any country with a long coastal line, coastal areas are of prime importance as they function as gateways to promote and boost the nation's economy, security, trade, and tourism-related activities. In recent years, coastal areas are under increasing threat and the contributing factors are both natural and anthropogenic. Natural attributes linked to climate change and global warming can cause an increase in sea level, a number of cyclonic, and other extreme events, which lead to coastal flooding and erosion in vulnerable areas. Similarly, anthropogenic activities like the construction of hard structures on the coast as part of coastal development, shore protection, port/harbour development, coastal sand mining (both offshore and nearshore), and beach sand mining, also, will have a detrimental effect on the stability of the coast and the long-term effects can be quite damaging resulting in the eventual loss of land, which in turn can affect the natural coastal ecosystem of the area. Hence, for any developing country like India with a vast coastline of about 7500 km length, the prime objective is to aim for sustainable development of the coast with an increasing focus on minimising the negative impacts by adopting proactive measures.

Over the years, the Indian coastline has been showing trends of degradation (Rajawat et al. 2015). Of the nine maritime states, four are experiencing nearly 40% erosion of the shoreline and if the trend continues, it will be disastrous for these states with a highly dense population in the coastal area and causing an alarming situation along the Indian coastline (Kankara et al. 2018). Continuous site-specific monitoring of the coast, particularly in vulnerable areas is essential for the collection of nearshore data for understanding the complex coastal processes, which exhibit both spatial and temporal variation. The data is extremely useful for achieving the objective of sustainable coastal development so that the shoreline can be preserved intact as timely corrective measures in the form of disaster mitigation or coastal protection can be adopted whenever required. Even though it is widely accepted that measurements/observations of the nearshore hydrodynamics by the direct deployment of instruments is the best possible method for data collection, it is not always feasible due to the huge cost and high risk involved in maintaining instruments in a hostile environment for long periods. In this context, the application of remote sensing methods to obtain high-resolution (both spatial and temporal) observations for continuous monitoring of the coastal dynamics and the associated changes in morphology gains importance as a technically viable and cost-effective solution. However, it is a known fact that some of these methods also have certain inherent limitations in adopting regular procedures for sampling with in situ sensors or satellite imagery/aerial photography; frequency at which images are available depending on the satellite revisit period (data may not be available during extreme events); availability and reliability of satellite data whenever cloud cover percentage is significant; gaps during maintenance/repairs, etc. With the advent of high-resolution digital cameras and the development of custom-made image processing tools, the use of land-based coastal video monitoring system has emerged as one of the most reliable and proven methods for continuous monitoring of the coast. The method is relatively cheap and has been in use since 1990, with the first ARGUS (Holman and Stanley 2007) station installed in the United States, and later adopted in other parts of the world mainly in England, Australia, and European countries like France, Portugal, Spain, and Italy. Recently, particularly during the last 3–5 years, the application of this method has spread to SE Asian countries like China, Korea, Japan, Singapore, the Philippines, Vietnam, Thailand, and Hong Kong.

The main advantage of adopting digital imaging technology through the installation of shore-based video stations is that it provides a cost-effective solution without compromising the quality of data, which is very important for providing valuable data for planning, design, and implementation of coastal management measures as well as other activities. ARGUS is the first ever developed coastal video monitoring system, pioneered by the Coastal Imaging Laboratory, US, in 1990 and over the years it has developed into a fully fledged coastal environmental monitoring system with proven capabilities for coastal morphological analysis (Holland and Holman 1997; Lippmann and Holman 1989; Coco et al. 2000; Quartel et al. 2006), bathymetry (Stockdon and Holman 2005; Aarninkhof et al. 2003), currents (Chickadel et al. 2003), wave height (Gal et al. 2013), shoreline analysis (Smit et al. 2007), and coastal zone management (Kroon et al. 2007; Davidson et al. 2007). Presently, the Argus System has a commercial status thereby limiting its application to other parts of the world particularly Asian regions/developing countries because of huge costs. This has subsequently paved the way for the development of other similar video monitoring systems and also video processing software for coastal monitoring across the world, namely, CAM-ERA (Coco et al. 2004), Beach keeper plus

(Brignone et al. 2012), ORASIS (Vousdoukas 2013), CASAGEC (http://www.casagec.fr/), COSMOS (Taborda and Silva 2012), HORUS (www.horusvideo.com), SIRENA (Zarruk et al. 2012; Nieto et al. 2010), and ULISES (Simarro et al. 2017). These systems which are at different stages of development are yet to attain the present status of ARGUS even though many of the systems have almost equal capabilities. A similar system has been introduced for the first time in India, with the installation of the pilot VBMS Station at Valiyathura (Thiruvananthapuram coast) SW coast of India by NCESS in 2016 (Ramesh et al. 2020). The pilot VBMS system is indigenously designed and with the experience gained from the first installation, further improvements are made for enhancing its performance.

Similar to the coastal video monitoring methods, there are two other remote sensing-based techniques/methods, namely, CoastSnap (Harley et al. 2018, 2019) and CoastSat (Vos et al. 2019) are relatively new (about 5 years) but are being used in other parts of the world. Of the two, CoastSnap is a community-based beach monitoring program that invites the public to take a photo and upload it to social media, using a specific hashtag. This simple method introduced in 2017, follows a citizen-science approach, wherein an average community member is elevated to a coastal scientist by taking pictures of the coastline with their smartphones and share to the database (Harley et al. 2019). Since this technique does not require expensive equipment or trained technical personnel to collect the data, it is a low-cost method but still provides reliable good quality data. The advantage is that the method can be adopted at any critical/vulnerable location which warrants immediate attention as it does not need sophisticated instruments/gadgets/infrastructure facilities except for a simple cradle stand for mounting the mobile phone. The third method discussed here is the CoastSat, which is a python-based toolbox developed using advanced algorithms to enable faster extraction of shorelines from satellite imageries freely downloadable utilising the google earth engine facility (Vos et al. 2019).

During the last 5 years, NCESS has been involved in developing a coastal monitoring system based on the application of a combination of remote sensing methods, namely, land-based Video Beach Monitoring Systems (VBMS), photographs from CoastSnap, and satellite images with CoastSat techniques. The aim is to have a standalone system comprising all three methods mentioned above, which can be customised for use at any location along the Indian coast. The monitoring systems once fully established can generate valuable data sets for understanding coastal erosion, sediment transportation, surf zone characteristics, temporal changes in shoreline (both short-term and long-term), etc., which vary both spatially and temporally. It is expected that the wealth of coastal data generated would be useful for unravelling many of the complex coastal processes which could not be explained earlier in a scientific way due to a lack of site-specific information. The long-term goal is to establish a network of cost-effective coastal monitoring systems for different locations in India (identified as vulnerable locations, tourist spots, geological and heritage sites, strategically important locations, ecologically sensitive areas, etc.) so that continuous data would be available for taking timely management and protection measures to ensure sustainable development of the coast.

26.2 Study Area

The southwest coast of India, particularly the Thiruvananthapuram coast, which is a high-energy coast has been showing a retreating shoreline trend since 2005 and the erosion rate has drastically increased during the last decade (Rajawat et al. 2015; Kankara et al. 2018; Mallik et al. 1987; Sreekala et al. 1998; Noujas and Thomas 2015). There also has been an increase in the number of flash flood events reported known as Kallakadal (Baba 2005; Kurian et al. 2009a, b). Hence, the high-energy Thiruvananthapuram coast has been chosen as the study area. It is an ideal location for the pilot study as it represents a coast with diverse environmental settings comprising low-lying beaches like Adimalathura (located at the southern tip of the Thiruvananthapuram coast); pocket beaches like Kovalam – a popular tourist destination (separated by rocky headlands), Valiyathura, Shanghumukam, and Veli in the middle; and a cliff beach comprising of three pocket beaches skirting the Varkala cliff (which is a national geological monument) at the northern end. Geographically, it lies between 8.299°N latitude, 77.082° E longitude (Southern boundary), and 8.766° N Lat., 76.666° E Long. (Northern boundary) and the shoreline length is 55 km (approx.). For the present study, data recorded by the two VBMS stations at Valiyathura (8.4636°N latitude, 76.9275°E longitude) and Varkala (8.7376° N latitude, 76.7017° E longitude); CoastSnap photographs shared from Adimalathura (8.3533° N latitude, 77.0153° E longitude) and CoastSat analyses carried for six coastal locations, namely, Adimalathura, Kovalam, Valiyathura, Shanghumukam, Muthalapoozhi, and Varkala are considered as the combined results of these three methods applied together, which can give an overall picture of the short-term beach state conditions of the Thiruvananthapuram coast. Figure 26.1 shows the map of the study area with the position of VBMS stations, CoastSnap, and CoastSat locations used for the study marked.

26.2.1 Installation and Setup of VBMS and CoastSnap Stations

26.2.1.1 VBMS – Valiyathura

It is the pilot VBMS station, installed in 2016 at Valiyathura beach, Thiruvananthapuram (SW coast of India). Being the first attempt in India, the prime objective was to have a basic understanding of the installation setup, methodology to be adopted, image processing requirements, etc., needed for continuous monitoring of a high-energy coast experiencing distinct seasonal variability in the morphology. The camera station comprises four cameras, out of which three are bullet cameras mounted on top of a building just behind the beach, at an elevation of 10.3 m above the MSL. The camera positions are such that they face the south, west, and north of the beach covering the entire stretch (approx. 250 m) of Valiyathura pocket beach and the extent of coverage of the nearshore region (western side,

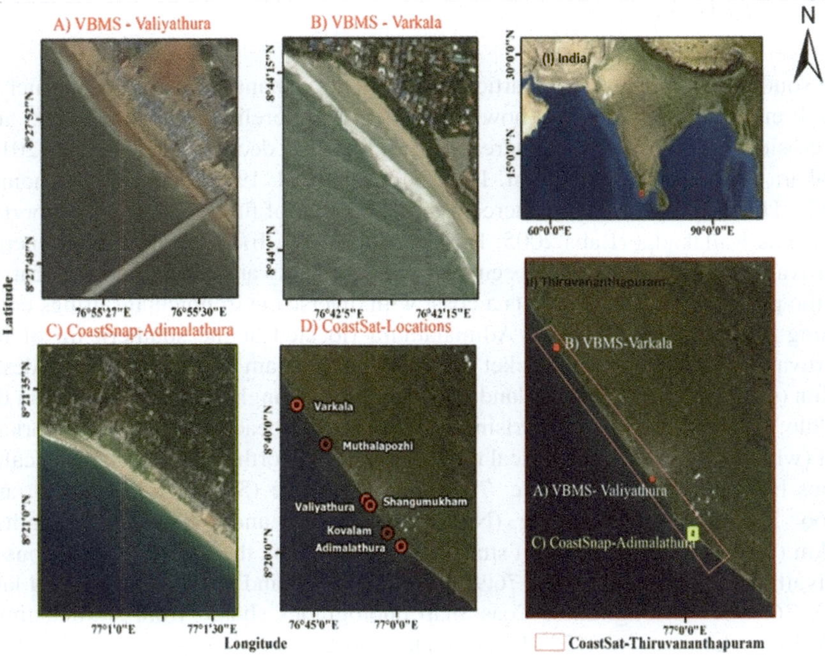

Fig. 26.1 Study area map: (**a**) VBMS – Valiyathura, (**b**) VBMS – Varkala, (**c**) CoastSnap – Adimalathura, and (**d**) CoastSat locations along Thiruvananthapuram

towards the sea) is about 500 m, which is sufficient to study the beach-surf zone dynamics for the high energy coast. In addition to the three cameras which together give a panoramic view of the beach, a fourth camera, which is a PTZ camera is used to give critical information on the nearshore wave characteristics and beach-surf zone dynamics during flooding events due to remote forcing and also other extreme events. The cameras are set to record videos daily during high tide, at a rate of 25 frames per second for 30 minutes. Figure 26.2 represents the VBMS station at Valiyathura, including the three views from cameras.

26.2.1.2 VBMS – Varkala

This is an advanced VBMS installed in 2020 with improvements made in the setup that includes solar panel installation for power supply, facility for automatic data transfer through FTP and also for monitoring during the night through a thermal camera. This station is situated on the famous Varkala cliff, a national geological monument, has the additional advantage of having a natural height of about 25 m above the MSL. Hence, it can cover a larger stretch of the beach (about 1 km) compared to that of the Valiyathura Station. Recording intervals are 06:00 AM to 06:00 PM for the optical camera and 06:00 PM to 06:00 AM for the thermal camera

Fig. 26.2 (**a**) Valiyathura VBMS station. (**b**) Camera 1: view towards the south of the beach. (**c**) Camera 2: view towards the west of the beach. (**d**) Camera 3: view towards the north of the beach

at 8 fps and 20 minutes recording for every hour. Figure 26.3 represents VBMS station at Varkala, including the views from optical and thermal cameras.

26.2.1.3 CoastSnap – Adimalathura

The CoastSnap station at Adimalathura, Thiruvananthapuram is operational since 2020. The installation consists of a simple camera cradle structure fixed on a pole at a suitable height to capture the full view of the beach as per requirement. By adopting this method, members of the local community/any tourist/visitor/individual are encouraged to take snapshots of the beach by mounting their smartphone on the camera cradle and contribute to the image database by uploading the same on social media platforms (using a hashtag identifier) or via other sources like email. A board printed with all the relevant instructions needed for taking and sharing the CoastSnap photographs is fixed just below the mobile cradle fixture on the pole. This method is considered as one of the cheapest and simplest methods that can be adopted for coastal monitoring as the data collection is entirely based on smartphone images

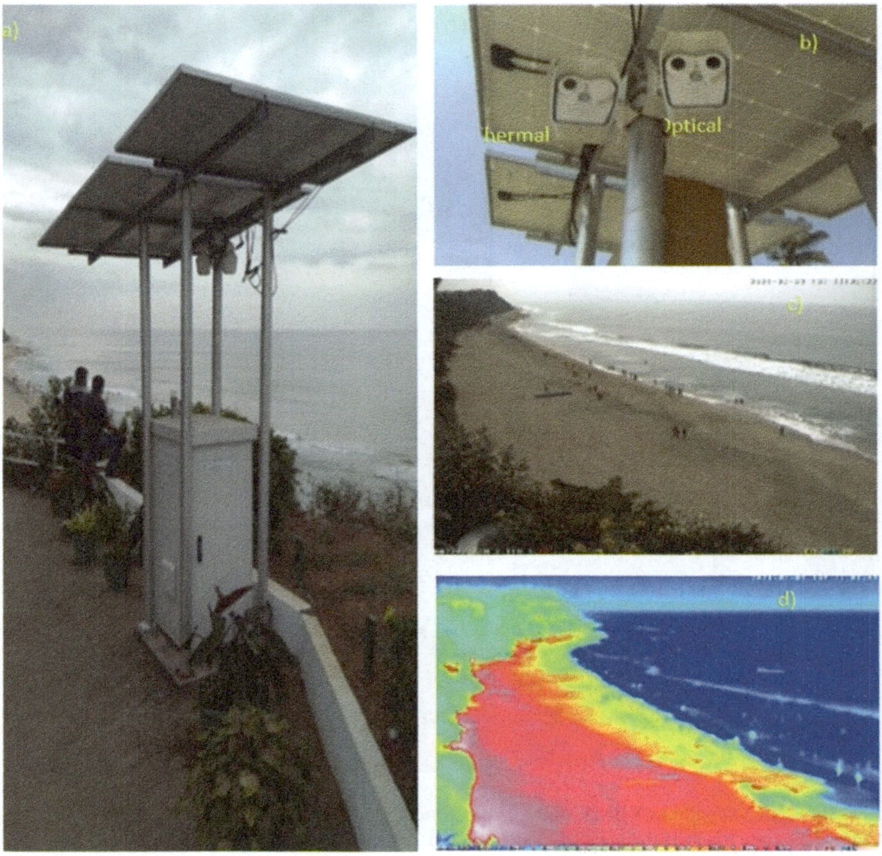

Fig. 26.3 (**a**) VBMS Varkala station. (**b**) Optical and thermal camera. (**c**) Optical camera image. (**d**) Thermal camera image

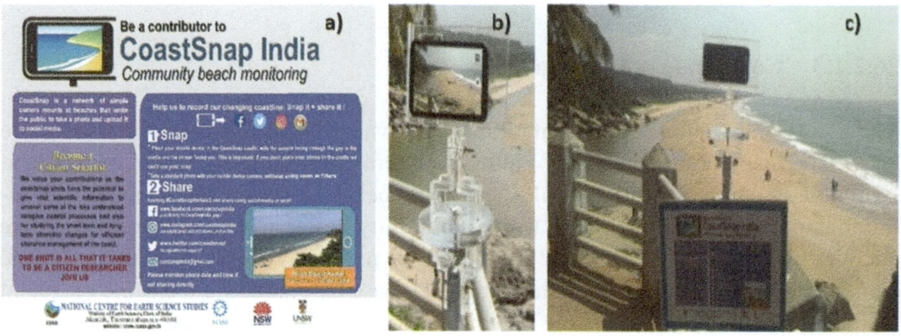

Fig. 26.4 CoastSnap. (**a**) Notice board with instructions. (**b**) Closeup view of the CoastSnap cradle. (**c**) CoastSnap Station at Adimalathura

Table 26.1 CoastSnap social media community account details

Account	Username
Gmail	coastsnapindia@gmail.com
Facebook	CoastsnapIndia (https://www.facebook.com/CoastsnapI)
Instagram	CoastsnapIndia (https://www.instagram.com/coastsnapindia/)
Twitter	CoastsnapIndia (https://twitter.com/CoastsnapI)

shared by the community by uploading on social media platforms, namely, Facebook, Instagram, WhatsApp, and Twitter. The CoastSnap station at Adimalathura installed by NCESS is shown in Fig. 26.4 and Table 26.1 provides the social media accounts being maintained by NCESS for sharing the data.

26.3 Data and Methodology

26.3.1 Data Products

The data products used for the study comprise video data collected from the two VBMS stations; CoastSnap photographs/snapshots of the coast taken and shared through various community platforms by individuals and downloaded satellite images from Sentinel2-A&B, which form the basic input data for CoastSat application. In addition, a set of ground control points (GCPs) established in the area of interest which falls within the field of view covered by the VBMS and CoastSnap stations are used for the rectification process of video imagery, and CoastSnap images. For validation of the coastal parameters derived from the rectified images, in situ field measurements like beach profiles and LEO observations are used.

26.3.2 Methodology

26.3.2.1 VBMS

The camera station is programmed to record the videos at pre-scheduled intervals with a defined duration and transfer the data to the processing station. Camera calibration (Bouguet 2007) and rectification are the pre-processing tasks carried out prior to the generation of rectified plan-view Timex images (Simarro et al. 2017). Later, in-house developed algorithms are used for deriving morphological parameters from the rectified plan view Timex images (Ramesh et al. 2020) and are archived in the database.

26.3.2.2 CoastSnap

All the images are initially registered to a control image using the auto-align function of the Adobe Photoshop application. This program has been selected based on its proven capability to register images from different smartphone models and lighting conditions at a wide range of image resolutions. Once the images are registered to control image, image processing techniques are applied to rectify the community-sourced images using intrinsic and extrinsic parameters from the control image. The images are rectified following the standard method (Harley et al. 2019) and the outputs are generated at a resolution of 0.5 m. Once the images are rectified, shoreline mapping and tidal correction are done using open-source tools (Harley et al. 2019). In order to map the horizontal position of the shoreline from the geo-rectified image, an edge detection algorithm is employed and the necessary tidal corrections are applied by using the tide data (generated from Tide toolbox in MIKE 21 software).

26.3.2.3 CoastSat

The use of satellite imagery to study both short-term and long-term shoreline changes is a proven method and is being widely adopted. Here, the CoastSat, which is an open-source software kit developed using Python provides fast and reliable tools to analyse and study the temporal variations in shoreline depending on the availability of high-resolution multi-dated satellite imageries. It is an open-source python-based advanced toolkit available for fast and efficient processing of satellite imageries downloaded from Google Earth Engine. The method involves downloading all the available satellite images for the user-defined region of interest, automatic detection of the land/water interface and mapping using a robust sub-pixel resolution shoreline detection algorithm (Vos et al. 2019).

26.4 Results and Discussion

26.4.1 Beach State Conditions from VBMS

26.4.1.1 VBMS – Valiyathura

Valiyathura VBMS station was installed in September 2016 and started full-fledged operation in December 2016 after completing the initial field trials (including testing, calibration, and validation) and continuous data was available till June 2017. This being the first station, nearly 2–3 months of field trials were carried out exclusively to understand the intricacies of operation and also gather additional knowledge based on practical experience to deal with issues related to installation, instrument settings, image processing requirements, a minimum height of installation, the influence of sunlight (intensity and direction) on the quality of image, sea spray on cameras, etc.

The ULISES tools were used for geo-rectification of the images recorded by the Valiyathura VBMS. The timex images are generated initially and are rectified and merged to create plan view images. Sample timex images recorded by the three cameras at Valiathura and the rectified plan view image obtained after merging are shown in Fig. 26.5a–d, respectively. The rectified timex images are then used to extract the beach surf zone parameters at daily intervals. Morphological changes in the beach-surf zone region like the formation of beach cusps, and rip channels, if present, can also be identified from these images. The variations in beach and surf zone widths from December 2016 to June 2017, are presented in Fig. 26.6. From the results, it is evident that the Valiyathura beach shows cumulative erosion with a reduction of 43 m in beach width during the fair season of 2016–17. This type of severe eroding trend is not observed from December to May period of previous

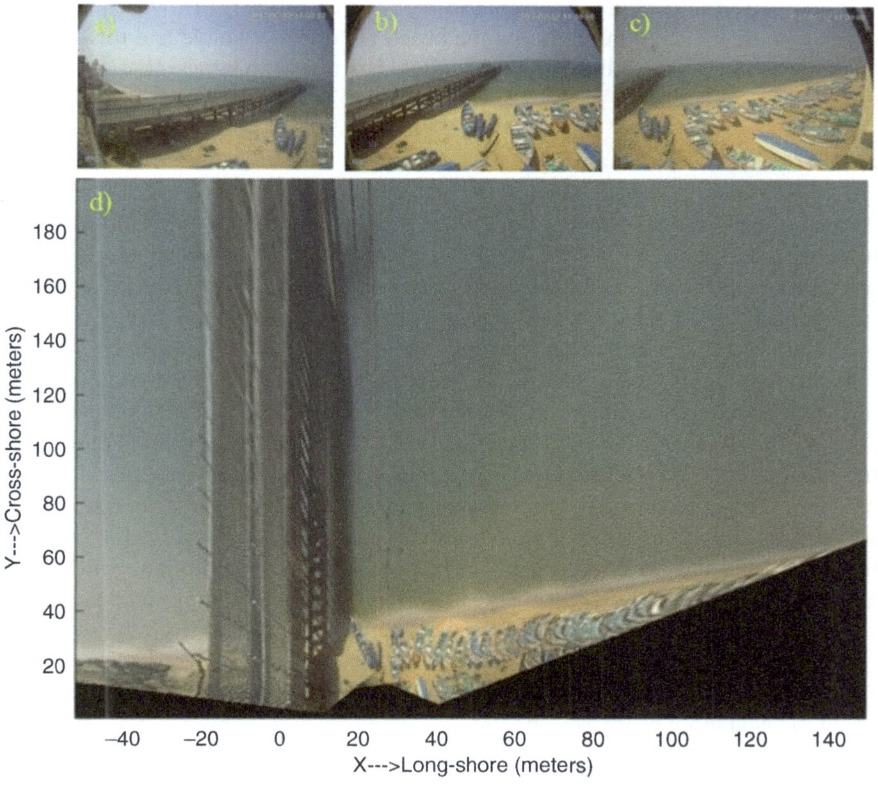

Fig. 26.5 Video generated timex image rectification – Valiyathura; (**a–c**) individual timex images, and (**d**) rectified planview image

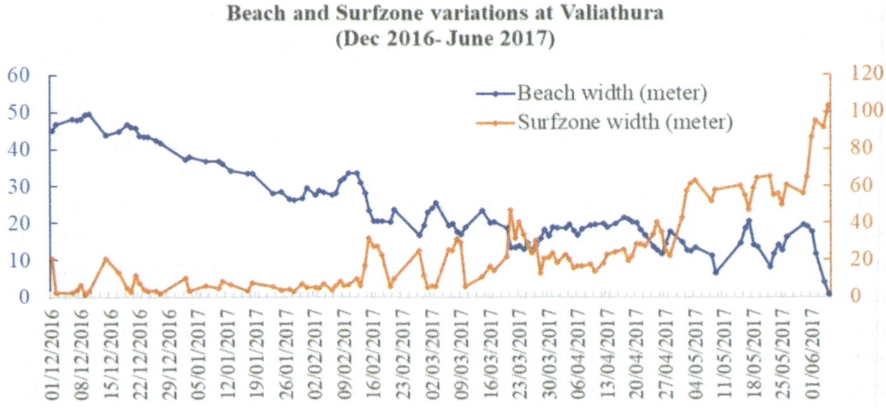

Fig. 26.6 Beach surf zone dynamics of Valiyathura

years, since, the fair season period from December–April is considered as a beach-building period in general and it used to be like that till 2015. However, from the present study, it can be inferred that the beach is no longer stable and this eroding tendency is likely to continue which definitely is a matter of concern.

26.4.1.2 VBMS – Varkala

As mentioned earlier, the Varkala VBMS is an upgraded system with improved features like solar power, advanced facilities for data recording, storage, and online transmission, aimed at further enhancing the overall performance of this indigenously designed system. Based on the learning experience gained from the installation of the first station, special care was taken to minimise the adverse influence of sunlight by adjusting the position of cameras by trial and error and further fine-tuning before fixing. Being situated on the cliff top, this station is devoid of the sea spray issue, which otherwise can affect normal functioning. For this station, the CIRN- UAV processing toolbox is being used to generate the rectified timex images which are subsequently used for deriving the morphological parameters. The rectified timex images, once generated are processed to estimate the parameters such as beach width, surfzone width, and wave runup. which gives a fairly good understanding of the nearshore processes including the beach-surfzone dynamics covering the inter-tidal area. Typical snapshots from the two cameras with the corresponding rectified timex images are presented in Fig. 26.7. Figure 26.8 shows the computed beach and surfzone widths, which gives an idea about the nearshore processes and the associated beach-surfzone dynamics during the observation period.

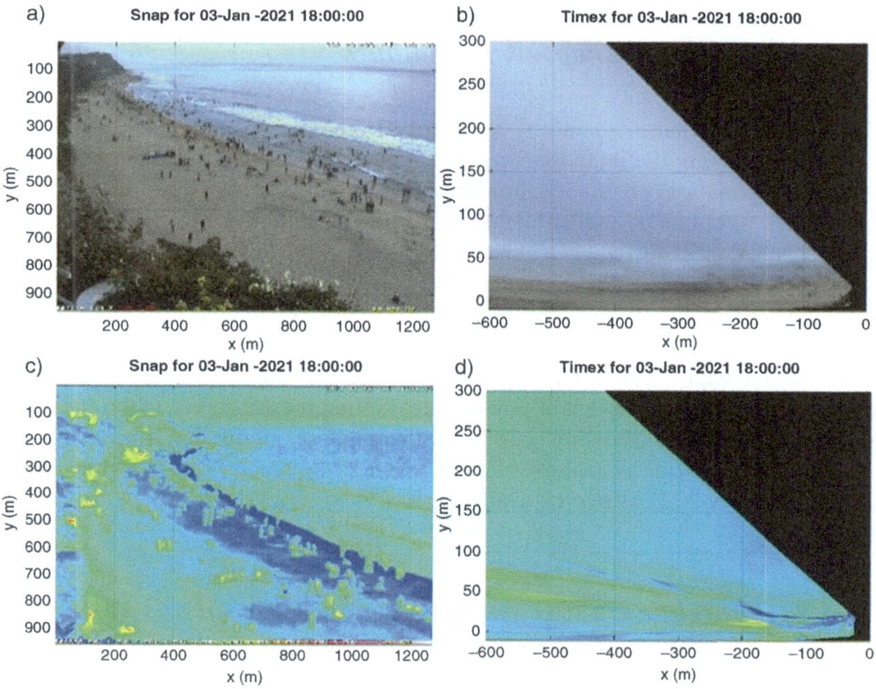

Fig. 26.7 Varkala (**a**) Snap image (optical), (**b**) Timex image (optical), (**c**) Snap image (thermal), and (**d**) Timex image (thermal)

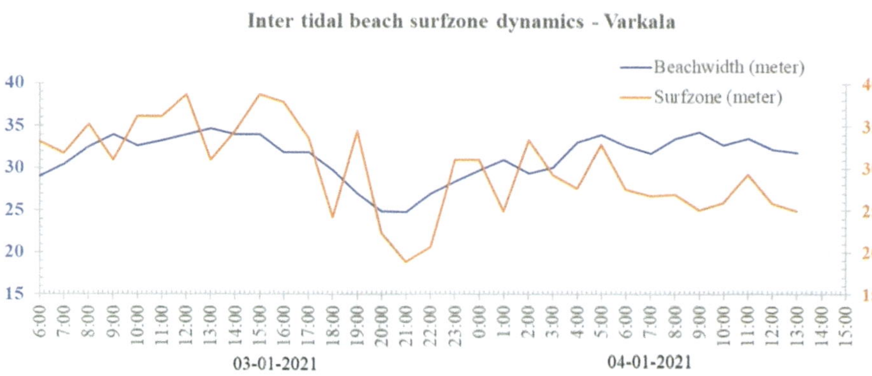

Fig. 26.8 Beach surf zone dynamics – Varkala

26.4.2 Shoreline Change Analysis from CoastSnap – Adimalathura

The CoastSnap images shared by the local community through social media are useful for studying the temporal variations in coastal sea state conditions and also getting information on the erosion/accretion trends linking with the observed changes in morphology, which includes changes in shoreline position, beach profiles, beach slopes, surf zone width, breaker height, type of wave breaking, etc. A sample of the rectified image obtained from the collected photograph is shown in Fig. 26.9. Figure 26.10 gives the plot of beach width computed using multi-dated photographs and the resulting shoreline change trend. The accuracy of the computed beach width from CoastSnap photographs is validated by comparing the derived beach width with that of physical field measurements taken at the site and it is observed that the correlation is 85%. Further analyses of the shoreline changes at Adimalathura indicate a declining trend of 0.73 m in the beach width from September 2020 to January 2021 as per the available data.

26.4.3 Shoreline Change Analysis Using CoastSat

In the present study, the 5-year (2016–2020) short-term shoreline changes at selected locations along Thiruvananthapuram coast have been analysed using the CoastSat tools. The results indicate that the southern part of the study area (Adimalathura to Kovalam) is in a dynamically stable state with monsoon erosion

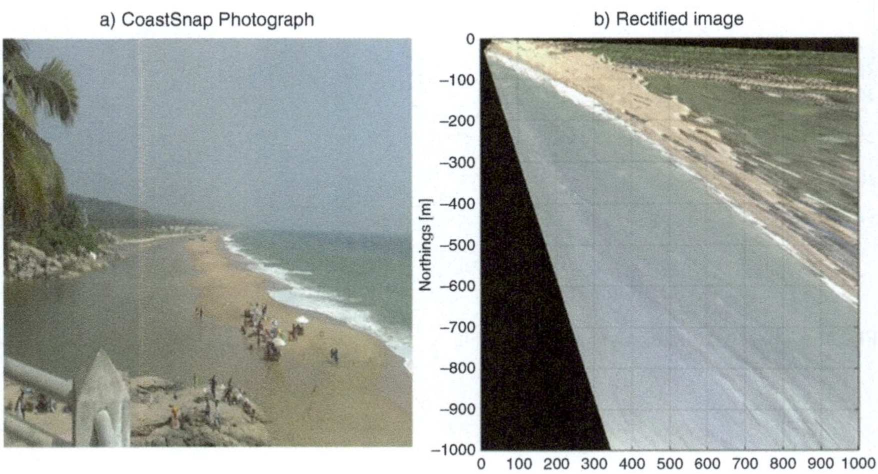

Fig. 26.9 CoastSnap – Adimalathura (**a**) original image (**b**) rectified image

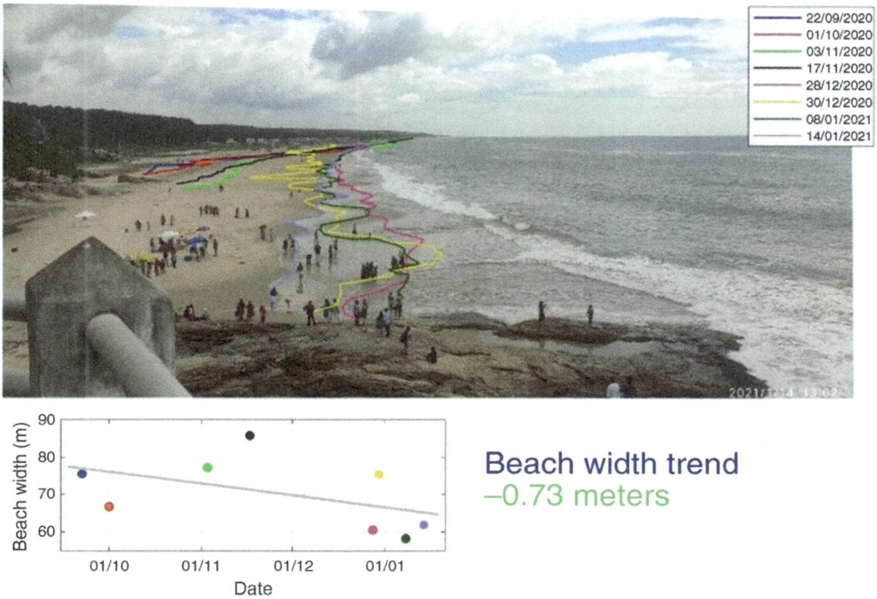

Fig. 26.10 Shoreline position extracted and the computed beach width variation from multi-dated CoastSnap photographs taken at Adimalathura

and recovery thereafter. However, the middle part comprising the Valiyathura-Shangumukham stretch shows an eroding trend with signs of becoming more critical during the last 2–3 years (since 2018). More than 40 m retreats in the shoreline have been observed, and this can be attributed to the influence of the cyclone OCKHI and the introduction of hard structures (particularly groin field) at about 3–5 km to the south of Valiyathura. Muthalapozhi and the Varkala locations which represent the northern part are more or less stable even though the Varkala cliff beach shows significant erosion during the monsoon. The variations in shoreline positions during the 5-year period of 2015–2020, at each of the six locations selected for the study are presented in Fig. 26.11.

26.5 Conclusions

The present study illustrates the application of three state-of-the-art image processing methods/techniques, namely, the VBMS, CoastSnap, and CoastSat, which have been introduced in India for the first time. The VBMS, even though being adopted in other parts of the world, in India it has been tried for the first time (a pilot project taken by NCESS) to study the morphological variations along the high-energy Thiruvananthapuram coast. The study area is a high-energy coast that provides good quality data for understanding the coastal processes at work, which includes the

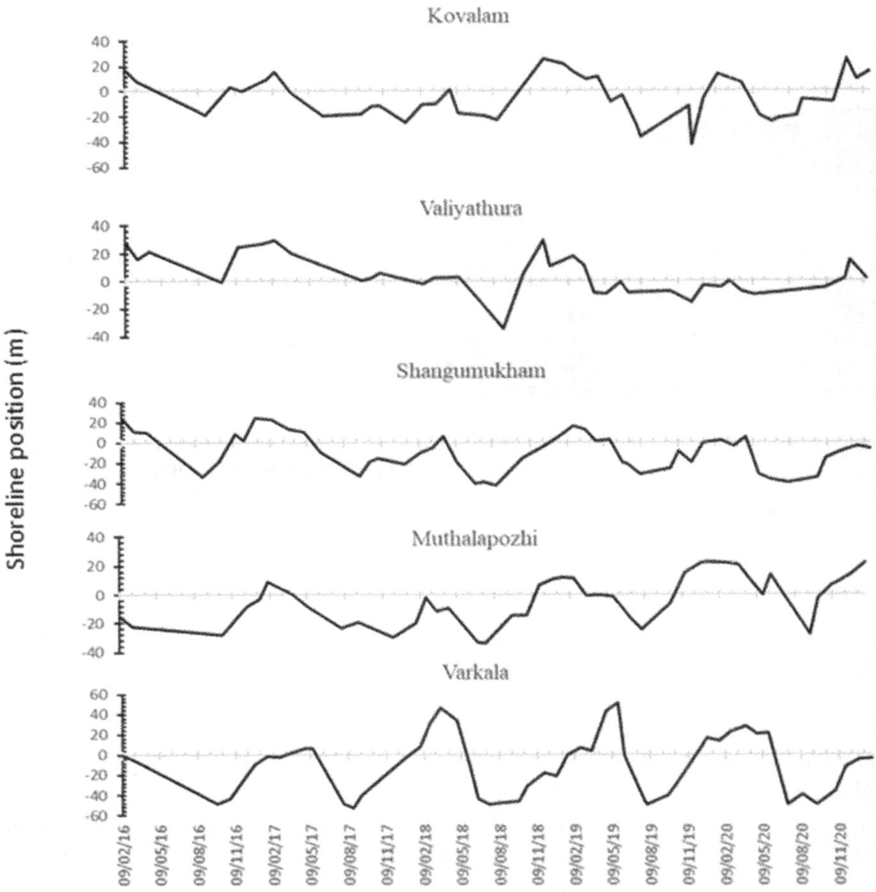

Fig. 26.11 Shoreline analysis at selected locations along Thiruvananthapuram

beach surf zone characteristics, the onset of beach development and its growth which is invariably linked to the nearshore hydrodynamics of the region. The advantage is that high-resolution continuous data is available for carrying out a micro-level study of the erosion/accretion processes, particularly in identifying the causative factors other than the high monsoonal wave activity (mostly westerly) during monsoon and the swells propagating from the Southern Indian Ocean, which are present throughout the year. In addition, morphological features like beach cusp, scarp, berm, nearshore bar formation and dissipation, which could not be studied earlier due to lack of site-specific continuous data are being studied to unravel the complex nearshore processes of a high-energy beach. CoastSnap is also a cost-effective tool to collect valuable and reliable first-hand information on the rapid variations in beach processes, especially at beaches with limited or no access, particularly during extreme events. Since it is a fairly simple method with the involvement of the local community/tourists/any layman visiting the area, the availability of real-time data just before/immediately after extreme events like storms will be useful to assess the damage due to coastal flooding, and high wave activity. Similarly, the CoastSat tools are highly advantageous to understand the short-term and long-term shoreline changes and impacts of extreme events on coastal morphological variations. The CoastSat is developed as a user-friendly practical toolkit that can be easily used to monitor and explore the coastlines at high resolutions. But the non-availability of satellite imageries during the monsoon season and whenever there is cloud cover is a matter of concern to analyse the micro-level temporal changes in shoreline during ephemeral events such as storms and coastal flooding due to remote forcing. However, a combination of all these techniques when applied at different locations can provide a wealth of information pertaining to the underlying coastal processes, sediment transport patterns and impact of structures that are present and also newly introduced if any. Apart from the scientific contribution, the site-specific short-term/long-term data collected can be used as a primary source by coastal engineers/managers for taking important decisions pertaining to planning and design of disaster mitigation and management measures; recommendation of appropriate shore protection measures, for selection of ideal location for the development of port/harbours, etc., without affecting the sustainability.

Acknowledgements The authors would like to thank Mr. Vighnesh R.S., Mr. Sameer V.K., and Mr Sreejith N. for their support during the installation and maintenance of VBMS stations and field data collections. The authors would like to thank the Director, NCESS, and the Ministry of Earth Sciences, India, for providing financial support to the research.

References

Aarninkhof SG, Turner IL, Dronkers TD, Caljouw M, Nipius L (2003) A video-based technique for mapping intertidal beach bathymetry. Coast Eng 49(4):275–289

Baba M (2005) Occurrence of 'swell waves' along the southwest coast of India from southern Indian ocean storm. J Geolog Soc India 66:248–249

Bouguet JY (2007) Camera calibration toolbox for MATLAB. http://www.vision.caltech.edu/bouguetj/calib_doc/

Brignone M, Schiaffino CF, Isla FI, Ferrari M (2012) A system for beach video-monitoring: Beachkeeper plus. Comput Geosci 49:53–61

Chickadel CC, Holman RA, Freilich MH (2003) An optical technique for the measurement of long-shore currents. J Geophys Res Oceans 108(C11):3364. https://doi.org/10.1029/2003JC001774

Coco G, Bryan K, Payne G (2004) The next era for cam-era. Coast News 26:22–24

Coco G, Bryan KR, Green MO, Ruessink BG, Turner IL, Van Enckevort IMJ (2000) Video observations of shoreline and sandbar coupled dynamics. In: Proceedings of coasts and ports, pp 471–476

Davidson M, Van Koningsveld M, de Kruif A, Rawson J, Holman R, Lamberti A, Aarninkhof S (2007) The CoastView project: developing video-derived Coastal State Indicators in support of coastal zone management. Coast Eng 54(6):463–475

Gal Y, Browne M, Lane C (2013) Long-term automated monitoring of nearshore wave height from digital video. IEEE Trans Geosci Remote Sens 52(6):3412–3420

Harley M, Kinsela M, Sánchez-García ES, Vos K (2018) CoastSnap: crowd-sourced shoreline change mapping using smartphones. In: AGU fall meeting abstracts., EP52D-2

Harley MD, Kinsela MA, Sánchez-García E, Vos K (2019) Shoreline change mapping using crowd-sourced smartphone images. Coast Eng 150:175–189

Holland KT, Holman RA (1997) Video estimation of foreshore topography using trinocular stereo. J Coast Res 13(1):81–87
Holman RA, Stanley J (2007) The history and technical capabilities of Argus. Coast Eng 54(6):477–491
Kankara RS, Ramana Murthy MV, Rajeevan M (2018) National assessment of shoreline changes along Indian coast- a status report for 26 years 1990–2016. NCCR Publication, p 69. http://www.nccr.gov.in
Kroon A, Davidson MA, Aarninkhof SGJ, Archetti R, Armaroli C, Gonzalez M, Spanhoff R (2007) Application of remote sensing video systems to coastline management problems. Coast Eng 54(6):493–505
Kurian NP, Rajith K, Shahul Hameed TS, Sheela NL, Ramana Murthy MV, Arjun S, Shamji VR (2009b) Wind waves and sediment transport regime off the south-Central Kerala coast, India. Nat Hazards 49:325–345
Kurian NP, Nirupama N, Baba M, Thomas KV (2009a) Coastal flooding due to synoptic scale, meso-scale and remote forcings. Nat Hazards 48:259–273
Lippmann TC, Holman RA (1989) Quantification of sand bar morphology: a video technique based on wave dissipation. J Geophys Res Oceans 94(C1):995–1011
Mallik TK, Samsuddin M, Prakash TN (1987) Beach erosion and accretion—an example from Kerala, Southwest Coast of India. Environ Geol Water Sci 10:105–110. https://doi.org/10.1007/BF02574668
Nieto MA, Garau B, Balle S, Simarro G, Zarruk GA, Ortiz A, Tintoré J, Álvarez-Ellacuría A, Gómez-Pujol L, Orfila A (2010) An open source, low-cost video-based coastal monitoring system. Earth Surf Process Landf 35:1712–1719
Noujas V, Thomas KV (2015) Erosion hotspots along southwest coast of India. Aquat Proc 4:548–555
Quartel S, Addink EA, Ruessink BG (2006) Object Oriented extraction of Beach Morphology from video images. Int J Appl Earth Obs Geoinf 8:256–269
Rajawat AS, Chauhan HB, Ratheesh R, Rode S, Bhanderi RJ, Mahapatra M, Kumar M, Yadav R, Abraham SP, Singh SS, Keshri KN (2015) Assessment of coastal erosion along the Indian coast on 1: 25,000 scale using satellite data of 1989–1991 and 2004–2006-time frames. Curr Sci 109:347–353
Ramesh M, Sheela NL, Ramachandran KK, Prakash TN (2020) Development of video monitoring system for coastal applications. J Coast Res 89:118–125
Simarro G, Ribas F, Álvarez A, Guillén J, Chic Ò, Orfila A (2017) ULISES: an open-source code for extrinsic calibrations and planview generations in coastal video monitoring systems. J Coast Res 33(5):1217–1227
Smit MWJ, Aarninkhof SGJ, Wijnberg KM, González M, Kingston KS, Southgate HN, Medina R (2007) The role of video imagery in predicting daily to monthly coastal evolution. Coast Eng 54(6):539–553
Sreekala SP, Baba M, Muralikrishna M (1998) Shoreline changes of Kerala coast using IRS data and aerial photographs. Indian J Geo-Marine Sci 27(1):144–148
Stockdon HF, Holman RA (2005) Estimation of wave phase speed and nearshore bathymetry from video imagery. J Geophys Res 105(C9):22,015–22,033
Taborda R, Silva A (2012) COSMOS: a lightweight coastal video monitoring system. Comput Geosci 49:248–255
Vos K, Splinter KD, Harley MD, Simmons JA, Turner IL (2019) CoastSat: a Google Earth Engine-enabled Python toolkit to extract shorelines from publicly available satellite imagery. Environ Model Softw 122:104528
Vousdoukas M (2013) ORASIS-a coastal video monitoring platform. In: EGU general assembly conference, vol 3036
Zarruk GA, Orfla A, Nieto MA, Garau B, Balle S, Simarro G, Ortiz A, Vizoso G, Tintore J (2012) SIRENA: an open source, low-cost video-based coastal zone monitoring system. Preprint submitted to Elsevier Preprint, 21p. http://costabalearsostenible.es/ATenpdf/1.1MA/4.OO/SIRENA/final/CMv09.pdf

Chapter 27
Coastal Morphodynamics and Environmental Variables of Ennore Creek: An Integrated Approach

M. Krishnaveni, K. Kalaivani, K. Vijaya Priya, and C. Jagadish

Abstract Creeks are coastal ecosystems and act as a habitat for flora and fauna which help to maintain biodiversity. The Ennore Creek located in North Chennai, Tamil Nadu, on the Southeast coast of India is the study area. The domestic wastes from Chennai City, industrial effluents from the industrial area, and coolant water from the Thermal Power Stations (NTPS) are disposed into Ennore Creek which affects the water quality of the creek. A survey is conducted to identify the major environmental issues faced by the creek. The temporal variation of creek mouth and water quality are the major environmental issues that pose threat to the economic and health well-being of the fisheries community. The creek mouth plays a role in diluting the concentration of pollutants before they are mixed with seawater carried in by the tides. The analysis of decadal variations of the creek mouth using remote sensing technique indicates that the width of the mouth varies between 32 and 223 m. The monthly variations of the shoreline changes around the creek are analysed during different months of the year 2014. The northern part of Ennore Creek is dominantly influenced by accretion and the southern part of the creek is predominantly influenced by erosion during pre-monsoon and vice versa during monsoon periods. The water quality parameters are evaluated at nine sampling locations, for the year 2014. The temporal variations of different water quality parameters, such as pH, temperature, salinity, dissolved oxygen (DO), biochemical oxygen demand (BOD), total suspended solids (TSS), ammonium (NH_4) and total nitrogen (TN), are analysed using the samples. The sampling results show the water quality of the creek is affected which may affect the aquatic life.

Keywords Coastal creek · Shoreline changes · Remote sensing · Water quality · Environmental issues

---

M. Krishnaveni (✉) · K. Kalaivani · K. Vijaya Priya · C. Jagadish
Institute for Ocean Management, Anna University, Chennai, Tamil Nadu, India
e-mail: mkveni@annauniv.edu

N. Jayaraju et al. (eds.), *Coasts, Estuaries and Lakes*,
https://doi.org/10.1007/978-3-031-21644-2_27

27.1 Introduction

Rivers and other freshwater sources meet the ocean in creeks and estuaries. The estuary is the world's most productive environment, a partially enclosed coastal body of brackish water with one or more rivers or streams flowing into it. The effect of waves, currents and tides are primarily the natural factors that influence the coast. The majority of the world's population (60%) lives along the coast, including creeks. The increase in urbanisation and industrialisation in the coastal zone results in adverse environmental impacts. Shoreline, the boundary between land and ocean keeps changing its shape and position continuously due to the dynamic environmental condition.

Human-induced morphological changes and discharge of untreated effluents adversely affect the fragile tidal creek ecosystem (Tanjina et al. 2018; Mushtaq et al. 2020). Many species of fish have adapted to spawning in marine coastal waters while juveniles migrate into estuaries or lagoons to feed and grow in their shelter (Hans 2011). The distribution of flora and fauna in a creek depends on the fluctuations in the physiochemical characteristics of water. Creek and marine environments are affected by a wide range of pollutants as a result of increased anthropogenic activities from rapidly increasing populations along the coastal zones. Estuaries, in particular, have served as major repositories for the disposal of industrial and municipal waste, sewage sludge and dredged material. Pollutants primarily enter estuaries and nearshore waters via pipeline discharges, disposal from vessels, riverside input, atmospheric deposition and nonpoint source runoff from land. The other major point sources of pollution are sediment from stormwater drains, nutrients from sewage outfalls and fish waste associated with aquaculture farms (Whitehead et al. 2010; Coughanowr et al. 2015; Oh et al. 2015). Figure 27.1 depicts the percentages of various pollutants entering the oceans. Eighty per cent of pollutants entering the marine environment are from terrestrial sources, both point and non-point. The most significant non-point source occurs as a result of runoff.

27.1.1 Creek Mouth and Water Quality

Creek mouth plays a role in diluting the concentration of pollutants before being mixed with seawater brought in by tides. Frequent monitoring of shorelines is essential to understanding the coastal dynamics. The Shoreline can be extracted from Google Earth pro for different periods to assess the status of Creek morphological changes. ArcGIS provides information on shoreline detection. The water quality also changes due to morphological changes in creek width. Also, The bio-physical and chemical characteristics of the creek are highly modified by various anthropogenic activities like the discharge of untreated municipal wastewater, industrial effluents, unsustainable land use and land cover modifications.

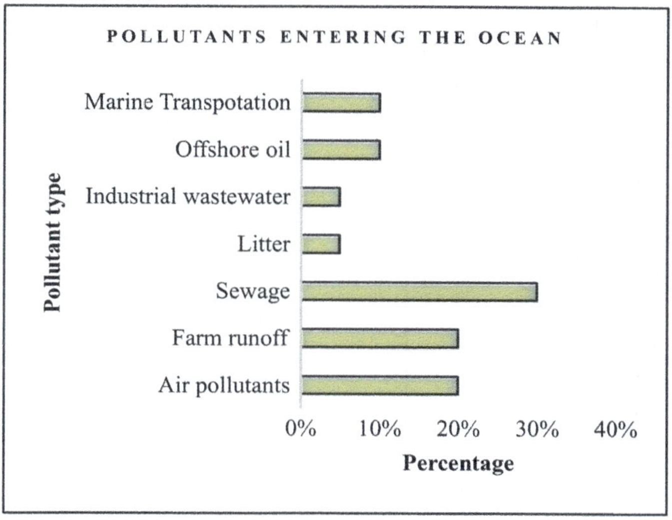

Fig. 27.1 Percentage of various pollutants entering the oceans. (Source: US Environmental protection agency, 1989)

27.2 Objectives

The main objectives of the present study are as follows:

- To assess the environmental issues of Ennore Creek through a socioeconomic survey
- To map the shoreline changes at the creek mouth through remote sensing data
- To analyse the water quality parameters for different periods as a consequence of the opening and closure of the mouth

27.3 Study Area

Ennore Creek is a backwater located in Ennore, Chennai, along the Coromandel Coast of the Bay of Bengal which is shown in Figs. 27.2 and 27.3. It is located between 13°16′ and 13°26′N latitudes and between 80°24′ and 80°35′E longitudes. Mangroves, reptiles, tortoises and unique fish used to live in abundance in this creek. The creek was once encompassed rich diversity of flora and fauna and constituted an excellent green belt, which has been almost completely obliterated by the disposal of untreated sewage and untreated/treated industrial effluents from several chemical industries located around the Creek. The once naturally beautiful creek has been reduced to a mere sewage channel and the biological productivity of the coast has drastically come down.

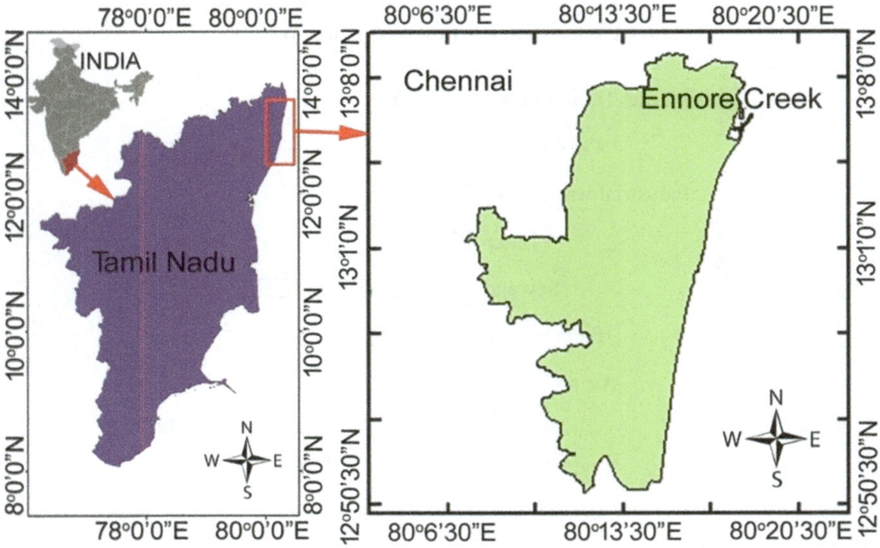

Fig. 27.2 Index map of the study area

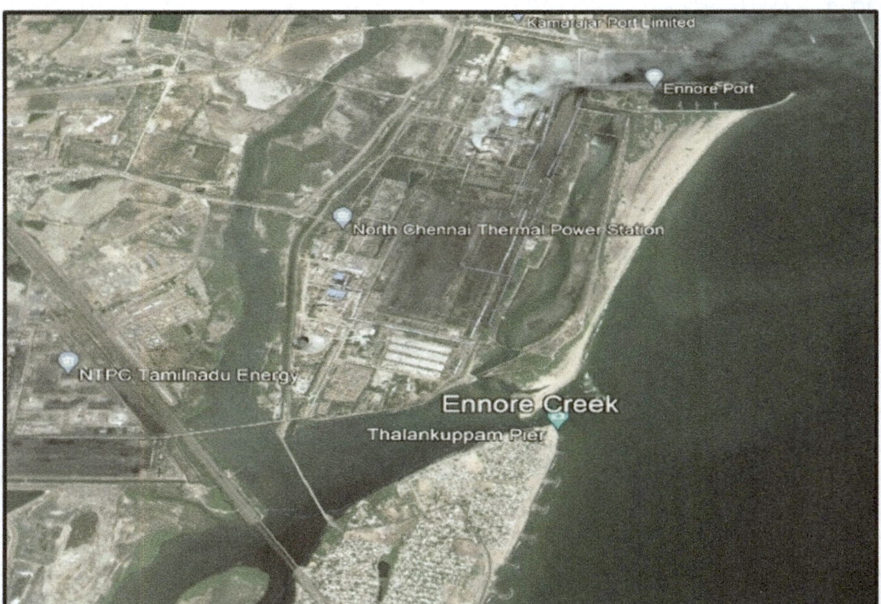

Fig. 27.3 Satellite view of the study area

The Ennore Creek mouth is changing frequently due to natural phenomena like cyclones, floods and tidal fluctuations and anthropogenic activities. During the northeast monsoon, the entire coast is affected by the cyclones. A high rate of shoreline change has been affecting the human settlement and coastal ecosystem along with the areas of the river mouth. The occurrence of tropical cyclones between October and November in the Bay of Bengal is a unique feature (Raj et al. 2019). On India's east coast, cyclones cause massive devastation. Currents circulate along the coast northbound from March to October and southbound from November to February. It affects the shoreline changes causing accretion and erosion (Waters 2004). In addition, the construction of jetties and breakwaters at Ennore Port has resulted in sand accretion. At present, Ennore Creek water is being used for industry, transport, recreation, mariculture and fish production. The thermal power plant utilises the creek water as a coolant, and it has been estimated that the hot water discharge into the creek has resulted in elevating the temperature of the creek and near-shore water by 5°–9 °C above the ambient seawater temperature (Buvaneshwari et al. 2014). The littoral drift is actively forming sandbars near the mouths, which is why regular dredging activities are carried out by state government agencies to keep the rinsing at the mouth of the creek. The study is focused mainly on a detailed analysis of shoreline changes in the mouth area using remote sensing.

27.4 Environmental Issues of the Ennore Creek

27.4.1 Socioeconomic Survey

The survey of the socioeconomic status of fishing and non-fishing communities and their opinion about the environmental issues are essential for the management of the creek ecosystem. The creek has been supporting the livelihood of many thousands of fishing families who are the original stakeholders in the adjacent villages and has been a source of sustenance for the traditional fishermen community. The survey is conducted among 70 households, 56 households belong to the fishermen community and the remaining belong to the non-fishermen community. Among the 70 respondents, 53 are men and 17 are women. Figure 27.4a depicts the age groupings of respondents in detail.

In Ennore Creek, 33% of the fishermen and 11.4% of the non-fishermen have yearly earnings of less than 60,000 rupees. The economic circumstances of respondents in the villages reveal that the vast majority of individuals are from the lower social classes and live below the poverty line. The main occupation of the people around the Ennore Creek is fishing, their income depends on fish catch. Local industries, thermal power plants, the Buckingham Canal source and the mouth closure are all obstacles that indirectly affect the availability of fish in the creek. According to respondents, the fish species like White Prawns (VellaiIral), Black Prawns (KaruppuIral), Sand Prawns (Mann Iral), Tiger Prawns, Green Crab, Irun Kezhuthi,

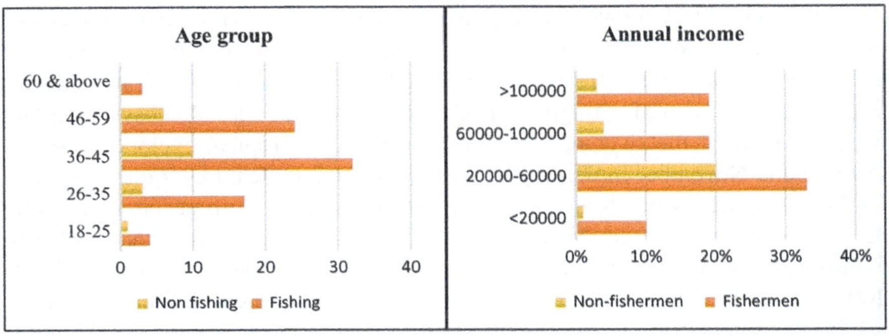

a.) Age group b.) Annual income

Fig. 27.4 Details of respondent's age group and annual income

Mullet (Madavai), Oodan, Kezhangan, Uppathi, Keechan, Kalvaan, Panna and Koduvaare were in abundance and float on top of the river before the 1960s, but after the construction of coastal infrastructures, the quality of water and the availability of fish reduced. Due to thermal power stations, industrial effluents and the closure of the mouth, the quality of water is reduced, and the people living around the creek face health issues. The deterioration of biodiversity caused by industrial effluents impoverishes the fishing community's livelihood and sanitary conditions. Figure 27.4b represents the annual income of the people living near the creek. About 20% of their income is spent on health issues.

27.4.1.1 Environmental Issues

Environmental issues faced by the Ennore Creek are shown in Fig. 27.5. The major environmental issues identified are the closure of the mouth and water pollution based on the survey. The closure of the mouth and the construction activities like jetties, breakwaters and port constructions change the shoreline morphology. Additionally, untreated domestic and industrial effluents are discharged directly through the Kosasthalaiyar River which affects the quality of water in the creek.

The Ennore Creek mouth closure is due to the siltation, shrinkage of the river, poor regulations of tidal inflow and sand bar formation. It faces intense erosion and irregular tidal flow conditions due to the morphological changes along the coast caused by the recent construction activities. According to the survey, the creek is highly polluted due to the domestic sewage, industrial effluents from the industrial belt, coolant water discharge from the ETPS and various other industrial effluents. The occurrence of algal blooms along the coast of Ennore is due to the release of untreated sewage water into the creek and high nutrient loads from agricultural and industrial activities in the surrounding catchment area. The decreased water quality affects the aquatic ecosystem of the Ennore estuary.

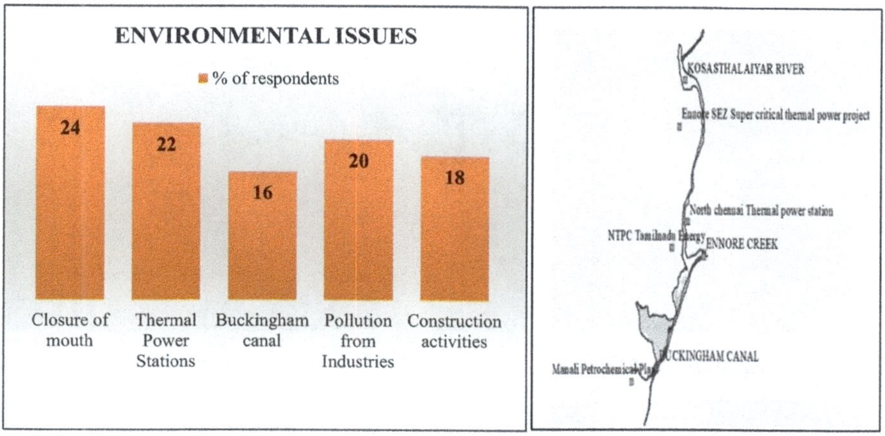

Fig. 27.5 Major issues of Ennore Creek

27.5 Creek Mouth Analysis Through Remote Sensing

27.5.1 Decadal Analysis of the Creek Width Changes

The decadal changes of Ennore Creek mouth are carried out for the years 2008 and 2015. High-resolution satellite images are obtained from Google Earth Pro and are collected for multiple years. November 2008, October 2009, August 2010, June 2011, February 2012, July 2013, March 2014, August 2014 and December 2015 are the months for which satellite images are collected using Google Earth Pro.

Remote sensing and GIS technology are used to evaluate the shoreline changes at the Creek. The mouth width changes range between 32 m and 223 m from 2008 to 2015. Due to erosion and accretion, the creek mouth width fluctuates regularly that shown in Fig. 27.6. The littoral drift, current, wave and wind play a major role in shoreline changes. The frequent closure of Ennore Creek mouth has resulted in insufficient tidal inflow and thus reduced the supply of water required for the thermal power stations.

27.5.2 Monthly Variations of the Ennore Creek Mouth

A decadal shoreline change analysis for 2008 to 2015 shows that shoreline changes from 32 to 223 m. For the detailed analysis of shoreline changes, the year 2014 is chosen. By using remote sensing and GIS Technology the shoreline drawn for March, May, July, September and November of 2014 is shown in Fig. 27.7. In this study, the shoreline change analysis is observed along the river mouth, to provide zone-wise erosion and accretion during different time periods. The creek mouth is

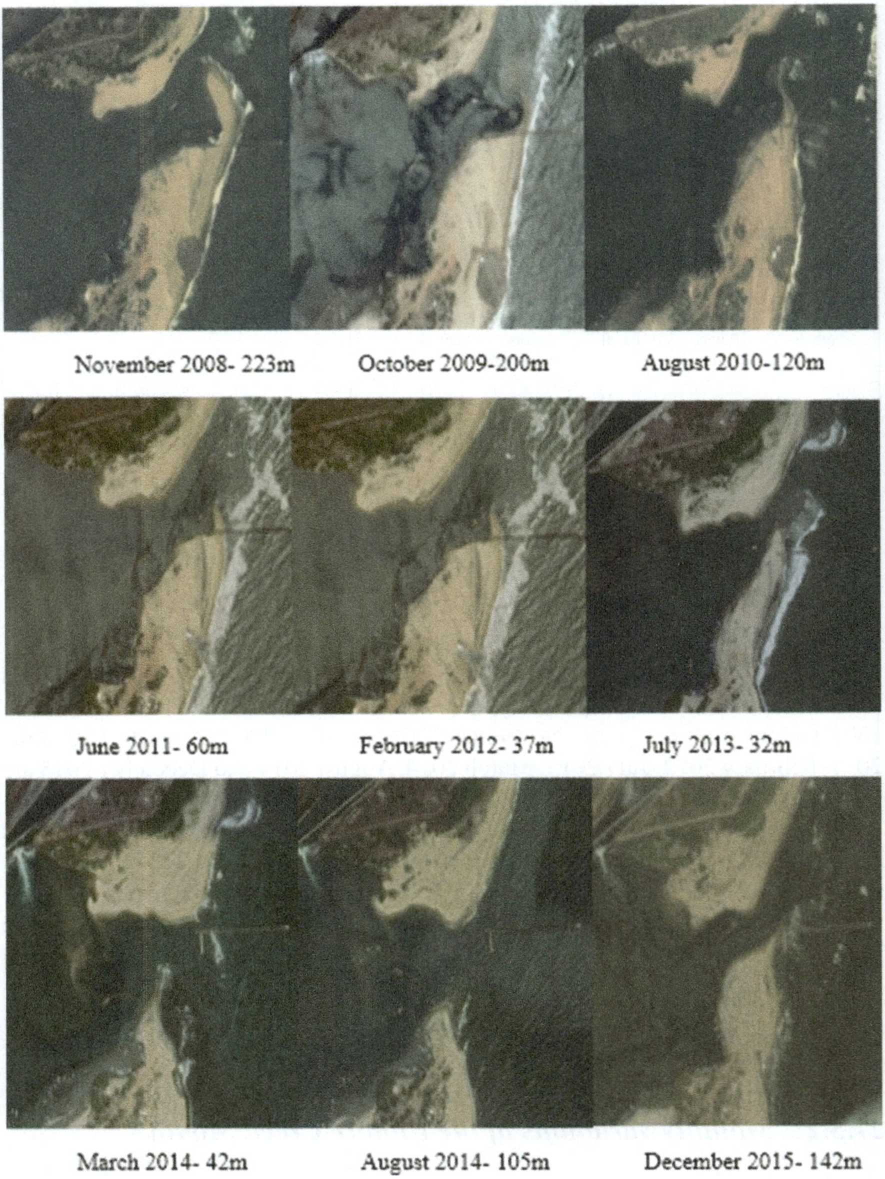

Fig. 27.6 Decadal mouth width variations of Ennore Creek

divided into two zones northern part is represented as zone 1 and the southern part is represented as zone 2 to find erosion and accretion.

The northern part of zone 1 is highly dominated by accretion during March, May and July and accordingly in the same period erosion are predominantly in the

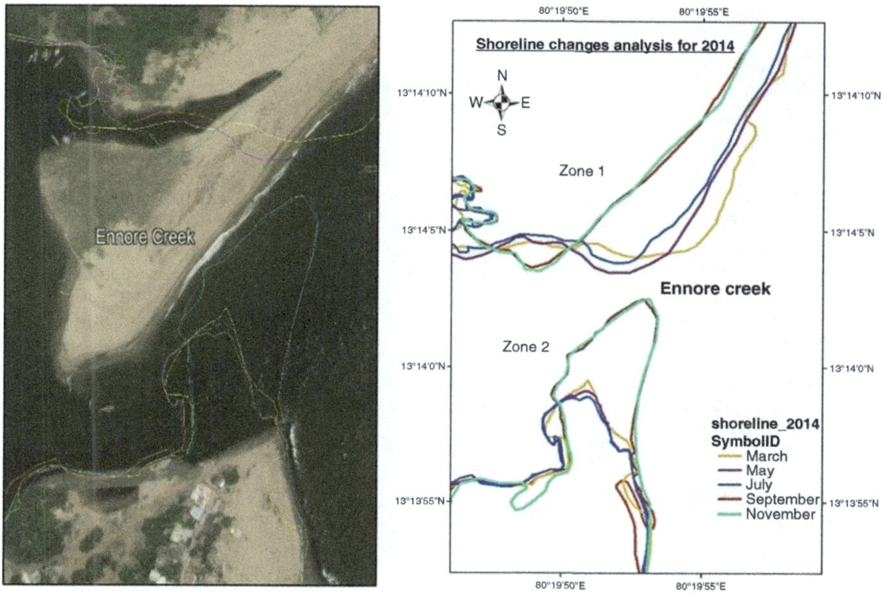

Fig. 27.7 Shoreline analysis in Ennore Creek for different months of 2014

southern part of zone 2. In September and November, erosion takes place in zone 1 and accretion takes place in zone 2 as shown in Fig. 27.7. This accretion and erosion change the coastal morphology, that is, accretion results in the closure of the mouth and erosion results in flooding. The accretion reduces the width of the creek mouth and prevents the river water from diluting with coastal water brought by tides. So the concentration of pollutants increases to a higher level which is enough to affect the aquatic ecosystem. To analyse the water quality of the creek, the water samples were collected in 2014 for three different months with different widths of the creek mouth and are discussed below.

27.6 Water Quality Analysis of the Creek

The shoreline changes often due to erosion and accretion which also affect the water quality. The physicochemical properties of Ennore Creek are analysed using samples collected from nine different stations during three different seasons are shown in Fig. 27.8. The physical properties of Ennore Creek water that are important in determining the well-being of the aquatic ecosystem include temperature, pH, salinity and total suspended solids (TSS). The important chemical properties of creek water are ammonium (NH_4), DO, BOD and Nitrogen.

The parameters analysed in situ are pH, salinity, and temperature pH and salinity were measured using WTW probes. The temperature is recorded

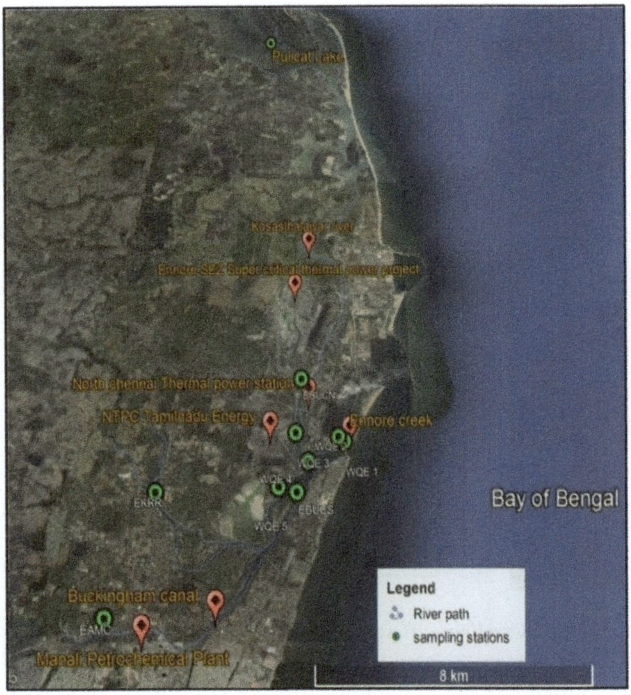

Sample stations	Locations
WQE-1	Bar mouth region
WQE-2	South of the railway bridge
WQE-3	North Buckingham Canal
WQE-4	D/S of Kosasthalaiyar river junction
WQE-5	D/SofAmullavoyal junction
EBUCS	South Buckingham Canal
EBUCN	D/S of Kosasthalaiyar River junction
EKRR	Kosasthalaiyar River
EAMC	Amullavoyal Canal

Fig. 27.8 Map showing the locations of samples collected for water quality

immediately using a digital thermometer of 0.01 °C accuracy. Winkler's method is used to determine dissolved oxygen. Biological oxygen demand (BOD) is determined using the DO method. The total suspended solids are estimated using the American Public Health Association method. Total nitrogen is determined using spectrophotometry. The standard range of brackish water quality parameters is shown in Table 27.1. If the value exceeds the standard range, aquatic life is under stress. The analysed sample result is shown in Fig. 27.9. During all the seasons, the observed temperature was high at WQE2. The maximum temperature of 35 °C was at WQE2, which is attributed due to the discharge of coolant water from the NCTPS. If the temperature of the discharged coolant exceeded 32 °C, enough to potentially impact aquatic life. Except for WQE2, all the other stations recorded less temperature.

The reduced DO level in two stations WQE2 and EBUCS are found. At high temperatures, fish consume more oxygen because their metabolic rates increase and respiration rate increases. A lower concentration of DO in EBUCS and WQE-5 stations increases the risk and stress of aquatic life. At two stations WQE5 and EBUCS, the DO level is very low. The salinity level also changes ranging from 2 to 34 ppt. Increasing salinity modifies the behaviour of the aquatic ecosystem and also limits reproduction and germination, lowering their fitness for survival in that environment. DO and temperature increases are enough to affect

Table 27.1 Water quality parameters in the year 2014

S. No	Parameters		March 2014		July 2014		December 2014		Recommended Value (CPCB std)
			Min	Max	Min	Max	Min	Max	
1.	Physical	Temperature°C	29	34	26	32	28	35	25–30 °C
2.		Salinity (ppt)	2	34	3	30	2	32	0.5–30 ppt
3.		pH	7.8	8.5	7.2	8.8	7.3	8.6	6.6–8.5
4.		TSS (mg/L)	6	81	4	69	2	61	10–20 mg/L
5.	Chemical	DO (ppm)	2	5.5	0.5	8	1.5	7	4–5 ppm
6.		BOD (mg/L)	1	38	1.9	31	2	70	3–20 mg/L
7.		NH_4(μmol/L)	2	149	1	130	5	145	0.2–2.0 mg/L
8.		TN (μmol/L)	25	310	40	320	30	500	–

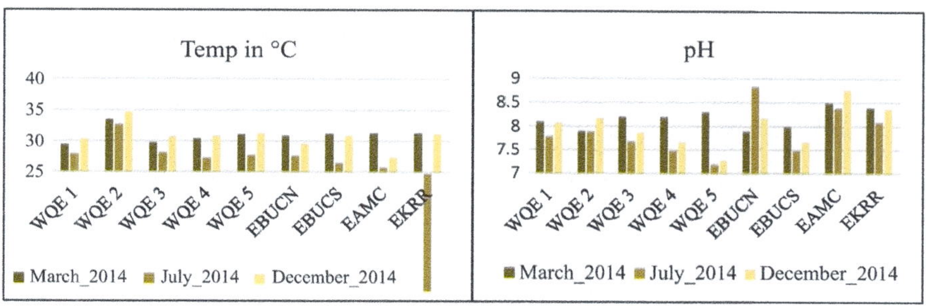

Fig. 27.9 Variations in physiochemical parameters for three different periods

aquatic life and increase stress. A fish's skin can be chemically burned by high-pH alkaline or basic substances.

27.7 Conclusions

Ennore Creek a healthy aquatic habitat formerly is well known for its biodiversity. A socioeconomic survey conducted indicates that the major environmental issues are the closure of the creek mouth and changes in water quality. Remote sensing and GIS technology are used to analyse shoreline changes. According to the decadal analysis of satellite images, the mouth width of Ennore Creek varies between 32 and 223 m. The monthly variation shows that erosion and accretion change the shoreline morphology around the creek throughout the year. The accretion results in the closure of the mouth and erosion results in flooding. Proper dredging is recommended to keep the creek mouth open and protect the aquatic habitat. The physicochemical properties of creek water at nine stations for three different seasons are evaluated and compared with the CPCB standard water quality parameters. The temperature is beyond the recommended value near the south railway bridge sampling location.

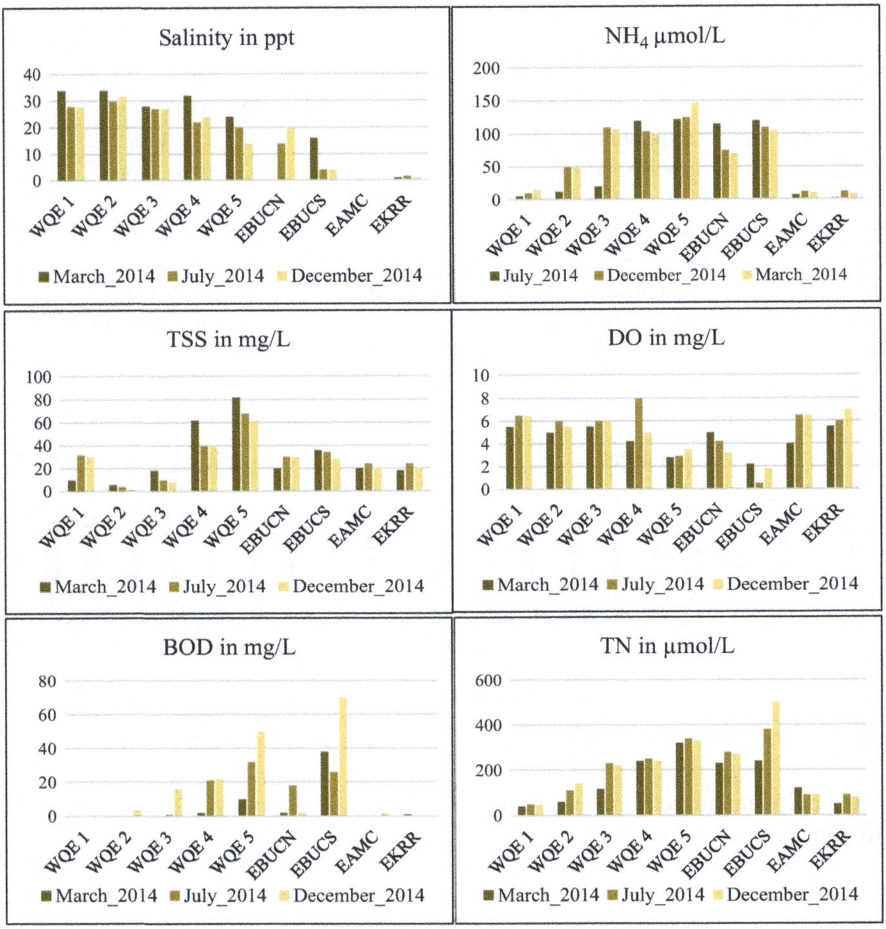

Fig. 27.9 (continued)

The increasing temperature may be due to coolant water from thermal power stations which is enough to threaten aquatic life. Extending the retention time of coolant water in thermal power plants will reduce the temperature of brackish water to safeguard aquatic life. It is recommended to increase the width of the creek mouth and its depth before the monsoon to ensure adequate dilution and to maintain the essential water quality of the creek.

References

Buvaneshwari S, Ravichandran V, Mudgal BV (2014) Thermal pollution modeling of cooling water discharge into a closed creek system. J Mar Sci 43(7):1–7

Coughanowr C, Whitehead S, Whitehead J, Einoder L, Taylor U, Weeding B (2015) State of the Derwent estuary: a review of environmental data from 2009 to 2014. Derwent Estuary Program, Hobart

Hans DW (2011) Globalisation and internationalisation of higher education. Int J Educ Technol High Educ 8:241–248

Mushtaq N, Singh DV, Bhat RA, Dervash MA, Hameed OB (2020) Freshwater contamination: sources and hazards to aquatic biota. In: Book of fresh water pollution dynamics and remediation. Springer, Singapore, pp 27–50

Oh ES, Edgar GJ, Kirkpatrick JB, Stuart-Smith RD, Barrett NS (2015) Broad-scale impacts of salmon farms on temperate macroalgal assemblages on rocky reefs. Mar Pollut Bull 98:201–209. https://doi.org/10.1016/j.marpolbul.2015.06.049

Raj N, Gurugnanam B, Sudhakar V, Francis PG (2019) Estuarine shoreline change analysis along with the Ennore river mouth, southeast coast of India, using digital shoreline analysis system. Geod Geodyn 10(3):205–212

Tanjina GN, Kabir A, Hossian MA (2018) Major environmental issues and problems of South Asia, particularly Bangladesh. In: Handbook of environmental materials management. Springer, Cham, pp 1–40

Waters, North Chennai Coastal (2004) Waste load allocation & waste assimilative capacity studies for Ennore Creek & North Chennai Coastal Waters

Whitehead J, Coughanowr C, Agius J, Chrispijn J, Taylor U, Wells F (2010) State of the Derwent estuary 2009: a review of pollution sources, load and environmental quality data from 2003–2009. Derwent Estuary Program, DPIPWE, Tasmania

Chapter 28
A Study on Dynamics of Krishna River Mouth, East Coast of India: A Geospatial Approach

B. Lakshmanna, N. Jayaraju, G. Sreenivasulu, T. Lakshmi Prasad, K. Nagalakshmi, M. Pramod Kumar, and B. Praveena

Abstract Because of the strategic location Krishna River mouth, research on the dynamics of the river mouth is crucial. The data were created by combining multi-temporal satellite images from IRS P6, LISS III, and Landsat 8 over an 8-year period, namely, 2009, 2012, 2013, 2014, 2015, 2016, 2017, and 2018. Change detection investigations on the river mouth width on its southern tip were regularly and eventually monitored. It has been discovered that the river mouth is rather dynamic. It is fluctuating on a regular and consistent basis. Obviously, there is a threat to the estuary. The river mouth is critical to mitigating, managing, monitoring, and protecting the existing width in order to keep the river biologically, ecologically, and economically vibrant and active. This work discovered a significant alteration in the river mouth as inferred by using geospatial tools such as Remote Sensing and GIS.

Keywords Krishna River mouth · IRS LISS III data · Geospatial tools · East coast of India

B. Lakshmanna
Department of Energy and Environmental Engineering, CSIR-IICT, Hyderabad, Telangana, India

N. Jayaraju (✉)
Department of Geology, Yogi Vemana University, Kadapa, Andhra Pradesh, India

G. Sreenivasulu
Department of Geology, Sri Venkateswara University, Tirupati, Andhra Pradesh, India

T. L. Prasad · M. P. Kumar · B. Praveena
Department of Earth Sciences, Yogi Vemana University, Kadapa, Andhra Pradesh, India

K. Nagalakshmi
Department of Geology, Government College (A), Anantapur, Andhra Pradesh, India

N. Jayaraju et al. (eds.), *Coasts, Estuaries and Lakes*,
https://doi.org/10.1007/978-3-031-21644-2_28

28.1 Introduction

The Krishna River drainage basin is geologically made up of Deccan basalts, Archean granites, Gneiss sand, and Precambrian sand Khondalites. The study area is surrounded by deltaic alluvium. The study area contains paleobeach beach ridge, ridges, swale complexes, mangroves, sand dunes, stabilised sand bars, long creeks, deep and shallow bays, estuaries, and intricate tidal flats. This region has several sets of paleo-morphological features (Nageswara Rao and Vaidyanadhan 1978).

28.2 Geology of the Study Area

Before reaching the estuary, the River Krishna divides into three distributaries, each one about 10 km, Gollumuttapaya, Nadimeru, and the main Krishna River. This Krishna River currently carries the majority of the discharge and empties into the Bay of Bengal. Shallow tidal channels and mudflats are noticed at the lower part of the delta. The study area is an embayment on the southern side of the Krishna delta. The coastline of the area is about 125 km. It is a shallow marine environment with a smooth bottom topography that generally slopes towards the open Bay (Fai et al. 2014; Sreenivasulu et al. 2016; Short and Nakamura 2000).

28.3 Shore Currents

During the SW monsoon season, longshore currents generally move from the Krishna delta's western margin in an N to NW direction, and the delta's eastern side in a W to NW direction. Longshore currents typically have speeds ranging from 15 to 32 cm/s (Mullick et al. 2019; Jayappa et al. 2003).

28.4 Waves

In this region, wave directions and heights are constantly changing. Waves propagate in the SW and NW, the south and SW, the south and SW, and the south. The wave heights range from 0 to 2.8 m (Michelles et al. 2003).

28.5 Sediments

The texture of the sediments ranges from fine to moderate and is very well sorted. They are often coarse to finely skewed and are leptokurtic in nature (Nageswara Rao et al. 2005).

28.6 Weather

Scientists use classification and theoretical models in order to know paleo and present climatic scenarios. Melting Arctic sea ice, a decrease in cloud cover and a lack of precipitation, depletion of vegetation and animals, and historical and archaeological evidence are all examples of physical evidence.

28.7 Efficacy of the Data (Table 28.1)

- It may be used as baseline data for the dynamics Krishna River mouth.
- The data generated maybe used to understand River dynamics between 2009 and 2018.
- The data are georeferenced and can be used in future works.
- It is needed by policymakers and researchers, who work on river mouth dynamics.
- It is beneficial to the socio-economic status of the river-dependent human community.

The data show the changes in river mouth width over the course of 8 years, namely, 2009, 2012, 2013, 2014, 2015, 2016, 2017, and 2018. The data is based on satellite images that were corroborated with ground realities. The key interpretation of satellite imagery and assessments of mouth dynamics are depicted (Tables 28.2

Table 28.1 Specification table

Subject area	Earth and planetary sciences
More specific subject area	Remote sensing
Type of data	Table, figures
How data was acquired	IRS6, LISS III satellite Imageries
Data format	Analysed
Experimental factors	Satellite imageries were georeferenced by using GIS and ERADAS imagine software
Experimental features	River mouth dynamics were calculated for the period 2009–2018
Data source location	Nizampatnam Bay and Lankevanidibba
Data accessibility	The data are available in this chapter

Table 28.2 Salient features of satellite data sets

Name of the satellite	Date of data acquisition	Sensor	Resolution	Path	Row	No. of bands
Landsat8	1 January 2018	OLI/ TIRS	30	102	050	4
Landsat8	18 April 2017	OLI/ TIRS	30	142	050	11
Landsat8	19 February 2016	OLI/ TIRS	30	142	050	11
Landsat8	21 April 2015	OLI/ TIRS	23.5	142	050	11
IRS-P6 (Resourcesat-1)	14 April 2014	LISS-III	23.5	102	063	4
IRS-P6 (Resourcesat-1)	3 March 2013	LISS-III	23.5	102	063	4
IRS-P6 (Resourcesat-1)	8 March 2012	LISS-III	23.5	102	063	4
IRS-P6 (Resourcesat-1)	9 October 2009	LISS-III	23.5	102	063	4

Table 28.3 The Krishna River mouth on the southern portion of the Sea mouth (measured through satellite imageries (km))

S. no	Date	Golumuttapaya (km)	Nadimieru (km)	Krishna River (km)
1	1 January 2018	1.93	1.23	Close
2	18 April 2017	1.47	1.25	"
3	19 February 2016	1.80	1.18	Close
4	21 April 2015	1.61	0.94	1.03
5	14 April 2014	1.55	1.29	1.02
6	3 March 2013	1.67	1.05	1.01
7	8 March 2012	1.71	0.69	0.94
8	9 October 2009	1.68	0.99	0.73

and 28.3). The area's location map, satellite images, and assessments demonstrating variations in mouth opening over time are provided (Figs. 28.1 and 28.2).

28.8 Methodology

The Krishna River has three entrances, before emptying into the Bay of Bengal satellite imagery shows that a spiti – like features is forming at the northern end (Sreenivasulu et al. 2016). It is critical to understand the estuary mouth's erosion or accretion phase because beaches are ideal for the survival of diverse biota due to the mixing of fresh and saltwater (Prabaharan et al. 2010). If the southern tip is closed by sediment, there will be an ecological imbalance. A variety of datasets, including

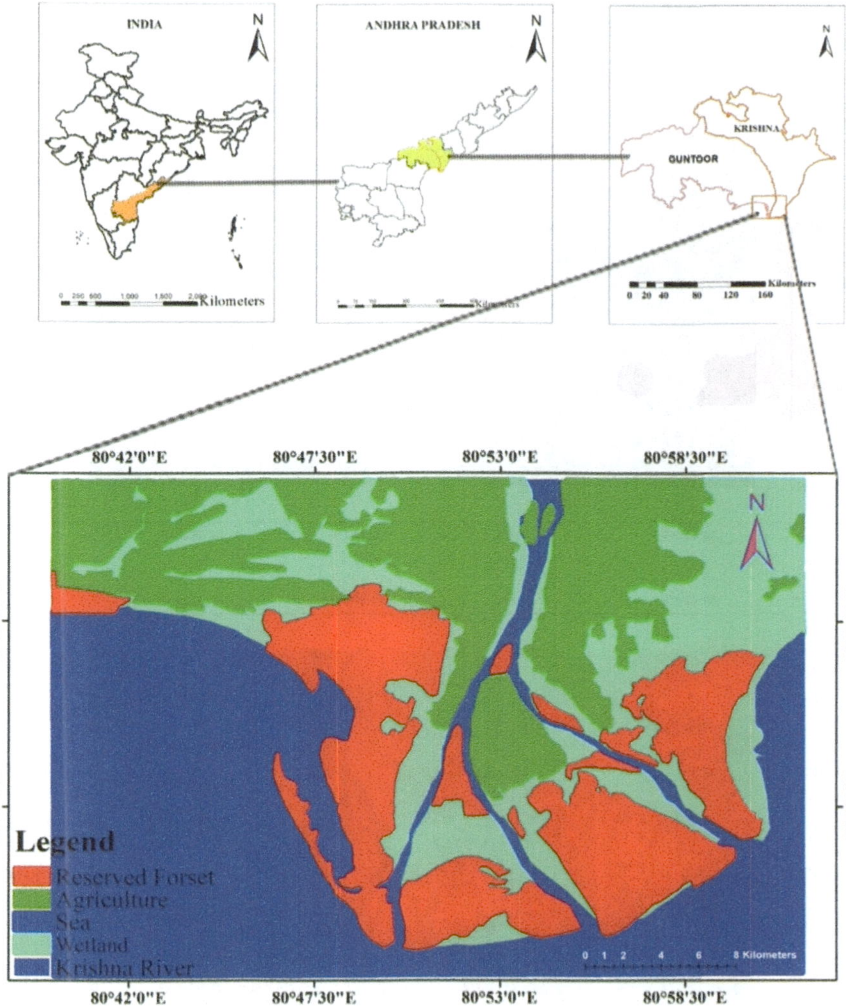

Fig. 28.1 Location map of the study area

Survey of India (SOI) toposheets at a scale of 1:50,000, IRS LISS-III at a scale of 1:50,000 imageries were used in this study. To begin, a baseline map was created using toposheets 66C/1. The datasets were obtained from the National Remote Sensing Centre (NRSC), Bhuvan and Earth Explorer websites from October 2009 to 2018. The satellite images were cropped to the Area of Interest (AOI) and geometric corrections were applied using ERDAS software (2014), (Mullick et al. 2019). The datasets thus processes were georeferenced by projecting them onto the referee projection, namely, UTM .WGS 1984 (Tay et al. 2013). The images then were visually and digitally interpreted using ArcGIS 10.3. The results thus obtained

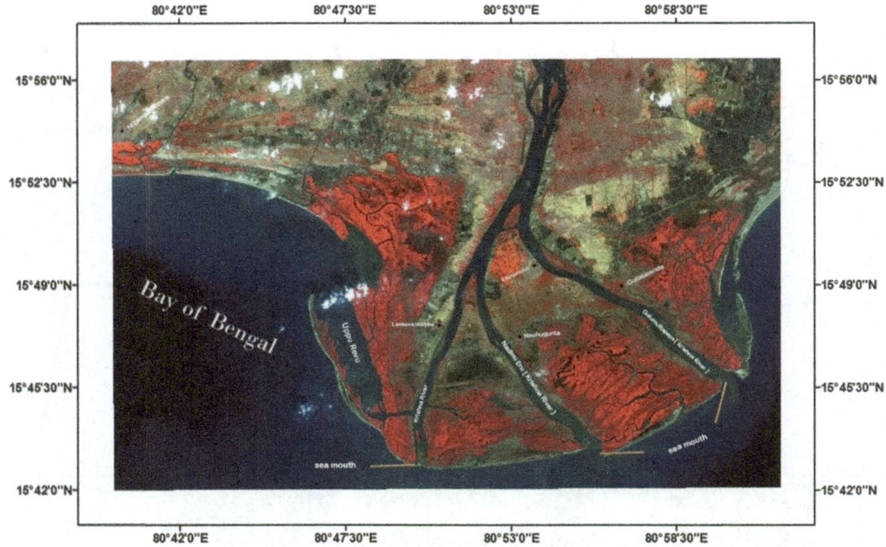

(a) 07.01.2018

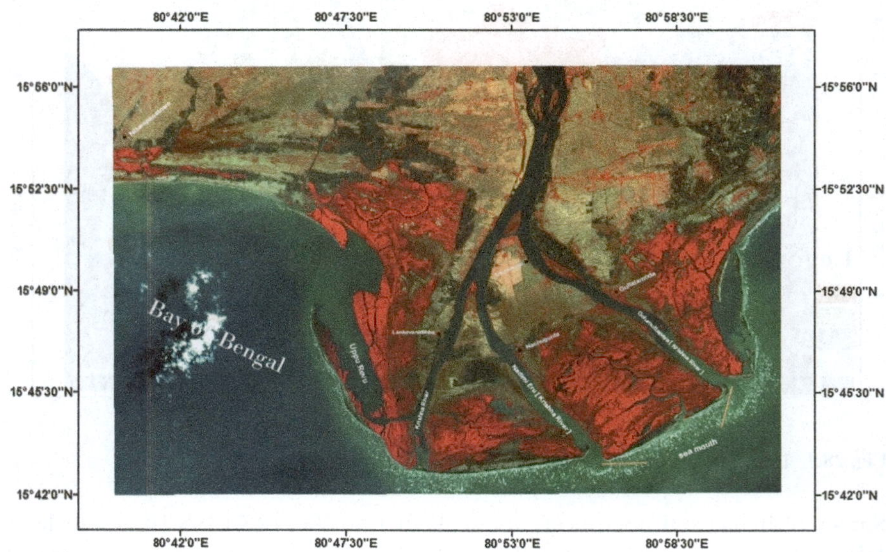

(b) 08.04.2017

Fig. 28.2 Satellite imageries showing the southern part of the mouth opening dynamics in different periods. (**a**) 1 January 2018 Landsat8, (**b**) 18 April 2017 Landsat8, (**c**) 19 February 2016 Landsat8, (**d**) 21 April 2015 Landsat8, (**e**) 14 April 2014 LISS-III, (**f**) LISS-III 3 March 2013, (**g**) 8 March 2012 LISS-III and (**h**) 9 October 2009 LISS-III

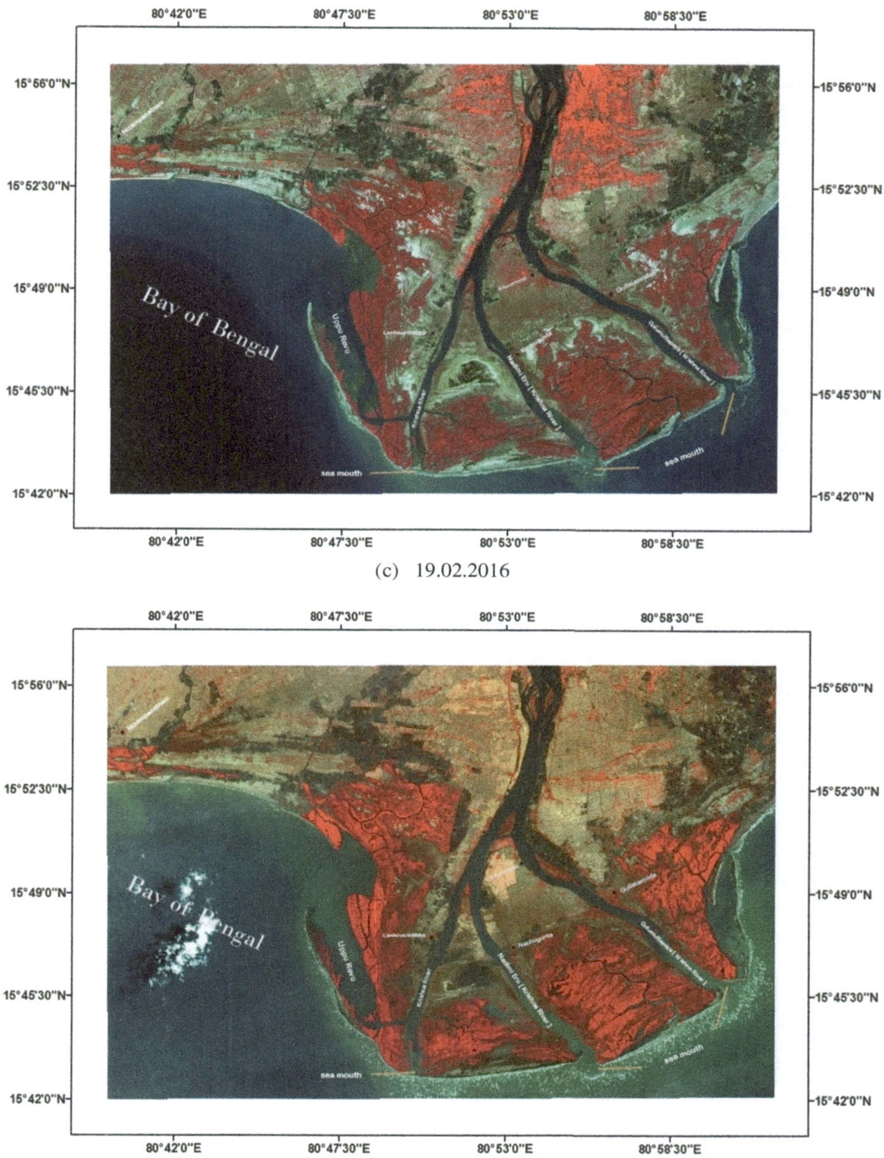

(c) 19.02.2016

(d) 23.05.2015

Fig. 28.2 (continued)

were cross-checked with ground reality, which included in-person field checks (Avinash et al. 2012; Kaliraj and Chandrasekar 2012; Kaliraj et al. 2013; Sreenivasulu et al. 2014; Nagalaksmi et al. 2017).

(e) 14.04.2014

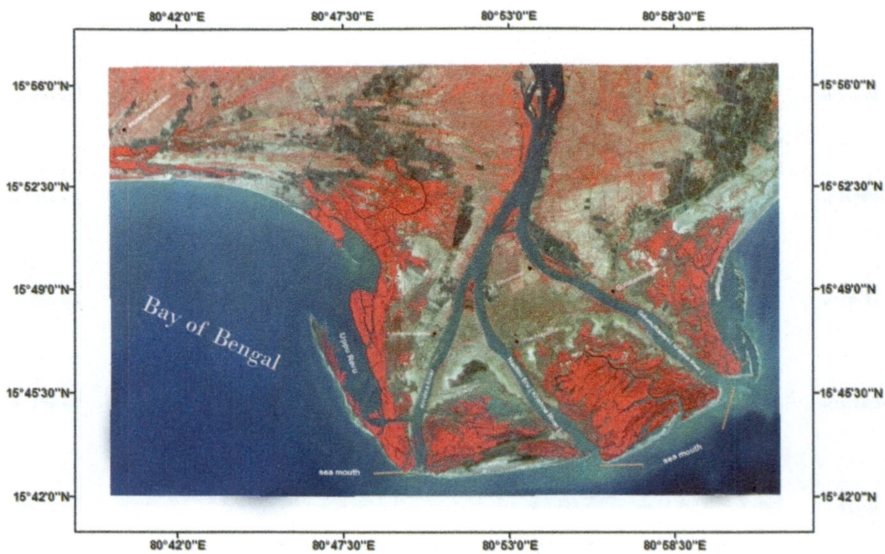

(f) 13.03.2013

Fig. 28.2 (continued)

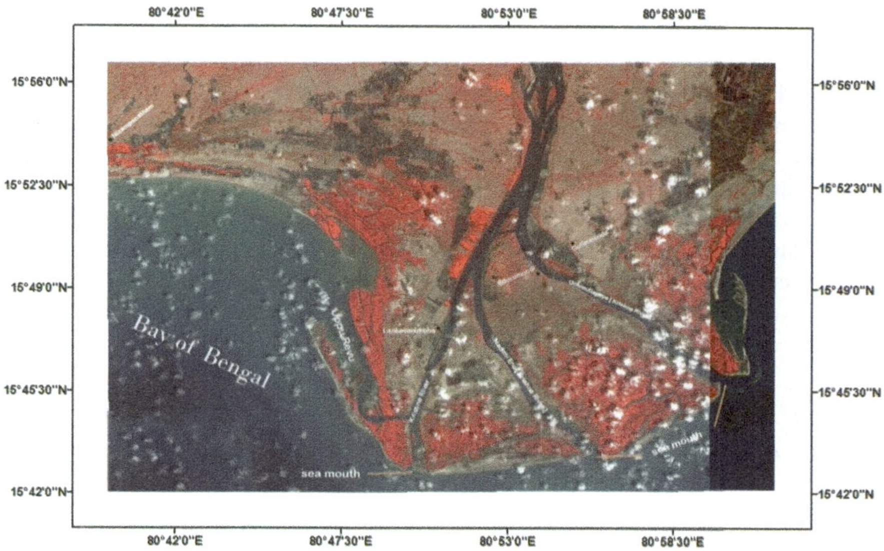

(g) 18.03.2012

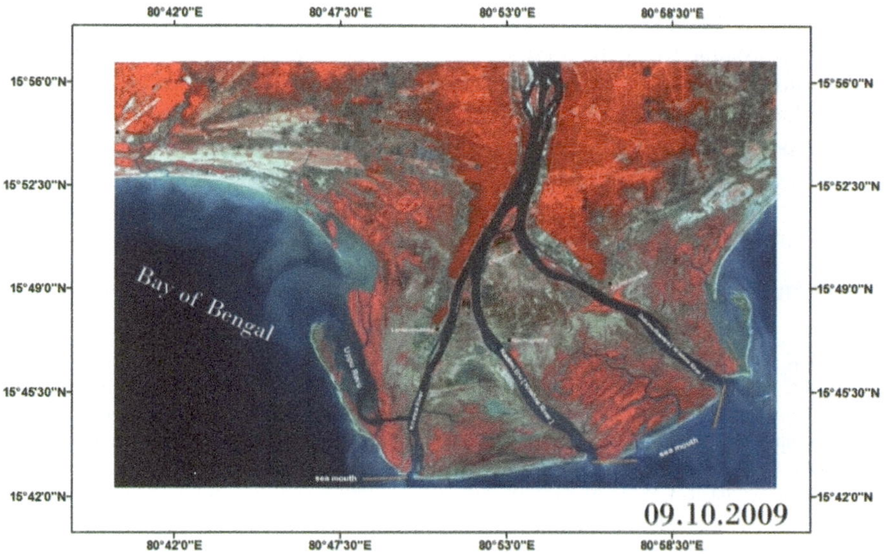

(h) 09.10.2009

Fig. 28.2 (continued)

28.9 Results and Discussion

Changes in the mouth of the Krishna River were observed by scaling the opening width using satellite imagery. The geospatial data collected from 2009 to 2018 show that the river mouth and the formation of sand bar-like structures are highly vibrant and dynamic (Nunes and Adams 2014). The river mouth opening width developed in 2009 was used as a reference point in this study (Mutti et al. 2000) (Fig. 28.2a–h). The width of the beach mouth was measured in relation to it in subsequent years (Bonilla et al. 2005) (Table 28.3). The data generated by employing satellite imagery were cross-checked by field checks. This was similar to coincided the previous values.

28.10 Conclusion

Fragile coastal environments are the result of the interaction of biological, marine, meteorological, and physical activities. Coastal dynamics are largely governed by the activities of currents, tides, and waves. Investigation of geospatial data and field studies inferred that the river mouth opening was enlarging and reducing periodically. It is assumed that the abrupt decrease in 2012 may be owing to prevailing conditions of low wave and current activities, and sediment accretion. It clearly shows the massive sediment deposition. Stilt roots, massive mangroves catch silt (mud deposits) at stream mouths, slowing down the Bay dynamic activities (waves, currents, etc.) allowing the sediments to deposit slowly but not short.

References

Avinash K, Jayappa KS, Vethamony P (2012) Evolution of Swarna Estuary and its impact on braided islands and estuarine banks, southeast coast of India. Environ Earth Sci 65:835–848

Bonilla C, Aubriot L, Perez MDC (2005) Influence of hydrology on phytoplankton species composition and life strategies in a subtropical coastal lagoon periodically connected with the Atlantic Ocean. Estuaries 28:884–895

Fai K, Lo A, Gunasiri CWD (2014) Impact of coastal land use change on shoreline dynamics in Yunlin County, Taiwan. Environments 1:124–136

Jayappa KS, Vijaya Kumar GT, Subrahmanya KR (2003) Influence of coastal structures on the beaches of southern Karnataka. India J Coast Res 19(2):389–408

Kaliraj S, Chandrasekar N (2012) Spectral recognition techniques and MLC of IRS P6 LISS III image for coastal landforms extraction along south west coast of Tamilnadu, India. Bonfring Int J Adv Image Proc 2:1–7

Kaliraj S, Chandrasekar N, Magesh NS (2013) Impacts of wave energy and littoral currents on shoreline erosion/accretion along the south-west coast of Kanyakumari, Tamil Nadu using DSAS and geospatial technology. Environ Earth Sci. https://doi.org/10.1007/s12665-013-2845-6

Michelles O, Crooks S, Williams PB (2003) Will restored tidal marshes be sustainable? San Franc Eastuary Watershed Sci 1:1–33

Mutti E, Tinterri R, di Biase D, Fava L, Mavilla N, Angella S (2000) Delta front facies associations of ancient flood-dominated fluvio-deltaic systems. Rev Soc Geol Espana 13(2):165–190

Nunes M, Adams JB (2014) Responses of primary producers to mouth closure in the temporarily open/closed Great Brak Estuary in the warm-temperate region of South Africa. Afr J Aquat Sci 39:387–394

Nagalaksmi K, Pramod Kumar M, Jayaraju N, Lakshmi Prasad T, Lakshmanna B, Sreenivasulu G (2017) Dynamics of Pulicat Lake mouth analysis using geospatial data, east coast of India: implications to socio-economic scenarios. Data Brief 15:142–147. https://doi.org/10.1016/j.dib.2017.09.016

Nageswara Rao K, Vaidyanadhan R (1978) Geomorphic features of the Krishna River delta and its evolution, Proc. Symp. on morphology and evolution of landforms. University of Delhi, pp 120–130

Nageswara Rao PV, Suryam RK, Ranga Rao V (2005) Depositional environment inferred from grain size parameters of the beach sediments between False Devi Point to Kottapatnam, Andhra Pradesh Coast. J Geol Soc India 65:317–324

Prabaharan S, Srinivasa Raju K, Lakshumanan C, Ramalingam M (2010) Remote sensing and GIS applications on change detection study in coastal zone using multi temporal satellite data. Int J Geomat Geosci 1:159–166

Mullick MRA, Ashraful Islam KM, Tanim AH (2019) Shoreline change assessment using geospatial tools: a study on the Ganges deltaic coast of Bangladesh. J Earth Sci Inform, Springer 13:299

Sreenivasulu G, Jayaraju N, Sundara Raja Reddy BC, Lakshmi Prasad T, Lakshmanna B, Nagalakshmi K, Prashanth M (2016) River mouth dynamics of Swarnamukhi estuary, Nellore coast, southeast coast of India. Geodesy Geodynam 7(6):387–395

Short DA, Nakamura K (2000) TRMM radar observations of shallow precipitation over the tropical oceans. J Clim. https://doi.org/10.1175/1520-0442(2000)013<4107:TROOSP>2.0.CO

Sreenivasulu G, Jayaraju N, Lakshmi Prasad T (2014) Land use and land cover change detection study at Pennar River Estuary, Nellore District, Andhra Pradesh, southeast coast of India J. Geotech Eng 1:1–9

Tay HW, Bryana KR, de Langea WP, Pilditchb CA (2013) The hydrodynamics of the southern basin of Tauranga Harbour. N Z J Mar Fresh Res. https://doi.org/10.1080/00288330.2013.778300

Chapter 29
Non-monsoonal Coastal Erosion Due to the Tropical Cyclone (OCKHI) and Its Impacts Along Thiruvananthapuram Coast, Southwest Coast of India – A Geospatial Approach

J. R. Princy, M. Ramesh, J. Jyothi, P. S. Swathy Krishna, and L. Sheela Nair

Abstract The tropical cyclone OCKHI of 2017 adversely affected the SW coast of India, even though it did not have a direct hit on the coast. The present study mainly focuses on the shoreline changes witnessed along the Thiruvananthapuram coast under the influence of the OCKHI cyclone and assesses the negative impact by comparing it with historical shoreline data. Landsat imagery for the past 43 years and Sentinel-2 imagery, depicting the pre and post-cyclonic events have been used to analyse the short-term impacts of cyclone OCKHI along the study area. Extreme events like cyclones/storm events pose major challenges to the fishing community as their socioeconomic (lively hood) and cultural lives are badly affected. As part of this work, the extent of the damage caused by the cyclone on the livelihoods of the coastal community along the Thiruvananthapuram coast is studied through local field data collection which involved observations of field signatures, physical survey and interviews, conducted in the highly affected coastal villages. Through this study, it is found that the long-term erosion rate which was about 45 per cent increased to 91 per cent along the Thiruvananthapuram coast subsequent to the passage of the cyclone OCKHI even though it did not have a direct hit on the coastal area. Some of the coastal sectors which used to be dynamically stable like the Puthenthoppu and Adimalathura beaches started showing eroding tendency. Cyclone OCKHI also caused significant changes in the coastal morphology and high-resolution satellite observations are very much useful to study the impacts of tropical cyclones on short-term coastal variation.

J. R. Princy · M. Ramesh (✉) · P. S. Swathy Krishna
National Centre for Earth Science Studies, Ministry of Earth Science Studies, Thiruvananthapuram, Kerala, India

Cochin University of Science and Technology, Ernakulam, Kerala, India
e-mail: ramesh.madipally@ncess.gov.in

J. Jyothi · L. Sheela Nair
National Centre for Earth Science Studies, Ministry of Earth Science Studies, Thiruvananthapuram, Kerala, India

N. Jayaraju et al. (eds.), *Coasts, Estuaries and Lakes*,
https://doi.org/10.1007/978-3-031-21644-2_29

Keywords OCKHI · Remote sensing · Thiruvananthapuram coast · Coastal erosion and societal impacts

29.1 Introduction

Coastal geomorphology is one of the most sensitive systems in nature, which responds spontaneously to tides, waves, currents, cyclonic storms and river/estuarine inputs (Nayak 2005; Leah and Veronica 2010). In addition to the natural factors, human interventions such as the construction of coastal structures that reduce/alter the sediment supply to the coastal zone, beach sand mining etc. can also affect the coastal morphology (Shamji 2011). Shoreline analyses conducted to study the shift/changes in shoreline positions are primary measurements for understanding the temporal and spatial changes in coastal geomorphology and images captured by remote sensing satellites facilitate frequent monitoring of coast with varying geomorphology (Woodroffe 2002; Davis 1972; Nayak 2002). Remote sensing techniques can be used for mapping of coastal landforms for providing detailed and robust information on various features such as shape, distribution and morphological status during the past and present (Smith et al. 2006; Bocco et al. 2001; Rao 2002; Sallema and Mahika 2004; Greeshma and Jayaraj 2014; Jayappa et al. 2006; Elkoushy and Tolba 2004; Abdulrahman 2010). In the case of the Indian coast, particularly the southwest coast of India, rapid changes in shoreline characteristics mostly in the coastal erosion/accretion patterns are reported (Mujabar and Chandrasekar 2011; Neelima et al. 2018; Noujas et al. 2017; Sreekala et al. 1997; L&T and INCOIS 2012). Previous studies reveal that the SW coast of India is under pressure and prone to erosion (Noujas and Thomas 2015; Chandrasekar et al. 2013). Natural factors along with artificial structures such as seawalls, groins, harbour breakwaters have significant influence on the shoreline dynamics (Noujas et al. 2019).

Coastal hazards like tropical cyclones also can have a profound influence on coastal morphology. In recent years there has been an increase in the number of extreme events reported and this can invariably be attributed to climate change and related events and the OCKHI cyclone of November 2017 can be cited as a typical example. The OCKHI cyclone which had its origin in the southwest BoB as a depression on 28 November moved westward and later emerged as a deep depression over the Comerin on 30 November further moving westward towards the Lakshadweep transforming itself into a very severe cyclonic storm (Table 29.1). Even though the OCKHI did not hit the SW coast directly its movement in the Arabian Sea off the SW coast of India had a significant influence on the SW coast of India. It had resulted in strong winds, widespread rainfall and high wave activity all along the coast. The present study mainly focuses on the morphological changes brought out by OCKHI along some of the badly affected coastal areas of the Thiruvananthapuram coastal sector on the SW coast of India. Since OCKHI struck the coast during the post-monsoon season which is considered as the beach-building

Table 29.1 Genesis and path of cyclone OCKHI

Date	Time (IST)	Stage of cyclone	Location
28.11.2017	08:30	Low pressure area	Southwest Bay off Sri Lanka
29.11.2017	08:30	Depression	Southwest Bay off Sri Lanka
30.11.2017	02:30	Deep depression	Comorin Area
	08:30	Cyclonic storm	Comorin Area
01.12.2017	05:30	Severe cyclonic storm	Lakshadweep
	14:30	Very severe cyclonic storm	Lakshadweep
04.12.2017	17:30	Severe cyclonic storm	East-central Arabian Sea
05.12.2017	08:30	Cyclonic storm	East-central Arabian Sea
	17:30	Deep depression	East-central Arabian Sea
	20:30	Depression	East-central Arabian Sea
06.12.2017	02:30	Well-marked low	North-east Arabian Sea
	05:30	Low pressure area	South Coastal Gujarat

period, it has affected the normal beach development process which was in progress and the present study aims to investigate the extent of the negative impact.

29.1.1 The OCKHI Cyclone – Origin, Growth and Dissipation

Cyclone OCKHI was first formed as a low-pressure area in the adjoining area of south Sri Lanka over the southwest Bay of Bengal and the equatorial Indian Ocean in the forenoon (08:30 IST) of 28 November 2017. A well-marked low-pressure area in the early morning (05:30 IST) of 29 November over the same region was formed and under favourable environmental conditions over the southwest Bay of Bengal off the coast of southeast Sri Lanka, it got transformed into a depression by the forenoon (08:30 IST) of 29th November. The depression moved across the Sri Lankan coast and appeared in the Comorin Sea on the evening (17:30 IST) of 30 November. It further moved towards the northwest direction and intensified into a 'Cyclonic Storm' in the forenoon (0830 IST) of 30th November over the Comorin Sea. The cyclone OCKHI rapidly intensified as a depression from its genesis and became a Cyclonic Strom within a period of 24 hours. Thereafter, OCKHI moved towards the west-northwest direction, further intensifying into a 'Severe Cyclonic Storm' over the Lakshadweep area in the early morning (05:30 IST) of 1 December 2017 and into a 'Very Severe Cyclonic Storm' over the southeast Arabian Sea, to the west of Lakshadweep in the afternoon (14:30 IST) of 1 December. The Cyclone then travelled towards the northwest direction and reached its maximum intensity of 150–160 km/h rising to 180 km/h on 2 December with a high central pressure of 976 hectopascal (hPa). The cyclone then moved towards the north-northwest direction for a short time and further in the north-northeast direction with a stable intensity until the early morning of 3 December and later slowly got dissipated while moving in the same direction. It eventually crossed the southern coast of Gujarat, between Surat and Dahanu, as a well-marked low-pressure trough around early morning (05:30 IST) of 6 December 2017 (Parliament of India 2018).

29.1.2 Societal Impacts of Cyclone OCKHI Along Southwest Coast of India

The impact of cyclone OCKHI was huge as it greatly affected the socioeconomic conditions and livelihood of the coastal community in Lakshadweep, the southern part of Tamil Nadu, Kerala and Gujarat. During the cyclone, isolated heavy rainfalls were reported over the states of Tamil Nadu, Kerala, Lakshadweep and Gujarat. A trail of destruction was observed in the areas close to the Western Ghats. Several houses and roads were damaged, power communication lines were mutilated, crops were uprooted and road and rail traffic was also hindered. The arrival of the OCKHI tropical cyclone badly affected the nearshore areas of the seabed. The strong waves triggered by the cyclone also damaged vast areas of the underwater ecosystem which had a negative influence on the post-monsoon fish harvest. The cyclone during its passage off the SW coast of India and the Lakshadweep carried huge quantities of plastic waste which eventually reached the nearshore region and this later became a threat near shore ecosystem and also the coastal community inhabiting the coastal areas. Table 29.2 provides the societal impacts of OCKHI along the Kerala, Tamilnadu and Lakshadweep coastal regions.

29.2 Study Area

The study area is the Thiruvananthapuram coast in Kerala state, India. It is geographically situated between Puthenthoppu (76°49′01.95″E 8°35′38.13″N) to pozhiyur (77°05′37.53″E 8°17′41.58″N) (Fig. 29.1). The study area has different geology, vegetation, geomorphology and other physical features and has a variant topography with different features in elevation, rock types, stability and resource potential. Thiruvananthapuram coastal stretch comprises distinct geomorphological features such as coastal plains, rocky coast to undulated lateritic terrain with occasional structural or residual hills. There are two promontories at Vizhinjam and Kovalam along the selected study area. The climate of the coast is monsoon controlled and has a steep bathymetry (Shaji 2014). The coast is highly dynamic in nature and subjected to erosion both by the influence of natural and anthropogenic factors. Several hard structures as part of coastal protection measures and harbour and port development are present along this stretch. There are four seasonal river inlets within this area, namely, Veil, Adimalathura, Poonthura and Poovar. The study area includes a total number of 27 coastal villages among which 16 villages are identified as highly eroding coasts and are protected with hard structures such as groynes (mostly short) and seawall. The total length of coastal stretch considered for the study is 42 km, which also includes a 6 km stretch comprising two rocky cliffs (premonitory) at Kovalam and an area which is being reclaimed as part of the Vizhinjam International Sea Port construction which is underway. Short-term shoreline change analysis along this particular 6 km stretch is not possible during the study period mainly because of the ongoing construction works. Hence this region

Table 29.2 Societal impacts of cyclone OCKHI

Impact	Tamilnadu	Kerala	Lakshadweep
No: of deaths	30	75	Nil
Livestock	7654	Nil	1691
No. of missing fishermen	203	141	Nil
Houses damaged	Huts damaged – 6262 Houses damaged – 101	Fully – 221 Partially – 3253	Fully – 87 Partially – 935
Infrastructure damaged	(i) Mechanised boats partially – 640, fully – 60 (ii) Fibre Reinforced Plastic (FRP) boats partially – 3407, fully – 3407 (iii) Electricity Board Poles – 15,858 Transformers – 95 (iv) Fallen Trees – 25,526 (v) 38 Breaches in tanks and 31 Breaches in channels/canals (vi) 103 Government buildings damaged (vii) Damage to 75.046 km State Highways, 98.93 km National Highways, 417.18 km Rural/ Urban Roads	(i) Boats fully damaged/ lost – 384 (ii) Loss of road – 41 km (iii) Damage to Pumps – 180 (iv) Damage to Supply Tanks 430	(i) Boats fully damaged/ lost – 12 Boats partially damaged – 25 (ii) Houses – fully damaged – 87 Houses partially damaged – 935 (iii) Government building – 340 (iv) Coconut trees – 32,747 (v) Other trees – 5514 NIOT drinking water plant was also damaged to some extent
Crop area affected	6625 hectares	7817.43 hectares	–

Source: PARLIAMENT OF INDIA, RAJYA SABHA, Department-Related Parliament Standing Committee on Home Affairs, report no: 211 (2018)

is excluded from the present analysis and it has been considered as the reference area for describing the other locations. In the study area, 11 km stretch extending from Chowara to Pozhiyur lies on the southern side of the excluded area and the remaining 25 km extends from Panathura to Puthenthope on the northern side. Figure 29.2 represents the important coastal villages which are considered as reference coastal regions for the detailed discussion of the results and analysis and an enlarged view of the excluded area is shown in the inset.

29.3 Data and Methodology

29.3.1 Satellite Data

Two sets of satellite imagery are used in the present study for the assessment of the short-term shoreline changes due to the effect of the OCKHI cyclone along the Thiruvananthapuram coast. Landsat images available for the period 1973–2015 were used to study the long-term shoreline changes and the short-term changes due to OCKHI were analysed using the high-resolution Sentinel 2 images for the period

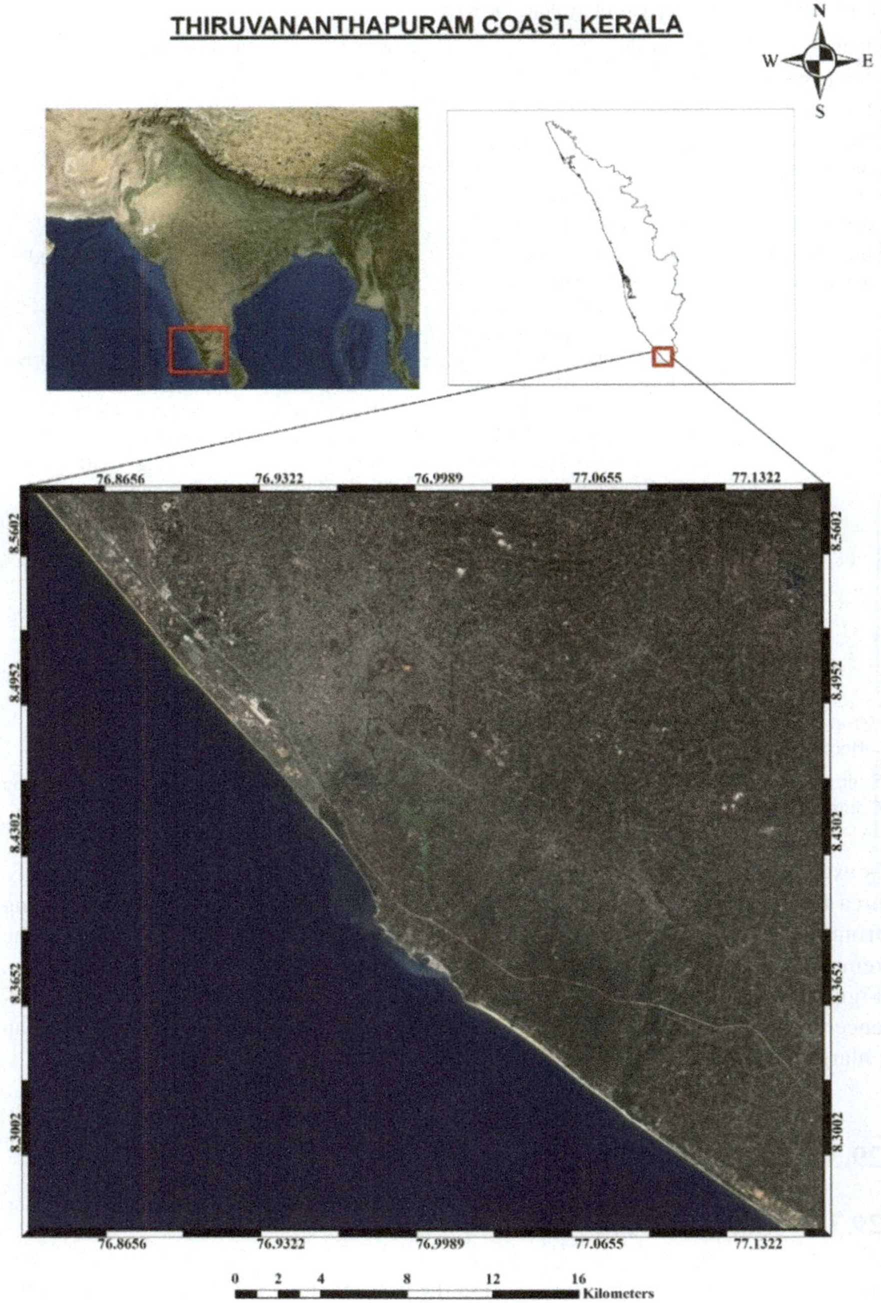

Fig. 29.1 Study area – Thiruvananthapuram coast, Kerala, India

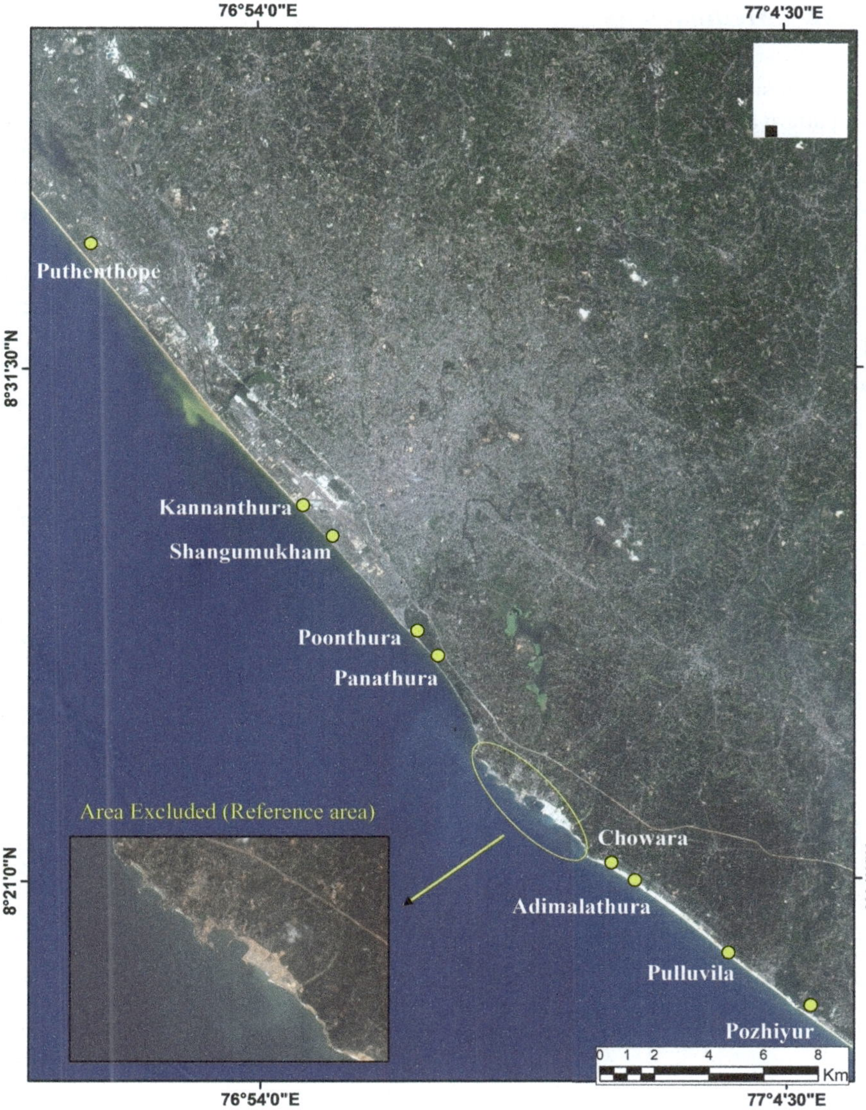

Fig. 29.2 Important coastal villages along the study area and excluded (reference) area in the inset

September 2017 (pre-cyclone) to March 2018 (post-cyclone period). Details of the Landsat and Sentinel data products used for the study are given in Table 29.3.

29.3.2 Methodology

Historical shoreline variations are studied to understand the coastal erosion/accretion pattern along the study region during the past 43 years and the data used is a set of sixteen Landsat images from 1973 to 2015. The Landsat images are processed for geometric and radiometric correction and classified using isodata clustering algorithm. The shorelines are then extracted after raster to vector conversion followed by digitisation. Survey of India toposheets of Thiruvananthapuram available at 1:50000 scale have been used to create baseline and shoreline analysis carried out using the ArcGIS DSAS (Himmelstoss 2009) tools. Transects are generated at 50

Table 29.3 Details of Landsat and Sentinel−2 imagery used for the study

Satellite		Sensor	Date of pass	Resolution	Row/path
Landsat-1		MSS	10-01-1973	80 m	156/052, 155/052
			23-01-1973	80 m	155/053
Landsat-3		MSS	11-01-1979	80 m	156/052, 155/052
			10-01-1979	80 m	155/053
Landsat-5		MSS	02-04-1989	30 m	145/052
			22-02-1989	30 m	144/053
Landsat-5		TM	13-01-1995	30 m	145/052
			04-02-1995	30 m	144/053
Landsat-7		ETM	03-03-2001	30 m	145/052
			28-01-2001	30 m	144/053
Landsat-7		ETM	05-03-2005	30 m	145/052
			10-02-2005	30 m	144/053
Landsat-7		ETM	03-03-2010	30 m	145/052
			08-02-2010	30 m	144/053
Landsat-8		OLI	27-03-2015	30 m	145/052
			16-01-2015	30 m	144/053
Sentinel 2A		MSI	06-09-2017	10 m	
			11-09-2017	10 m	
			01-10-2017	10 m	
			16-10-2017	10 m	
			05-12-2017	10 m	
			25-12-2017	10 m	
			04-01-2018	10 m	
			14-01-2018	10 m	
			29-01-2018	10 m	
			03-02-2018	10 m	
			13-02-2018	10 m	
			28-02-2018	10 m	
			05-03-2018	10 m	
			13-03-2018	10 m	
			30-03-2018	10 m	

meters spacing with 500-meter length from the baseline and linear regression rate (LRR) statistics are calculated. Based on the LRR criteria (Table 29.3) coastal regions are classified into five classes for demarcating the eroding and accreting areas. Similarly, high-resolution (spatial 10 m, temporal ~15 days) sentinel 2A satellite images have been collected for a period of 6 months from September 2017 to March 2018 that is three months prior to and after the cyclone to assess the impacts of the cyclone on shoreline variations along the study area. A total of 15 images have been collected for the study period, but it is observed some of the images, particularly during the 2-month period of October to November 2017 are fully covered with clouds due to cyclonic impact making it not suitable for processing. Similarly, there was a well-marked low-pressure region formed over the Indian Ocean on 12 March 2018, as a backdrop of the OCKHI cyclone and hence, the satellite image of 13 March 2018, also has huge cloud coverage which makes it unsuitable f it. Therefore, out of fifteen, only eleven images are fit enough to be used for the analysis. The selected eleven images are processed and shorelines are extracted by ArcGIS digitisation methods. Transects are generated at 50 meters spacing with a 500-meter length from the baseline for shoreline change analysis using DSAS tools. LRR statistics are calculated and based on the pre-defined LRR criteria (Table 29.4), coastal regions are classified into erosion and accretion classes. In addition, the data collected as part of several field surveys and local interviews have been used to study the impacts of cyclone OCKHI on the socioeconomic conditions of the fishing villages. In addition, physical measurements and photographs taken during the field visits to the affected fishing villages are used as signatures to provide vital information on the changes in coastal morphology as well as the beach-building processes, subsequent to the passage of the OCKHI cyclone.

29.4 Results and Discussion

29.4.1 Short-Term Shoreline Change Analysis Using Satellite Imagery

The study was carried out to understand the short-term shoreline variations subsequent to the passage of the cyclone OCKHI off the Thiruvananthapuram coast, using remote sensing, GIS and linear regression analysis methods. The result shows

Table 29.4 LRR criteria used for classification

Class	LRR criteria (Sentinel)	LRR criteria (Landsat)
High erosion	<−5	<−2
Erosion	−5 to 0.5	−2 to −1
Stable beach	−0.5 to 0.5	−1 to 1
Accretion	0.5 to −5	1 to 2
High accretion	>5	>2

that almost all the beaches along the Thiruvananthapuram district experienced severe erosion except a few. The increase in the rate of erosion due to the cyclone OCKHI along coastal villages is assessed by the comparison of erosion/accretion trends between the historical shoreline data and the shoreline data after the cyclone. Rapid erosion with an increasing trend was observed on the northern side of the reference excluded area whereas considerably low rates of erosion prevailed to the south of the reference excluded area which was showing accreting trend earlier, that is, before the cyclone. The northern side coastal stretch that is from Panathura to Shangumukham has been identified as a perennially eroding stretch from historical shoreline analysis and the coastal stretch from Kannanthura to Puthenthope seems to be a stable beach during the last 43 years. The stretch to the south of the excluded area comprises Adimalathura beach, which is an accreting beach and the coastal stretch up to Pozhiyoor seems to be stable for the last 43 years. The erosion/accretion patterns along the Thiruvananthapuram coast from 1973 to 2015 are shown in Fig. 29.3a. The erosional/accretional trends after the cyclone OCKHI (Fig. 29.3b), indicate that almost all the open beaches along the Thiruvananthapuram coast have undergone erosion and this includes the stable beach stretches in the northern and southern region of the study area which extends from Puthenthope to Shankhumuhgam and from Pulluvila to Poovar respectively, the Adimalathura beach being an exception.

Further, the variations in the shoreline inflicted by OCKHI are studied by comparing and analysing the LRR rates of two cases, that is LRR statistics of past shoreline positions (1973–2015) with that of the post-OCKHI shoreline positions (2017–2018). Figure 29.4 represents the trends of historical LRR statistics (LRR1) and the post-OCKHI LRR statistics (LRR2) along the study area. From Fig. 29.4, the LRR1 trend denotes the accretion pattern at most of the beaches along the study area, which are stable and accreting in the past. That is, beaches from Puthenthope to Shankhumugham represented by transect numbers 1–214, are stable with very little erosion whereas beaches along the Shankhukmugham – Poonthura stretch, represented by transect numbers 214–276 are mostly eroding. On the southern side, Adimalathura beach represented by transect numbers 277–348 shows an accreting trend whereas further south, the Chowara to Pozhiyur sector, represented by transect numbers 348–392 appears to be stable. The LRR2 statistics, on the other hand, indicate an erosive trend all along the study area. High erosion is reported on the northern side of the study area that is the area between Puthenthope to Panathura represented by transect numbers 1–294 and the Chowara-Pozhiyur sector towards the south represented by transect numbers 344–394 also indicates erosion. The Adimalathura beach represented by transects 295–343 is the only sector where accretion is observed. However, the accretion rates have not reached the previous level.

The present study clearly shows that the majority of the erosional sites lie on the northern side of the excluded area (Vizhinjam seaport under construction and Kovalam cliff), whereas the southern part of the study area (i.e. from Pozhiyoor to Adimalathura with a coastline of the 11 km length) shows a mixed trend with both erosion and accretion. The analysis results point out that the rate of erosion in the earlier erosion-prone areas has increased considerably after the OCKHI cyclone.

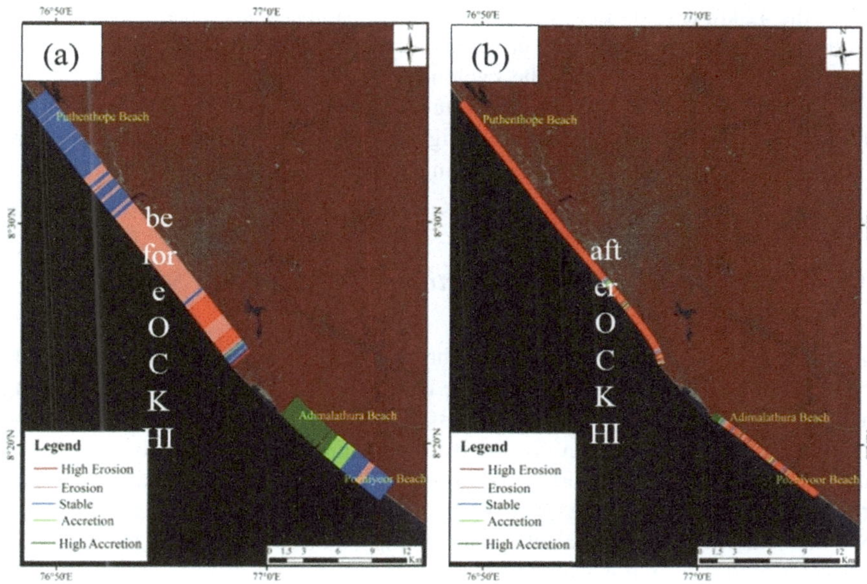

Fig. 29.3 (**a**) Landsat classified image and (**b**) Sentinel classified image for the analysis of coastal erosion before and after OCKHI

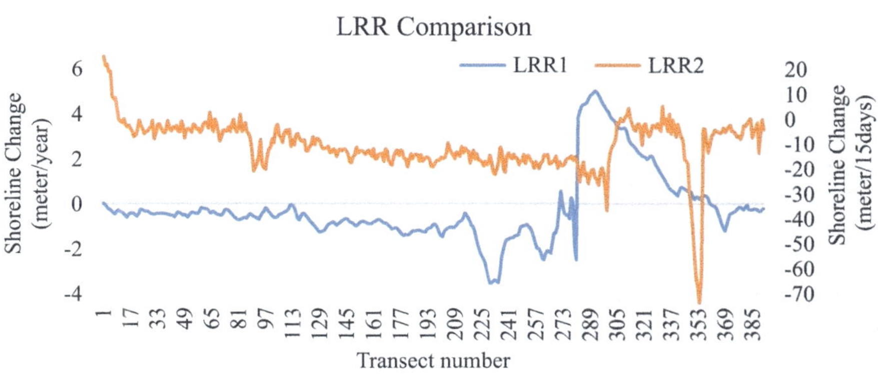

Fig. 29.4 Comparison of the rate of shoreline changes – historical versus OCKHI effected: LRR1 is Landsat-based classification for the past 43 years and LRR2 is Sentinel-based classification after OCKHI

Table 29.5 Change statistics of coastal erosion along Thiruvananthapuram due to OCKHI

Class	Before OCKHI (%)	After OCKHI (%)	Change (%)
High erosion	8.65	66.87	+58.22
Erosion	36.90	24.91	−11.99
Stable	37.40	3.54	−33.86
Accretion	4.58	2.53	−2.05
High accretion	12.47	2.15	−10.32

Out of the 42 km length of the study area, high erosion is observed along 66.87% of the stretch followed by moderate erosion which is 24.91%. About 91% of the study area experienced erosion after the cyclone and out of the remaining 9%, 3.54% remains stable, 2.53% shows moderate accretion and high accretion 2.15% (Table 29.5). In effect, the percentage of highly eroding stretch along the Trivandrum coast has gone up by 58% and the accreting area has reduced by 12%.

29.4.2 Site-Specific Data Collection Through Field Survey

Interviews and surveys were conducted throughout the fishing villages in the study area to obtain primary data regarding the experience of the local fishermen during the cyclone period. It is noted that almost 90% of the men are engaged in fishing activities and 70% of women are fish vendors, and about 30% of the people are engaged in fishing-related works. The experience of the indigenous traditional fishermen (native fishermen) is expected to provide a profound knowledge of the history and stability of the coastal stretch being studied. According to the native fishermen, under normal seasonal conditions, the sand removed by the high-intensity monsoonal waves will be gradually brought back to the shore by the end of September and this continues throughout the fair season (i.e. till April to May). If normal conditions prevail during the post-monsoon season which is considered as a conducive environment for beach building, about 80% of the eroded material gets re-deposited on the shore by the end of December. If it is a fairly stable beach, the stability will be restored by the end of January and it continues to be in a dynamically stable state from February to May, that is, till the onset of the monsoon. However, it is observed that even along some of the stable stretches of the coast, the beaches are no longer able to sustain their stability mainly because of the negative influence imparted due to episodic events like cyclones, other storm events and anthropogenic interventions which interfere with the natural coastal processes at work.

In the present study, we are giving the details of the survey conducted at Poonthura, one of the coastal villages identified as an erosion hotspot. The Poonthura coastal stretch has an open beach of 1.6 km in length, with 8 groynes in the east-west direction and also a seawall which was built in 2008. The open beach is being used by the local fisherman for launching their boats to venture into the sea for fishing (except during monsoon season). Since this stretch of the coast was exhibiting an overall erosional tendency which was deteriorating year after year, measures were taken to protect the Panathura and Poonthura locations (identified as perennial erosion hotspots) by constructing groynes and seawalls on these sites. But as per the field observations, these structures after being constructed extended the problem of erosion to the adjacent stretches which gradually got extended to the Shangumukham beach. The survey reports also confirmed that the normal beach building process which used to take place till April–May was disrupted soon after the OCKHI and the presence of hard structures like groins and seawall also contributed to the deterioration of the coast as sand deposition could take place only on the upstream side of the groins (southern side due to the predominant northerly transport during the post-monsoon season). But in December 2017 soon after the OCKHI, there was hardly any beach as erosion was severe due to the impact of the cyclone. It took several days for the beach building process to get initiated and it was not

Fig. 29.5 Photographs collected from Poonthura from January 2018 to March 2018

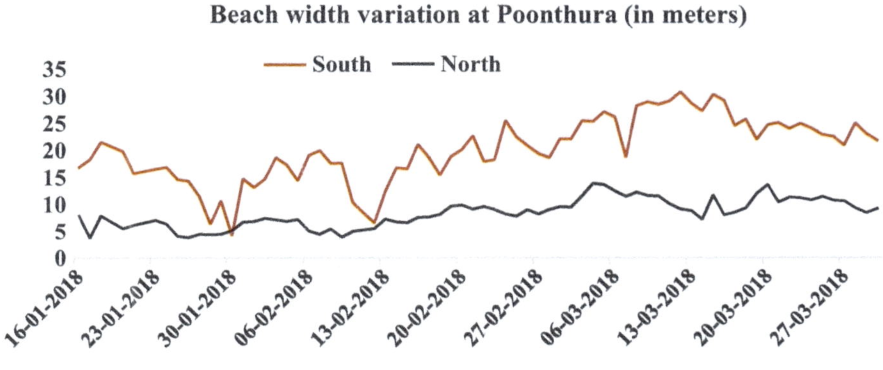

Fig. 29.6 Beachwidth variations at centre groyne from field survey

encouraging as the original beach width as in the previous year could not be regained before the onset of the next, that is, the 2018 monsoon in June. The downdrift side of the groin field was badly affected and also many of the open beaches experienced a drastic reduction in the beach width during the fair season. The photographs taken from the Poonthura village soon after the passage of the OCKHI cyclone are presented in Fig. 29.5. Post-cyclone beach width variations at both sides of the centre groyne field in Poonthura are represented in Fig. 29.6. From Fig. 29.6, it is evident that both protected areas (by groin field and sea wall) and the open beaches were subjected to erosion during the OCKHI cyclone.

29.5 Discussion and Conclusions

The present study has analysed the varying trends of coastal erosion along Thiruvananthapuram by using historical shoreline position data from LANDSAT imagery and post-cyclonic shoreline position data from Sentinel-2 imagery. The study shows that 45% of the study area was eroding (8.65% high erosion and 36.90 erosion), 38% of the study area is stable and the remaining 17% of the study area is accreting for the past 43 years, that is, 1973 to 2015. From the long-term shoreline change analysis using historical data, it can be inferred that the overall stability of the coast was showing more or less the same trend till 2015, with more than 50% of the total stretch falling under the stable and accreting beaches category. However, a drastic change in the trend pattern is observed during the year 2018, that is, immediately after the OCKHI cyclonic event and this certainly can be attributed to OCKHI. A close examination of the shoreline changes analysis results reveals that most of the stable and accreting beaches showed eroding tendency by the end of the post-monsoon season of 2017–2018, which is a clear indication of the negative influence of OCKHI cyclone during the beach building stage. From the study, it is seen that more than 90% of the study area had undergone drastic erosion, which is quite evident from the increase in erosion rate from 45% to 91% with the maximum erosion rates reported at the monsoon erosion hotspots identified along the Thiruvananthapuram coast. The accreting stretch reduced from 17% to 5%, which indicates that 70% of the earlier accreting beaches became erosive beaches which is a clear indication of the impact of OCKHI. The impact of OCKHI had caused distinct changes in the coastal morphology which certainly was quite different from the season changes witnessed earlier.

It is observed that the high waves during the OCKHI, the event had even affected the beaches which were considered as stable. Another important point worth noting is that the accretion/beach-building process after the event was very slow the stable areas of the coast did not reach their normal state before the onset of the monsoon in 2018 and this has aggravated the erosion all along the study area during the successive monsoon seasons and also during extreme events /stormy events, the frequency of which has increased considerably in recent years. From the study, it is clear that the high-energy Thiruvananthapuram coastal stretch was seriously affected by the impact of the OCKHI cyclone and it is evident that even after a couple of years after the incident, the coast is yet to regain its normalcy. The presence of hard structures and other anthropogenic activities like river sand mining, construction of dams along with natural storm events have drastically changed the coastal processes at work and made them even more complex.

Based on the study carried out it can also be concluded that even though the OCKHI cyclone did not have a direct hit on the SW coast of India, its passage off the SW coast was powerful enough to cause widespread coastal erosion and damage to fishing boats, nets, tools and other accessories thereby affecting the lively hood of the coastal community. It also had an adverse impact on the socioeconomic conditions of the coastal community which comprises mostly fishermen and people who are involved in fishing-related activities. As per the reports more than 216 fishermen were missing. The post-cyclone field visits and interviews carried out in the cyclone-affected areas have made a clear picture of the need for scientific analysis and understanding of the cyclonic impact on coastal morphology. In this context, the need for conducting a comprehensive scientific study aimed at a better understanding of the complex coastal processes considering both anthropogenic and natural factors along with the spatial and temporal changes in coastal morphology as well as the influence of climate change such as sea level rise is essential.

Through this study, the use of high-resolution (both spatial and temporal) Sentinel-2 satellite imagery to observe the short-term shoreline variations due to episodic events like a cyclone is illustrated. Application of advanced remote sensing methods and tools for understanding the changes in shoreline as well as coastal morphology will be extremely useful for scientific studies related to coastal erosion which forms part of disaster management and mitigation as they can be used along with the field data collected/observations to understand the processes and also for validation of the numerical modelling results.

Acknowledgements The authors would like to thank Mr. Akhildev and Mr. Sarankumar S.G. for their support during field data collection. Also, the authors would like to thank the Director, NCESS, and the Ministry of Earth Sciences, India, for providing financial support to the research.

References

Abdulrahman KA (2010) Remote sensing, ministry of higher education and scientific research. Republic of Iraq 3:3–4

Bocco G, Mendoza M, Velázquez A (2001) Remote sensing and GIS-based regional geomorphological mapping – a tool for land use planning in developing countries. Geomorphology 39(3–4):211–219. https://doi.org/10.1016/S0169-555X(01)00027-7

Chandrasekar N, Joe Viviek V, Saravanan S (2013) Coastal vulnerability and shoreline changes for southern tip of India- remote sensing and GIS approach. J Earth Sci Climate Change 4(4). https://doi.org/10.4172/2157-7617.1000144

Davis JL (1972) Geographical variation in coastal development. Oliver and Boyd, Edinburgh, pp 204–212

Elkoushy AA, Tolba ERA (2004) Prediction of shoreline change by using satellite aerial imagery. Western Indian Ocean. J Mar Sci 3(1):1–10
Greeshma MS, Jayaraj PG (2014) Coastal vulnerability assessment along Kerala coast using remote sensing and GIS. Int J Sci Eng Res 5(7):228–234
Himmelstoss EA (2009) DSAS 4.0 installation instructions and user guide. U.S. Geological Survey Open-File Report
Jayappa KS, Mitra D, Mishra AK (2006) Coastal geomorphological and land-use and land-cover study of Sagar Island, Bay of Bengal (India) using remotely sensed data. Int J Remote Sens 27:3671–3682
Leah P, Veronica T (2010) Tasmanian coastal works manual: a best practice management guide for changing coastlines. Department of Primary Industries. Parks Water and Environment, Tasmania, p 428
Mujabar SP, Chandrasekar N (2011) Dynamics of coastal landform features along the southern Tamilnadu of India by using remote sensing and geographic information system. Geocarto Int 27(4):347–370
Nayak S (2002) Use of satellite data in coastal mapping. Indian Cartograph 22:147–156
Nayak GN (2005) Indian Ocean coasts – coastal geomorphology. Encyclopaedia of Coastal Science Springer International, pp 554–557
Neelima T, Noujas V, Thomas KV, Kurian NP (2018) Coastal morphology and beach stability along Thiruvananthapuram, south-west coast of India. Nat Hazards 90:1–23
Noujas V, Thomas KV (2015) Erosion hotspots along southwest coast of India. Aquat Proc 4:548–555
Noujas V, Thomas KV, Ajeesh NR (2017) Shoreline management plan for a protected but eroding coast along the south west coast of India. Int J Sedim Res 32:495–505
Noujas V, Kurian NP, Thomas KV (2019) Comparison of coastal hydrodynamics in different energy regime coasts along southwest coast of India. J Earth Syst Sci 128:93
Rao DP (2002) Remote sensing application in geomorphology. Trop Ecol 43(1):49–59
Report: L&T-Rambol Private Limited and INCOIS (2012) Assessment of long-term shoreline changes in and around proposed Vizhinjam port, Kerala. Government of Kerala. pp- ii+25
Report: Parliament of India Rajyasabha (2018) The Cyclone Ockhi-its impact on fishermen and damage caused by it. Department- Related Parliament Standing Committee on Home Affairs, report no: 211
Sallema VR, Mahika C (2004) Monitoring shoreline change using remote sensing and GIS: a case study of Kunduchi area. Western Indian Ocean J Marine Sci 3(1):1–10
Shaji J (2014) Coastal sensitivity assessment for Thiruvananthapuram west coast of India. Nat Hazards 73:1369–1392
Shamji VR (2011) Studies on beach morphological changes using numerical models. Research thesis, Marine Science Division, Centre for Earth Science Studies
Smith MJ, Rose J, Booth (2006) Geomorphological mapping of glacial landforms from remotely sensed data: an evaluation of the principal data sources and an assessment of their quality. Geomorphology 76:148–165
Sreekala SP, Baba M, Muralikrishna M (1997) Shoreline changes of Kerala coast using IRS data and aerial photographs. Indian J Marine Sci 27(1997):144–148
Woodroffe CD (2002) Coasts: form, process and evolution. Cambridge University Press, Cambridge, pp 1–26

Index

N. Jayaraju et al. (eds.), *Coasts, Estuaries and Lakes*,
https://doi.org/10.1007/978-3-031-21644-2

The manufacturer's authorised representative in the EU is Springer Nature Customer Service Centre GmbH, Europaplatz 3, 69115 Heidelberg, Germany. If you have any concerns regarding our products, please contact ProductSafety@springernature.com

Printed and bound by CPI Group (UK) Ltd, Croydon, CR0 4YY
30/09/2026
02225244-0018